Lecture Notes in Artificial Intelligence   3554

Edited by J. G. Carbonell and J. Siekmann

Subseries of Lecture Notes in Computer Science

Anind Dey   Boicho Kokinov
David Leake   Roy Turner (Eds.)

Modeling and Using Context

5th International and Interdisciplinary Conference
CONTEXT 2005
Paris, France, July 5-8, 2005
Proceedings

Volume Editors

Anind Dey
Carnegie Mellon University, School of Computer Science
Human Computer Interaction Institute
5000 Forbes Ave, Pittsburgh, PA, 15206, USA
E-mail: anind.dey@cs.cmu.edu

Boicho Kokinov
New Bulgarian University
Central and East European Center for Cognitive Science
Department of Cognitive Science and Psychology
21, Montevideo Str., Sofia 1638, Bulgaria
E-mail: bkokinov@nbu.bg

David Leake
Indiana University
Computer Science Department
Lindley Hall 215, 150 S. Woodlawn Ave, Bloomington, IN 47405-7104, USA
E-mail: leake@cs.indiana.edu

Roy Turner
University of Maine
Department of Computer Science
Orono, Maine 04469, USA
E-mail: rmt@umcs.maine.edu

Library of Congress Control Number: 2005928169

CR Subject Classification (1998): I.2, F.4.1, J.3, J.4

ISSN   0302-9743
ISBN-10   3-540-26924-X Springer Berlin Heidelberg New York
ISBN-13   978-3-540-26924-3 Springer Berlin Heidelberg New York

This work is subject to copyright. All rights are reserved, whether the whole or part of the material is concerned, specifically the rights of translation, reprinting, re-use of illustrations, recitation, broadcasting, reproduction on microfilms or in any other way, and storage in data banks. Duplication of this publication or parts thereof is permitted only under the provisions of the German Copyright Law of September 9, 1965, in its current version, and permission for use must always be obtained from Springer. Violations are liable to prosecution under the German Copyright Law.

Springer is a part of Springer Science+Business Media

springeronline.com

© Springer-Verlag Berlin Heidelberg 2005
Printed in Germany

Typesetting: Camera-ready by author, data conversion by Scientific Publishing Services, Chennai, India
Printed on acid-free paper   SPIN: 11508373   06/3142   5 4 3 2 1 0

Preface

Context is of crucial importance for research and applications in many disciplines, as evidenced by many workshops, symposia, seminars, and conferences on specific aspects of context. The International and Interdisciplinary Conference on Modeling and Using Context (CONTEXT), the oldest conference series focusing on context, provides a unique interdisciplinary emphasis, bringing together participants from a wide range of disciplines, including artificial intelligence, cognitive science, computer science, linguistics, organizational science, philosophy, psychology, ubiquitous computing, and application areas such as medicine and law, to discuss and report on context-related research and projects.

Previous CONTEXT conferences were held in Rio de Janeiro, Brazil (1997), Trento, Italy (1999, LNCS 1688), Dundee, UK (2001, LNCS 2116), and Palo Alto, USA (2003, LNCS 2680). CONTEXT 2005 was held in Paris, France during July 5–8, 2005.

There was a strong response to the CONTEXT 2005 Call for Papers, with 120 submissions received. A careful review process assessed all submissions, with each paper first reviewed by the international Program Committee, and then reviewer discussions were initiated as needed to assure that the final decisions carefully considered all aspects of each paper. Reviews of submissions by the Program Chairs were supervised independently and anonymously, to assure fair consideration of all work. Out of the 120 submissions, 23 were selected as full papers for oral presentation, and 20 were selected as full papers for poster presentation. These outstanding papers are presented in this proceedings.

In addition to the papers presented here, the conference featured a rich program including invited talks and panel sessions, posters of late-breaking work, system demonstrations, a doctoral consortium, and workshops whose results are published in a separate volume.

We would like to thank the members of the Program Committee and all of our additional reviewers for their careful work in assuring the quality of the technical program. We would also like to thank the members of the Organizing Committee, and especially Patrick Brezillon, Organizing Committee Chair, for their tireless efforts; Elisabetta Zibetti, Workshops Chair, for organizing an exceptional workshops program; Chiara Ghidini for putting together an inaugural Doctoral Consortium; Leslie Ganet, Denesh Douglas and Christine Leproux for the wonderful website; and the Steering Committee for their guidance. Special thanks are due to Laure Leger and Mary Bazire for their preparation of this proceedings.

In addition, we would like to thank all the chairs, co-chairs and the many volunteers involved in making CONTEXT 2005 a successful conference. The conference is also grateful to the AAAI, the Association Francaise pour le Contexte, CNRS, the Cognition et Usages Laboratory, FLAIRS, the Laboratory of Computer

Sciences, Paris 6, Nokia, Paris 8 University, Pierre and Marie Curie University, RISC, Supelec, and The University of Maine for their generous assistance to the conference.

We hope that the contributions reported here will illustrate the rich range of current contributions to context, and will provide a foundation for drawing on the best work from many areas in the next generation of research on context.

July 2005

Anind Dey
Boicho Kokinov
David Leake
Roy Turner

Organization

CONTEXT 2005 was hosted at the Pierre and Marie Curie University, Paris 6
http://www.context-05.org/

Conference Chair	Roy Turner (University of Maine, USA)
Program Co-chairs	Anind Dey (Carnegie Mellon University, USA)
	Boicho Kokinov (New Bulgarian University, Bulgaria)
	David Leake (Indiana University, USA)
Managing Editors	Mary Bazire (University of Paris 8, France)
	Laure Léger (University of Paris 8, France)

Organizing Committee

Chairs Patrick Brézillon (University of Paris 6, France)
Jean-Charles Pomerol (University of Paris 6, France)
Charles Tijus (University of Paris 8, France)

Members
  Web Site Management
    Christine Leproux (University of Paris 8, France)
    Leslie Ganet (University of Paris 8, France)
    Denesh Douglas (University of Paris 6, France)
  Communication Management
    Chantal Perrichon (University of Paris 6, France)
    Virginie Amiache (University of Paris 6, France)
    Hind Boumlak (University of Paris 8, France)
  Workshop Management
    Elisabetta Zibetti (University of Paris 8, France)
    Isabel Urdapilleta (University of Paris 8, France)
    Patrick Yeu (University of Paris 8, France)
  Doctoral Consortium Management
    Chiara Ghindini (ITC-IRST, Italy)
    Bich-Liên Doan (France)
  Poster and Demonstration Session Management
    Anind Dey (Carnegie Mellon University, USA)
    Jean-Marc Meunier (University of Paris 8, France)
  Volunteers Management
    Jean-Marc Meunier (University of Paris 8, France)
    Lionel Médini (University of Lyon 1, France)
  Publicity Chair
    Aline Chevalier (University of Paris 10, University of Paris 8, France)

Program Committee

Aamodt Agnar (NTNU, Norway)
Akman Varol (Bilkent U., Turkey)
Ala-Siuru Pekka (VTT, Finland)
Andonova Elena
  (New Bulgarian U., Bulgaria)
Arló Costa Horacio (CMU, USA)
Barnden John (U. Birmingham, UK)
Barsalou Larry (Emory U., USA)
Bauer Travis
  (Sandia Laboratories, USA)
Bazzanella Carla (U. Torino, Italy)
Beigl Michael (TeCO, Germany)
Bell John (U. London, UK)
Benerecetti Massimo
  (U. Naples, Italy)
Bermudez Jose Luis
  (Washington U. St. Louis, USA)
Bouquet Paolo (U. Trento, Italy)
Brézillon Patrick (U. Paris 6, France)
Budzik Jay (Northwestern U., USA)
Campbell Roy (UIUC, USA)
Canas Alberto (IHMC, USA)
Castefranchi Cristiano (CNR, Italy)
Cavalcanti Marcos
  (Fed. U. Rio de Janeiro, Brazil)
Chen Guanling (Dartmouth, USA)
Cheverst Keith (Lancaster U., UK)
Christoff Kalina (Cambridge U., UK)
Conlon Tom (U. of Edinburgh, UK)
Corazza Eros (U. Nottingham, UK)
Coutaz Joelle (INRIA, France)
de Paiva Valeria (Xerox PARC, USA)
de Roure David
  (U. Southampton, UK)
Dichev Christo
  (Winston-Salem U., USA)
Ebling Maria (IBM, USA)
Edmonds Bruce
  (Manchester Inst. Tech., UK)
Fernando Tim
  (Trinity College Dublin, Ireland)
Fetzer Anita (U. Stuttgart, Germany)
Finin Tim (UMBC, USA)
French Robert (U. of Liege, Belgium)

Gellersen Hans-Werner
  (Lancaster U., UK)
Ghidini Chiara (ITC-irst, Italy)
Giboin Alain (INRIA, France)
Giunchiglia Fausto (U. Trento, Italy)
Goker Mehmet
  (PricewaterhouseCoopers, USA)
Gonzalez Avelino
  (U. Central Florida, USA)
Graf Peter
  (U. British Columbia, Canada)
Grinberg Maurice
  (New Bulgarian U., Bulgaria)
Guha Ramanathan
  (IBM Research, USA)
Hobbs Jerry (USC/ISI, USA)
Hoffman Robert (IHMC, USA)
Hong Jason (CMU, USA)
Horvitz Eric (Microsoft, USA)
Humphreys Patrick
  (London School of Econ. &
  Poli. Sci., UK)
Indurkhya Bipin
  (Tokyo U. of Ag. & Tech., Japan)
Jameson Tony (DKFI, Germany)
Kahana Michael (Brandeis U., USA)
Korta Kepa
  (U. the Basque Country, Spain)
LaPalme Guy (U. Montreal, Canada)
Lieberman Henry
  (MIT MediaLab, USA)
Mantaras Ramon
  (Spanish Sci. Res. Council, Spain)
McClelland James (CMU, USA)
Mostafa Javed (Indiana U., USA)
Nixon Paddy (Strathclyde, UK)
Nossum Rolf (Agder U., Norway)
Pavel Dana (Nokia, USA)
Pimentel Maria
  (U. of Sao Paulo, Brazil)
Pu Pearl (EPFL, Switzerland)
Rissland Edwina (NSF, USA)
Roth-Berghofer Thomas
  (U. Kaisers-lautern, Germany)

Salber Daniel (The Netherlands)
Schmidt Albrecht
　(U. Munich, Germany)
Serafini Luciano (ITC, Italy)
Shafir Eldar (Princeton U., USA)
Sierra Carles
　(Spanish Sci. Res. Council, Spain)
Smyth Barry (UCD, Ireland)
Staab Steffen
　(U. of Karlsruhe, Germany)
Steenkiste Peter (CMU, USA)
Stock Oliviero (ITC-IRST, Italy)
Sowa John
　(VivoMind Intelligence, Inc., USA)
Sumi Yasuyuki (Kyoto U., Japan)
Ter Meulen Alice
　(U. Groningen, The Netherlands)
Thomason Rich (U. Michigan, USA)
Turner Roy (U. Maine, USA)
Turner Elise (U. Maine, USA)
van Benthem Johan
　(U. Amsterdam, The Netherlands)
Wilson David (UNC Charlotte, USA)
Wilson Deirdre (UCL, UK)
Winograd Terry (Stanford, USA)
Young Roger (U. Dundee, UK)

Additional Reviewers

Aihe David (USA)
Christopouloum Eleni (Greece)
Gena Cristina (Italy)
Gerganov Encho (Bulgaria)
Holleis Paul (Germany)
Hristova Penka (Bulgaria)
Joshi Hemant (USA)
Kranz Matthias (Germany)
Laine Tei (USA)
Matsuka Toshihiko (USA)
Merler Stefano (Italy)
Pleuss Andreas (Germany)
Reichherzer Thomas (USA)
Riezler Stefan (USA)
Rukzio Enrico (Germany)
Sow Daby (USA)
Stamenov Maxim (Germany)
Terrenghi Lucia (Germany)
Trossen Dirk (USA)
Van den Bergh Jan (Belgium)

Acknowledgments

We would particularly like to thank the FLAIRS society, and especially Susan Haller, John Kolen and Zdravko Markov for their authorization to use the FLAIRS-04 Website as initial version for the CONTEXT-05 Website.

　We also thank Brigitte Bricout, graphic designer at the University of Paris 6 for the realization of the poster of the conference, and A. Jeanne-Michaud for the photos on the poster.

　We thank very much Nina Runge and Daphne Batamio for their enthusiastic support during the Web site development from June to September 2004.

　We thank finally the volunteers who helped us enthusiastically before, during and after this event.

Sponsoring Institutions

PIERRE ET MARIE CURIE UNIVERSITY
http://www.upmc.fr/

PARIS 8 UNIVERSITY
http://www.univ-paris8.fr/

COGNITION & USAGES LABORATORY
http://www.cognition-usages.org/

LABORATORY OF COMPUTER SCIENCES, PARIS 6
http://www.lip6.fr/

NOKIA
http://www.nokia.com/

THE UNIVERSITY OF MAINE
http://www.umaine.edu/

ASSOCIATION FRANÇAISE POUR LE CONTEXTE
http://www-poleia.lip6.fr/~brezil/AFC/index.html

Association Française pour le Contexte

CNRS
http://www.cnrs.fr/

AAAI
http://www.aaai.org/

RISC
http://www.risc.cnrs.fr/

FLAIRS
http://www.flairs.com/

SUPÉLEC
http://www.supelec.fr/

RITZ
http://www.ritzparis.com/

MINISTERE DE L'EDUCATION NATIONALE, DE L'ENSEIGNEMENT SUPERIEUR ET DE LA RECHERCHE
http://www.education.gouv.fr/

Table of Contents

Modelling the Context of Learning Interactions in Intelligent Learning Environments
Fabio N. Akhras 1

Contextual Modals
Horacio Arló Costa, William Taysom 15

Understanding Context Before Using It
Mary Bazire, Patrick Brézillon 29

Epistemological Contextualism: A Semantic Perspective
Claudia Bianchi, Nicla Vassallo 41

Task-Realization Models in Contextual Graphs
Patrick Brézillon 55

Context-Dependent and Epistemic Uses of Attention for Perceptual-Demonstrative Identification
Nicolas J. Bullot 69

Utilizing Visual Attention for Cross-Modal Coreference Interpretation
Donna Byron, Thomas Mampilly, Vinay Sharma, Tianfang Xu 83

Meaning in Context
Henning Christiansen, Veronica Dahl 97

Descriptive Naming of Context Data Providers
Norman H. Cohen, Paul Castro, Archan Misra 112

Quotations and the Intrusion of Non-linguistic Communication into Utterances
Philippe De Brabanter 126

Mobile Phone Talk in Context
Mattias Esbjörnsson, Alexandra Weilenmann 140

Unsupervised Clustering of Context Data and Learning User Requirements for a Mobile Device
John A. Flanagan 155

Identification of Textual Contexts
Ovidiu Fortu, Dan Moldovan 169

Investigation of Context Effects in Iterated Prisoner's Dilemma Game
Evgenia Hristova, Maurice Grinberg 183

Context-Aware Configuration: A Study on Improving Cell Phone Awareness
Ashraf Khalil, Kay Connelly 197

Context-Aware Adaptation in a Mobile Tour Guide
Ronny Kramer, Marko Modsching, Joerg Schulze, Klaus ten Hagen 210

Contextual Factors and Adaptative Multimodal Human-Computer Interaction: Multi-level Specification of Emotion and Expressivity in Embodied Conversational Agents
Myriam Lamolle, Maurizio Mancini, Catherine Pelachaud, Sarkis Abrilian, Jean-Claude Martin, Laurence Devillers 225

Modeling Context for Referring in Multimodal Dialogue Systems
Frédéric Landragin 240

Exploiting Rich Context: An Incremental Approach to Context-Based Web Search
David Leake, Ana Maguitman, Thomas Reichherzer 254

Context Adaptative Self-configuration System Based on Multi-agent
Seunghwa Lee, Heeyong Youn, Eunseok Lee 268

Effect of the Task, Visual and Semantic Context on Word Target Detection
Laure Léger, Charles Tijus, Thierry Baccino 278

Ontology Facilitated Community Navigation – Who Is Interesting for What I Am Interested in?
Nils Malzahn, Sam Zeini, Andreas Harrer 292

Contextual Information Systems
Carlos Martín-Vide, Victor Mitrana 304

Context Building Through Socially-Supported Belief
Naoko Matsumoto, Akifumi Tokosumi 316

A Quantitative Categorization of Phonemic Dialect Features in Context
Naomi Nagy, Xiaoli Zhang, George Nagy, Edgar W. Schneider 326

Context-Oriented Image Retrieval
*Dympna O'Sullivan, Eoin McLoughlin, Michela Bertolotto,
David Wilson* 339

An Approach to Data Fusion for Context Awareness
*Amir Padovitz, Seng W. Loke, Arkady Zaslavsky, Bernard Burg,
Claudio Bartolini* 353

Dynamic Computation and Context Effects in the Hybrid Architecture AKIRA
Giovanni Pezzulo, Gianguglielmo Calvi 368

Goal-Directed Automated Negotiation for Supporting Mobile User Coordination
*Iyad Rahwan, Fernando Koch, Connor Graham, Anton Kattan,
Liz Sonenberg* 382

In Defense of Contextual Vocabulary Acquisition – How to Do Things with Words in Context
William J. Rapaport 396

Functional Model of Criminality: Simulation Study
*Sarunas Raudys, Aini Hussain, Viktoras Justickis, Alvydas Pumputis,
Arunas Augustinaitis* 410

Minimality and Non-determinism in Mutli-context Systems
Floris Roelofsen, Luciano Serafini 424

'I' as a Pure Indexical and Metonymy as Language Reduction
Esther Romero, Belén Soria 436

Granularity as a Parameter of Context
Hedda R. Schmidtke 450

Identifying the Interaction Context in CSCLE
*Sandra de A. Siebra, Ana Carolina Salgado, Patrícia A. Tedesco,
Patrick Brézillon* 464

Operational Decision Support: Context-Based Approach and Technological Framework
*Alexander Smirnov, Michael Pashkin, Nikolai Chilov,
Tatiana Levashova* 476

Threat Assessment Technology Development
  Alan N. Steinberg 490

Making Contextual Intensional Logic Nonmonotonic
  Richmond H. Thomason 501

Modeling Context as Statistical Dependence
  Sriharsha Veeramachaneni, Prateek Sarkar, George Nagy 515

Robust Utilization of Context in Word Sense Disambiguation
  Xiaojie Wang 529

Understanding Actions: Contextual Dimensions and Heuristics
  Elisabetta Zibetti, Charles Tijus 542

Applications of a Context-Management System
  Andreas Zimmermann, Andreas Lorenz, Marcus Specht 556

Author Index 571

Modelling the Context of Learning Interactions in Intelligent Learning Environments

Fabio N. Akhras

Renato Archer Research Center, Rodovia Dom Pedro I, km 143,6
13082-120 Campinas, São Paulo, Brazil
fabio.akhras@cenpra.gov.br

Abstract. The aim of this paper is to present an approach to modelling and using the context of learning in intelligent learning environments. The approach developed allows the modelling of the content and dynamics of learning situations and the way situations develop as a consequence of interactions. Situation models are used to support two aspects of intelligent learning environments: evaluation of learning, taking into consideration the context of learning activity, and adaptation of the learning environment to the learner's perceived needs, taking into consideration the affordances of learning contexts. The main focus of the paper is on the description of the formalism developed to support context modelling in intelligent learning environments. An application of the approach in a system to support learning of software engineering concepts is briefly described.

1 Introduction

Contemporary theories of learning stress that learning requires the development of active experiences in authentic contexts. According to these theories, learning cannot be separated from activity and an essential part of what is learned is the situation in which learning takes place, which refers to the physical as well as to the social context in which the learner is engaged in activity [1], [2], [3].

The implication of this view to the design of systems to support learning is that the computational learning environments must provide interactive situations for the learner in which the conceptual, physical and social aspects of these contexts are addressed in ways that create possibilities for learning.

Moreover, in order that an intelligent system can reason about the learning experiences developed in a situation, so that it can adapt the learning opportunities to the particular needs of the learner, the characteristics of the situation as well as the information about the interaction developed in the situation must be modelled and explicitly represented in the system.

Therefore, a formalism to represent the context of learning interactions and the interactions that happen in this context is needed. This formalism can inherit some characteristics of formalisms developed to represent context in other artificial intelligence research in which contextual representations are sought for different purposes, such as research in natural language understanding, or research on the frame problem. There might be, however, issues in the formalisation of contexts of learning interactions that might require particular approaches, also depending on the particular focus of the intelligent system to support learning, which might have an influence on the contextual aspects that are emphasised and in the way they are represented and used.

In this paper we explore some of the issues involved in the formalisation of contexts of learning interaction, and describe an approach that we have developed to model the characteristics of situations and of interactions that develop in situations. The main aim of the work developed is to support reasoning about the context of learning interactions in Intelligent Learning Environments (ILEs). The approach has been applied in the development of INCENSE, an ILE to support learning of software engineering concepts [4], which will be briefly described later in the paper.

2 Formalising Context

Formalising context has been a central issue of research in artificial intelligence and related areas [5]. Among the approaches that have been developed, situation theory [6], [7] was particularly influential in the development of our approach to formalising the context of learning interactions. In addition, situation calculus [8], histories [9] and other related work [10], [11], have also been considered. Below we briefly point to some of the main issues that were relevant to the development of our approach.

According to situation theory, situations are structured parts of the world that constitute the context for the behaviour or communication of agents. Situations are defined intentionally and are related to the infons that hold in the situation, which represent items of information about the world. Infons are represented by ordered sets denoted by $<<R, a1, ..., an, i>>$, where R is a n-place relation; $a1, ..., an$ are the arguments of R; and i is the polarity. The polarity can assume the values 1 or 0, to indicate whether the relation does or does not hold.

In order to represent actions and the changes in situations that may be caused by the occurrence of actions, Ohsawa and Nakashima [12] have proposed an approach in which actions are represented by pairs of sets of infons (the first set of infons corresponding to the action precondition and the second to the action postcondition). A problem associated with reasoning about changes in situations caused by actions is the frame problem [8]. As a way of addressing it, Reiter [11] has defined a logical theory to specify the effects of actions on states of the world, which includes: action precondition axioms, to specify the conditions for the occurrence of an action, and successor state axioms, to specify the ways in which actions affect the states of the world. These axioms, along with a set of general axioms, allow inference of the facts that hold in a new situation after the occurrence of an action.

The fact that in the situation calculus situations are spatially unbounded has been addressed by Hayes [9], who proposed the notion of histories as an alternative to situations that take chunks of space-time as components, and denote snapshots of the universe that are extended through time and spatially bounded.

Our approach to modelling context is based on the definition of situation types, which are abstract entities defining a context in which learning interactions can occur, and situations, which are instances of situation types occurring at particular times. Before we proceed defining the primitives needed to represent these and other entities of our approach, we will briefly discuss some issues of learning situations that have been considered.

3 What Is in a Learning Situation

Concerned with the environmental conditions in which learning experiences are developed, researchers have investigated the differences between learning activity in school and in everyday real life and work situations, and have suggested that in order that students become able to think with and about the entities of a domain, rather than just learning what these entities are, they need to learn how these entities are generated and how they work in authentic contexts and activities [13], [1], [14], [2], [15].

Therefore, the computational setting of a learning situation may display aspects of the authentic physical settings that characterise real life situations in the domain, and may embody the social and cultural aspects related to these situations, reflected, for example, in the practices of members of the community, their common views, and their tools. These aspects may appear in the situation in the form of:

- *Physical entities,* representing physical aspects of phenomena.
- *Conceptual entities,* representing ideas, concepts, and information.
- *Living entities,* representing people or other living things.

In addition, in order to provide ways for learners to act in the situation and learn, several kinds of opportunities for activity must be implemented in a learning situation, which may appear in the situation in the form of:

- *Action events,* representing several ways of accessing, using and creating information, or other kinds of entities of a domain, in the circumstances provided by a situation.
- *Communication events,* representing several ways of communicating with other learners or living entities, in the circumstances provided by a situation.

Therefore, through activity learners interact in a situation taking the physical or conceptual entities of the situation as objects of their actions and producing new or transformed entities. In this process they may also communicate to each other or to the living entities designed to perform particular roles in the learning situation.

4 Modelling Learning Situations

Taking into consideration the issues involved in formalising the context of learning interactions, we have developed an approach for modelling learning situations which allows the modelling of structural information about learning situations as well as temporal information associated with the way situations develop. The formalism is a many-sorted first-order predicate theory whose entities address the following aspects of contexts of learning interaction: situation types, content of situation types, dynamics of situation types, situations and situation development. Examples of the use of the formalism will be given in the section that describes the application developed.

4.1 Situation Types

Situation types are abstract entities that are concerned with possibilities for the development of interaction and are not located in any particular time. They denote contexts for the development of learning interactions, comprising many kinds of entities and holding various possibilities for action. A *situation* is an instance of a situation type occurring at a particular time.

A situation type is defined in terms of two kinds of entities: entities that denote the content of the situation type, and entities that denote the dynamics of the situation type (the way learners interact with the content of the situation type). An entity x, of content or dynamics, is defined as part of a situation type s, using the notation *define(x, s)*.

4.2 Content of Situation Types

The units of content of situation types are the *objects* which denote physical or conceptual entities that are part of a situation type and are represented by n-place predicates. *Relations* between objects are represented as *relation(r(object1, object2))*, where r is the type of the relation which is a 2-place predicate that take objects as arguments.

Two special kinds of relations that are treated as particular kinds of entities are properties and states. To denote *properties* of objects we use the notation *property(x, pt(pv))*, where x is an object, pt is the property type, and pv is the property value. Sometimes, in modelling properties of entities of varied levels of abstraction, it may be necessary to attribute only a *type of property* to an object. We use the notation *prop-type(x, pt)* for this purpose, where x is an object and pt is the property type.

As for *states* of objects, they are denoted by *state(x, st)*, where x is an object and st is a state. In addition to actual states, we represent the states in which objects might be in by types of states, and the transitions of states that objects might go through, i.e., their state graphs, by types of state transitions. *Types of states* are denoted by state-type(x, st), where x is an object and st is a state. To denote *types of state transitions* for objects we use the notation *tran-type(x, st1, st2)*, where x is an object and $st1$ and $st2$ are states.

In modelling the content of learning situations, two hierarchical relations that are useful are: generalisations, that characterise is-a relations between sub-class entities and super-class entities, and aggregations, that characterise part-of relations between component entities and aggregate entities. To denote relations of *generalisation* between objects we use the notation *kind(x1, x2)*, and for *aggregation* we use *part(x1, x2)*, where $x1$ and $x2$ are objects.

In our theory, situation types may also appear as constituents of other units of content. Therefore, most of the entities introduced above can also take a situation type as argument. The form of the argument that corresponds to a situation type *is sit-type(s)*, where s is a situation type. This form can be used in the same places where objects are used.

Of particular interest is the specification of hierarchies of *generalisation of situation types*, for which we have defined a simple mechanism of inheritance that allow sub-class situation types to inherit the content of super-class situation types. So, for example, the generalisation *kind(sit-type(s1), sit-type(s2))* lead to the content of *s1* being inherited by *s2*, unless otherwise stated in *s2*. In the applications that we have developed, this mechanism was used to facilitate the modelling of sets of situation types that have common characteristics.

Table 1 presents a summary of the entities defined in our theory to model the content of situation types.

4.3 Dynamics of Situation Types

Agents can interact with the content of a situation type, and with each other, by means of events. *Events* are entities of situation types that represent the potential behaviour of agents in learning situations. To denote events we use the notation *event(a, e)*, where *a* is an agent and *e* is an event type

Table 1. Entities of content of situation types

Basic sorts:		
o	object	n-place predicate with constants or variables as arguments
pv	property value	constant or variable
pt	property type	constant or variable
st, st1, st2	state	constant or variable
s	situation type	constant
Derived sorts:		
r relation		2-place predicate with objects as arguments
p property expression		1-place predicate with property value as argument
Nonlogical symbols and their sorts:		
relation(r)		
property(o|s, p)		
prop-type(o|s, pt)		
state(o|s, st)		
state-type(o|s, st)		
tran-type(o|s, st1, st2)		
kind(o|sit-type(s), o|sit-type(s))		
part(o|sit-type(s), o|sit-type(s))		

Obs: The symbol | means "or".

The conditions of activation of an event are stated by the preconditions defined for the event type. *Preconditions* are denoted by *pre(e, x, pa)*, where *e* is an event type, *x* is any content entity and *pa* is the participation of *x* in the precondition. The participation can assume the values 1 or 0 and indicate whether *x* must hold or not to satisfy the precondition. For example, *pre(e, x, 0)* means that the precondition for *e* is *not(x)*.

The changes in a learning situation caused by the occurrence of an event are stated by the effects defined for the event type. To denote *effects* we use the notation *effect(e, x, pa)*, where *e* is an event type, *x* is any content entity and *pa* is the participation of *x* in the effect.

In the definition of preconditions and effects of event types, the arguments of content entities can be variables. We will see in the next section how the variables in the preconditions and effects are instantiated when events occur and their actual values are associated with the occurrence of the event.

In addition, an event can be related to other content of a situation type, which may represent a particular aspect of the context of the event (such as, a background for the event, an aspect of its authentic setting, etc.). The *context of an event*, is denoted by *context(e, x)*, where *e* is an event type and *x* is a content entity that characterises the context of the event type in the situation type.

Two final notions that are relevant for modelling the dynamics of more complex learning situations are generalisation and aggregation of events. As usual, generalisations are used to model is-a relations between events whereas aggregations model part-of relations. To denote these relations we use the same notation introduced for content entities: *kind(e1, e2)* for generalisations, and *part(e1, e2)* for aggregations, where *e1* and *e2* are events. In addition, the kind of an event can be an object.

Table 2 presents a summary of the entities defined in our theory to model the dynamics of situation types.

Table 2. Entities of dynamics of situation types

Basic sorts:		
e	event type	constant
a	agent	constant
pa	participation	1 or 0
Other sorts:		
x	any content entity	
Nonlogical symbols and their sorts:		
event(a, e)		
pre(e, x, pa)		
effect(e, x, pa)		
context(e, x)		
kind(event(e, a)|o, event(e, a))		
part(event(e, a), event(e, a))		

4.4 Situation Development

A situation is defined by the pair *(situation type, time)* and denote the state of a situation type at a certain time. This state is changed by the occurrence of events which gives rise to new situations.

The occurrence of events in situations is denoted by *occurs(e, a, s, t)*, where e is an event type, *a* is the agent that activated the event, and *(s, t)* is the situation in which the event occurs. The holding of a content entity in a situation is denoted by *in(x, s, t)*, where *x* is any content entity, and *(s, t)* is the situation in which the content entity is present.

The way situations develop from the occurrence of events in a situation type depends on the preconditions and effects defined for these events. The preconditions determine the content entities that must hold in a certain situation for the activation of an event. When these preconditions hold and the event is activated by an agent, then the effects defined for the event determine the content entities that will hold in the resulting situation.

This is made precise by a set of axioms of situation development which allow to track the changes that occur in situations caused by the occurrence of events. Some of these axioms, are effect axioms, which are presented below.

SD-1: $occurs(e, a, s, t) \wedge define(pre(e, x, 1), s) \wedge define(effect(e, y, 1), s) \wedge in(x, s, t) \Rightarrow in(y, s, t+1)$

SD-2: $occurs(e, a, s, t) \wedge define(pre(e, x, 0), s) \wedge define(effect(e, y, 1), s) \wedge \neg in(x, s, t) \Rightarrow in(y, s, t+1)$

SD-3: $occurs(e, a, s, t) \wedge define(pre(e, x, 1), s) \wedge define(effect(e, y, 0), s) \wedge in(x, s, t) \Rightarrow \neg in(y, s, t+1)$

SD-4: $occurs(e, a, s, t) \wedge define(pre(e, x, 0), s) \wedge define(effect(e, y, 0), s) \wedge \neg in(x, s, t) \Rightarrow \neg in(y, s, t+1)$

The effect axioms allow us to determine the effects of events that occur but they do not help in the general problem of determining the content entities that hold in a certain situation, which is an instance of the frame problem. This is determined in our theory by taking into consideration the histories of participation of content entities in the preconditions or effects of events that have occurred, as described below.

In the previous section we have introduced preconditions and effects as entities of dynamics that denote the preconditions and effects that are defined for an event, and which may contain variables to be instantiated in time of event occurrence. These entities characterise abstract preconditions and effects, as they contain variables. When the event occurs, the variables are instantiated, and actual preconditions and effects will hold in the situations of the beginning and of the end of the event.

As well as content entities that can hold in situations, actual preconditions and effects of events can also hold in situations. This is denoted by *in(pre(e, x, pa), s, t)* or *in(effect (e, x, pa), s, t)*, with *x* being an instantiated content entity, as above, and *pa* assuming 0 or 1. These actual preconditions and effects of events characterise points in the histories of participation of content entities in the preconditions and effects of events that occur. These histories are used in determining the content entities that hold in situations.

To allow this kind of inference we have formulated two history axioms and one state axiom, which help to determine the content entities that hold in situations. These axioms are presented below.

SD-5: $define(pre(e, x, pa), s) \wedge occurs(e, a, s, t) \Rightarrow in(pre(e, x, pa), s, t)$

SD-6: $define(effect(e, x, pa), s) \wedge occurs(e, a, s, t) \Rightarrow in(effect(e, x, pa), s, t+1)$

SD-7: $[in(pre(e, x, 1), s, t) \lor in(effect(e, x, 1), s, t)] \lor$
$[(\exists t1)[t1<t \land in(x, s, t1)] \land (\forall t2)[t1<t2<t \land \neg in(effect(e, x, 0), s, t2)] \lor$
$[define(x, s) \land (\forall t1)[t1<t \land \neg in(effect(e, x, 0), s, t1)] \Rightarrow in(x, s, t)$

As we have described, entities of content, *x*, are normally placed as arguments of entities of dynamics, such as in *pre(e, x, pa)*, and as arguments of entities of situation development, such as in *in(x, s, t)*. In addition, we have just shown that some entities of dynamics, such as *pre(e, x, pa)*, can also be placed as arguments of entities of situation development, such as in *in(pre(e, x, pa), s, t)*.

Now, a further feature of our theory is that some entities of situation development such as *in(x, s, t)* or *occurs(e, a, s, t)* can also be placed as arguments of entities of content and of dynamics, such as in the entity *part(x, occurs(e, a, s, t))*. This is useful when entities that denote aspects of situation development have a relevance to the the content of situations and, therefore, may appear as preconditions or effects of events.

We have seen in a previous section that situation types characterise open contexts for learning interactions. However, within the context of a situation type the interactions are limited by the nature of the content and dynamics defined for that situation type. Therefore, it is expected that productive learning interactions extend over several situation types. In this way, sequences of situations of a single or various types will be formed, giving rise to the notion of course of interaction.

A *course of interaction* is defined by a sequence of situations and is denoted by *course(s1, t1, ..., sn, tn)*, where *(s1, t1)* and *(sn, tn)* are any two situations, and for which $n>=2$ and $tn>t1$. A particular case of course of interaction is the course of two situations *course(s1, t1, s2, t2)*, which was used in the definitions of our application. Note that courses of interaction are not necessarily contiguous, as there may be other situations located between *(s1, t1)* and *(s2, t2)*. Therefore, courses of interaction can overlap in many ways, and a situation can appear in more than one course of interaction.

So far we have focused our formalism on two aspects of contexts of learning interactions: the internal structure of these contexts (in terms of content and dynamics), and interaction development in these contexts (in terms of occurrence of events, entities that hold, and situations and courses of interaction that develop). A further aspect has to do with changing the context of learning interactions, and involves consideration of spaces of interaction, entering contexts, and leaving contexts.

The set of situation types that are available for an agent to interact with, at a certain time, defines the space of interaction of that agent at that time. To denote that a situation type *s* is in the space of interaction of an agent *a* at time *t*, we use the notation *space(s, a, t)*.

When a situation type is in the space of interaction of an agent, then the agent can enter the context provided by the situation type and start interacting with it. To denote the engagement of an agent in a situation type, meaning that the agent enters the context for interaction provided by the situation type, we use the notation *enters(a, s, t)*, where *t* is the time in which the agent *a* enters the situation type *s*.

Besides entering contexts, agents can also leave contexts. To denote when an agent leaves a situation type we use the notation *leaves(a, s, t)*, where *t* is the time in which the agent *a* leaves the situation type *s*.

The information represented by these two predicates (enters and leaves) is used to mark when agents change their contexts of interaction. Therefore, there is no increase in the time when agents enter or leave contexts.

Finally, in Table 3 we present a summary of the entities of situation development defined in our theory.

5 The Role of Situation Models in ILEs

Following the issues emphasised by contemporary theories of learning in relation to the role of context in learning, and according to the approach to modelling the context of learning interactions presented in the previous sections, two main roles for situation models in ILEs have been explored. First, situation models support perception of interaction in situations and are the basis for the definition of units of analysis of interaction that take into consideration the context in which the interaction occurs. Second, they support the analysis of the affordances of potential situations which are used to create opportunities for new learning. These two roles are discussed next.

Table 3. Entities of situation development

Basic sorts:
t, t1, tn	time	natural number
s, s1, sn	situation type	constant
e	event type	constant
a	agent	constant

Other sorts:
xc	any content entity
x	any content entity or precondition or effect

Nonlogical symbols and their sorts:
sit-type(s)
define(xc, s)
occurs(e, a, s, t)
in(x, s, t)
course(s1, t1, ..., sn, tn)
space(s, a, t)
enters(a, s, t)
leaves(a, s, t)

5.1 Perception of Interaction in Situations and Reasoning About Learning Interactions

According to the views of learning discussed earlier, interacting in situations is inherent to learning. It follows that learning depends on three aspects of interactions: the situation in which the interaction occurs, the cognitive structures of the learner involved in the interaction, and the nature of the activity that is developed by the learner in the situation. This indicates the need for units of analysis that take into consideration the various ways in which these three aspects can relate in learning interactions.

Modelling the contexts in which interactions occur and the nature of these interactions allow the modelling of this sort of units of analysis of interaction, which we call patterns of interaction, and are defined in terms of the more basic entities previously modelled, such as the occurrence of events, or the presence of entities in situations.

Modelling patterns of interaction and representing them in an ILE will allow the ILE to reason about learning interactions in situations in order to recognise the occurrence of regularities that are meaningful in terms of understanding the learning that occurs. Some types of regularities involving context that we have modelled are:

Regularities involving context and activity, which relate aspects of the context in which a learning experience is developed, to aspects of the activity that takes place in the learning situation.

Regularities involving context and cognitive structures, which relate aspects of the context in which a learning experience is developed, to aspects of the cognitive state of the learner involved in the activity.

Regularities involving different contexts, which relate aspects of different contexts in which learning experiences are developed.

An example is the definition of the pattern *utilises* which relates a learner's action to the situation in which it happens.

Definition (Utilises). *A learner a utilises an entity x through an event e in a situation (s, t), iff the event e, defined in situation type s as part of the alphabet of the learner a, occurs in (s, t), and x is a precondition of e that holds in (s, t).*

$define(event(a, e), s) \wedge occurs(e, a, s, t) \wedge in(pre(e, x, pa), s, t)$
$\Leftrightarrow utilises(a, x, e, s, t)$

In addition to the patterns of interaction, in order to address issues related to the temporal extension of the learning process we can define higher-order regularities that involve sequences of interactions, and relate patterns of interaction that hold in different situations of a course of interaction. These regularities are defined in terms of properties of courses of interaction. An example is presented below.

Definition (Cumulative). *A course of interaction course(s1, t1, s2, t2) is cumulative with respect to a content entity x for a learner a, if situations (s1, t1) and (s2, t2) share the entity x, and the learner a utilises x in (s1, t1) or generates it in (s1, t1-1), and further utilises x in (s2, t2) or generates it in (s2, t2-1).*

$share(s1, t1, s2, t2, x) \wedge$
$[utilises(a, x, e1, s1, t1) \vee generates(a, x, e1, s1, t1-1)] \wedge$
$[utilises(a, x, e2, s2, t2) \vee generates(a, x, e2, s2, t2-1)] \wedge t2>t1$
$\Rightarrow cumulative(course(s1, t1, s2, t2), a, x)$

The definition of the property of cumulativeness tries to capture part of the idea stressed by learning theorists that repetition of similar experiences in different contexts may enable access to prior knowledge, as current experiences are interpreted in the light of previous ones.

These and other similar definitions can be used by an ILE to obtain an interpretation of how learners have been interacting with the situations of their learning environments in terms of expected patterns of interaction and properties of courses of interaction (for more examples of definitions of patterns of interaction and properties of courses of interaction see [4]).

Therefore, the first role of situation models in ILEs, according to our perspective, is to support the evaluation of the learning process. Initially, through the perception of basic aspects of interactions in situations, then through reasoning to obtain higher-order interpretations of these interactions, which ultimately provide an evaluation of learning according to the view of learning being addressed.

The second role of situation models in ILEs, presented next, is to support the adaptation of the learning environment according to the needs of the learner. This involves two aspects: (a) reasoning about potential situations to be offered to the learner, determining the affordances of these situations to the learner according to the evaluation of the learning process, and (b) changing the environment in order to introduce situations that create opportunities for new learning.

5.2 Reasoning About Affordances of Situations and Adapting the Learning Environment

Following Gibson [16], the affordances of an environment are what it offers to an organism, such as the opportunities for actions or the dangers that exist in an environment for an organism. Affordances are located neither in the environment nor in the organism. Instead, they capture units of analysis that refer to both the environment and the organism in a complementary way.

In our theory, according to our modelling of situation types, we can say that the occurrence of events whose preconditions can be satisfied are basic kinds of affordances of a situation type to a learner. In addition, consider the occurrence of an event e at time t, such that:

occurs(e, learner, s, t)
in(pre(e, x, 1), s, t)

This indicates, according to definition of the pattern of interaction *utilises*, given in the previous section, the development of the pattern:

utilises(learner, x, e, s, t)

Therefore, a situation type may also afford to a learner the development of particular patterns of interaction. In the example above, we can say that the situation type s affords to the learner utilising the entity x through the event e. As an example, the definition of the affordance for the pattern *utilises* is presented below.

Definition (Affords utilises). *A situation type s affords to a learner a utilising an entity x through an event e, iff the event e is defined in situation type s as part of the alphabet of the learner a, and x is a precondition of e defined in s, and x is defined or can be generated in s.*

define(event(a, e), s) ∧ define(pre(e, x, pa), s) ∧
[define(x, s) ∨ affords(s, generates, al, x, el)]
 ⇔ affords(s, utilises, a, x, e)

In addition to what a situation type can afford to a learner in terms of the development of patterns of interaction, a situation type can also afford to a learner the development of courses of interaction that exhibit particular properties. As an example, the definition of the affordance for the property of cumulativeness is presented below.

Definition (Affords cumulative). *A situation type s affords to a learner a the development of a course of interaction from situation (s_i, t_i) that is cumulative for the learner with respect to a content entity x, if the learner utilises x in (s_i, t_i) or generates it in (s_i, t_i-1) and the situation type s affords to the learner a situation that shares the entity x with situation (s_i, t_i) and also affords to the learner utilising or generating x.*

$[utilises(a, x, e_i, s_i, t_i) \lor generates(a, x, e_i, s_i, t_i\text{-}1)] \land$
$affords(s, share, s_i, t_i, x) \land$
$[affords(s, utilises, a, x, e) \lor affords(s, generates, a, x, e)]$
$\Rightarrow affords(s, s_i, t_i, cumulative, a, x)$

As a result of modelling affordances, an ILE can determine the set of situation types that are most beneficial to a learner at a certain moment, based on what they afford to that learner. This helps the ILE in individualising learning, offering particular situations that hold particular opportunities for interaction to particular learners, at particular times of their learning processes.

6 Application

Our approach to modelling and reasoning about learning situations and interactions has been implemented in INCENSE [4], an ILE in the domain of software engineering which is capable of analysing interactions in situations in order to evaluate the learning that occurs and to build a model of the affordances of potential situations to the learner, adapting the learning environment accordingly.

In INCENSE, the situation types describe settings of a software engineering laboratory in which the learner can develop models of software engineering processes or apply these models to the development of particular software projects. For example, in situation types for modelling software engineering processes, the content of the situation types include various kinds of information about software engineering and about the learning situation, software engineering concepts to be used in constructing the models, and the models constructed. The dynamics of these situation types include ways of accessing the various kinds of information and of constructing software engineering models. Some examples are:

content:
concept(customer needs)
concept(define the project scope)

dynamics:
event(learner, create-process)
pre(create-process, concept(X), 1)
effect(create-process, process(X), 1)

event(learner, create-material)
pre(create-material, concept(X), 1)
pre(create-material, process(Y), 1)
effect(create-material, relation(material(concept(X), process(Y), 1)

Let us imagine that a learner develops a short course of interaction in this situation type, in which she or he is constructing a model of a software project planning process, having created the process 'define the project scope' and the material 'customer needs' as input to this process. The events that occur and the changes in the content of the situations developed are represented by:

situation development:
occurs(create-process, learner, s, t1)
in(pre(create-process, concept(define the project scope), 1, s, t1)
in(effect(create-process, process(define the project scope), 1, s, t2)
in(process(define the project scope), s, t2)
occurs(create-material, learner, s, t2)
in(pre(create-material, concept(customer needs), 1, s, t2)
in(pre(create-material, process(define the project scope), 1, s, t2)
in(effect(create-material, relation(material(concept(customer needs),
    process(define the project scope))), 1, s, t3)
in(relation(material(concept(customer needs),
    process(define the project scope))), s, t3)
course(s, t1, s, t2, s, t3)

On the basis of these events, and of the definitions of patterns of interaction and properties of courses of interaction, the system can reason about the interactions developed in the situations of the course of interaction to determine the patterns of interaction and properties of course of interaction developed. It can also, on the basis of the definitions of affordances, reason about the situations available for interaction to determine the affordances of these situations to the learner, as discussed in the previous section. Then, the system can adapt the learning environment, providing situations that enable the occurrence of courses of interaction that lead to the development of the desirable properties.

7 Conclusion

In this paper we have presented an approach to modelling and using the context of learning in intelligent learning environments. The main aim of intelligent learning environments and intelligent tutoring systems is to individualise learning, adapting the learning resources available to the learner's perceived needs. To do so, they rely on models that inform the system about the learner (what has been learned and how) and about the opportunities for learning at particular times. In order to address issues related to the role of the context in learning, which has been emphasised by recent research on learning and instruction, we have developed ways of modelling the context of learning and of taking it into consideration in evaluating learning and in adapting the learning environment to the learner's needs.

The approach developed allows the modelling of the context of learning in terms of situation types, their content and dynamics, situation development and situations. Modelling situations in this way makes possible the analysis of interactions in situations in evaluating learning, and the analysis of potential situations in adapting the learning opportunities to the learner.

An implementation of the approach was developed in INCENSE, an intelligent learning environment in the domain of software engineering. In INCENSE, learners learn by interacting in situations that represent particular kinds of contexts in which software engineering activities are developed. The system evaluates learning and adapts the learning environment taking into consideration the context in which learning interactions develop, making use of models of situations and affordances.

Acknowledgements. This work has been partially financed by CNPq, Brazil.

References

1. Brown, J. S., Collins, A. and Duguid, P. (1989). Situated cognition and the culture of learning. *Educational Researcher*, 18(1), pp. 32-42.
2. Piaget, J. and Garcia, R. (1991). *Toward a Logic of Meanings*. Hillsdale, NJ: Lawrence Erlbaum.
3. von Glasersfeld, E. (1989). Cognition, construction of knowledge, and teaching. *Synthese*, 80, pp. 121-140.
4. Akhras, F. N. and Self, J. A. (2000). System intelligence in constructivist learning. *International Journal of Artificial Intelligence in Education*, 11(4):344-376.
5. Akman, V. and Surav, M. (1996). Steps toward formalizing context. *Artificial Intelligence Magazine*, Fall, pp. 55-72.
6. Barwise, J. and Perry, J. (1983). *Situations and Attitudes*. Cambridge, MA: MIT Press.
7. Devlin, K. (1991). *Logic and information*. Cambridge University Press.
8. McCarthy, J. and Hayes, P. J. (1969). Some philosophical problems from the standpoint of artificial intelligence. In: Mitchie, D. (ed.), *Machine Intelligence 4*, Edinburgh: Edinburgh University Press, pp. 463-502.
9. Hayes, P. J. (1985). The second naive physics manifesto. In: Hobbs, J. R. and Moore, R. C. (eds.), *Formal Theories of the Commonsense World*, Norwood, NJ: Ablex, pp. 1-36.
10. Davis, E. (1990). *Representations of Commonsense Knowledge*. San Mateo, CA: Morgan Kaufmann.
11. Reiter, R. (1991). The frame problem in the situation calculus: A simple solution (sometimes) and a completeness result for goal regression. In: Lifshitz, V. (ed.), *Artificial Intelligence and Mathematical Theory of Computation: Papers in Honor of John McCarthy*, Academic Press, pp. 359-380.
12. Ohsawa, I. and Nakashima, H. (1991). *Actions in situation theory*. Technical Report TR-91-13, Electrotechnical Laboratory, Tsukuba, Japan.
13. Greeno, J. G. (1989). A perspective on thinking. *American Psychologist*, 44(2), pp. 134-141.
14. Resnick, L. B. (1987). Learning in school and out. *Educational Researcher*, 16(9), pp. 13-20.
15. von Glasersfeld, E. (1995). *Radical Constructivism: a way of knowing and learning*. London: The Falmer Press.
16. Gibson, J. J. (1979). *The Ecological Approach to Visual Perception*. Boston: Houghton Mifflin.

Contextual Modals

Horacio Arló Costa[1] and William Taysom[2]

[1] Carnegie Mellon University
hcosta@andrew.cmu.edu
[2] Institute for Human and Machine Cognition
wtaysom@ihmc.us

Abstract. In a series of recent articles Angelika Kratzer has argued that the standard account of modality along Kripkean lines is inadequate in order to represent context-dependent modals. In particular she argued that the standard account is unable to deliver a non-trivial account of modality capable of overcoming inconsistencies of the underlying conversational background. She also emphasized the difficulties of characterizing context-dependent conditionals. As a response to these inadequacies she offered a two-dimensional account of contextual modals. *Two* conversational backgrounds are essentially used in this characterization of contextual modality.

We show in this paper that Kratzer's *double relative* models (with finite domains) are elementary equivalent to well known neighborhood models of normal modalities originally proposed by D. Scott [?] and R. Montague [?]. We also argue that neighborhood models can be also used to represent some (non-normal) graded modalities that are difficult to represent in her framework (like 'it is likely that' or 'it is highly probable that', etc). Finally we show that an extension of the neighborhood semantics of conditionals is able to capture some of her proposals concerning dyadic modals. DR models with infinite domains can be shown to be pointwise equivalent to neighborhood models, but they are not guaranteed to have relational counterparts. So DR models surpass the representational power of relational (Kripkean) models. Neighborhood representations are, nevertheless, always possible, making clear as well that the central feature of double relative modals is that they are capable of encoding two central aspects of context: its propositional content, and its dynamic properties (which in Kratzer's models are represented via an *ordering source*).

1 Introduction

Phrases like *what the law provides*, *what we know*, or *what we presuppose* are usually employed to introduce what semanticists and logicians call *conversational backgrounds*. And such types of contexts are usually encoded as world-dependent sets of propositions. In other words conversational backgrounds (or context sets) are usually represented as functions which assign to every member w of a universal set of worlds W a subset $f(w)$ of the power set of W.

Then modals like, 'must', 'might', etc. can be analyzed in terms of this given background as follows: 'It must be the case that P' is true at w with respect to $f(w)$ if and only if the proposition expressed by P follows from everything one knows at w. By the same token: 'It might be the case that P' is true at w with respect to $f(w)$ if and only if the proposition expressed by P is compatible with everything one knows at w.

This is the simple analysis of contextual modals along Kripkean lines that Kratzer proposes in various writings ('the standard model' of modalities according to her terminology). The surface description of these standard models for modalities might not sound nevertheless familiar to readers used to relational semantics. But it is easy to show that models of these type can be straightforwardly presented as neighborhood models obeying certain constraints, and that, in turn, these neighborhood models are always pointwise equivalent to relational models.[1] Parenthetical references below point to [Ch].

Definition 1 (Ch Def. 7.1). *A model* $M = \langle W, f, P \rangle$ *is a* neighborhood model *iff:*

1. W is a set.
2. f is a function which maps each world to a set of propositions (i.e. $f : W \to \wp(\wp(W))$).
3. $P : N \to \wp(W)$.

Definition 2 (Ch Def. 7.2). *Let w be a world in a neighborhood model* $M = \langle W, f, P \rangle$. *Then:*

1. $\models^M_w \Box A$ *iff* $||A||^M \in f(w)$.
2. $\models^M_w \Diamond A$ *iff* $W - ||A||^M \notin f(w)$.

This is the basic presentation of modalities in terms of neighborhood models. Kratzer's models can be presented as the following particular case:

Definition 3. *Let w be a world in a Kratzer model* $M = \langle W, f, P \rangle$. *Then:*

1. $\models^M_w \Box A$ *iff* $\bigcap f(w) \subseteq ||A||^M$.
2. $\models^M_w \Diamond A$ *iff* $||A||^M \cap \bigcap f(w) \neq \emptyset$.

In this particular case the value of the function f at w encodes the propositional content of the context at w. The propositions in $f(w)$ are the ones that are presupposed, or known, or the propositions that encode what the law provides, etc. There is, of course, the question of how structured this body of information should be. Should it be closed under logical consequence? Should it be closed

---

[1] A neighborhood model M is pointwise equivalent to a relational model N, when for any sentence A and any world w, A is true at w in M iff A is true at w in N. One can show, as suggested above, that for every relational model, there is a pointwise equivalent neighborhood model obeying some constraints, and vice versa.

under conjunctions? Kratzer is right, nevertheless in proceeding gradually from the less structured situation (where f is unconstrained) to the adoption of constraints that fit models capable of handle modalities along Kripkean lines. The following definitions will help to clarify which are these assumptions.

A neighborhood function f is *supplemented* if and only of if $p \in f(w)$ and $q \in f(w)$, when $p \cap q \in f(w)$. In addition a neighborhood model is *augmented* if and only of if it is supplemented and for every world $w \in W$, $\bigcap f(w) \subseteq f(w)$.

A Kratzer model can consist of a frame containing a function f encoding finite information ($f(w)$ might have finite cardinality). The augmentation of such a frame is obtained by taking the intersection of $f(w)$ and closing under supplementation. We can see the augmentation ($f!$) of a function f as the representation of the logical consequences of the information encoded in $f(w)$ for every w in the domain. It is also easy to see that the augmentation of the frame of a Kratzer model validates exactly the same modal sentences that the Kratzer model itself. Moreover it is well known that augmented neighborhood models are pointwise equivalent to standard relational models.[2]

So, Kratzer's 'standard account' is indeed relatable to the standard relational account of modalities. But as Kratzer correctly observes, the standard account cannot tolerate a weakly inconsistent context (i.e. a case where $\bigcap f(w)$ is empty). In this case every proposition is both necessary and impossible. Kratzer also argues (correctly, we think) that the standard account cannot handle correctly *graded* modalities and that its treatment of dyadic modality is also poor.

2 Double Relative Models

Kratzer's solution to the former problem is the adoption of models dependent on two conversational backgrounds rather than one:

Definition 4. *A model $\langle W, f, g, P \rangle$ is doubly relative (a DR model) iff:*

1. *W is a set.*
2. *f and g are conversational backgrounds.*
3. *$P : N \to \wp(W)$.*

The background f is called the *modal base* for the model; g is called the *ordering source*. The role of the modal base is slightly different than in the Kratzer models presented above. Its central function is to determine for each world, which worlds are accessible from it. A conversational background e is *empty* just in case $e_w = \emptyset$ at every world w. We conventionally use e to denote the empty conversational background. When the DR model $\langle W, f, g, P \rangle$ is understood and we have fixed a world w, we let, for every world u, $\gamma(u) = \{p \in g_w | u \in p\}$.

Definition 5. *The ordering source g is used to fix a typicality ordering $\leq_{g(w)}$ such that for all $w, u, v \in W$, $v \leq_{g(w)} u$ iff $\gamma(u) \subseteq \gamma(v)$.*

---

[2] See [Ch], page 221.

So, for each world the second conversational background induces a (partial) ordering of the set of worlds accessible from that world (where this accessible set is determined by the first background). Truth conditions in this model function in a simple way, only complicated by technical concerns abut infinity. The basic idea is that a proposition is necessary if and only of it is true in all accessible worlds which come close to the ideal established by the ordering source. The formal definition is as follows;

Definition 6. *Let w be a world in a DR model $M = \langle W, f, g, P \rangle$:*

1. $\models_w^M \Box A$ *iff for all $u \in \bigcap f_w$, there is a $v \in \bigcap f_w$ such that $v \leq_{g(w)} u$ and for all $z \in \bigcap f_w$ if $z \leq_{g(w)} v$, then $z \in ||A||^M$.*
2. $\models_w^M \Diamond A$ *iff $\models_w^M \neg \Box \neg A$.*

The definition for $\models_w^M \Box A$ is a bit difficult to parse. The idea is that $\models_w^M \Box A$ just in case for every world u if we look at a series of worlds each progressively more typical than u, there comes a point in the sequence v such that for any world z more typical than v, A is true at z. Or in other words, $\models_w^M \Box A$ when no matter what world u you choose, you can find a world more typical than v such that A is true in every world more typical than v. The only difference between what has just been expressed and the full formal definition is that in the definition the domain of quantification for u, v, and z is restricted to $\bigcap f_w$. Also observe that when the typicality ordering has most typical worlds in $\bigcap f_w$, the truth conditions for a sentence A can be simplified: $\models_w^M \Box A$ just in case $||A||^M$ is a superset of the set of most typical worlds in $\bigcap f_w$. When this happens, we say that a doubly relative model is reduced.

We introduce some terminology and show that any partial ordering can be a typicality ordering.

Definition 7. *Let $\leq$ be a partial ordering on U (i.e. $\leq \subseteq U \times U$ that is reflexive, antisymmetric, and transitive). If $V \subseteq U$, then the ordering of $\leq$ restricted to V is the ordering $I(\leq, V) = \{\langle u, v \rangle \in \leq | u, v \in V\}$.*

Definition 8. *A point u is minimal in an ordering $\leq$ iff for every $v \leq u$, $v = u$. An ordering $\leq$ of U has a bottom iff for every $u \in U$, there is a minimal point $m \in U$ such that $m \leq u$. If $\leq$ does not have a bottom, $\leq$ is bottomless. If $\leq$ has a bottom, then $bot(\leq)$ denotes the set of minimal points in $\leq$. If $\leq$ is bottomless, $bot(\leq)$ is undefined.*

A minimal point is one which has has no points less than it. Suppose $\leq$ is an ordering on U. If every element in U is greater than or equal to a minimal point of U, then $bot(\leq)$ is the set of minimal points of U. On the other hand, if there is some element u of U such that there is no minimal point m of U with $m \leq u$, then $bot(\leq)$ is undefined.

Theorem 1. *If $\leq$ is a partial ordering of W, then there is a DR model $\langle W, f, g, P \rangle$ such that $\leq = \leq_{g(w)}$.*

This theorem shows that in general, $\leq_{g(w)}$ can be any partial ordering of W.

2.1 Relating DR Models and Augmented Models

For some DR models, the doubly relative condition DRC can be simplified substantially. We call the simplified condition the reduced doubly relative condition (abbreviated RDRC). If a DR model can be characterized by RDRC, we say that it is a reduced doubly relative model (or RDR model for short). We show that the class of RDR models is pointwise equivalent to the class of augmented models. Furthermore, a DR model is pointwise equivalent to an augmented model just in case the DR model is reduced.

Definition 9. *A DR model $M = \langle W, f, g, P \rangle$ is reduced (alternatively, a reduced doubly relative model or RDR model) just in case $bot(I(\leq_{g(w)}, \bigcap f_w))$ is defined.*

Given a sentence A, we call the condition $bot(I(\leq_{g(w)}, \bigcap f_w)) \subseteq ||A||^M$ the *reduced doubly relative condition* (RDRC for short). Intuitively, The reduced doubly relative condition just says after restricting $\leq_{g(w)}$ to $\bigcap f_w$, the resulting restricted ordering has a bottom. So a DR model is not reduced, if the restriction of $\leq_{g(w)}$ to $\bigcap f_w$ is bottomless.

Theorem 2. *The class of RDR models is pointwise equivalent to the class of augmented neighborhood models.*

The final theorem of this section shows that the class RDR models characterizes those DR models which are pointwise equivalent to augmented minimals.

Theorem 3. *A DR model is reduced iff it is pointwise equivalent to an augmented minimal $N = \langle W, N, P \rangle$ such that for all $w \in W$, $\bigcap N_w = ||B||^N$ for some sentence B.*

2.2 A Concrete Example

Kratzer offers a concrete example, which we can analyze here in order to illustrate some of the main constructions. The example uses a set $\{M, G, \neg G\}$ containing background information about laws passed in a hypothetical country. 'M' stands for 'Murder is a crime' and 'G' for Owners of goats are liable for damage caused by them'. The idea is that two courts in different parts of the country have passed contradictory laws - the intuition is that different geographical conditions justify different jurisprudence.

With this example in mind, Kratzer believes that the following sentences should be true in M at a given world w:

1. $\Box M$ – "Murder is necessarily a crime."
2. $\Diamond G$ – "Owners of goats are possibly liable for damage caused by their goats."
3. $\neg \Box G$ – "Owners of goats are not necessarily liable for damage caused by their goats."

But these conditions, though feasible, are not the only ones that fit with the scenario of several courts and conflicting judgments. The way Kratzer has it, any conflict between the courts mutes the precedent. Conflicting rulings essentially cancel each other out. Yet the judicial system might work differently. Suppose that each ruling of a court is binding individually but that the normative force of independent rulings is not combined. So each ruling serves to establish precedent, but the precedent of two or more rulings cannot be conjoined. Such cases need to be decided individually. Then, the following sentences are true in M at a given world w:

1. $\Box(G)$
2. $\Box(\neg G)$
3. $\neg\Box(G \wedge \neg G)$

No DR model gives us these truth conditions. A class of minimal models, on the other hand, does.

In what follows, we show how Kratzer uses a DR model for her truth conditions. Then we show that there is no DR model for the modified scenario. We close by showing that there are minimal models which capture both Kratzer's intuition and the alternative.

2.3 How a Doubly Relative Model Handles Kratzer's Intuition

Let $M = \langle W, f, g, P \rangle$ be a DR model such that the modal base f is empty and the ordering source $g_w = \{||M||^M, ||G||^M, ||\neg G||^M\}$ for some world w. Let $W = \{a, b, c, d\}$. Let $||M||^M = \{a, b\}$. And let $||G||^M = \{a, c\}$. We show that the three sentences mentioned above are all true in M at w (i.e. $\models^M_w \Box M$, $\models^M_w \Diamond G$, and $\models^M_w \neg \Box G$).

By definition, a conversational background f is empty just in case $f_w = \emptyset$ for every world w. Recall that f_w is a set of propositions and that a proposition is just a set of worlds. So $\bigcap f_w$ is the set of worlds such that each $u \in \bigcap f_w$ is in every proposition $p \in f_w$. Now, since f_w is empty, there are no propositions in f_w. Therefore, every world is in every proposition in f_w vacuously. So, $\bigcap f_w = W$. Thus, $I(\leq_{g(w)}, \bigcap f_w) = \leq_{g(w)}$. We show that R is defined. We determine the value of R, and then use RDRC to show the three sentences are true.

Now, consider the typicality ordering $\leq_{g(w)}$ induced by the ordering source g_w. We know that $v \leq_{g(w)} u$ iff $\gamma(u) \subseteq \gamma(v)$ where $\gamma(x) = \{p \in g_w | x \in p\}$ for any world x. The following table shows the value of $\gamma(x)$ for each $x \in W$:

x	$\gamma(x)$
a	$\{||M||^M, ||G||^M\}$
b	$\{||M||^M, ||\neg G||^M\}$
c	$\{||G||^M\}$
d	$\{||\neg G||^M\}$

Hence, $\leq_{g(w)}$ contains $\langle a, b \rangle$ and $\langle b, c \rangle$ in addition to the reflective pairs $\langle x, x \rangle$. The following diagram shows, graphically, the ordering:

It is clear from the diagram that $bot(\leq_{g(w)}) = \{a, b\}$. Therefore, if the proposition expressed by a sentence is a super set of $\{a, b\}$ it is necessary in M at w. Clearly, $||M||^M$ is a super set of $\{a, b\}$. Therefore, $\models^M_w \Box M$.

Since $\Diamond$ is the dual of $\Box$, $\models^M_w \Diamond G$ iff $\models^M_w \neg\Box\neg G$ iff it is not the case that $\models^M_w \Box\neg G$ iff $||\neg G||^M$ is not a super set of $\{a,b\}$. So since $||\neg G||^M = \{b,d\}$, $\models^M_w \Diamond G$.

Likewise, since $||G||^M = \{a,c\}$ is not a superset of $\{a,b\}$, it is not the case that $\models^M_w \Box G$. Therefore, $\models^M_w \neg\Box G$. Thus, we see that the DR model has the truth condition that Kratzer intends.

2.4 Why No Doubly Relative Model Handles the Alternative Intuition

Suppose there is a doubly relative model that handles the alternative intuition. Then there is a pointwise equivalent filter $M = \langle W, N, P\rangle$ that also handles the alternative intuition. Thus for every world w, $\models^M_w \Box G$, $\models^M_w \Box\neg G$), and $\models^M_w \neg\Box(G \wedge \neg G)$. So by definition, $||G||^M \in N_w$ and $||\neg G||^M \in N_w$. Then since M is closed under finite intersections, $||G||^M \cap ||\neg G||^M \in N_w$. Therefore, $\models^M_w \Box G \wedge \neg G$ counter to the intuition.

2.5 Minimal Models Capture Both Intuitions

Previously, we showed that every RDR model is propositionally equivalent to an augmented minimal model. We apply the construction used in proving the theorem to find the minimal model which captures Kratzer's intuition.

Let $M = \langle W, f, g, P\rangle$ be the DR model we showed to capture Kratzer's intuition. Then according to the theorem, we let $N = \langle W, N, P\rangle$ be an augmented minimal such that $\bigcap N_w = bot(\leq_{g(w)})$. By definition, $\models^N_w \Box A$ iff $||A||^N \in N_w$ iff $\bigcap N_w \subseteq ||A||^N$ iff $bot(\leq_{g(w)}) \subseteq ||A||^N$ iff $\models^M_w A$. So, clearly, a sentence is true in M at w iff it is true in N at w.

With Kratzer's intuition out of the way, we return to the alternative intuition. The alternative intuition is that residents of New Zealand are obligated by the consequences of each ruling individually, but that combinations of rulings do not establish obligations. In terms of neighborhoods, closure under logical consequence corresponds to the idea of supplementation.

Let $N = \langle W, N, P\rangle$ be a minimal model such that for all w, N_w is supplementation of $\{||M||^M, ||G||^M, ||\neg G||^M\}$. Then since $||G||^N \in N_w$, $||G||^M \in N_w$. Therefore, $\models^N_w \Box G$. Likewise, $\models^N_w \neg G$. But since the empty set is not a superset of any set, it is not N_w. Therefore, $\models^N_w \neg\Box(G \wedge \neg G)$.

2.6 Relating DR Models and Filters

A neighborhood model which is supplemented and in addition is closed under finite intersections and contains the unit is called a *filter*.

Theorem 4. *The class of DR models is pointwise equivalent to the class of filters.*

Proof outline. This is the main theorem of the paper. Presenting its detailed proof would take too much space here (we would need new technical notions, which only have instrumental value for the completion of the proof). So, we will only present an outline with the main ideas.

In the first part half of the proof, we show that for any DR model we can construct a pointwise equivalent filter. We accomplish the construction by defining the neighborhood of a minimal model to validate just those sentences which are valid in the DR model. Thus pointwise equivalence follows trivially. The rest of this part is dedicated to showing that the minimal model is indeed a filter.

The second part requires much more work. We take advantage of the fact that every filter has a nesting.[3] Using one such nesting, we construct the ordering source for the DR model. (We just let the modal base be empty.) The construction is accomplished by slicing the intersection of the defined nesting out of the nesting itself and letting the ordering source be the sliced nesting[4] together with the intersection, which was removed.

We prove four properties about the constructed ordering source. These give us enough properties to show that the DR model validates a sentence iff the filter does. We split the proof into two cases. First, we show modal sentences validated by the filter are validated by the DR model. Then we show the converse. This concludes the proof. ●

The theorem shows that the DR models go beyond relational models. In fact there are examples of filters which are not augmented and that therefore have no relational (Kripkean) counterparts. The following example is adapted from David Lewis [?]. Suppose that two particles are separated by a distance of just over an inch. Let the set W of possible worlds be a continuum expressing possible distances between the two particles. So, $16 \in W$ is the possible world in which the particles are 16 inches apart. Relative to W, propositions (sets of worlds) allow is to specify ranges for the distance separating the particles. For instance, the assertion that, "the two particles are separated by a distance of less than 3 inches" is represented with the proposition, $\{x \in W | x < 3\}$.

Let w be a world in which our original assertion – that the two particles are separated by a distance of just over an inch – is true. Let g_w be a nest of propositions compatible with this assertion so that a proposition $p \in g_w$ just in case p is an interval $p = (a, b)$ such that $a = 1$ and $1 < b$. Now, choosing to call this set g_w is no coincidence: g_w as viewed as an ordering source, produces a doubly relative model for the world w in which the particles are just over an inch apart. (Let the ordering source f be empty.) To see this, just consider the typicality ordering $\leq_{g(w)}$ induced by g_w. Let $u \in W$. If $u < 1$, then u is in none of the propositions of g_w (i.e. $\gamma(u) = \emptyset$) – we know that the particles are more than an inch apart, so any world where they are less than an inch a part isn't typical at all. If $1 < u$, on the other hand, then $\gamma(u) = \{(1, x) | u \leq x\}$. Therefore, if $v \in W$ such that $1 < v < u$, then $\gamma(u) \subseteq \gamma(v)$. Hence, $v \leq_{g(w)} u$. Furthermore, there is no most typical world under $\leq_{g(w)}$, since for all $u > 1$, there is a $v > 1$

---

[3] A set **C** of sets is a *nest* if and only if for every pair of sets U, V in **C**, either $U \subseteq V$ or $V \subseteq U$.

[4] This sliced nest is the nest obtained from the original nest by subtracting the intersection of the original nest from each element of the original nest.

such that $v < u$. The typicality ordering is bottomless, hence the DR model is not reduced.

Notice that for use in natural language, it doesn't really matter whether or not particles or points or that space is continuous. The very fact that we can consider an example of this sort shows that people can talk about points and continuous spaces. Notice further, that a filter equivalent to this model will not be augmented since there is no most typical world under the typicality ordering. Therefore there is no Kripke model pointwise equivalent to it.

3 Going Beyond Double Relative Models

Even when Double Relative models can go beyond relational semantics when the underlying domain is infinite, they are guaranteed to have Kripkean counterparts when the domain is finite. In order to see this consider;

$$B_w = \{||A||^M : M, w \models \Box A\}$$

It is a consequence of Kratzer's definition that if W is finite then B_w is guaranteed to be consistent in the sense that $\bigcap B_w$ is non-empty. But many of the graded modals that she or other people have considered do not have this feature. For example, the modality 'highly probable' lacks this feature. Or the modality 'legally obligatory' also lacks this feature when the underlying legal context encodes jurisprudence containing contradictory norms. Kratzer seems to confine her attention to a particular class of neighborhood models, namely those conforming filters. This seems to be limitative from a representational point of view.

Say that you are reasoning about an assembly line implementing certain standard for quality control. The background information is determined by high probability judgments about the line. Given the standards for quality control enforced in the line we have that it is highly probable for each piece fabricated in the line that it is non-defective (a piece is defective when its length is less than a threshold *min* and when its length is more than a threshold *max*). At the same time it is highly probable that some piece is defective. It seems that one should be able to draw sound inferences from this background. That one can conclude for example for lengths t' between *min* and *max* that: 'If the length is t' then the piece is non-defective'; or 'It is not (seriously) possible that there will not be defective pieces' ('It is expected that there will be defective pieces'); or 'It is likely that there will be some defective pieces', etc.

This background cannot be represented directly as a modal base. Apparently in a situation of this kind Kratzer needs to proceed indirectly by assuming a space of points (in the propositional representation) each one of which encodes a situation where some number of pieces are defective, and where there is information about the length of each piece. Then an ordering is constructed for this space. The points where certain small number n of pieces (for a threshold value n) are defective and the rest sane will be in the bottom of the ordering (maximally probable). Points where all pieces are defective will be maximally improbable, and so on. Any proposition that is a superset of the bottom of the ordering will be considered highly probable. This gives us: 'it is highly probable that there is some defective piece'; but it does not give us for each piece p, 'it is highly probable that p is defective'. We can have at most: 'it is possible that piece p is defective'.

A situation of this sort can be straightforwardly represented via neighborhood models. For each w we can directly use $f(w)$ as in the standard case. The propositions in $f(w)$ are the propositions receiving high probability. What is a possibility in this model? We can extrapolate from the standard case. In this case A is possible if $||A||^M$ is compatible with $\bigcap f(w)$. In this new type of scenario we can define that A is possible if $|A|$ intersects each proposition in $f(w)$. Then we can add rationality conditions to the neighborhood representation that fit the modality in question. For example in this case one would like to add supplementation and the constraint establishing that neighborhoods contain the unit. But it is clearly unreasonable to require closure under finite intersections.

So, it seems that one can do with neighborhoods all that one can do with Double Relative Models and more. But it is also true that some of the basic insights in neighborhood semantics can be supplemented with some of the ideas that inspired Double Relative models. Kratzer's analysis of contextual conditionals offers a clear illustration of this possibility.

4 Conditionals

Kratzer's basic idea about conditionals is that the function of the antecedent in evaluating an if-clause at a world w is to restrict the set of worlds which are accessible from w. This is done formally by adding the proposition expressed by the antecedent to the modal base.

This proposal (based in David Lewis's ideas about the semantics of counterfactuals) has various well-known problems representing conditionals of various kinds. One of the them is the validation of an axiom establishing that $A \wedge B$ entails $A > B$, for all propositions A, B. It is easy to find counterexamples to this axiom for various types of conditionals, both of ontic and epistemic type. Stochastic counterexamples are the simplest that one can exhibit. Suppose that at time t and with respect to some determinate chance set-up (officiating as background) a coin is tossed (T) and that it lands heads (H). Does one want to infer from that that $T > H$ (i.e. 'If the coin is tossed at time t in this chance set-up, it lands heads')? This does not seem reasonable. Other types of counter-examples abound.

There are remedies for this kind of problem. We will propose a possible solution, and we will show that the solution in question can be implemented in an extension of the neighborhood semantics for conditionals proposed in [Ch].Our intention is to show that the central semantic ideas defended by Kratzer can all be accommodated in a slight extension of the classical presentation of neighborhood semantics. Neighborhood semantics has the adequate resources and the representational flexibility needed to implement a two dimensional analysis of modalities (both monadic and dyadic) that is both sensitive to context and capable of tolerating weakly inconsistent background information.

4.1 Neighborhood Models for Conditionals

Neighborhood models of conditionals are offered in chapter 10 of [Ch]. In comparison with other kind of semantic account of conditionals they remain relatively unexplored. Perhaps due to the fact that these models for conditionals were proposed independently of models for unary modalities, and with different intended applications, the two accounts of modality remain also unrelated. One of the virtues of Kratzer's models is that she presents an integrated account of both modalities.

Definition 10 (Ch Def. 10). *A model* $M = \langle W, f, P \rangle$ *is a* neighborhood conditional model *iff*:

1. W *is a set.*
2. f *is a function which maps pairs of worlds and propositions to a set of propositions (i.e.* $f : W \times \wp(W) \to \wp(\wp(W))$ *).*
3. $P : N \to \wp(W)$.

The truth conditions for conditionals proceed along familiar lines:

$\models^M_w A > B$ iff $||B||^M \in f(w, ||A||^M)$.

Models where the conditionals neighborhoods form filters have been recently considered in order to represent non-monotonic notions of consequence - see, among other papers [?]. Nevertheless in an integrated representation one would like to establish some lawful connection between the content of $f(w)$ and the conditional neighborhood $f(w, ||A||^M)$. Intuitively the latter encodes the result of supposing that A is the case with respect to the conversational background given by $f(w)$. And one can capture this intuition by having a fixed partial ordering which regulates suppositional background changes. Obviously the more unstructured is the content of $f(w)$ the more difficult is to establish the desired relation. We will consider here only cases already analyzed by Kratzer, where the conversational background $f(w)$, for every world w, forms at least a filter, and we will restrict our analysis to finite domains. The extension to the infinite case is technically more complicated but it preserves the central ideas of the model. The extension for backgrounds that do not manage to form filters requires modifications that are beyond the scope of this note. It should be noted in passing though that the use of a partial, rather than a total ordering, extends some of the most common analysis that one can find in the literature on conditional logic, while preserving most of the basic intuitions advanced by Kratzer.

Given a fixed partial ordering $\leq$ we can define some useful notions. $C(\leq)$ is a *downward cone* for $\leq$ if and only if for every z in $C(\leq)$ we have that for every x, such that $x \leq z$, that $x \in C(\leq)$. We can restrict here our attention to backgrounds $f(w)$ obtained by taking supersets of a given downward cone of $\leq$ for $f(w)$.

A chain of worlds will be called minimal if and only if it has a minimal point as endpoint. The intuitive idea is that the background f is given by an augmented neighborhood whose intersection is determined by taking the union of a set of minimal chains of worlds. At least this is so if the domain is finite. Otherwise the notion of downward cone might also include infinite descending chains of worlds.

So, what is supposing that A is the case with respect to $f(w)$? When $C(\leq)$ is compatible with the proposition expressed by A we can just take $C(\leq) \cap ||A||^M$. Otherwise we can implement the idea that supposing A requires opening our mind with respect to A first and then inputing A in this suppositional scenario. This requires contracting both A and its negation from $C(\leq)$ – this idea is first presented by Isaac Levi in [?]. This requires, in turn, an explanation of how can we contract A from a view that affirms A by appealing to the fixed ordering $\leq$. This can be done in two steps. First we identify a halo of $\neg A$ points with respect to $C(\leq)$. This can be done via the following operation;

$h(\neg A) = \{w \in ||\neg A||^M :$ for some $z \in C(\leq)$, $z \leq w$ and for all y such that $w > y$, $y \in ||A||^M \}$

As a second step we can define the contraction in question as follows:

$||C(\leq) \div A||^M = \{w \in ||A||^M : w \leq z$ for $z \in h(\neg A) \}$

With the help of these preliminary definitions we can introduce now a conditional neighborhood model as follows:

Definition 11. *A model* $M = \langle W, s, P \rangle$ *is a conditional neighborhood model iff:*

1. *W is a set.*
2. *s is a function which maps pairs of world and propositions to propositions (i.e. $f : W \times \wp(W) \to \wp(W)$).*
3. *$\leq$ is a partial ordering on W.*
4. *$P : N \to \wp(W)$.*

The selection function s is defined in terms of the contraction operation presented above;

$s(w, ||A||^M) = ||(C(\leq) \div A) \div \neg A||^M \cap ||A||^M$.

Definition 12 (Ch Def. 7.2). *Let w be a world in a neighborhood model* $M = \langle W, f, P \rangle$. *Then:*

1. $\models_w^M A > B$ *iff* $s(w, ||A||^M) \subseteq ||B||^M$

Basically we have introduced one more selection function in a model that contains the double relative construction of Kratzer. W officiates here as the modal base of Kratzer. The ordering has the role of her ordering source. And the selection function is constructed by performing a more sophisticated manipulation in the ordering than just taking a restriction with respect to the antecedent. Of course, when $C(\leq)$ entails the proposition expressed by A the result of $s(w, ||A||^M)$ need not yield $C(\leq)$ back. One might then select a strictly larger set of worlds as the representative of the supposition with A.

5 Conclusions and Related Work

Kratzer's account of modalities is based on two central contextual parameters given by the modal base and the ordering source. Formally these conversational backgrounds provide two main tools: an accessibility relation and a partial ordering defined over it. Kratzer manages to show that these two parameters can be used advantageously in order to actually go beyond the common approach to modality based on the use of relational semantics (what she calls the 'standard approach'. We argued above that neighborhood semantics can be used in order to represent most of what she is interested to represent, and that actually neighborhood representations can go beyond the expressive power of her approach when infinity matters (and when one wants to represent genuine conflict in the background).

But we also argued that some of the insights deployed in Double Relative models can be transposed into an enriched version of neighborhood semantics. The central idea is to enrich neighborhood models with a partial ordering. This ordering can play an explanatory role in determining the behavior of selection functions in modeling conditional reasoning.

But the use of an ordering both in hypothetical reasoning and in fixing the content of background information should be familiar from its use in neighboring disciplines. We are thinking about Economics as a paradigmatic example. It is true nevertheless that it is only recently that economists began to think about rationality without the help of a *total* ordering. And the use of partial orderings in encoding typicality or entrenchment is also relatively recent in philosophy and mathematical psychology. But it is clear that abandoning the assumption of completeness of an underlying ordering helps to obtain more realistic descriptions useful in all these disciplines.

Notice that in the previous section we followed Kratzer in distilling augmented neighborhoods from a given partial ordering. Our $C(\leq)$ is a generalization of Kratzer's use of the bottom of the partial ordering in order to determine the content of neighborhoods (used in the semantics of monadic modal operators). But an approach in terms of neighborhoods is more flexible and allows for more scattered representations of epistemic context. For example, we can follow the lead of Amrtya Sen in [?] and we can construct neighborhoods from partial orderings as follows:

$Min(\leq) = \{P \in 2^W : \text{for every } x, y \text{ in } P, x = y \text{ and for no } z \in P, r \in W\ z > r\}$

$Min(\leq)$ picks the propositions composed by (comparable and equi-preferred) worlds that fail to dominate any other world. If we take the supplementation of $Min(\leq)$ there is no guarantee that the corresponding neighborhood is even weakly consistent (its intersection can be empty). Nevertheless, we can define monadic and dyadic modalities with respect to this neighborhood by appealing to minor variants of the techniques used above. So, for example, the propositions included in the neighborhood obtained by taking the supplementation of $Min(\leq)$ might indicate what is known 'all things considered' - if the ordering is interpreted as a typicality ordering.

Moreover as Sen arguend in [Sen] (by appealing to ideas first developed by Stig Kanger in modal logic) it is also reasonable to make the partial ordering dependent on the menu of preferences as well as dependent on the chooser. As he argued in [Sen] much of the tools used in maximization (minimization) remain useful in this case, while the representation gains in realism.

References

[AC]   Arlo Costa, H., First order extensions of classical systems of modal logic:The role of the Barcan schemas, forthcoming in *Studia Logica* 71 (2002) 87-118.

[HAC]   Arlo Costa, H., and Pacuit, E. First order classical modal logic, Technical Report No. CMU-PHIL-164, Carnegie Mellon University (2004). An abbreviated version is forthcoming in TARK X.

[Ben]   Ben-David, S., and Ben-Eliyahu, R., A Modal Logic for Subjective Default Reasoning". *Artificial Intelligence*, Volume 116 (2000) 217-236.

[Ch]   Chellas, B. *Modal logic: An introduction*, Cambridge UP, Cambridge (1980).

[K]   Kratzer, A Modality. In: von Stechow, A. and Wunderlich, D. (eds.). *Semantik. Ein internationales Handbuch der zeitgenossischen Forschung* Walter de Gruyter, Berlin (1991) 639-650.

[K2]   Kratzer, A. What 'must' and 'can' must and can mean. In: *Linguistics and Philosophy*, 1(3) (1977) 337-356.

[Le]   Levi, I.: *For the sake of the argument: Ramsey test conditionals, Inductive Inference, and Nonmonotonic reasoning*, Cambridge University Press, Cambridge (1996).

[L]   Lewis, D. *Counterfactuals*, Blackwell (1973).

[M]   Montague, R. Pragmatics, *Contemporary Philosophy*, La Nuova Italia Editrice (1968) 101-121.

[Sen]   Sen, A. Non-Binary Choice and Preference, in *Rationality and Freedom*, Harvard University Press, Cambridge, Mass. (2003).

[S]   Scott, D. Advice on modal logic. *Philosophical Problems in Logic*, (1970) 143-173.

Understanding Context Before Using It

Mary Bazire[1] and Patrick Brézillon[2]

[1] Laboratoire Cognition & Usages, CNRS FRE 2627,
2 rue de la Liberté 93523 Saint-Denis Cedex
mary.bazire@cognition-usages.org
[2] LIP6, Case 169, Université Paris 6,
8 rue du Capitaine Scott, 75015 Paris, France
Patrick.Brézillon@lip6.fr

Abstract. This paper presents an attempt to point out some problematic issues about the understanding of context. Although frequently used in cognitive sciences or other disciplines, context stays a very ill-defined concept. Our goal is to identify the main components of the context on the basis of the analysis of a corpus of 150 definitions coming mainly from the web in different domains of cognitive sciences and close disciplines. We analyzed this corpus of definitions through two methods, namely LSA [1], [2] and STONE [3], [4], and we conclude that finally the content of all the definitions can be analyzed in terms of few parameters like constraint, influence, behavior, nature, structure and system.

Keywords: Context, Definition, Contextual elements.

1 Introduction

Etymologically, context of a given utterance (= co-text) is composed by the part of text before [5] and after the utterance (cf. textual linguistics). The meaning of the term knew an evolution towards a larger acceptation and now the meaning generally accepted is that context is the set of circumstances that frames an event or an object. This concept is increasingly used in a large number of disciplines like psychology, especially since the emergence of situated cognition theories [6], those theories considering cognition in its natural context [7].

However, it is difficult to find a relevant definition satisfying in any discipline. Is context a frame for a given object? Is it the set of elements that have any influence on the object? Is it possible to define context a priori or just state the effects a posteriori? Is it something static or dynamic? Some approaches emerge now in Artificial Intelligence (e.g. see [8]). In Psychology, we generally study a person doing a task in a given situation. Which context is relevant for our study? The context of the person? The context of the task? The context of the interaction? The context of the situation? When does a context begin and where does it stop? What are the real relationships between context and cognition?

Hereafter, the paper is organized in the following manner. Section 2 shows the key role plays by context in psychology. Section 3 presents the importance to elaborate a robust of context for an effective use of this concept. Section 4 introduces briefly the two techniques of semantic analysis (the Latent Semantic Analysis [1], [2] and STONE [3], [4]) that have been applied to a corpus of definitions of context for identifying more clearly the concept of context. Finally, Section 5, proposes a first model taking into account every component of the context and presents an efficient definition for addressing psychological problems.

2 Context, a Fundamental Notion in Psychology

There are a number of concepts in Psychology that are considered to be close to context, such as the notions of *situation* and *field* (e.g. in the Witkin theory on cognitive styles [9]), the notion of *milieu* (e.g. in animal psychology), the notion of *distractor* (e.g. in perception psychology), *background* (e.g. in the gestalt theory). All these notions are considered as parts of the immediate context of a stimulus. References on context exist in all the different branches of psychology. This is certainly the most used concept [10]. In general psychology, context is often used in the sense of "set of situational elements in which the object (i.e. stimulus considered above) being processed is included" [5]. Context has multiple effects of interest in Psychology: studies on memorization and lexical access processes point out some priming phenomenon (the primer belongs to the context and is a stimulus that affects previously the processing of the current one; for example, categorization [11] and attentional capacities [12] are affected by context.

Bastien [13] distinguishes two opposite views about the role of context in human cognition. The first view considers cognition as a set of general processes that modulate the instantiation of general pieces of knowledge by facilitation or inhibition. In the second view (in the area of situated cognition [6], [14], context has a more central role as a component of cognition by determining the conditions of knowledge activation as well as the limit of knowledge validity. These two opposite views underline that context may have an internal nature or an external one. On the one hand, context is an external object relative to a given object, when, on the other hand, context belongs to an individual and is an integral part of the representation that this individual is building of the situation where he is involved. According to this second viewpoint, "context cannot be separated from the knowledge it organizes, the triggering role context plays and the field of validity it defines" [13].

Those two opposite views could be reconciled in a prototype model [15]. Those seem to be two points of attraction "around which the various notions of context seem to converge: (1) a local point which is related to the structural environment. It is activated and constructed in the ongoing interaction as it becomes relevant [16], and is eventually shared by interactants; and (2) a global point, which refers to the given external components of the context. It includes knowledge and beliefs, and the general experience resulting from the interplay of culture and social community" [17].

Thus, context has an explanatory power in Psychology. The question then is to identify the main dimensions of context. We address this question in the next section.

3 Need of a Consensual Definition

Previous sections show that context is almost a buzzword in psychology, as in a number of other ones, such as computer science or linguistic. However, like some words as "concept" or "system", the word "context" either is not defined or it is possible to find as many definitions (in an ad hoc manner) as authors. A reason is that this word is used supposing that every body knows its meaning, otherwise the author wants to delineate the particular meaning he gives at this word. This epistemological problem [18], [19] will be not discussed in this paper. Moreover, one notes an increasingly use of this word on the Web: 5% in 1997 and 15% in 2004 of Web pages used this word, clearly not always in a useful sense (e.g. "In the context of this situation, we think that...") but more frequently now. When the word "context" is used in a meaningful sense, there is no real consensual definition, mainly because definitions of context are too much dependent of their own contexts (e.g. the discipline in which the definition is taken but also on both the kind and the goal of a given text). However, we think that few parameters may leads to an objective view (decontextualized view) of the definitions of context.

3.1 The Corpus of Definition

Brézillon [8], [18], [19] and Brun [20] collected a set of definitions of context, mainly on the Web. Initially the goal was to develop an effective model of context to be used in a knowledge-based system. After a while it appeared that it was not possible to develop in isolation a model of context because context, knowledge and reasoning [21] are strongly intertwined. Thus, the idea was to design and develop rather a software based on a context-based formalism of representation of knowledge and reasoning, formalism called Contextual Graphs [21]. For Brézillon [18], the lack of an operational definition explains several failures noted in knowledge based system use because (1) users and their contexts are not taken into account, (2) out of its context of validity, there is an incorrect use of the knowledge, (3) with the infinite number of contexts, it is not possible to endow prior its use a system with all the needed knowledge, and (4) computer systems have not the means to identify the context in which a user's request must be interpreted.

We now have a database of more than 150 definitions of context pick up on the Web from various disciplines such as computer science, philosophy, economy, business, HCI, etc. Each definition is entered in the database according to a frame containing:

- the definition
- the object about which context is defined
- the domain from which the definition has been found
- the reference (URL or bibliographical reference)
- eventual comments or complements to the definition.

For example:

> **Definition 2:**
>   Context is what constrains a step of a problem solving without intervening in its explicitly
> Item:
>   A step of a problem solving
> Domain:
>   Artificial Intelligence, Decision Support System
> Source:
>   Brézillon P. (1999) Context in problem solving: A survey. The Knowledge Engineering Review, 14(1): 1-34.
> Comments:

The goal of this collection of definitions relies on the extraction of consensual, necessary or dependent features related to a given domain of knowledge. For instance, the definition "context is what constrains a problem solving without intervening in it explicitly" above is similar to the definition "All that may influence a given process whom first causes are known" [22]. By abstracting key elements of the definitions we thought to develop a resulting definition able to cover a large class of definitions. Among the different methods available, we chose to extract main components of the definitions with the software called STONE [3], [4]. In our approach we have taken only half of the corpus of definitions (indeed, 66 definitions) in order to use the second half for validation of our model (Selection of definitions has been made by classical statistical means).

However, most of the definitions in our initial corpus mostly belongs to cognitive sciences. The most represented areas in the database are: artificial intelligence (39), documentation (27), cognitive ergonomics (9), cognitive psychology (6), business (4), philosophy (2), linguistic (4) and one in education sciences, another in medicine and the last one in neuroscience (some definitions are multidisciplinary).

The definitions are collected in order to avoid redundancy in order to have, as much as possible, each definition original and unique. We have been obliged to put apart definitions coming from some highly specialized domains, such as logic or programming, because such definitions resulted in too much abstract, formal or specific definitions. The corpus presents nevertheless a sufficient variety: "Context is what constrained a problem solving without intervening in it explicitly" [18]; "Something that growths and change as a function of the time , user's requests and the growth intelligence of the system" [23]; "An ordered set of modules which can change during the process of proving a formula" [24], etc. This variety arises from the fact that there is no absolute context, context being relative to something. As a consequence, context must be studied according to its use.

3.2 The Latent Semantic Analysis (LSA)

Latent Semantic Analysis (LSA, in free access at http://lsa.colorado.edu) is a theory (a memory model) and a method (a mathematical model) used for extracting and representing the meaning of words in their context of use by applying statistical computations to very large corpus of texts [1], [2]. The representation of the meaning of a word or a sentence in LSA is given by a vector in a multi-dimensional space, as an average of the meanings of all the paragraphs where the word occurs (this point is detailed in [25]). The distance between two vectors coding two words or paragraphs represents a semantic distance between two terms. By choosing this method, we hypothesize that clusters of definitions semantically close could emerge.

The application of LSA to our corpus gives a correlation matrix that represents the semantic proximity between each paragraph (a definition in our corpus). The lower correlation was 0.27 for two definitions corresponding to "a set of constraints" and "a shared ontology which constitutes a shared vocabulary." The higher correlation was 0.93 for definitions including "what constrains each step of a problem solving without intervening in its explicitly." and "The set of the properties that are associated to an entity according to the environment in which is the entity." The two later definitions were the closest according to the LSA model. The correlation matrix being too large to treat manually, we apply then a cluster analysis.

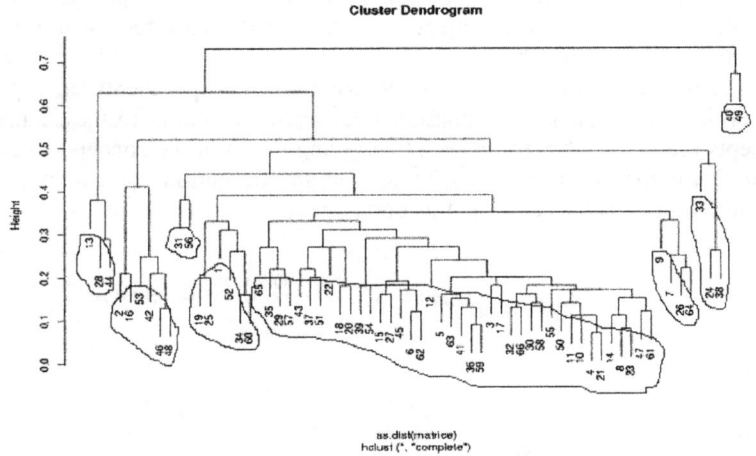

Fig. 1. Cluster analysis of 66 definitions of context applied to the correlation matrix

However, the cluster analysis does not provide relevant results for pointing out some particular clusters. The method allows to identify successive fitting and the marginality of some definitions as shown in Figure 1. First, it is possible to isolate two definitions, namely:

<u>Definition 40</u>: "Context is any identifiable configuration of environmental, mission-related, and agent-related features that has predictive power for behaviour." and
<u>Definition 49</u>: "A shared goal between instructor and students."

Marginality could appear from a too large specificity. Indeed, the two definitions are in-depth and are domain-dependent with respect to the knowledge. There is a problem of granularity and this opens the question of the management of different levels of generality in the corpus. Some definitions are at a high level of generality and could be applied directly across disciplines, when other definitions are specific of a given situation in a given domain, generally given as ad hoc definitions.

Therefore, it is possible to explain some groups of definitions in Figure 1. A first group concerns definitions around the idea of a psychological process of representation building. For Bastien [5], one of the main role of context is to remove ambiguity or to interpret situation. A second group meets together definitions about the spatial dimension of context. This is generally in this acceptance of the term that the popular speech uses the term of context: the context of an object is the set of physical elements in which the object is embedded. A third group gathers some definitions taken from linguistic. The other groups are difficult to identify.

The lack of clear result by the LSA approach leads us to try a second approach called STONE [3], [4]. It is a tool for the analysis by a binary questionnaire, i.e. the answering modalities must fit presence or lack of selected features. This method allows the representation of inter-modality associations by a set of implication relations, such as either a tree (with respect to the answering modality) or a Galois lattice (with respect to the set of answers given by one person. In such a representation, nodes represent object categories and edges represent relations of semantic inclusion [4], [3], [26].

A first step of our work was to extract manually concepts identified explicitly in the definitions with their logical implications. Then, according to the occurrence of the concepts found, STONE provides a tree giving the fit of the concepts as presented in Figure 2 for the component "Behavior" in the definitions of context, with the number of definitions concerned by this component.

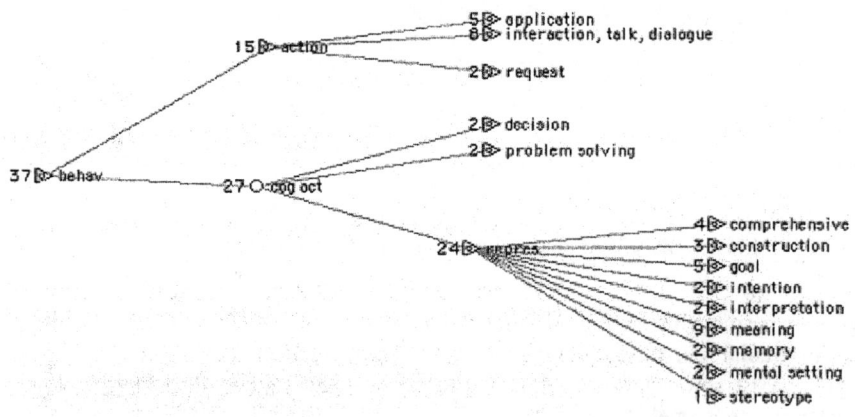

Fig. 2. An example of the context component "behavior"

The analysis of our corpus by STONE provides the following results. Six main components are identified--constraint, influence, behavior, nature, structure and system—and are then refined (for 4 of them) into more specific parts. The object concerned by a context can be an action or a cognitive activity. Its nature is external, internal (see details in [13]) or conditional. Context can be either a process (and thus dynamic) or an information (a concept, a data, a document, an entity, a piece of knowledge, a proposition, a set of propriety or a stimulus). Finally, context can have a structure of set or frame such as a network or an ontology, whose constituting elements share some relations or not.

More than half of the 66 definitions (37 definitions) concern the context of a behavior, the behavior being an action (applied to an application software, a language act or a request), or a cognitive activity (decision, problem solving, or representation construction). Such contexts are not external and objective physical contexts of an object because they intervene for the understanding, the goal determination, the intention, the construction of an interpretation, a meaning or a stereotype and for the memorization.

From these trees, STONE builds the Galois lattice for representing the implication relations shared by the definitions. For example, see Figure 3 the lattice corresponding to the context component "behavior".

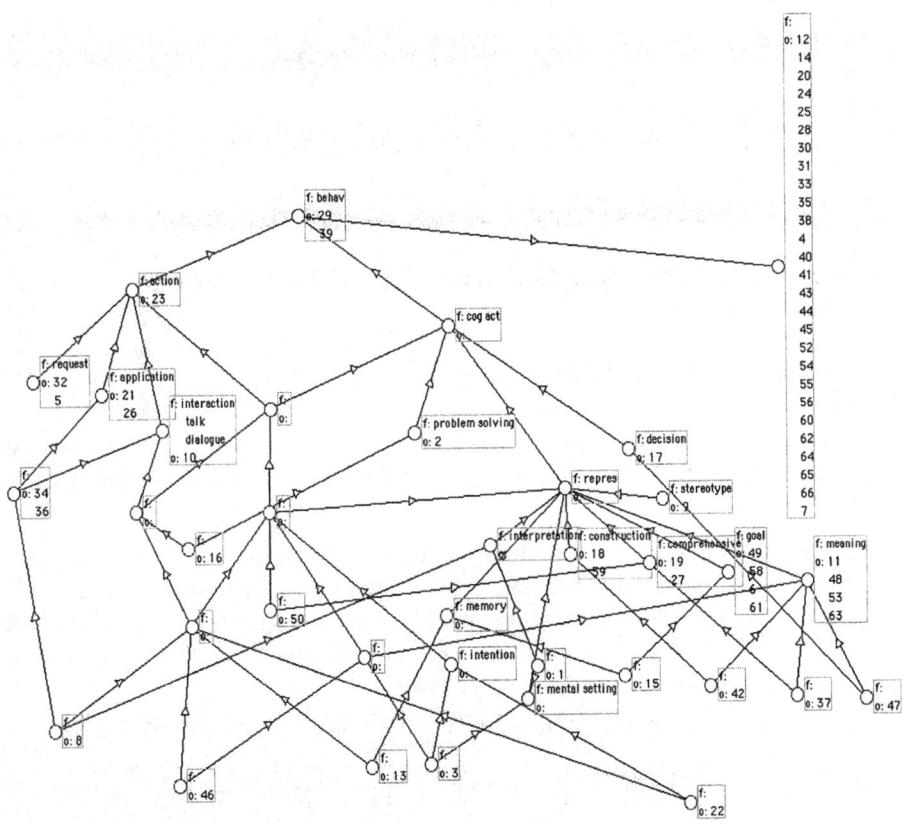

Fig. 3. Example of lattice: the context component "behavior"

In such an analysis, the identified groups are characterized by the fact that some definitions share concepts in a more or less direct way. Thus, seven definitions of our corpus of 66 definitions mention the notion of meaning as a sort of representation, a sort of cognitive activity, or a sort of behavior. Nodes without neither label nor definition represent categories without any example. Such nodes are generally intermediate and with more than one super-ordinate node. Moreover, the lattice defines relations as implication relations between definitions: nodes representing categories (action, cognitive activity, decision, for instance) and edges representing relations of semantic inclusion. Consequently, the following two definitions:

Definition 34: "Context is any information that characterizes a situation related to the interaction between humans, applications and the surrounding environment", and

Definition 36: "Context is any information that can be used to characterize the situation of an entity, where the entity is a person, place, or object that is considered relevant to the interaction between a user and its application, including the user and the application themselves.."

belong to the same category and their subordinate categories are the categories "application" and "interaction, dialogue, talk", themselves, subordinates to "action." which is a kind of behavior. Some categories make obviously a choice between the two main sub-categories (the observed behavior is either an action or a cognitive activity), whereas other definitions, such as the definition:

Definition 46: "Context is the interrelated conditions in which something exists or occurs, and the parts of a discourse or treatise that precede and follow a special passage and may fix its true meaning"

takes the two alternatives into account. Nevertheless, it happens that the lattice become very complex if a definition mentions a great deal of proprieties. For example, the definition:

Definition 8: "Context is any information that can be used to characterize and interpret the situation in which a user interacts with an application at a certain time."

may be too explicit because it mentions the ideas of application software, interaction, cognitive activity, problem solving, interpretation and representation. Such an accuracy in the definition make the lattice complex. Conversely, excluding such definitions could show only the significant relations, notably relations that concern more than one definition. There is a compromise to find here.

The two analysis used (LSA and STONE) reveal the diversity (almost a heterogeneity) showed by the collected definitions. Indeed, the mixing up of the lattices and the progressive exclusion of definitions in the cluster analysis illustrates this situation.

Note that the grouping in clusters of linguistic definitions confirms that a definition of context can be very dependent on the discipline it belongs to (see the grouping in clusters of the linguistic definitions).

4 A Working Model for Context

Our conclusions of the corpus study lead us to build a model of context representing the components of a situation (where the context is taken into account) and the different relations between those components, with the hypothesis that the reason why definitions diverge is that they don't put their focus of attention on the same topics.

Then, the topic on which definitions put the focus on may allow discriminating definitions. A situation could be defined by a user, an item in a particular environment, and eventually an observer (according to certain definitions). Context interferes with each of these elements as illustrated in Figure 3. Some definitions focus on certain relations, when others concern other relations.

We separate context and environment because we consider that the physical environment is not totally relevant in a task running and we propose that context represents all that it is significant at a given moment but can belong to any of the terms of our model, depending to the task goal.

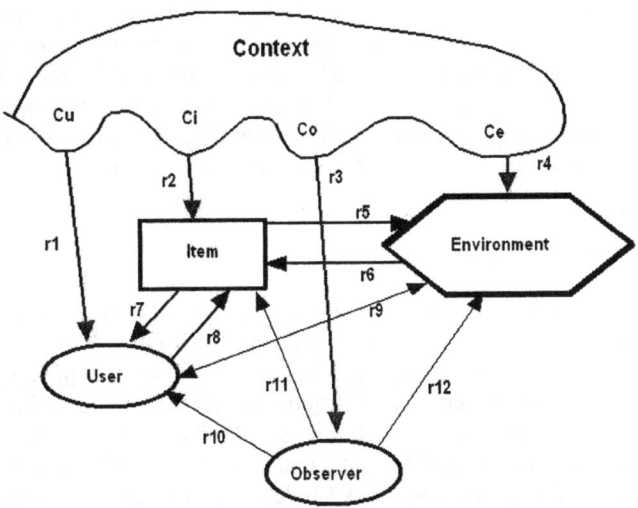

Fig. 4. Our proposal for a model of context

For example, consider the two following definitions.

<u>Definition 2</u>: "Context is what constrains each step of a problem solving without intervening in its explicitly."

The focus is here on the item (a step of a problem solving) and the goal is to solve the problem at this step. In Figure 4, the definition takes into account the context of the item (Ci), the user and the environment (the other steps of the problem solving), eventually the relations R6 and R8. The other topics are not considered. The definition:

Definition 57: "Context has two dimensions: (1) Ecology: aspects of the school that are not living, but nevertheless affect its inhabitants (resources available, policies and rules, and size of the school); and (2) Culture to capture the informal side of schools."

focuses on life in a school (the item in Figure 4). The goal is here to integrate a school in the neighborhood of a city. Here, the important topics are the user, the contexts of the item and of the environment and the relations R5, R6, R2, R4, R7, R8 and R9.

We hope that this model will allow us to build groups regarding to the topics they focus on. Then, in order to work on a definition of context, we would concentrate our work first on the definitions that focus on context itself. We hypothesize also that the differences between domains could be explain in those terms: by the topics they single out.

Another study, lead concurrently concerned the necessary components of a situation in order to be easily understood and agreed with [27], allows us to consider that context can be specified for a given situation by the answering of the following questions : Who? What? Where? When? Why? and How? "Who" indicates the subject, the agent of the action. "What" represents the object, the patient that sustains the action. "Where" and "When" give information about the spatio-temporal location of the considered action. "Why" gives the intentions, the goal (and eventually the emotions) of the subject. At last, "How" makes explicit the procedure needed to realize the action. Our study mentioned above showed that the optimal order of presentation is Action-Object-Agent-Location, which gives a primordial role to action that confirms the necessity to consider the goal as a part of context.

5 Conclusion

Our analysis of definitions of context collected on the Web shows that context could be analyzed through six essential components. The context acts like a set of constraints that influence the behavior of a system (a user or a computer) embedded in a given task. We discuss the nature and structure of context in the definitions. There is no consensus about the following questions: Is context external or internal? Is context a set of information or processes? Is context static or dynamic? Is context a simple set of phenomenon or an organized network? Indeed, the two analysis used in this study (LSA and STONE) point out the fundamental role played by the context in the process of representation construction. Finally we can say that context occurs like what is lacking in a given object for a user to construct a correct representation. We state that a definition of context depends on the field of knowledge that it belongs to.

Concurrently to these two analysis, an other one, more qualitative (by hand), leads us to extract some determining factors in a definition of context: the entity concerned by the context, its focus of attention, its activity, its situation, its environment and eventually, an observer.

Now, we plan to validate our results on the whole corpus of 150 definitions. However, this study offers a contribution on the cohesiveness of the different approaches met in the literature (e.g. for modeling reasoning, ubiquitous computing and in context-aware applications). In this sense, we think to have bring a new insight. A possible extension of this work would be to take into account the dynamic dimension of context through the definitions of context because we are convinced that if context is often considered in a passive way, context has a real active nature (its dynamical properties are not systematically shown). We hope that this work can be useful as a beginning for the community to discuss the meaning of the term.

References

1. Landauer, T. & Dumais, S. (1997) A solution to Plato's problem: The Latent Semantic Analysis Theory of Acquisition, Induction and Representation of Knowledge. *Psychological Review.* 104: 211-240.
2. Landauer, T., Foltz, P. & Laham, D. (1998) An introduction to Latent Semantic Analysis. *Discourse Processes.* 25 : 259-284.
3. Poitrenaud, S. (2001) Complexité cognitive des interactions Homme-Machine. Paris: L'Harmattan.
4. Bernard, J.-M. & Poitrenaud, S. (1999) L'analyse implicative bayésienne multivariée d'un questionnaire binaire: quasi-implications et treillis de Galois simplifié. *Math. Inf. Sci. hum.* 147: 25-46.
5. Bastien, C. (1998) Contexte et situation *In* Houdé, O., Kayser, D., Koenig, O., Proust, J. & Rastier, F., *Dictionnaire des Sciences Cognitives.* Paris: PUF.
6. Clancey, W. (1994) Situated Cognition : How representations are created and given meaning, *In* Lewis, R. and Mendelsohn, P., Eds. *Lessons from learning.* Amsterdam. pp 231-242.
7. Seifert, C. (1999) Situated cognition and learning. *In* Wilson, R. & Keil, F. *The MIT encyclopedia of the cognitive sciences.* Cambridge: The MIT Press.
8. Brézillon, P. (2003) Representation of procedures and practices in Contextual graphs. The Knowledge Engineering Review, 18(2): 147-174.
9. Witkin, H.A., Goodenough, D.R. & Oltman, P.K. (1979) Psychological differenciation current status, *J. Person. Soc. Psychol.*, 37 : 1127-1146.
10. Péreira, F. & Quelhas, A. C. (1998) Cognition and context: introductory notes. *In* Quelhas, A. C. & Péreira, F., *Cognition and Context.* Analise psicologica: Lisbonnes.
11. Barsalou, L. (1982) Context-independent and context-dependent information in concepts. *Memory & Cognition.* 10(1): 82-93.
12. Sabri, M., Melara, R. & Algom, D. (2001) A confluence of contexts: Asymmetric versus global failures of selective attention to Stroop dimensions. *Journal of experimental psychology: Human perception and performance.* 27(3): 515-537.
13. Bastien, C. (1999) Does context modulate or underlie human knowledge? *In* Quelhas, A.C. & Péreira, F., *Cognition and Context.* Analise psicologica: Lisbonnes.
14. Newell, A. & Simon, H. (1972) *Human Problem Solving.* Prentice-Hall, Inc. : Englewood Cliffs.
15. Rosch, E. H. (1978) Principles of categorization. In: E. Rosch & B. Lloyd, eds., *Cognition and Categorization.* Hillsdale, N.J.: Erlbaum Associates. 27-48.
16. Sperber, D. & Wilson, D. (1986). Relevance: Communication and Cognition. Cambridge, MA: MIT Press.
17. Akman V., Bazzanella C. (2003) The complexity of context, in Varol Akman, Carla Bazzanella(eds.) *On Context, Journal of Pragmatics*, special issue, 35, 321-329
18. Brézillon, P. (2002) Modeling and using context: Past, present and future. *Rapport de recherche interne LIP6*, Paris.

19. Brézillon, P. (1999) Context in Problem solving: A Survey. *The Knowledge Engineering Review.* 14(1): 1-34.
20. Brun, V. (2002) Mémoire de maîtrise: Université de Paris 8.
21. Brézillon, P., (2005) Task-realization models in Contextual Graphs. Modeling and Using Context (CONTEXT-05), A. Dey, B.Kokinov, D.Leake, R.Turner (Eds.), Springer Verlag, LNCS (This volume).
22. Denis, M. & Sabah, G. (1993) In Modèles et concepts pour la science cognitive. Hommage à J.-J. Le Ny. (Grenoble: PUG).
23. Minsky, M. (1986) *The Society of Mind*, Picador Publisher.
24. Bielikova, M. & Navrat, P. (1997) A multilevel knowledge representation of strategies for combining modules. *Proceedings of the Seventh International Conference on AI and Information-Control Systems of Robots.* World Scientific, Singapore: 155-168.
25. Bazire, M. (2003) *Effets des composantes du contexte sur la compréhension.* Mémoire de DEA: Université de Paris 8.
26. Tijus, C. (2001) Contextual categorization and cognitive phenomena, P.B.V. Akman, R. Thomason, R.A. Young (Eds) *Modeling and Using Context.* Third International and Interdisciplinary Conference, CONTEXT, 2001.
27. Jang, S. & Woo, W. (2003) Ubi-UCAM : A unified context-aware application model. Proceedings of the 4th international and interdisciplinary conference CONTEXT 2003. Stanford : P. Blackburn et al. (Eds). pp.178-189.

Epistemological Contextualism: A Semantic Perspective

Claudia Bianchi[1] and Nicla Vassallo[2]

[1] University San Raffaele, Faculty of Philosophy, Palazzo Borromeo,
I-20031 Cesano Maderno (MI), Italy
[2] University of Genova, Department of Philosophy, via Balbi 4,
I-16126 Genova, Italy
{claudia, nicla}@nous.unige.it

Abstract. According to epistemological contextualism, a sentence of the form "S knows that p" doesn't express a complete proposition. Different utterances of the sentence, in different contexts, can express different propositions: "know" is context-dependent. This paper deals with the *semantic* contextualist thesis grounding *epistemological* contextualism. We examine various kinds of linguistic context dependence, which could be relevant to epistemological contextualism: ambiguity, ellipsis, indexicality, context-sensitivity of scalar predicates, dependence on standards of precision. We argue that only an accurate analysis of the different varieties of context sensitivity secures us a better understanding and a clearer evaluation of the contextualist approach.

1 Introduction

Consider the two following scenarios.

Case A: It is Friday afternoon. Nicla and I are walking around in the town. We stop in front of my travel agency. I would like to collect my plane ticket for a trip to Paris in two weeks. But I realize that the agency is too crowded and I hate the crowds. I tell Nicla: "Tomorrow I will come back to collect my ticket". She says: "It is better to do it now. Perhaps tomorrow the agency is closed. Several travel agencies are closed on Saturday". I reply: "I know that the agency will be open tomorrow. It is open on Saturday. I personally saw it two weeks ago".

Case B: It is Friday afternoon. Nicla and I are walking around in the town. We stop in front of my travel agency. I would like to collect my plane ticket for a trip to Paris on Sunday. But I realize that the agency is too crowded and I hate the crowds. I tell Nicla: "Tomorrow I will come back to collect my ticket". She says: "It is better to do it now. Perhaps tomorrow the agency is closed. Several travel agencies are closed on Saturday". I reply: "I know that the agency will be open tomorrow. It is open on Saturday. I personally saw it two weeks ago" She retorts: "You have to collect your ticket, otherwise we can't go to the Context Conference in Paris. The agency might have changed its opening days during the last two weeks. Do you really know that it will be open tomorrow?". I admit: "Perhaps I do not know. It is better to ask which days the agency is open".

Epistemological contextualism is the doctrine that the truth-conditions of knowledge ascribing and knowledge denying sentences vary according to the context in which they are uttered. According to the classic view in epistemology, knowledge is justified true belief. Invariantism[1] claims that there is *one and only one* epistemic standard for knowledge. Therefore it is wrong to claim – for the same cognitive subject S and the same proposition p – that

(1) S knows that p[2]

is true in one context, and is false in another context. On the contrary, contextualism admits the legitimacy of *several* epistemic standards that vary with the context of use of (1); it is right to claim – for the same cognitive subject S and the same proposition p – that (1) is true in one context, and is false in another context. The contextualist thesis is quite interesting and appealing because, compared to invariantism, promises epistemological theories more compatible with our everyday epistemic practices and solutions more alluring to the problem of skepticism. In what follows we focus mainly on the semantic issues of epistemological contextualism, trying to provide a better formulation of the *semantic* contextualist thesis grounding *epistemological* contextualism: our goal is then to assess differences and similarities between "know" and context-sensitive terms in natural language.

This paper is structured as follows. In section 2 we briefly present a standard version of epistemological contextualism. In section 3 we sketch the contextualist response to skepticism. In section 4 we examine various kinds of context dependence, which could be relevant to epistemological contextualism: ellipsis, ambiguity, indexicality, context-sensitivity of scalar predicates, dependence on standards of precision. In the conclusion we argue that only an accurate analysis of the different varieties of context sensitivity secures us a better understanding and a clearer evaluation of the contextualist approach, and of its response to the skeptic.

2 Contextualism in Epistemology

The idea that there are two senses of "know" – a weak or ordinary sense and a strong or philosophical sense – is not new: it has been defended by Descartes, Locke, Hume[3], and more recently by Malcom.[4] Those authors maintain that there are *two* different epistemic standards for knowledge attributions. This surely is a contextualistic thesis, but mild, since contemporary contextualism allows that there are *several* standards. To say that there are several standards of knowledge is nothing but to recognize the indisputable fact that we apply different standards in different conversational contexts, so that it happens that we are willing – for the very same proposition p and the very same subject S – to attribute knowledge in contexts where

---

[1] The term is due to [39].
[2] Or "S is justified in believing that p": on this point, cf. [1].
[3] Cf. [16] Part IV, [27], Book IV, Ch. XI, par. 3, and [22].
[4] Cf. [28] 183.

low standards count and to deny knowledge in contexts where *high* standards count. As Keith DeRose defines it, contextualism "refers to the position that the truth-conditions of knowledge ascribing and knowledge denying sentences (sentences of the form "S knows that p" and "S doesn't know that p" and related variants of such sentences) vary in certain ways according to the context in which they are uttered. What so varies is the epistemic standards that S must meet (or, in the case of a denial of knowledge, fail to meet) in order for such a statement to be true".[5]

According to contextualism, then, the truth-values of knowledge attributions vary on the basis of certain characteristics of the conversational context.[6] Contextualism allows the possibility of truly asserting

(1) S knows that p

in one context and

(2) S does not know that p

in another context, identical to the previous one in all features relevant for the determination of indexicals or usual contextual expressions: different contexts call for different epistemic standards – lower or higher, weaker or stronger – that S must satisfy.

Let's go back to the two cases presented in the Introduction. From a contextualistic perspective, the sentence

(3) Claudia knows that the agency will be open tomorrow

is true in case A, and false in case B.[7] Or, to put it in an equivalent way, (3) is true in case A and

(4) Claudia doesn't know that the agency will be open tomorrow

is true in case B – where (4) is the denial of (3). While according to the invariantist, it is the strength of Claudia's epistemic status that changes, according to the contextualist, Claudia has the same epistemic position in case A and B, but there is a variation in what semantically counts as "knowing".

3 Skepticism

Contextualism has been often developed in order to face the skeptic's challenge[8]. Hume surely was a precursor: the Humean suggestion is that we must distinguish between a philosophical, or skeptical, context and an ordinary one. The skeptical hypotheses are normally raised in the former, while in the latter they are so obliterated

---

[5] [14] 187-188.
[6] For *subjective* contextualism, they are the characteristics of the context of the cognitive subject, while for *attributive* contextualism they are the characteristics of the context of the attributor. We will not address the question of the difference between subjective and attributive contextualism here: on this point see [42] and [43].
[7] Taking however for granted that the three traditional conditions are satisfied in both cases, i.e. that: it is true that the travel agency will be open tomorrow; Claudia believes that the travel agency will be open tomorrow; Claudia's belief is justified.
[8] For contextualist approaches to skepticism cf. [17], [18], [19], [40], [7], [8], [9], [12], [13], [25], [26], [44], [45]. For criticisms of these approaches cf. [33] and [37].

that they appear cold and ridiculous, once we come back to philosophical reflection.[9] Contemporary contextualism elaborates a claim analogue to the Humean one: the skeptic modifies by her hypotheses the ordinary epistemic standards, and, in particular, raises them in order to create a context in which we cannot truly attribute knowledge to ourselves and to others. Once the standards are raised, or strengthened (even if aberrantly[10]), we must admit that we feel all the force of skepticism, and concede that we accept its conclusion: we don't know what we ordinarily claim to know.

Let us examine the skeptical case by considering the sentence

(5) S knows that she has hands:

(5) is true iff S has hands, S believes it and she is in a good enough epistemic position with respect to "S has hands". If I am not a philosopher, but a normal person, I may truly assert that (5) is true, if S has hands, S believes it and she is in a certain epistemic position: for example, if her perceptual faculties are well functioning, and there is no special reason to believe that any potential defeater obtains. In such an ordinary context the epistemic standards are Low (or Easy). What does the skeptic do? She mentions a skeptical hypothesis, and so confers relevance to it, compelling us to consider it.[11] She changes the context: now we are in a skeptical context. S's position is not judged good enough anymore: the standards are High (or Tough[12]), and so in this case I may say that (5) is false. In order truly to state I now know something, I must rule out the skeptical hypothesis. But, I cannot: therefore we must admit the triumph of skepticism.

Once we are back in more ordinary conversational contexts, we apply more relaxed standards and realize that we can truly attribute knowledge to ourselves and to the others. In DeRose's words: "As soon as we find ourselves in more ordinary conversational contexts, it will not only be true for us to claim to know the very things that the skeptic now denies we know, but it will also be *wrong* for us to *deny* that we know these things".[13] The fact that the skeptic employs high standards in her context cannot show at all that we do not satisfy the weaker standards of ordinary contexts. So there is not any contradiction between saying that we know and that we do not know: the skeptical negation of knowledge is perfectly compatible with ordinary knowledge attributions.

---

[9] Cf. [22] 316: "Most fortunately it happens, that since reason is incapable of dispelling these clouds, nature herself suffices to that purpose, and cures me of this philosophical melancholy and delirium, either by relaxing this bent of mind, or by some avocation, and lively impression of my senses, which obliterate all these chimeras. I dine, I play a game of backgammon, I converse, and am merry with my friends; and when after three or four hour's amusement, I wou'd return to these speculations, they appear so cold, and strain'd, and ridiculous, that I cannot find in my heart to enter into them any farther".

[10] In Goldman's words, "the skeptic… exercises an aberrant pattern of possibility exploration": [20] 148.

[11] Lewis speaks of "Rule of Attention", Cohen of "Rule of Salience", DeRose of "Rule of Sensitivity".

[12] The terms "Easy" and "Tough" are due to [33].

[13] Cf. [14] 194.

The obvious advantage of contextualism over invariantism is its compatibility with everyday epistemic practice. In fact it is evident that, contrary to what invariantism claims, we legitimately apply different epistemic standards in different contexts. Contextualism theorizes this point and bridges the gap between epistemological reflection and ordinary practice. It recognizes the validity of skepticism and, at the same time, the validity of our ordinary knowledge attributions by explaining why we face an alleged paradox.

4 Context Dependence

Broadly speaking, the semantic thesis grounding epistemological contextualism is that a sentence of the form (1) doesn't express a complete proposition. Different utterances of (1), in different contexts of utterance, can express different propositions. The proposition expressed by a knowledge attribution is determined in part by the context of use: we must add in information about the context in order to determine the proposition expressed by a sentence of the form (1). If we fill in the gaps by appealing to low epistemic standards in case A, (3) will be evaluated as expressing a true proposition; if we fill in the gaps by appealing to high epistemic standards in case B, (3) will be evaluated as expressing a false proposition. Little attention has been paid to a precise formulation of the semantic contextualist thesis grounding epistemological contextualism.[14] Our goal is then to examine various kinds of linguistic context dependence, which could be relevant to epistemological contextualism: ellipsis, ambiguity, indexicality, context-sensitivity of scalar predicates, dependence on standards of precision. One point, before starting, to clarify our overall project. We will see that contextualist supporters of the different semantic theories of context dependence agree on the semantic value of (1) in the different contexts – that is, on its truth-conditions. The disagreement, then, doesn't concern the semantic interpretation of (1), but what features of the context have a bearing on its semantic interpretation – and in particular the semantic mechanisms explaining how context affects its semantic interpretation.[15]

4.1 Ellipsis

It has sometimes been suggested to draw a parallel between "know" and standard cases of linguistic ellipsis, as in
 (7) Jack has finished eating; Jill has finished too.
To be evaluated, the second sentence must be supplemented with some linguistic material, drawn from the first sentence. We will then obtain:
 (8) Jack has finished eating; Jill has finished eating too.
In a similar vein, we can postulate for (3) some supplemented linguistic material, allowing to completely determining the sentence truth conditions. One might say that

---

[14] With the noteworthy exceptions of [33], [35], [36], [31], and [15].
[15] The distinction between semantic values and what in the context makes it the case that an utterance has the semantic value it has, is an instance of a well-established distinction within semantics: the one between *descriptive* and *foundational* semantics: cf. [34] 535.

there is an elliptical "according to standard Low" in case A, and an elliptical "according to standard High" in case B, as in

(9) According to standard Low, Claudia knows that the agency will be open tomorrow

and

(10) According to standard High, Claudia doesn't know that the agency will be open tomorrow.

However, there is a widespread agreement that one should in general avoid to try to explain such a contextual variation in terms of some kind of ellipsis: as a matter of fact, one should appeal to ellipsis in a systematic way, for every occurrence of the predicate "know". Moreover, as we mentioned in § 2, it is possible to multiply the standards of justification, implausibly extending the list of linguistic material that should be supplemented in each case.

4.2 Ambiguity

Traditional semantics postulates that two or more conventional meanings are associated with an ambiguous expression, and two or more sets of truth conditions are associated with an ambiguous sentence, such as

(11) Jack went to the bank.

To account for A and B cases, contextualists could devise an ambiguity approach, postulating for "know" different conventional senses: $know_1$ meaning "to know relative to standard Low" and $know_2$ meaning "to know relative to standard High". Sentence (3), then, would have two alternative meanings, corresponding to two alternative logical forms:

(12) Claudia knows$_1$ that the agency will be open tomorrow

and

(13) Claudia knows$_2$ that the agency will be open tomorrow.

The role of context is just to help us determine which of the two (or more) sentences it is uttered in case A and in case B.

Such an approach should be ruled out. Again, it is possible to multiply the standards of justification, extending the list of senses conventionally associated with the predicate. This is a choice hardly plausible and certainly not economical for any semantic theory. Moreover, as Jason Stanley rightly points out, claiming that "know" is ambiguous would amount to showing that there are languages in which the different senses are represented by different words.[16]

4.3 Indexicals

A strategy in terms of some kind of indexicality seems far more plausible. Indexicals are referential expressions depending, for their semantic value, on the context of utterance. Context determines a contextual parameter that fixes the value of an indexical expression: "know" is an indexical expression like "I", "here" or "now". The interpretation of a sentence containing an indexical depends on the characteristics of

---

[16] Cf. [36] 139.

the context in which it is uttered: the interpretation varies with context of use. The sentence

(14) I am French,

for example, is true if uttered by Claudine (who is a French), while it is false if uttered by Claudia (who is an Italian). Language conventions associate with an indexical a rule (a Kaplanian *character*) fixing the reference of the occurrences of the expression in context. The semantic value of an indexical (its *content*, its truth conditional import) is thus determined by a conventional rule and by a contextual parameter, which is an aspect of the utterance situation.[17] The character of an indexical encodes the specific contextual co-ordinate that is relevant for the determination of its semantic value: for "I" the relevant parameter will be the speaker or the agent of the utterance, for "here" the place of the utterance, for "now" the time of the utterance, and so on.

Sentences containing "know" are considered in the same way as sentences containing indexicals. The truth of (1), then, "is relative to the attributor's context, but the notion of truth is preserved by treating knowledge claims as having an indexical component".[18] The character of (1) may be expressed as "S knows that p relative to standard N", while its content would be, in case A, "S knows that p relative to standard Low" and, in case B, "S knows that p relative to standard High".[19] Now, what happens in the skeptical case (sentence (5))? The character of (5) is always constant: "S knows that she has hands relative to standard N". The content varies with the context of the attributor, and, in particular, with the epistemic position she requires for the cognitive subject. If I am not a philosopher, but a normal person, I may truly say that (5) is true, if S knows it relative to standard Low. Of course in a philosophical context, such a position is not judged good enough: in this case I may say that (5) is false. Again it is claimed that the way in which the truth conditions of (5) vary with context is not different at all from the way in which the truth conditions of (14) vary with context.

Let's further examine the analogy between "know" and indexicals. The meaning of an indexical expression is a *function* from contextual factors (such as speaker, place and time of the utterance) to semantic values. By applying the functional conception to examples (1) and (3), we generalize the idea that the conventional meaning of "know" is a function. The function will have the following disjunctive form:

- "know" = *know relative to standard x* if "know" (in context A) or *know relative to standard y* if "know" (in context B) or *know relative to standard z* if "know" (in context C) or *know relative to standard w* in all the other cases.

where, for example, context A is an everyday context with no urgent practical concerns, B is an everyday context with urgent practical concerns, C is a skeptical context, etc. This approach has the valuable benefit of maintaining a stable conventional meaning associated with "know": there is only one function associated

---

[17] Cf. [23]. On Kaplan, and, more in general, on indexicality, cf. [2] and [3].
[18] Cf. [6] 648.
[19] Cf. [33] 326-328. According to DeRose, the character of (1) is roughly the following: "S has a true belief that p and is in a *good enough* epistemic position with respect to p". Its content is "how good an epistemic position S must be in to count as knowing that p", and this shifts from context to context: cf. [11] 922.

with the predicate, and all its different values depend on the different arguments the function takes (context A, B, C, etc.).

We must however reject the functional strategy: it is, in fact, conceivable to obtain for (3) in context A the interpretation of "know" which is normally obtained for (3) in context B. Let's see. Once the context of utterance is fixed, the linguistic rules governing the use of the indexicals determine completely and automatically their reference, no matter what the speaker's intentions are. If, for example, Nicla utters (14) with the intention of referring to Catherine Deneuve (if, for example, she believes she is Catherine Deneuve), she will nonetheless express the (false) proposition "Nicla is French". Analogously, according to the indexical strategy, context fixes the epistemic standards – no matter what the knowledge attributor's intentions are. If Bea, the knowledge attributor, utters (3) in context A, with the intention of expressing the proposition

- Claudia knows that the agency will be open tomorrow relative to standard High, she will nonetheless express the proposition
- Claudia knows that the agency will be open tomorrow relative to standard Low.

In other words, there is no way for her to express, in a context where there are no particular practical concerns, that Claudia knows something according to high standards – which is a very common case (we may suppose, for example, that, in context A, Claudia has recently checked the opening hours of the agency for some other reasons).

4.4 Demonstratives

In the case of an analogy of "know" with a demonstrative, the situation is quite different: a difference, to our knowledge, never correctly underlined.[20] Demonstratives can take an indefinite number of senses depending on the context of use. The meaning of a demonstrative, like "she" in the sentence

(15) She is French,

by itself doesn't constitute an automatic rule for identifying the referent of the expression in a given context. The semantics of "she" cannot unambiguously determine its reference: if, for instance, in the context of utterance of (15) there is more than one woman, the expression "she" can equally identify any woman. In sophisticated versions of traditional semantics, demonstratives (expressions like "he", "she", "this", "that", etc.) are given a different treatment from the indexical one. According to Kaplan, the occurrence of a demonstrative must be supplemented by a *demonstration*, an act of demonstration like pointing, or "the speaker's directing intention". There is no automatic rule of saturation: the semantic value of a demonstrative is fixed according to the speaker's directing intentions.[21] The reference of "I" is the object satisfying, in a given context, the condition coded in its own character: "'I'

---

[20] Schiffer's critique to hidden-indexicality doesn't account for that difference, which we view as a mistake: cf. [33] 326-328.

[21] There are only constraints on possible referents: "he" refers to a male individual who is neither the speaker nor the addressee, "she" to a female individual who is neither the speaker nor the addressee, and so on.

refers to the speaker or to the agent"; while the rule associated with a demonstrative is "an occurrence of 'she' refers to the object the speaker intends to refer to".[22]

In the same way, we could say that there is a variable hidden in the syntactic structure of the predicate "know" (a variable for the epistemic standard): we must specify that variable for every occurrence of the predicate, in every context, in order to have complete truth conditions. The rule associated with (1) would then be: "S knows that p relative to standard N". "Relative to standard N" is now a free variable for epistemic standard: the variable must be saturated according to the context, but there is no automatic rule of saturation, no function from a contextual parameter to a semantic value. Its value depends *on the knowledge attributor's intentions*. Let us examine again the two possible contexts for (3): in case A the set of truth conditions is: *Claudia knows that the agency will be open tomorrow relative to standard Low* – and (3) is true. In case B, the set of truth conditions is: *Claudia knows that the agency will be open tomorrow relative to standard High* – and (3) is false.

In our opinion, the interpretation of "know" as a demonstrative, and not as an indexical, offers a way out from the puzzle mentioned in § 4.3. Suppose again we are in context A modified: in this context there are no particular practical concerns, but Claudia happens to know that the agency will be open tomorrow according to high standards (she has recently checked the opening hours of the agency for some other reasons). Now Bea, the knowledge attributor, may utter (3) in context A, with the intention of expressing the proposition

– Claudia knows that the agency will be open tomorrow relative to standard High,

and succeed in expressing it. The variable hidden in the syntactic structure of the predicate is not fixed automatically by the context (as for indexicals like "I"), but saturated according to Bea's directing intentions.

The analogy between "know" and demonstratives seems promising; but there is a powerful argument against it. As many contextualists have pointed out, in every context there is only one epistemic standard: the epistemic standards for "know" do not shift within a single sentence. Failure to respect such a rule amounts to the formation of what DeRose's calls *Abominable Conjunctions* – sentences such as

(16) S doesn't know she is not a bodiless brain in a vat, but S knows she has hands.[23]

Moreover, not only the contextual parameter corresponding to epistemic standards cannot shift within a clause, but also, once standards have been raised, it is not possible to lower them again in the next sentence.[24] The context-dependence of the predicate "know" seems tied not to the expression itself, but to the whole discourse. This is not the case of many contextual expressions – and in particular this is not the case of demonstratives: demonstratives shift internal to a single sentence, as in

---

[22] See [23] and [24]; cf. [2] 74-76 and [32] 56-57.
[23] Cf. [12] 28. For a different view on this kind of sentences, see [30].
[24] Cf. [25] 247: "the rule of accommodation is not fully reversible. For some reason, I know not what, the boundary readily shifts outward if what is said requires it, but does not so readily shift inward if what is said requires that".

(17) She is French and she is not French

(uttered with two different demonstrations, or two different referential intentions):[25] the interpretation of the relevant contextual parameter can change within a sentence. Notice that while demonstratives do allow for shifts within a clause, indexicals do not:[26] the sentence

(18) I am French and I am not French

is contradictory, while (17) is not. Relative to this particular feature, pure indexicals like "I" behave as "know" does.

4.5 Scalar Predicates

The most promising approach views the predicate "know" as a predicate like "flat", "happy", "rich", "empty" and so on: they are all context-dependent terms in need of "precisification", namely specification of a relevant comparison class. As Cohen claims, "Many, if not most, predicates in natural language are such that the truth-value of sentences containing them depends on contextually determined standards, e.g. 'flat', 'bald', 'rich', 'happy', 'sad'... For predicates of this kind, context will determine the degree to which the predicate must be satisfied in order for the predicate to apply simpliciter. So the context will determine how flat a surface must be in order to be flat".[27] An attribution of flatness is sensitive to a contextually salient scale of flatness. Even for the very same cognitive subject, the very same surface can be judged flat or bumpy, depending on the context: for everyday aims (eating outside, sunbathing, playing volley-ball, and so on) I can judge flat a certain lawn; the same judgement may be plainly false if Wimbledon's tennis tournament is going to be played on that lawn. In a similar vein, an attribution of knowledge is sensitive to a contextually determined epistemic parameter.

Again, we must draw an important distinction here. Following Barbara Partee (and *contra* Jason Stanley and Stewart Cohen), we must distinguish between scalar adjectives like "tall"[28] and absolute adjectives like "flat", "bald", and "certain":[29] while "tall" is a gradable term, "know" is not[30] – exactly like "flat", "bald", and "certain". Rather than abandon contextualism, as Stanley suggests, we claim that the analogy should be maintained between "know" and absolute, context-dependent adjectives.

Does this strategy solve the problem of the context-sensitivity of "know" to discourse, rather than to the term itself? In other words, do absolute context-dependent adjectives allow for standard-shifts within a single sentence? If they do –

---

[25] Cf. [36] 134: "Contextualists typically speak as if there is one contextual standard in a context for all context-sensitive expressions in a discourse... But this is not in general a good description of how context-sensitive expressions work. Rather, the context-sensitivity is usually linked to the term itself, rather than the whole discourse".
[26] Neither Stanley nor Partee in her comments acknowledge this fact.
[27] [16] 60. On "flat", cf. [25] and [19.
[28] "Tall" is Stanley's main target in his critique of contextualism: cf. [36].
[29] Absolute adjectives like "flat" admits "absolutely" and "perfectly" as modifiers: cf. [31]. On "certain", cf. [38] ch. 2.
[30] Cf. [36] 124-130.

and it is usually claimed that they do - the analogy should be rejected. As Stalnaker points out, "Many binding examples of these are familiar:

(19) Many of the animals in the zoo are old,

where it is understood that an elephant is old if it is old for an elephant, a boa constrictor is old if it is old for boa constrictors, etc.".[31] Stanley proposes similar examples, like

(20) That butterfly is small, and that elephant is small

where a butterfly is small if it is small for a butterfly, and an elephant is small if it is small for an elephant.[32] We maintain that intuitions are unclear: for example, what about

(21) Bill Gates is rich and Claudia is rich

or

(22) My desk is flat and Holland is flat?[33]

Is it understood that Bill Gates is rich if he is rich relative to a high standard, and Claudia is rich if she is rich relative to a (much) lower standard? Is it understood that my desk is flat if it is flat relative to a high standard, and Holland is flat if it is flat relative to a lower standard? Remember that once standards have been raised, it is not possible to lower them again: (21) and (22) look as Abominable Conjunctions to us. The same goes for

(23) Bob is tall and Shaq is tall

(where Bob is tall according to some low standards) or

(24) Many people in this room are tall,

if Shaq and Bob are in the room, or

(25) Many people in this room are rich,

if Bill Gates and Claudia are in the room (and Claudia is rich according to some low standards), or

(26) Yul Brinner is bald and Berlusconi is bald.

In our opinion, absolute context-dependent predicates do not shift internal to a single sentence. Is it really obvious that examples like (20) are acceptable? Intuitions are far from stable. If one is hesitant about her intuitions, as we are, then the analogy between "know" and absolute context-dependent predicates still holds and deserves attention and further inquiry.

But there is an alternative solution to the puzzle mentioned in § 4.4.

4.6 Standards of Precision

The crucial point is the possibility, or not, of allowing context shifts internal to a single sentence. Contextual expressions, it is claimed, appeal to different contexts in different parts of the same sentence: sentences (19) and (20) are examples of such a variation. Conversely, "know" does not allow for changes in the context within a sentence: sentence (16) is an example of such impossibility. In other words, we cannot accept sentences like

---

[31] [35] 111.
[32] [36] 134.
[33] Stanley's example is actually "That field is flat, and this rock is flat".

(27) S knows that p and S doesn't know that p

by claiming that the epistemic standards have shifted from Low in the first occurrence of "know" to High in the second one. In this respect, standards of precision are an interesting analogy for knowledge attributions. As it is well known, according to Lewis, once you have fixed some conversational standard of precision, by saying, for example

(28) Italy is boot-shaped,

then you may truly assert

(29) France is hexagonal;

"But if you deny that Italy is boot-shaped, pointing out the differences, what you have said requires high standards under which 'France is hexagonal' is far from true enough".[34] The essential point is that standards of precision are not tied with the expressions "boot-shaped" and "hexagonal": they are part of the conversational score, and are therefore associated with the whole discourse. The same goes for (27). If you say

(3) Claudia knows that the travel agency will be open tomorrow,

and, as Lewis puts it, "get away with it", then (assuming that Claudia and Nicla are in the same epistemic position) you can truly assert

(30) Nicla knows that the travel agency will be open tomorrow.

Low standards are required, hence (31) is true. But if you deny (3), then you cannot assert (30): you have raised the epistemic standards and neither Claudia's nor Nicla's epistemic positions are good enough. The epistemic standards for (3), or for its denial, are not associated with the expression "know": that explains why raising the standards for (3) will affect the standards for (30).

5 Conclusion

In this paper we have focused on the semantic issues raised by epistemological contextualism. Our aim was to provide a better formulation of the semantic thesis grounding epistemological contextualism. We have underlined differences and similarities between "know" and context-sensitive terms in natural language, and distinguished various kinds of context dependence: ellipsis, ambiguity, dependence on the context of utterance of indexicals, demonstratives, and scalar predicates, dependence on standards of precision. We have argued that only an accurate analysis of the different varieties of context sensitivity secures us a better understanding and a clearer evaluation of the contextualist approach, and of its response to the skeptic. More specifically we have identified a crucial question: while "know" does not allow for changes in the epistemic standards within a sentence, contextual expressions appeal to different contexts in different parts of the same sentence. Those remarks usually suggest adopting a strategy in terms of standards of precision, drawing on Lewis and Partee's proposals. Yet, we have claimed that intuitions about absolute context-dependent predicates like "flat" and "bald", and about the way they allow for context shifts, are far from stable. In our opinion the analogy still holds and deserves attention and further inquiry.

---

[34] [25] 245.

References

1. Annis D.B.: A Contextualist Theory of Epistemic Justification. American Philosophical Quarterly 15 (1978) 213-229
2. Bianchi C.: Context of Utterance and Intended Context. In V. Akman et al. (eds.): Modeling and Using Context. Third International and Interdisciplinary Conference, CONTEXT '01, Dundee, Scotland, Proceedings. Springer, Berlin (2001) 73-86
3. Bianchi C.: How to Refer: Objective Context vs. Intentional Context. In P. Blackburn, et al. (eds.): Proceedings of the Fourth International and Interdisciplinary Conference on Modeling and Using Context (CONTEXT'03), Lecture Notes in Artificial Intelligence, vol. 2680. Springer, Berlin (2003) 54-65
4. Bianchi C.: Nobody loves me: Quantification and Context. Philosophical Studies (2005) forthcoming
5. Bianchi C. (ed.): The Semantics/Pragmatics Distinction. CSLI, Stanford (2004)
6. Brower B.W.: Contextualism, Epistemological. In Routledge Encyclopedia of Philosophy, Routledge, London (1998) 646-650
7. Cohen S.: Knowledge, Context, and Social Standards. Synthese 73 (1987) 3-26
8. Cohen S.: How to be a Fallibilist. Philosophical Perspectives 2 (1988) 91-123
9. Cohen S.: Contextualist Solutions to Epistemological Problems: Scepticism, Gettier, and the Lottery. Australasian Journal of Philosophy 76 (1998) 289-306
10. Cohen S.: Contextualism, Skepticism, and the Structure of Reasons. In Philosophical Perspectives, vol. 13, Epistemology. Blackwell, Oxford (1999) 57-98
11. DeRose K.: Contextualism and Knowledge Attributions. Philosophy and Phenomenological Research LII (1992) 913-929
12. DeRose K.: Solving the Skeptical Problem. The Philosophical Review 104 (1995) 1-52
13. DeRose K.: Relevant Alternatives and the Content of Knowledge Attributions. Philosophy and Phenomenological Research LVI (1996)
14. DeRose K.: Contextualism: An Explanation and Defence. In Greco J., Sosa E. (eds.): Epistemology. Blackwell, Oxford (1999) 187-205
15. DeRose K.: The Ordinary Language Basis for Contextualism and the New Invariantism. The Philosophical Quarterly forthcoming
16. Descartes R.: Discours de la méthode. Leida (1637)
17. Dretske F.: Epistemic Operators. Journal of Philosophy 67 (1970) 1007-1023
18. Dretske F.: Conclusive Reasons. Australasian Journal of Philosophy 49 (1971) 1-22
19. Dretske F.: The Pragmatic Dimension of Knowledge. Philosophical Studies 40 (1981) 363-378
20. Goldman A.I.: Psychology and Philosophical Analysis. Proceedings of the Aristotelian Society 89 (1989) 195-209. Now in Goldman A.I.: Liaisons. Philosophy Meets the Cognitive and Social Science. MIT Press, Cambridge, Ma (1992) 143-153
21. Hale B., Wright C. (eds.): A Companion to the Philosophy of Language. Blackwell, Oxford (1997)
22. Hume D.: A Treatise of Human Nature (1739-40). Penguin, London (1969)
23. Kaplan D.: Demonstratives (1977). In Almog J., Perry J., Wettstein H.K. (eds): Themes from Kaplan. Oxford University Press, Oxford (1989)
24. Kaplan D.: Afterthoughts (1989). In Almog J., Perry J., Wettstein H.K. (eds): Themes from Kaplan. Oxford University Press, Oxford (1989)
25. Lewis D.: Scorekeeping in a Language Game. Journal of Philosophical Logic 8 (1979) 339-359. Reprinted in D. Lewis: Philosophical Papers, vol. I. Oxford University Press, New York (1986) 233-249

26. Lewis D.: Elusive Knowledge. Australasian Journal of Philosophy 74 (1996) 549-567
27. Locke J.: An Essay concerning Human Understanding London (1690)
28. Malcolm N.: Knowledge and Belief. Mind 51 (1952) 178-189
29. Moore G.E.: Philosophical Papers. Collier Books, New York (1962)
30. Nozick R.: Philosophical Explanations. Harvard University Press, Cambridge, Ma (1981)
31. Partee, B.: Comments on Jason Stanley's 'On the linguistic basis for contextualism'. Philosophical Studies 119 (2004) 147-159
32. Recanati, F.: Literal Meaning. Cambridge University Press, Cambridge (2004)
33. Schiffer S.: Contextualist Solutions to Scepticism. Proceedings of the Aristotelian Society XCVI (1996) 317-333
34. Stalnaker, R.: Reference and Necessity (1997). In [21] 534-554
35. Stalnaker, R.: Comments on 'From contextualism to contrastivism'. Philosophical Studies 119 (2004) 105-117
36. Stanley J.: On the linguistic basis for contextualism. Philosophical Studies 119 (2004) 119-146
37. Stroud B.: Epistemological Reflection on Knowledge of the External World. Philosophy and Phenomenological Research 56 (1996) 345-358
38. Unger P.: Ignorance. Clarendon Press, Oxford (1975)
39. Unger P.: Philosophical Relativity. University of Minnesota Press, Minneapolis (1984)
40. Unger P.: The Cone Model of Knowledge. Philosphical Topics 14 (1986) 125-178
41. Vassallo N.: Teorie della conoscenza filosofico-naturalistiche. Angeli, Milano (1999)
42. Vassallo N.: Contexts and Philosophical Problems of Knowledge. In V. Akman et al. (eds.): Modeling and Using Context. Third International and Interdisciplinary Conference, CONTEXT '01, Dundee, Scotland, Proceedings. Springer, Berlin (2001) 353-366.
43. Vassallo N.: Teoria della conoscenza. Laterza, Roma-Bari (2003)
44. Williams M.: Unnatural Doubts. Epistemological Realism and the Basis of Scepticism. Blackwell, Oxford (1991)
45. Williams M.: Skepticism. In Greco J, Sosa E. (eds.): Epistemology. Blackwell, Oxford (1999) 35-69

Task-Realization Models in Contextual Graphs

Patrick Brézillon

LIP6, Case 169, University Paris 6, 8 rue du Capitaine Scott,
75015 Paris, France
Patrick.Brezillon@lip6.fr

Abstract. Enterprises develop procedures to address focuses in any case. However, procedures result often in sub-optimal solutions for any specific focus. As a consequence, each actor develops his own practice to address a focus in a given context, focus and its context being particular and specific. The modeling of practices is not an easy task because there are as many practices as contexts of occurrence. This paper proposes a way to deal practically with practices. Based on our definition of context, we present a context-based representation formalism for modeling task accomplishment by users called contextual graphs and its interest for the tasks of incremental acquisition, learning and explanation. Contextual graphs are discussed on a modeling in information retrieval.

1 Introduction

Context plays an important role since a long time in domains where reasoning such as understanding, interpretation, diagnosis, etc. intervenes. This activity relies heavily on a background or experience that is generally not make explicit but gives a contextual dimension to the knowledge and the activity. Thus, context is always relative to something: context of the reasoning, context of an action, context of an object, etc., something that we call focus in this paper.

In this paper, we present contextual graphs that are used in several domains such as medicine, ergonomics, psychology, army, information retrieval, computer security, road safety, etc. The common factor in all these domains is that the reasoning is described by procedures established by the enterprise. These procedures are adapted by actors that take into account the context in which they have to deal with the focus and thus actors create practices as contextualizations of the procedures. Livet [10] shows that for building a robot that put a nut into a bolt, the perfect splinned trajectory (i.e. the procedure) is unrealistic because any small error in the positioning of the bolt with respect to the not will irremediably blocks the operation. There is clearly a need to introduce capacity of light variations around the theoretical trajectory (i.e. allowing practices), to introduce some play for an adaptation to error of positioning of any kind. A practical reasoning is not a logical and theoretical reasoning for which the action leads to a conclusion. The practical reasoning has more a status of inductive probabilistic reasoning: the conclusion cannot be detached (i.e. take a meaning) from the premises.

Procedures are collections of secure action sequences developed by the enterprise to address a given focus in any case. These procedures are decontextualized for covering a large class of similar focuses (generally differing by their contexts of occurrence), not a narrow focus. As a consequence, actors prefer to plan their action in real time rather than to rely on procedures for two main reasons. Firstly, the procedure is not perfectly adapted to the situation at hand and can lead to improper actions or sub-optimal incident resolution strategies. Secondly, the actor can miss some important facts and notice them too late to adequately solve the incident. Thus, the modeling of actors' reasoning is a difficult task because they use a number of contextual elements. These pieces of knowledge, which are not necessarily expressed, result in more or less proceduralized actions that are compiled in comprehensive knowledge about actions. However, procedures are considered as useful guidelines for actors to be adapted for particular focuses. Thus, each actor develops his own practice to address a focus in a given context, and one observes almost as many practices as actors for a given procedure because each actor tailors the procedure in order to take into account the current context, which is particular and specific.

This is a general way to reach the efficiency that decision makers intended when designing the task [2, 3]. Such know-how is generally built up case by case and is complemented by "makeshift repairs" (or non-written rules, "magic book", etc.) that allow the actors to reach the required efficiency. This is a way of getting the result whatever the path followed.

If it is relatively easy to model procedures, the modeling of practices is not an easy task because they are as many practices as contexts of occurrence of a given focus. Moreover, procedures cannot catch the high interaction between the task at hand and the related tasks that are generated by the task itself.

Hereafter, the paper is organized in the following way. Section 2 presents our view on context with a presentation of a general view and our definition. Section 3 discusses the different aspects of the building of the proceduralized context that plays a central role in our approach. Section 4 presents the context-based representation called contextual graphs that relies on the ideas develop in the previous section, with its main characteristics, an example, the way in which the proceduralized context is managed in contextual graphs, and some other properties with a more general impact.

2 Context Characteristics

2.1 Context and Focus

We cannot speak of context out of its context. Context surrounds a focus (e.g. the task at hand or the interaction) and gives meaning to items related to the focus. The context guides the focus of attention, i.e. the subset of common ground that is pertinent to the current task. Indeed, context acts more on the relationships between items in the focus than on items themselves, modifying their extension and surface.

As a consequence, the context makes the focus explicit and the focus defines the relevant pieces in the context. On the one hand, the focus determines what must be contextual knowledge and external knowledge at a given step. For example, a focus on software development implies contextual knowledge such as the programming language, the constitution of the designer team, etc., i.e. knowledge that could eventually be used when the focus evolves. Some knowledge from the designers' individual context could also be considered such as a previous experience with a given piece of software. On the other hand, the context constrains what must be done in the current focus. This could correspond to the choice of a specific method at a given step of a task. A software programmer will focus his/her programming activity in defining classes and methods when in an object-oriented project, he/she will define modules and functions if the project uses the functional paradigm. Indeed, some contextual elements are considered explicitly, say for the selection of the method and thus can be considered as a part of the way in which the problem is solved at the considered step.

The focus evolves along the execution of a series of actions. According to this dynamics of the focus, there is a dynamics of its context (some external events may also modify the context of the focus): focus and its context are intertwined.

2.2 A Definition of Context

For a given focus, Brézillon and Pomerol [3] consider context as the sum of three types of knowledge. First, there is the part of the context that is relevant at this step of the decision making, and the part that is not relevant. The latter part is called **external knowledge**. External knowledge appears in different sources, such as the knowledge known by the actor but let implicit with respect to the current focus, the knowledge unknown to the actor (out of his competence), contextual knowledge of other actors in a team, etc. The former part is called **contextual knowledge**, and obviously depends on the actor and on the decision at hand. Here, the focus acts as a discriminating factor between the external and contextual knowledge. However, the frontier between external and contextual knowledge is porous and evolves with the progress of the focus.

Second, a sub-set of the contextual knowledge is proceduralized for addressing the current focus. We call it the **proceduralized context**. The proceduralized context is a part of the contextual knowledge that is invoked, assembled, organized, structured and situated according to the given focus and is common to the various people involved in decision making.

It could arrive that it is not possible to build the right proceduralized context for a given focus because some information is missing. Once identified from the distance between the focus and the built PC, this missing knowledge can be acquired from the external knowledge and introduced in the sub-set of contextual knowledge on which is built the proceduralized context. This is a process of incremental knowledge acquisition that is intertwined with a process of practice learning as discuss in the next section.

2.3 Conditions of the PC Building

When an element of the contextual knowledge moves in the proceduralized context, this means that we consider explicitly its current instantiations. For example, when I go to my Lab., I look outside to the weather (contextual element of my travel). If it is raining (instance of the contextual element), then I take my umbrella (the action realized with the instantiation of the contextual element) because the subway is not close from my home (other contextual element). With the identification of the contextual element concerned and its specific instantiation, the actor has executed an action (or method) different of the action known by the system (I take my coat if it is not raining in the example).

We consider the process of proceduralized context building (PC building) in two situations, namely a building by a unique actor or by at least two actors. The focus is associated with the contextual knowledge (CK), part of the actor's context that could be mobilized for the current focus; the other part of the context being the external knowledge (EK). A sub-set of CK is explicitly consider to address the focus. Elements of this CK sub-set are assembled, organized, compiled as a proceduralized context (PC) to address the focus.

Sometimes, the building of a proceduralized context fails for a given focus and new (external) knowledge is needed. Associated with the proceduralized-context building, there are simultaneously an incremental acquisition of new contextual elements and the learning of a new practice. Acquisition and learning occur in a specific context that is acquired and learned jointly with the new knowledge. Indeed, first, learned practice and acquired knowledge are more properly two aspects of the learning process, and, second, learning is more an assimilation process rather than an accommodation process because this corresponds generally to a process of refinement of contexts. Second, if the addition of a piece of external knowledge in the PC correspond to a knowledge acquisition, it is nevertheless a learning process because the piece of external knowledge is not simply added to the PC, but assembled and connected to the construction already existing. Thus, it is possible to explain and justify each practice or item in the contextual graph later.

This triple aspect—context growth by integration of external knowledge in the PC building, by integration of a new "chunk of knowledge" in the contextual knowledge, and context change by the movement between the body of contextual knowledge and proceduralized contexts—gives a dynamic dimension to context [2]. This dynamic component is generally not considered in the literature and explains why making context explicit in an application is a difficult task, except if we restrict context at what can be obtained by sensors like in context-aware applications.

In a usual situation, a sub-set of contextual knowledge (CK) pieces are chosen, assembled and structured in a proceduralized context (PC) for addressing the current focus. Once this PC has satisfied the focus, this chunk of knowledge goes back in CK. Thus, CK contains the PC, the initial sub-set of CK elements and the way in which the PC has been built. This allows relevant explanation on why the PC was necessary for the focus, how the PC has been built and what the PC contains.

2.4 Learning of Practices

Learning is a social process that involves building connections: connections among what is being learned and what is important to the actor, connections among what is being learned and those situations in which it is applied, and connections among the actor and other actors of the work group (e.g. see [1]).

2.4.1 Learning by Assimilation of External Knowledge

The Proceduralized Context (PC) built from the contextual knowledge sometimes does not correspond to the focus. As a consequence, there is a need to add some pieces of External Knowledge (EK) in the proceduralized-context building. EK is the (contextual) knowledge not represented because never needed up to now. An EK piece is acquired when a new practice is introduced in the graph (learned by the system). This addition occurs once the discrepancy between the focus and the PC built is known and the missing knowledge identified. This generally occurs because a contextual element, which was not considered before, had the same instantiation all along the practice development and in the particular situation at hand takes another instantiation. With this specific instantiation of a contextual element, a new item (action, activity, method) has been used. The main point here is that an EK piece is introduced first in the sub-set of contextual knowledge used for the PC building, not directly in CK. Our approach brings a solution to the problem of the infinite dimension of context evoked by McCarthy [11]. As a consequence, each element, and especially a contextual element, in a contextual graph has been used at least once.

This situation corresponds to an incremental process of (1) acquisition of new knowledge pieces, and (2) learning of a new knowledge structure as a proceduralized context. If the addition of an EK piece in the PC correspond to a kind of incremental knowledge acquisition, it is also a learning process because the piece of external knowledge is not simply added to the PC, but assembled and connected to the construction already existing and corresponds to the learning of a new way to build a PC as discussed in Section 3.3 and Figure 2 below.

2.4.2 Experience Learnt After PC Use

Once a PC allows the evolution of the focus, the PC is not lost and goes into the body of contextual knowledge from which are coming its elements. This is not only the product that is stored, but also all the way in which this PC has been built, the reasons behind the choices, the alternatives abandoned, etc. The PC is totally integrated in the body of contextual knowledge and the learning is here an accommodation process. The PC that is stored could be recalled later either as a whole (as a part of a new proceduralized context, a chunk of knowledge a la Schank, [14] or the way in which it has been built will be reused for the new proceduralized context. This is a type of learning by structuration of the contextual knowledge, and the more a person is experimented, the more the person possesses available structured knowledge (i.e. chunks of contextual knowledge).

3 Representation by Contextual Graphs

3.1 Characteristics

A contextual graph is a context-based representation of a task execution. Contextual graphs are oriented without circuits, with exactly one input and one output, and a general structure of spindle. A path (from the input to the output of the graph) represents a practice (or a procedure), a type of execution of the task with the application of selected methods. There are as many paths as practices Different solutions can be associated with the unique output, like in the following example chosen in information retrieval: abandon, copy, or save a page before to close the window, but all of them lead to the same conclusion: end of the exploration of the page. A contextual graph is an acyclic graph because user's tasks are generally in ordered sequences. For example, the activity "Make the train empty of travelers" is always considered at the beginning of an incident solving on a subway line, never at the end of the incident solving. A more drastic divergence in the type of output (e.g. the execution of the task is stopped like "Error 104" in information retrieval) must be considered at a upper level in which the contextual graph at hand is a branch of an alternative (a contextual element such as "Are the conditions required for the task execution present? If yes got to the contextual graph otherwise does not consider this contextual graph).

Elements of a contextual graph are: actions, contextual elements, sub-graphs, activities and parallel action groupings.

- An **action** is the building block of contextual graphs. We call it an action but it would be better to consider as an elementary task. An action can appear on several paths (see the example of the action A5 in the example below). This leads us to speak of instances of a given action, as one speaks of the instantiation of a contextual element because an action (e.g. action A5 in Figure 2), which appears on several paths in a contextual graph, is considered each time in a specific context.
- A **contextual element** is a couple of nodes, a contextual node and a recombination node; A contextual node has one input and N outputs (branches) corresponding to the N instantiations of the contextual element already encountered. The recombination node is [N, 1] and shows that even if we know the current instantiation of the contextual element, once the part of the practice on the branch between the contextual and recombination nodes corresponding to a given instantiation of the contextual element has been executed, it does not matter to know this instantiation because we do not need to differentiate a state of affairs any more with respect to this value. Then, the contextual element leaves the proceduralized context and (globally) goes back to the contextual knowledge.
- A **sub-graph** is itself a contextual graph. This is a method to decompose a part of the task in different way according to the context and the different methods existing. In contextual graphs, sub-graphs are mainly used for obtaining different displays of the contextual graph on the graphical interface by some mechanisms of aggregation and expansion like in Sowa's conceptual graphs [15].
- An **activity** is a particular sub-graph (and thus also a contextual graph by itself) that is identified by actors because appearing in several contextual graphs. This recurring sub-structure is generally considered as a complex action. Our definition of activity is close from the definition of scheme given in cognitive ergonomics [9]. Each scheme organizes the activity around an object and can call other schemes to complete specific sub-goals.
- A **parallel action grouping** expresses the fact (and reduce the complexity of the representation) that several groups of actions must be accomplished but that the order in which action groups must be considered is not important, or even could be done in parallel, but all actions must be accomplished before to continue. The parallel action grouping is for context what activities are for actions (i.e. complex actions). This item expresses a problem of representation of items at a too low level of granularity. For example, the activity "Make train empty of travelers" in the SART application [4, 12] accounts for the damaged train and the helping train.

There is no importance to empty first either the damaged train or the helping train or both in parallel. This operation is at a too low level with respect to the general task "Return back rapidly to a normal service" and would have otherwise to be detailed in three paths in parallel (helping train first, damage train first, both in parallel) leading to the same sequence of actions after.

3.2 An Example

Figure 1 gives an example of contextual graph (the definition of the symbols are given in the Table 1) and Figure 2 presents the Activity-1 in the contextual graph represented in Figure 1. This contextual graph represents the different practices that can be used during an information retrieval when one clicks on a link. Square boxes represent actions, circles represent contextual elements (large circles for contextual nodes and black circles for recombination nodes). Rectangular boxes represent activities that can been opened as a sub-graphs. (There is no parallel action grouping in this example.) A path is followed from the left to the right and corresponds to the crossing of a series of elements.

The structure size of such a contextual graph is easily controlled and the consideration of a new contextual element will add few elements and do not increase drastically the size such as in a tree representation where the addition of a contextual element leads often to double the size of the tree for very few changes. A contextual graph represents a specific procedure and all its variants (the practices) at a given level of representation. The number of practices is not infinite and thus the size of the contextual graph is controlled easily because we consider changes only at the same level of representation of the problem solving (otherwise parallel action grouping are introduced in the contextual graph). Moreover, a new practice is generally introduce in a contextual graph as a variant of an existing practice differing from the previous one by a contextual element that was not initially taken into account because its instantiation did matter any more, and by an action or an activity. As such, a contextual graph appears as the corporate memory of the task execution represented by this contextual graph.

Table 1. Exploration of a link target

Contextual element	
C1	What is the link target ?
C2	Is there a HTML version ?
C3	Is the page interesting ?
C4	Are there figures to retrieve ?
C5	Have I time now ?
C6	Is it for a course ?
C7	Can the content of the page be retrieved?
C8	Duration of the download ?
C9	Explore the whole site ?
C10	Is the site already known?
C11	Is the whole page interesting?
C12	Which type of search is it?

Action	Definition
A1	Copy the slide
A2	Paste the slide in my ppt document
A3	Note the idea for a future exploitation
A4	Explore the rest of the presentation
A5	Close the window
A6	Save the page at the html format
A7	Copy and paste the text in a Word document
A8	Open the target in a new window
A9	Download the document
A10	Store the document in the « To read » folder
A11	Open the document
A12	Look for keywords on a colored background
A13	Search where are keywords in the page
A14	Go the home page of the site
A15	Go to the next slide
A16	Look for new stuffs
A17	Select the interesting part of the text on the page
A18	Look for searched item (e.g. reference of a read paper)
A19	Search management (key words or search engine)
A20	Select the text of the whole page

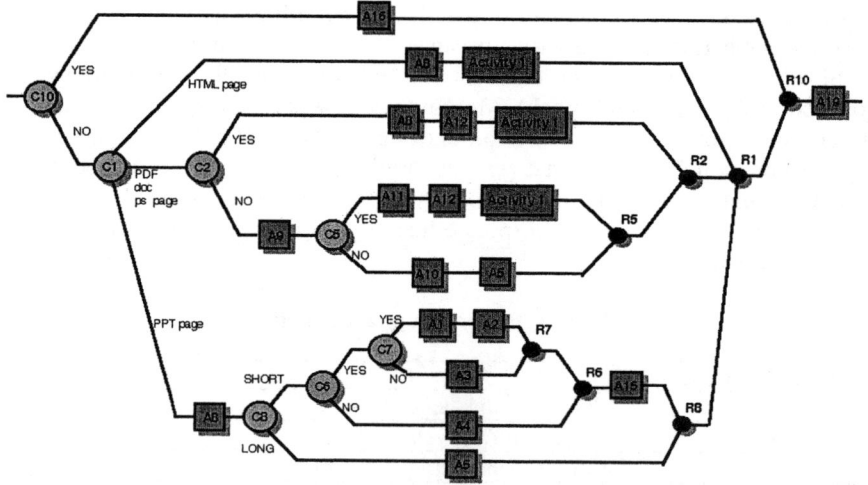

Fig. 1. Exploration of a link target found on a Web page

Contextual elements existing in the contextual graph belong to the contextual knowledge or the proceduralized context depending on the current focus. The context of an action, say action A6 at the top of Figure 2, is described in a static way with a fixed number of pieces of contextual knowledge (C9 and C11), and of an ordered sequence of instantiated contextual elements of the proceduralized context (C8 with the value "Short", C12 with the value "Exploratory", C3 with the value "Yes", and C4 with the value "Yes"). Note that an action such as A5 that appears in several paths (i.e. practices) in Figure 2 is associated with different contexts according to its position in the graph.

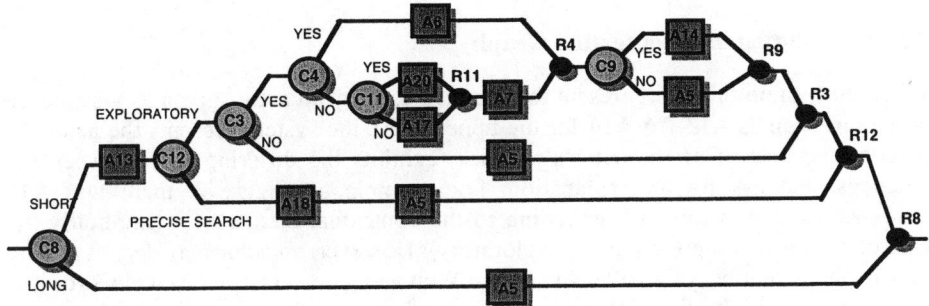

Fig. 2. Information exploitation on a link target (Activity 1 in Figure 1)

The context of a practice (say the upper path including the action A6 in Figure 2), is described in a dynamic way along the progress of the application of the practice (the current focus). For example, the practice represented by the upper path in Figure 2 goes through the steps described in Table 2.

Table 2. Status evolution of context pieces Ci during practice progress (with CK and PC standing respectively for contextual knowledge and proceduralized context with instantiations of the C in the PC)

Step	C8	C12	C3	C4	C9	C11
0	CK	CK	CK	CK	CK	CK
1	PC (short)	CK	CK	CK	CK	CK
2	PC (short)	PC (Exploratory)	CK	CK	CK	CK
3	PC (short)	PC (Exploratory)	PC (yes)	CK	CK	CK
4	PC (short)	PC (Exploratory)	PC (yes)	PC (yes)	CK	CK
5 (= 3)	PC (short)	PC (Exploratory)	PC (yes)	CK	CK	CK
6	PC (short)	PC (Exploratory)	PC (yes)	CK	PC (yes)	CK
7 (= 3)	PC (short)	PC (Exploratory)	PC (yes)	CK	CK	CK
8 (= 2)	PC (short)	PC (Exploratory)	CK	CK	CK	CK
9 (= 1)	PC (short)	CK	CK	CK	CK	CK
10 (= 0)	CK	CK	CK	CK	CK	CK

The dynamics of the context during the application of a practice corresponds here to the movement of contextual elements between the contextual knowledge and the proceduralized context. Note also that the context of the practice at step 4 corresponds to the context of the action A6.

3.3 PC Building in a Contextual Graph

When the system fails to represent the practice used by an actor (given as a sequence of actions such as A13-A6-A14 for the upper path), the system presents the actor the practice the nearest of the actor's practice, exhibits the differing part between the practices, and ask for an explanation. For example, in Figure 2, there was A13 followed, say, by action A7 according to the contextual element C3 (implicitly the type of search was supposed to be exploratory). However, an actor may decide to look for the phone number of a colleague on the Web and gives very precise keywords and has his answer in the first link provided by the browser. The actor then explains that the contextual element C12 "types of search?" must be introduced because the treatment is different from the previous one, i.e. an exploratory search. (The numbering of the contextual elements in Figures 1 and 2 is made in the order in which elements were introduced when we have modeled this task and represented it by this contextual graph).

The proceduralized context is knowledge that is explicitly used at the current focus (e.g. the action A6 in Figure 2 is executed because the contextual element C4 is instantiated with the value "Yes" that is thus explicitly considered in the focus. A proceduralized context is represented as an ordered set of contextual elements that fit into each other like a nest of dolls (Russian dolls). Figure 3 gives a context-based representation of the contextual graph in Figure 2 where the instantiated contextual elements (contexts C8, C12, C3, etc.) are replaced by oval for symbolizing the different contextual elements during their instantiation, and a path crossing different contexts along the numbered dots. A path entering a context means that the corresponding contextual element is instantiated (enters the proceduralized context as at dots 01, 02, 03, 04, 05 and 08 and thus enter the focus too) or decontextualized when leaving a context (as dots 06, 07, 09, 10, 11, 12 when leaving the proceduralized context). As a consequence, the practice leaves first the last contextual element entered. Thus, what is important is not so much the collection of contextual elements but the way in which they are ordered to allow the execution of a given action or a practice.

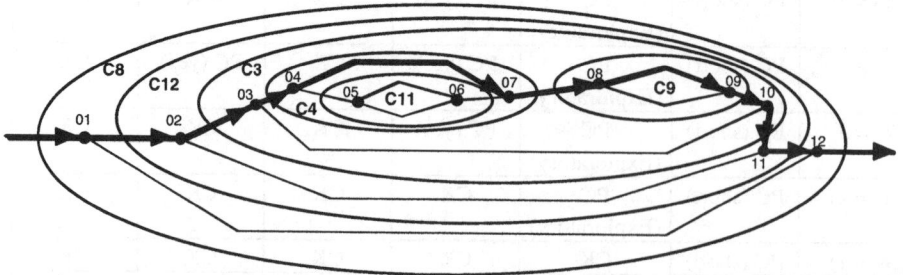

Fig. 3. A context-based representation of the graph in Figure 2 (See legend in the text)

This work, yet to be developed, is reminiscent of the work of Giunchiglia's team (e.g. see [6]). This is a new path to explore to developing contextual graphs as a formalism of representation of the knowledge and the reasoning.

3.4 Implementation Status

Right now, there is a software for visualizing contextual graphs and incrementally acquiring new practices. There are already a number of functions (zoom, handling of parts of a graph, change of language, etc.) for handling contextual graphs and the current development concerns the introduction of explanations. The use of this software in a number of applications shows that if the software itself cannot be separated from its interface (because of the incremental acquisition capability), the software is really independent of the applications already developed. A next step will be the integration of this piece of software in an intelligent support system.

Now, we study the re-writing of a graph in an operational programming language as, say, production rules. For example, there are two ways to write the path uses in the previous example (with the action A6 in Figure 2). The first one is to write the whole path as a unique rule, something like:

```
IF      C8      = (short)
        A13 is executed
        C12     = (exploratory)
        C3      = (yes)
        C4      = (yes)
THEN Execute action A6.
```

The second way is to have as many production rules as contextual elements, rules like:

```
RULE-11
        IF      C11     = (yes)
        THEN    Execute action A7
        ELSE    Execute action A17.
RULE-C4
        IF      C4      = (yes)
        THEN    Execute action A6
        ELSE    Check RULE-11.
```

This is only one way to obtain an operational representation of a contextual graph. Indeed, it would be also possible to use Bayesian networks, influence graphs, etc. for representing a contextual graph because we distinguish a model of a domain from its representation in a given formalism.

Contextual elements appear early in contextual graphs because actors want first to have a maximum of information. Then actors try to retrieve as much as possible known intermediary situations in order to reuse known strategies. A contextual graph is a kind of corporate memory and a return from experience. This could be the basis for the building of a more robust procedure (based on experience).

4 Conclusion

We propose contextual graphs for a uniform representation of elements of reasoning and contextual elements at the same level. This is different from the view of Campbell and Goodman in [5] for example (and Hendrix, [7] before for semantic networks) that consider context as a way to partition a graph. Moreover, context in our formalism intervenes more at the levels of the links between actions than actions themselves. Contextual elements being organized in contextual graphs in the spirit of "nest of dolls", we have not a hierarchy of context because a given contextual element is itself contextualized and can appear encompassed in different other contextual elements. Rather, a contextual element is a factor of knowledge activation.

We show that contextual issues cannot be addressed in a static framework only and that eliciting and sharing contextual knowledge in a dynamic way is a key process in addressing and understanding context problems.

Moreover, one finds in the literature the management of a local context in reference to a global context. The literature on context-aware systems distinguishes two types of context: (1) the "local" context that is close of the focus of attention and highly detailed, and (2) the "distant" context that is general (with less details). This approach is also found in other domains. For example, van Dijk [16] presents a very similar position on political discourses with:

- A local or micro context (called situation), defined by a specific setting and specific participants, and
- A global or macro context, informally defined in terms of higher level societal structures, involving, e.g., groups, group relations (such as power and inequality), organizations, institutions or even whole states and nations.

van Dijk embeds this view by representing context as a mental model. Contextual graphs propose a continuous view on context instead of such a dichotomy micro versus macro, local versus global, internal versus external, etc. We plan to extend this notion of granularity in situation of collaborative work where we distinguish different types of context at different levels. Movement from one context to another one is ensure by a proceduralization of a part of the knowledge from the first context to the second context. It is a way to manage different users' context (far or global context) and the collaborative work context (the local context).

The introduction of the item "Parallel Action Grouping" (PAG) simplify the representation of contextual graphs. However, if an activity is assimilated to a complex action, a PAG is more than a complex contextual element. An activity sums up a complexity at the same level of representation. A parallel action grouping generally represents (as a simplification) a complex entanglement of contextual elements corresponding to a low level of description of the problem solving modeled in the contextual graph. In the popular example of the coffee preparation given in UML manuals, it is said that we must take the coffee and the filter in one order or the other (or in parallel). However, according to the type of coffee machine (e.g. we must put it apart to fill the reservoir with water), the piece of the coffee machine where must be put the filter can be independent of the coffee machine, mobile on the coffee machine or fixed into the coffee machine. Each situation would be considered independently, but all situations will conclude on a unique action: "Put the coffee in the filter." Thus, instead of making complicated a contextual graph for representing its (natural) complexity, which is at a lower level of detail, we use parallel action groupings.

Information can be drawn from a contextual graph, such as the way in which it has been developed, which actors has developed a given part of the contextual graph. It is possible to have an evaluation of the distance between two practices (i.e. two paths in the contextual graphs). Contextual graphs are a formalism of representation allowing the description of decision making in which context influences the line of reasoning (e.g. choice of a method for accomplishing a task). This formalism has been already used in different domains such as medicine, incident management on a subway line, road sign interpretation by a driver, computer security, psychology, cognitive ergonomics, usual actions in a house (preparing hard-boiled eggs, change of an electric bulb, etc.). The extensions that will be given to this work concerns: (1) its introduction in an intelligent assistant system for providing suggestions to the actor, (2) the management of the database (operations on the items, regrouping contextual elements, etc.) in order to produce robust procedure; statistic on the development, use of a given path; and (3) the introduction of a module of explanation generation of different types and at different levels of details.

References

[1] Barab, S.A., Scheckler, R. and MaKinster, J.: Designing System Dualities: Building Online Community. American Educational Research Association in Seattle, WA (April, 2001).
[2] Brézillon, P.: Representation of procedures and practices in contextual graphs. The Knowledge Engineering Review, 18(2) (2003) 147-174.
[3] Brézillon, P. and Pomerol, J.-Ch.: Contextual knowledge sharing and cooperation in intelligent assistant systems. Le Travail Humain, 62(3), Paris: PUF, (1999) pp 223-246.
[4] Brézillon, P., Cavalcanti, M., Naveiro, R. and Pomerol, J-Ch.: SART: An intelligent assistant for subway control. Pesquisa Operacional, Brazilian Operations Research Society, 20(2) (2000) 247-268.
[5] Campbell, B. and Goodman, J.: HAM: A General Purpose Hypertext Abstract Machine. Commun. ACM, 31(7) (1988) 856–861.
[6] Serafini, L., Giunchiglia, F., Mylopoulos, J. and Bernstein, P.: Local relational model: A logical formalization of database coordination. In: P. Blackburn, C. Ghidini, R. Turner, F. Giunchiglia (Eds.), Modeling and Using Context, Springer Verlag, LNAI 2680, pp. 286-299 (2003)
[7] Hendrix,, G.: Expanding the utility of semantic networks through partitioning. Proceedings of the Fourth IJCAI, pp. 115-121 (1975).
[8] Karsenty, L. and Brézillon, P.: Cooperative problem solving and explanation. International Journal of Expert Systems With Applications 4 (1995) pp 445-462.
[9] Leplat, J. and Hoc, J.-M.: Tâche et activité dans l'analyse psychologique des situations. Cahiers de Psychologie Cognitive, 3 (1983) 49-63.
[10] Livet, P.: La Communauté Virtuelle: Action et Communication, Editions de l'Eclat (1994).
[11] McCarthy, J.: "Notes on formalizing context", Proceedings of the 13th IJCAI (1993) Vol. 1, 555-560.

[12] Pasquier, L., Brézillon, P. and Pomerol, J.-Ch.: Chapter 6: Learning and explanation in context.
[13] Pook, S, Lecolinet, E, Vaysseix, G and Barillot, E, 2000. Context and interaction in Zoomable User Interfaces *AVI 2000 Conference Proceedings (ACM Press)* (2003) pp 227-231.
[14] Schank, R.C.: *Dynamic memory, a theory of learning in computers and people* Cambridge University Press (1982).
[15] Sowa, J.F.: Knowledge Representation: Logical, Philosophical, and Computational Foundations. Brooks Cole Publishing Co., Pacific Grove, CA (2000).
[16] Van Dijk, T. A.: Cognitive Context Models and Discourse. In Maxim Stamenov ed *Language Structure, Discourse and the Access to Consciousness* Amsterdam: Benjamins pp 189-226 (1998).

Context-Dependent and Epistemic Uses of Attention for Perceptual-Demonstrative Identification

Nicolas J. Bullot

Institut Jean Nicod (CNRS/EHESS/ENS), 1 bis avenue de Lowendal,
75007 Paris, France
nicolas.bullot@college-de-france.fr

Abstract. Object identification via a perceptual-demonstrative mode of presentation has been studied in cognitive science as a particularly direct and context-dependent means of identifying objects. Several recent works in cognitive science have attempted to clarify the relation between attention, demonstrative identification and context exploration. Assuming a distinction between '(language-based) demonstrative reference' and '*perceptual-*demonstrative identification', this article aims at specifying the role of attention in the latter and in the linking of conceptual and non conceptual contents while exploring a spatial context. First, the analysis presents an argument to the effect that selection by overt and covert attention is needed for perceptual-demonstrative identification since overt/covert selective attention is required for the *situated cognitive access* to the target object. Second, it describes a hypothesis that makes explicit some of the roles of attention: the hypothesis of *identification by epistemic attention* via the control of perceptual routines.

1 Introduction

Demonstrative identification has been studied in philosophy as a particularly direct mode of identification of individual objects, since it requires direct perceptual discrimination. In spite of pioneering intuitions of Peirce [1], Russell [2] or Dretske [3], researches in the fields of the philosophy of demonstrative identification/reference and the psychology of attention have, until recently, evolved independently of each other. In contrast to this trend, several recent works [4-10] have attempted to clarify the relation between attention, demonstrative identification and context exploration. Within this collaborative tradition between philosophy and psychology, this article aims at specifying the role of attention in the preparation and performance of demonstrative identification while exploring a particular context. This analysis intends:

(i) To present an argument to the effect that attentional selection is needed for demonstrative identification since focal attention is required for the *situated cognitive access* to the target object in the explored context; in this approach, selective attention is required for mediating our relationships with any explored real context.

(ii) To describe a hypothesis that makes explicit some of the epistemic uses of attention, the hypothesis of *identification by epistemic attention* based on perceptual routines.

2 Perceptual-Demonstrative Identification by Means of Perceptual Selective Attention

According to a plausible assumption, the characteristics of demonstrative identification can be clarified and specified by studying the roles of focal/selective attention[1] during perceptual identification of objects in the currently explored context. One of the main arguments for this claim relates to the concept of cognitive access. It can be expressed by means of the following premises:

Premise 1. The perceptual-demonstrative identification of a token physical object x requires (necessarily) to have a situated and cognitive access to x's properties.

Premise 2. To have a situated and cognitive access to the object x's properties, any agent has to ($2i$) bring x into at least one of his or her sensory fields by overt movements (i.e., namely by *overt* attention) and ($2ii$) select x by *covert* attention to analyze its properties (and more generally to the cues to which x is related).

Conclusion. Hence, selecting x by one sensory system and by attention is necessarily required to identify x via a perceptual-demonstrative mode of presentation.

The justification for the first premise is derived from a classical (although not universally accepted) characterization of demonstrative identification, conceived of as perceptual-demonstrative identification. Classically, demonstrative identification of a token object indeed corresponds to the act of identification achieved (and which can only be achieved) at the time of the occurrence of the perception of the target object.[2] This act of identification is frequently achieved by, or is simultaneous to, its (egocentric) localization in a perceptual field [13], and in the peripheral space. For instance, I consider primary epistemic seeing in F. Dretske's [3: pp. 72-93] sense and object perceptual re-identification [12: pp. 31-36, 17] as paradigmatic cases of perceptual-demonstrative identification of object (which can occur in perception and thought, outside linguistic communication). This notion is distinct from the general concept of 'demonstrative reference' defined as any act of reference performed by the use during communication of a demonstrative term (cf. footnote 3). On this debatable issue, I will take for granted that perceptual modes of presentation are mandatory for

---

[1] The expression 'focal/selective attention' traditionally refers to the faculty which is responsible for the different types of selection operating in perception, such as in Treisman's sense [11]. The hypothesis below will propose a further specification of this concept.

[2] This is accepted in the tradition that studies 'knowledge by acquaintance', according to Russell's expression [2: pp. 127-74] – see e.g. P. F. Strawson [12: pp. 18-20], Evans' [13: pp. 143-203] notion of an 'information-link' with an object, Peacocke [14], McDowell [15], Millikan [16: pp. 239-56], Clark [6: pp. 130-63] and Campbell [4: pp. 84-113].

perceptual (re-)identification, but not for all kinds of demonstrative reference in language and thought.³

By definition, the act of perceptual-demonstrative identification cannot thus be effective in the absence of the genuine perception of the target object. In addition, to perceive an object in a veridical or correct way – keeping in view the classical analysis of the concept of veridical perception – requires a perceptual access to the categorical (or causal, intrinsic) properties of the object. This perceptual access (founded on an act of identification) has a *cognitive value* in the sense that it is linked to the cognitive significance of the representation of the object [e.g. 4: pp. 84-113].

The perceptual access is thus at the service of the agent's *epistemic* objectives, such as queries or inferences relating to the target object. Moreover, such an access is *situated* because to preserve the informational connection with the object (so that certain cues or properties dependent on the object are presented in the receiving fields of the agent's perceptual system), the agent must be roughly located in the same spatio-temporal *context* as the object (e.g. the same room or landscape at the same moment), so as to ensure an informational connection. According to this usual type of analysis, demonstrative identification of an object thus requires a cognitive and situated access to the properties of the object.

Premise 2 is more problematic and essential for our assumption. It relates to the problem of explaining the situated and cognitive access or tracking of the object. This problem is studied by several groups of theories, which are often conflicting. The *conceptualist* theories [15, 23-26] and *intentional* theories [27, 28] mainly concentrate on the concept-related and intention-related conditions of the cognitive relation to the object – for example the role of the sortal concepts and speaker intentions in spatio-temporal delineation of the referents. Another class of theories [5, 6, 29-31] favor the study of the *non-conceptual* conditions of the relation to the object – like the non-

---

³ A reviewer of an earlier draft of this article raised the objection that the evaluation of the argument was rendered difficult because of the notion of 'demonstrative identification'. This reviewer remarked that the phrase 'demonstrative identification' could not be equated with the notion of 'demonstrative reference' as understood in semantics and in Kaplan's tradition [18-21]. This difficulty is clarified by the distinction between 'demonstrative reference' and 'perceptual-demonstrative identification'. In traditional semantic theory, the notion of 'demonstrative reference' is usually (explicitly or implicitly) defined as an act of reference performed by the use of a demonstrative (or indexical) expression (such as the noun phrase "this/that" and the complex demonstrative "this/that *F*"). As specified above, the notion of 'perceptual-demonstrative identification' refers primarily to the acts/procedures of perceptual/direct identification of objects, which can ground the linguistic use of perceptual demonstratives but cannot be explained (at least according to the present account) merely by referring to linguistic processes. Here I agree with those [4-6, 9, 10] who consider that puzzles about demonstrative contents may sometimes be resolved by a closer examination of perceptual and attentional abilities. The argument explained in the present article mainly intends to start specifying what Evans [13: pp. 143-151] described as an 'information-link' (after Russell's 'acquaintance relation') on the basis of a theory of the epistemic uses of overt/covert attention. Regarding '(language-based) demonstrative reference' in general, one may use a demonstrative or a complex demonstrative without being or having been in perceptual contact with the referent. For instance, as argued by King [22] (but see [21]), the use of the 'that' phrase in a sentence such as 'That guy who scored hundred out of hundred in the exam is a genius' introduces a use of 'that' which does not seem to require perceptual contact with the referent.

conceptual contents and the peripheral mechanisms providing sensorimotor access (such as Pylyshyn's [5, 9, 29] visual indices). If an analysis of the cognitive access is completely unaware of the conditions studied by one or the other of these two classes of theories, it encounters the risk of leading to a circular analysis of the cognitive access (which is in this case conceived like a purely conceptual process or purely sensorimotor account). The interest of research in overt/covert attention is to connect the two types of capacities studied by these two classes of theories [4, 5: 160-61, 19], for there are reasons to postulate that attention is a *mediating* faculty whose function is precisely to articulate non-conceptual conditions (such as the control of eye saccades and sustained fixations [32, 33] and the construction of a visual object file [17, 34] of the target) and the conceptual conditions (such as the reasoning about the target) of cognitive access and perceptual tracking.

Premise 2 formulates this last assumption (mediating attention) by distinguishing *two types of access* relating to the faculty of attention: (2*i*) to bring the target into at least one sensory receiving field (by moving the body, the limbs and the sensory captors in space) – it is a question of motor preparation of the cognitive access; and (2*ii*) to select the target by covert (epistemic) attention to analyze its properties – it is a question of application of attentional analysis capacities to the properties of the object made available by selection. The next sections develop two complementary arguments, providing grounds for components 2*i* and 2*ii* of Premise 2.

3 Part 2*i* of Premise 2: Spatial Navigation, Motor Preparation and the Control of Perceptual Organ

The argument relating to the motor preparation assumes the following form: to have a perceptual access to a distal object, an agent must establish a sensory relationship to the properties of the object/agent or the cues depending relating to this object/agent. However, this relation continuously requires the displacement and the setting of parameter of the perceptual sensors in a manner which is adapted to the spatial and dynamic characteristics of the object (e.g. the occupied place, the shape of the object, its velocity, the type of tracks or clues left by the object) as well as of its context (e.g. the conditions of lighting, acoustic medium, load factor of the space configuration to which it belongs). Consequently, the cognitive access to an object requires us to have recourse to movements of approach which aim at including appropriately the object and its properties in the relevant sensory receiving fields (cf. 2*i*).

This argument is based on the need for the perceptual access (and the acts of identification that it allows) of various classes of body movements. They belong to the behavior directed towards the object, frequently named '*overt*' – i.e. publicly observable, as opposed to '*covert*'.[4] The representations of these movements can be classified according to the reference frames/systems[5] allowing their description.

The motor preparation of the perceptive and cognitive access sometimes requires displacements in space at long distances, which can be represented by reference frames known as allocentric or environmental. For example, an agent can move to go

---

[4] Cf. e.g. Findlay & Gilchrist [33] and Spence [35: 232].
[5] Cf. e.g. Milner & Goodale [36: pp. 88-92].

to seek a stationary token object being prompted by the episodic memory which he or she has of having handled it in another area of space than that which it occupies at present. In this case, the description of his or her movement in space is usually made according to allocentric reference frames – e.g. by means of a topographic representation mentioning salient landmarks, as in the case of cities' charts or crossed areas. The initial phase of his or her research corresponds to the movements of the bodies made to reach a still imperceptible target (especially if it is about an object which does not emit light). This class of movements unquestionably belongs to the motor preparation of the perceptual access to the target individual, because it corresponds to the combining of the sensory sensors (and effectors) of the target.

When the agent is sufficiently close to the place occupied by the target object, and has a perceptual access to the properties of this object, the representations of his or her body movements relate initially to an egocentric frame of reference. They are movements such as the capacity to move and adjust the orientation of the sensory sensors (e.g. the eyes, hands, ears) in order to scan/scrutinize the properties of the target object, as in the case of the saccades and ocular fixations. For example, the vision of human beings continuously uses more the mobility of the eyes as compared to the other parts of the body, so that the movements of repositioning of the visual axis occur several times in a second, by saccadic movements, in the majority of ordinary actions [32, 33, 37].

In addition to the initiation of certain movements, the motor preparation of the cognitive access rests moreover on the inhibition of certain movements (suspension of the locomotion, modification of breathing). This class of preparatory movements, related to the parameter setting of the sensory sensors relative to an egocentric reference frame, whose goal is to establish the informational connection with the target, depends on attentional capacities – like the ones that have been described since the end of the 19th century – since attentional selection has many motor implications (related as well to the selection of some gestures as to the inhibition of gestures without relevance). For example, James [38: pp. 434-38] supports that the attentional processes include "the accommodation and the adjustment of the sensors". This suit on the motor and overt (i.e., residing in publicly observable body movements) consequences of the attention is found in other authors from the end of the 19th century. In the same way, Sully [39: p. 82] describes attention as an active mode of consciousness that one can expect to find in certain motor process. Ribot [40: p. 3] writes that the mechanism of attention "is primarily motor, i.e. [*attention*] always acts on muscles, mainly in the form of a stop" [40: p. 3]. Among the contemporary theories of this tradition in the field of vision, one finds for example the motor (or pre-motor) theories of attention which analyze spatial covert attention as a process of (overt) movements of ocular saccades preparation [e.g. 41].

Various classes of movements, which are often *necessary* conditions of perception, thus prepare and optimize the perceptual and situated access to an object and consequently its perceptual-demonstrative identification. The component 2*i* of Premise 2 must thus be accepted. However, these motor conditions although necessary are not sufficient to explain cognitive access, because they do not account for the *epistemic* uses of attentional selection. This is why an analysis of the demonstrative identification must also consider the second side of the premise according to which the attentional *analysis* (epistemic) of indices is necessary (and even sufficient) for situated cognitive access.

4 Premise 2*ii* – Need for the Covert Attentional Selection (in Addition to the Overt Selection)

The second component of Premise 2 (maintains) states that (selection leading to) covert attentional analysis is necessary for *cognitive* access (the so-called selection 'for further processing'). This means that the covert attentional capacities (distinct from the overt behavior related to the orientation of the sensors) are selective and epistemic components also constitutive of the *access* to the target for its identification. In spite of the complementary nature of the two components – 2*i* and 2*ii* – it is necessary to distinguish them for the following reasons.

An argument is related to the fact that the presence of the object in a sensory field is a necessary but *not a sufficient* condition for identification. Contrary to the motor preparation via movements and overt attention (e.g. saccades and fixations) which can be partly carried out in the absence of the target, the achievement of perceptual-demonstrative identification requires accessing current information about the target object (and its causal/intrinsic properties) once it is included and tracked *in* a sensory field. One can account for this in a theory of visual attention by saying that one cannot demonstratively identify the target object as long as it is not *indeed* selected by covert attention for further processing. Conditions of stimulation such as presence of the object in the sensory field and supraliminal stimulations of the sensor, by themselves, are not enough to involve the perception and the identification of a token object/stimulus. Indeed, a sense organ can be directed very precisely towards a distal object (providing 'supra-liminal' stimulations) without this object consciously being noticed or identified, like in phenomena such as 'attentional blink' [42], 'change blindness' [43, 44], or 'inattentional blindness' [45] – which we will now consider.

The phenomenon of '*inattentional blindness*' [45], sometimes described as 'sighted blindness' [45: p. 61] refers to the failure to detect a theoretically entirely "visible" stimulus (for example a red square or a moving bar satisfying the sensory conditions of visual perception) when this stimulus is presented in the area of fixation (2° of visual arc). The expression 'inattentional blindness' was selected because this failure seems to be a direct consequence of the fact that the subjects had not paid attention' to the stimulus. According to the authors, this phenomenon could indicate that there cannot be conscious perception without a preliminary selection by attention [45: p. 61]. This is actually the most radical interpretation of the phenomenon, and it remains a matter of debate and controversy [46]. The discovery of Mack and Rock rests on an experimental protocol whose principle consists in putting the observer in a position where he would not pay attention to, nor would he expect to see, the object concerned – called the 'critical stimulus' – but, at the same time, would look at the area inside where it would be presented (figure below).

The experiments within this paradigm rest on the realization of a non critical attentional task during several tests. This task consists in presenting to the subject a cross, one segment of which is longer than the others, and asking the subject to determine which is the longest segment. The cross is briefly presented, and it is either centered at the point of fixation, or in the parafoveal area (figure above). At the third or fourth performance of these tests (this is the critical test), the critical stimulus is presented without any prior indication along with the cross. If the cross is centered at the point of fixation, the stimulus appears in one or more quadrants of the cross and is in a parafoveal area. Conversely, if the cross is centered in the parafoveal area, the critical stimulus is presented at the point of fixation (figure above).

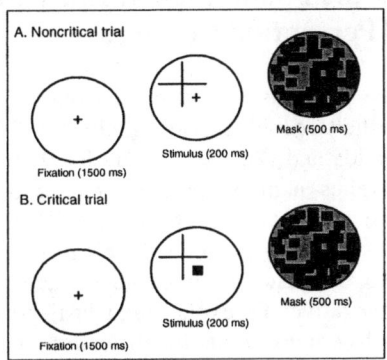

Fig. 1. The experimental protocol related to the discovery by Mack and Rock of inattentional blindness

One of the experimental outcomes was interpreted as an effect characteristic of the phenomenon of inattentional blindness. If the critical stimulus was presented at the point of fixation and the cross was centered in one of the four parafoveal areas previously occupied by the critical stimulus (figure above), inattentional blindness increased considerably. For example, a colored spot, which is seen in approximately 75% of the cases when it is presented in a parafoveal way in a quadrant of the cross, is seen only in 35% of the cases or even lesser when it is presented at the point of fixation whereas the cross is located in the parafoveal area.

With other phenomena such as 'change blindness' and 'attentional blink', in addition to the classical experiments in 'covert spatial orienting' [47], inattentional blindness illustrates a case in which directing a sensory organ (in the above example, the eyes and the fovea) towards an object or a property does not imply conscious detection or identification of this object or this property (or more carefully expressed, does not imply the capacity to submit a verbal report on its identification). This relative independence indicates the need to arrive at the demonstrative identification, to have an additional component, which is called here *(focal) attention*.

This work on attention is relevant for the theory of demonstrative identification. It helps to clarify Premise 2 of the general argument, i.e. it helps to specify the nature of the necessary conditions for a *cognitive situated access* to any elements in a context containing several elements. In the example of the experiment in inattentional blindness stated above, the cognitive situated access mainly relates to the properties of the cross to the detriment of those of the critical stimulus. To have x in one of one's sensory fields (x determining a supraliminal, durable and relatively stable stimulation) is thus not sufficient to have the benefit of a cognitive situated access to x. Consequently, the presence of x in a sensory field cannot even be sufficient to identify x in a demonstrative way, since the demonstrative identification depends on the cognitive situated access.

5 A Procedural Theory of the Epistemic Uses of Attention: The Hypothesis of the Foundation of Epistemic Attention on the Control of Perceptual Routines

Here is a summary of the previous argument. To obtain a situated and cognitive access to the properties of the target object, an agent must (2*i*) introduce x into at least one of his or her sensor fields and (2*ii*) select x by focal attention (covert attention). The justification for 2*ii* relies namely on empirical arguments that relate to the distinction between overt attention and covert attention. A number of examples show that an overt attentional attitude – such as looking toward x – does not imply conscious noticing of this x or epistemic processing of x's properties. However, this argument remains purely negative. To explain and justify the role of focal/selective attention in x's perceptual-demonstrative identification, one needs to specify the nature of the epistemic uses of attentional analysis – for the essential characteristic of the perceptual-demonstrative identification is to contribute to the *knowledge* available to the agent of his/her spatial environment (and the individual objects displayed in space). Something that deserves to be called *epistemic attention* may constitute demonstrative identification, but it leaves us the task to make explicit what epistemic attention precisely is. (Here the adjective 'epistemic' is akin to Dretske's use, when he refers to 'primary epistemic seeing' [3] and 'epistemic/meaningful perception' [48] – although Dretske [3, 48] does not explicitly study the contribution of covert attention to epistemic perception.)

The theories that appear to me as being relevant for the study of epistemic attention are the 'procedural theories' – that is, the theories according to which attention is a system that controls the performance of cognitive *procedures* and strategies for exploring the peripheral spatial context. Among the class of procedural theories, I shall include namely works by Miller & Johnson-Laird [49], Ullman [50] on visual routines, by Land et al. [32, 51, 52] on eye movements, by Campbell [4] on attention and reference, and by Pylyshyn [5] on visual reference. In the context of research on attention, one can use the term 'procedural' for referring to the theories that view attentional capacities as being coincident with the exercise/use of epistemic procedures or pragmatic procedures that are dependent on a situation/context of use and a strategy to obtain context- and task-relevant information.

According to this type of analysis, attention uses strategic and exploratory operations that enable the agent to obtain information (typically) on a particular target object or cues related to one object (and often also to constitute a singular representation of this target object). One can give an account of this strategic structure by analyzing focal attention as being dependant on two main components: (i) a set of instructions for the control of bodily or mental events, which can be called *epistemic* queries and pragmatic queries – and (ii) a set of elementary operations called *routines* that allow, according to varied and context dependent combinations, to give an answer to or to satisfy the epistemic and pragmatic queries. In this framework, the concept of epistemic query refers to the cognitive requests relating to objects (or contents) concerned with the problem which is currently under discussion. The concept of *pragmatic query* relates to the requests (of motor acts) relating to the objects concerned with the action under progress. An epistemic query aims thus at solving a particular problem (related to the action on token/individual objects or the knowledge about token objects), expressed in accordance with a particular spatio-temporal situation and particular goals (cognitive or pragmatic). The concept of *routine* refers to the perceptual or motor elementary procedures that can be used to satisfy or solve the queries (epistemic or pragmatic). I will present an analysis that suggests that attention, as a mediating faculty, seems to be the capacity that organizes the relations between conceptual and nonconceptual routines for identification.

Thus the central assumption of a procedural theory of epistemic attention is that attentional processes can (at least partly) be conceived of as a capacity for effecting epistemic routines, which can be expressed as follows:

Hypothesis of the foundation of epistemic attention on the control of perceptual routines: To select by (epistemic) attention a token physical object x implies carrying out epistemic and pragmatic routines that relate to x's properties – i.e. to carry out strategic procedures analyzing cues (or properties) that relate to x.

According to this assumption, the theory of attention as routines (mental) effector connects the attentional systems to the control of cognitive operations, (operated) performed at the time of interactions with distal targets, or more precisely on the cues that the distal target offers. The assumption suggests that attentional strategies can be described as a combination of distinct cognitive routines (having a bearing on those cues). Lastly, routines and their combinations are assumed to be necessary in order to constitute a singular representation of the entity involved. (The expression "epistemic routines" refers to the accomplished mental operations starting from information obtained by the mediation of an informational link with the target entity.)

Although the discussion of this point is outside of the scope of the present article, this hypothesis may help build a mediation view [4, 7, 19] based on the reconciliation of the *contextualist* views [53, 54] and the *intentional* views [27, 28] of demonstrative reference. For (i) the (operations) performance of the perceptual and motor routines are strictly context-dependant and conducted for having a bearing on publicly accessible features/cues of the context and (ii) the agents' intentions play a central role for the constructing of attentional strategies, i.e. strategies that aim at tracking down and identifying the target.

In the next paragraphs I will try to show that the appealing character of this assumption comes from the possibility of describing in a coherent way, along with its contents or its predictions, particularly precise mental operations such as perceptual verifications and probing. These two points can be further explained by considering precise examples analyzed according to this assumption. I will consider the example of perceiving abstract and geometric diagrams.

Shimon Ullman [50] develops a procedural theory according to which spatial perception of one or several objects and of their spatial relationships among them and their parts is achieved by sophisticated visual routines which are built up from more elementary operations, called 'primitive routines'. Ullman's discussion of visual routines focuses on the processes that might be used in analyzing complex properties of incoming visual information, but the concept can be easily extended to include the processes that operate on information provided by other sensory modalities or by memory systems. Ullman [50] suggests that the perception of shape and spatial relations among various parts is supported by sophisticated visual routines that are constructed from a set of primitive routines or elemental operations. These primitives, as well as useful sequences of them which are programmed into more complex routines, are stored in procedural visual memory and used in identifying objects and performing many other specific visual tasks. Ullman suggests five plausible elemental operations and gives a computational analysis of how they might be carried out:

Shifting the focus of processing [50: pp. 123-28]. This operation allows all visual routines to be applied to any location in the visual field simply by shifting the center of processing. It corresponds to changing the position of focal attention.

Indexing [50: pp. 129-35]. Indexing involves selection of a location where something is "different," as in visual pop-out and various pull cues that summon attention.

Bounded activation (or "coloring") [50: pp. 135-39]. Coloring is used to find the interior of a region by spreading activation within its boundaries.

Boundary tracing [50: pp. 139-46]. This operation is used to determine whether two locations are on the boundary of the same object.

Marking [50: pp. 146-52]. Marking a location is an operation that designates it as one to be remembered so that is can be accessed quickly at a later time.

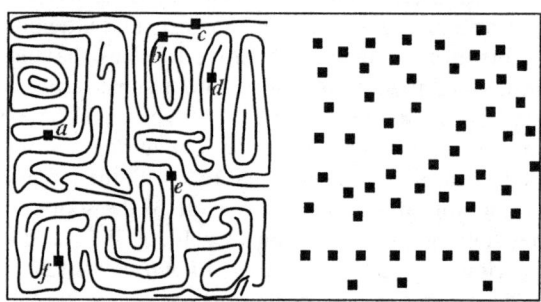

Fig. 2. Left part, Problem of the doubly secant curves: Is one and the same curve the secant of two black squares? Right part, To give a solution to the following epistemic query: How many black squares are being aligned?

Let us consider the problem 'Is the same curve a secant of two squares?' for the left part of Figure 2 (analyzed in the conceptual framework introduced by Ullman). The visual individuation of the six black squares (let us name them a, b, c, d, e, f) depends on primitive routines which segment salient elements. A complex routine is necessary to determine if a same curve is the secant of two black squares. This routine must make it possible to evaluate as true or false a property such as $C_1\text{-}Secant(e, f)$, meaning that the curve c_1 is the secant of the elements e and f. For example, on the basis of the area occupied by an individual square, the complex routine can include the visual tracking and activation of the two opposite segments of the secant curve of this square. In this way, it is possible to determine if one meets or not a second square on the same curve. This operation of activation, whose latency is longer than that of the visual individualization of a square, is imposed by the density of the grid, the length and the tangle of the curves because these factors prevent the fast individualization of the individual lines and supports their confusion.

These examples make it possible to specify why the study of epistemic attention – conceived as effector of perceptual routines (visual, auditory and others) – should contribute to the comprehension of the relation between the conceptual contents and the non-conceptual capacities and contents of perception during target demonstrative identification. Initially, there is no reason to doubt that the preceding epistemic queries can be expressed by conceptual contents, and can be communicated by linguistic statements like those stated above for presenting the problem. The understanding of the problem requires for example the possession of the concept of secant curve. The question under discussion relates to the status of the routines and contents which make it possible to check by visual experience of the image a predicate such as C_1-$Secant(e, f)$. Let us accept for the sake of argument the description of the proto-propositional contents given by Peacocke [55: pp. 74-90]. Within the conceptual framework of Peacocke's theory [55], the assembled routine enabling us to evaluate the predicate C_1-$Secant(e, f)$ is probably an operation constitutive of the proto-propositional contents of the visual experiment. Indeed, this routine requires the capacities consisting in tracking visually and continuously a curved line, to detect the distinctive properties of the squares and intersection between the line and a square (e.g. occlusion of the line by another form, presence of two 'T' junctions). This assembled routine thus seems able to take into account at the same time the difference between two specimens of individuals (this square here and this square there, this secant curve) and of the relevant properties to detect some of their relations (to be a square, to be black, to be a long curve, to be secant of) – the latter characteristics are specific to the proto-propositional contents of Peacocke's view [55].

Lastly, it is plausible to consider that this assembled routine operates by means of elementary routines which are not conceptual, as for example the routine which makes it possible to carry out the continuous tracking of a single curve without the error of 'sliding' onto another curve (this capacity is controlled by the spatial properties of a singular layout and the taking into account of such spatial properties does not depend on the possession of the general concept of curve). From the assumption of these hypotheses, the following conceptualization is thus plausible. The evaluation of the predicate C_1-$Secant(e, f)$ requires that we make use of the non-conceptual routines which are at the service of the resolution of an epistemic query formulated (or capable of being formulated) conceptually. To explain how the perceptual experience (or information processing performed by a perceptual system) gives a solution to a conceptual query, or verifies an observational proposition, is to understand how the conceptual query is articulated with non-conceptual routines which provide access to the relevant information to answer the query.

Similar analyses could be developed for the right part of Figure 2, from the moment when one seeks to solve the epistemic requests mentioned – to count aligned squares. The ability to count the collinear elements is supposed to determine if the property of colinearity applies to a class of elements selected at a given time. But to evaluate this property, the possession of the conceptual definition of colinearity as "located on the same line" is not enough, it is necessary in addition to lay out "see" and select the elements of which one seeks to know if they are located on the same line and for this reason it is necessary to plot by imagining a straight line or detect the totality of elements through which it can pass. Similar analyzes can also be developed for the perceptual analysis of spatial arrangements of 3D-objects. For instance, experimental researches on eye movements have been conducted with similar ideas by Ballard et al. [37] and Land et al. [32, 51, 52].

6 Conclusion

This article has described some of the epistemic uses of attention in the preparation and performance of context-exploration and perceptual-demonstrative identification. It has been shown that selective attention is required to obtain context-dependant knowledge about individuals. Attentional selection is needed for perceptual-demonstrative identification since selection by overt and covert attention is required for the situated cognitive access to the target object. In the present approach, selective attention is required for mediating our relationships with any explored context or situation – metaphorically, demonstrative identification seems to be based on the 'attentional navigation' through the spatial layout of physical objects. Finally, the procedural theories of perceptual and selective attention may be appropriate for the understanding of the epistemic uses of attention in context exploration.

Acknowledgments

I would like to thank N. Gangopadhyay and three anonymous reviewers for helpful comments on an earlier draft of this paper. This research has been made possible in part by the European Network of Excellence 'Enactive Interfaces'.

References

1. Peirce, C.S., Collected Papers of Charles Sanders Peirce, Vols. I-VI, ed. C. Hartshorne and P. Weiss. 1931-35, Cambridge, MA: Harvard University Press.
2. Russell, B., Logic and Knowledge, Essays 1901-1950 (ed. by R. C. Marsh). 1956, London: George Allen & Unwin.
3. Dretske, F.I., Seeing and Knowing. 1969, Chicago: The University of Chicago Press.
4. Campbell, J., Reference and Consciousness. Oxford Cognitive Science Series. 2002, Oxford: Clarendon Press.
5. Pylyshyn, Z.W., Seeing and Visualizing: It's Not What You Think. 2003, Cambridge, MA: MIT Press.
6. Clark, A., A Theory of Sentience. 2000, Oxford: Clarendon Press.
7. Siegel, S., The role of perception in demonstrative reference. Philosophers' Imprint, 2002. 2(1): p. 1-21.
8. Clark, A., Sensing, objects, and awareness: Reply to commentators. Philosophical Psychology, 2004. 17(4): p. 553-579.

9. Clark, A., Feature-placing and proto-objects. Philosophical Psychology, 2004. 17(4): p. 443-469.
10. Young, R.A., Demonstratives, reference, and perception, in Modeling and Using Context, 4th International Interdisciplinary Conference CONTEXT 2003, Standford, CA, USA, June 23-25, 2003, Proceedings, P. Blackburn, et al., Editors. 2003, Springer-Verlag: Berlin Heidelberg New York. p. 383-396.
11. Treisman, A., Strategies and models of selective attention. Psychological Review, 1969. 76(3): p. 282-299.
12. Strawson, P.F., Individuals: An Essay in Descriptive Metaphysics. 1959, London: Methuen.
13. Evans, G., The Varieties of Reference. 1982, Oxford: Oxford University Press.
14. Peacocke, C., Demonstrative content: a reply to John McDowell. Mind, 1991. 100: p. 123-133.
15. McDowell, J., Peacocke and Evans on demonstrative content. Mind, 1990. 99(394): p. 255-266.
16. Millikan, R.G., Language, Thought, and Other Biological Categories. 1984, Cambridge, MA: MIT Press.
17. Treisman, A., Perceiving and re-perceiving objects. American Psychologist, 1992. 47: p. 862-875.
18. Kaplan, D., Demonstratives, in Themes from Kaplan, J. Almog, J. Perry, and H. Wettstein, Editors. 1989, Oxford University Press: Oxford. p. 481-563.
19. Reimer, M., Three views of demonstrative reference. Synthese, 1992. 93: p. 373-402.
20. Recanati, F., Direct Reference: From Language to Thought. 1993, Oxford: Blackwell Publishers.
21. Corazza, E., Complex demonstratives qua singular terms. Erkenntnis, 2003. 52(2): p. 263-283.
22. King, J.C., Are complex 'that' phrases devices of direct reference? Noûs, 1999. 33(2): p. 155-182.
23. Wiggins, D., Sortal concepts: A reply to Xu. Mind and Language, 1997. 12(3/4): p. 413-421.
24. Wiggins, D., Sameness and Substance Renewed. 2001, Cambridge: Cambridge University Press.
25. Xu, F., From Lot's wife to a pillar of salt: Evidence that 'physical object' is a sortal concept. Mind and Language, 1997. 12(3/4): p. 365-92.
26. McDowell, J., Mind and World, Second Edition. 1996, Harvard: Harvard University Press.
27. Kaplan, D., Afterthoughts, in Themes from Kaplan, J. Almog, J. Perry, and H. Wettstein, Editors. 1989, Oxford University Press: Oxford. p. 556-614.
28. Donnellan, K.S., Reference and definite descriptions. The Philosophical Review, 1966. 75(3): p. 281-304.
29. Pylyshyn, Z.W., Visual indexes, preconceptual objects, and situated vision. Cognition, 2001. 80: p. 127-158.
30. Gunther, Y.H., ed. Essays on Nonconceptual Content. 2003, MIT Press: Cambridge: MA.
31. Cussins, A., Content, conceptual content, and nonconceptual content (1990), in Essays on Nonconceptual Content, Y.H. Gunther, Editor. 2003, MIT Press: Cambridge: MA. p. 133-163.
32. Land, M.F., N. Mennie, and J. Rusted, The role of vision and eye movements in the control of activities of daily living. Perception, 1999. 28: p. 1311-1328.
33. Findlay, J.M. and I.D. Gilchrist, Active Vision: The Psychology of Looking and Seeing. 2003, Oxford: Oxford University Press.

34. Kahneman, D., A. Treisman, and B.J. Gibbs, The reviewing of object files: Object-specific integration of information. Cognitive Psychology, 1992. 24(2): p. 175-219.
35. Spence, C., Crossmodal attentional capture: A controversy resolved?, in Attraction, Distraction and Action: Multiples Perspectives on Attentional Capture, C.L. Folk and B.S. Gibson, Editors. 2001, Elsevier: Amsterdam. p. 231-262.
36. Milner, A.D. and M.A. Goodale, The Visual Brain in Action. 1995, Oxford: Oxford University Press.
37. Ballard, D.H., et al., Deictic codes for the embodiment of cognition. Behavioral and Brain Sciences, 1997. 20(4): p. 723-767.
38. James, W., The Principles of Psychology. 1890, New York: Dover Publications.
39. Sully, J., Outlines of Psychology [1884]. 1898, London: Longmans, Greens & Co.
40. Ribot, T., Psychologie de l'Attention (Dixième édition). 1908, Paris: Félix Alcan.
41. Rizzolatti, G., L. Riggio, and B.M. Sheliga, Space and selective attention, in Attention and Performance XV: Conscious and Nonconscious Information Processing, C. Ulmità and M. Moscovitch, Editors. 1994, MIT Press: Cambridge, MA. p. 395-420.
42. Shapiro, K.L. and K. Terry, The attentional blink: The eyes have it (but so does the brain), in Visual attention, R.D. Wright, Editor. 1998, Oxford University Press: Oxford. p. 306-329.
43. Rensink, R.A., J.K. O'Regan, and J.J. Clark, To see or not to see : the need for attention to perceive change in scenes. Psychological Science, 1997. 8(5): p. 368-373.
44. Simons, D.J., Attentional capture and inattentional blindness. Trends in Cognitive Sciences, 2000. 4(4): p. 147-155.
45. Mack, A. and I. Rock, Inattentional blindness: perception without attention, in Visual attention, R.D. Wright, Editor. 1998, Oxford University Press: Oxford. p. 55-76.
46. Driver, J., et al., Segmentation, attention and phenomenal visual objects. Cognition, 2001. 80: p. 61-95.
47. Posner, M.I., Orienting of attention. Quarterly Journal of Experimental Psychology, 1980. 32: p. 3-25.
48. Dretske, F.I., Meaningful perception, in An Invitation to Cognitive Science: Visual Cognition, Second Edition, S.M. Kosslyn and D.N. Osherson, Editors. 1995, MIT Press: Cambridge, MA. p. 331-352.
49. Miller, G.A. and P.N. Johnson-Laird, Language and Perception. 1976, Cambridge, MA: Harvard University Press.
50. Ullman, S., Visual routines. Cognition, 1984. 18: p. 97-159.
51. Land, M.F. and S. Furneaux, The knowledge base of the oculomotor system. Philosophical Transactions: Biological Sciences, 1997. 352(1358).
52. Land, M.F. and M.M. Hayhoe, In what ways do eye movements contribute to everyday activities? Vision Research, 2001. 41: p. 3559-3565.
53. McGinn, C., The mechanism of reference. Synthese, 1981: p. 157-186.
54. Wettstein, H., How to bridge the gap between meaning and reference. Synthese, 1984. 58: p. 63-84.
55. Peacocke, C., A Study of Concepts. 1992, Cambridge, MA: MIT Press.

Utilizing Visual Attention for Cross-Modal Coreference Interpretation

Donna Byron, Thomas Mampilly, Vinay Sharma, and Tianfang Xu

The Ohio State University,
Department of Computer Science and Engineering,
2015 Neil Ave, Columbus, Ohio, 43210, USA
{dbyron, mampilly, sharmav, xut}@cse.ohio-state.edu

Abstract. In this paper, we describe an exploratory study to develop a model of visual attention that could aid automatic interpretation of exophors in situated dialog. The model is intended to support the reference resolution needs of embodied conversational agents, such as graphical avatars and robotic collaborators. The model tracks the attentional state of one dialog participant as it is represented by his visual input stream, taking into account the recency, exposure time, and visual distinctness of each viewed item. The model correctly predicts the correct referent of 52% of referring expressions produced by speakers in human-human dialog while they were collaborating on a task in a virtual world. This accuracy is comparable with reference resolution based on calculating linguistic salience for the same data.

1 Introduction

A challenging goal in computational linguistics is understanding all of the ways context modulates the meaning of linguistic forms. One contextual effect that has been observed across multiple experimental disciplines is the use of ambiguous referring expressions for entities that are salient in the context. Speakers use underspecified nominal expressions, especially pronouns such as *this* and *he* but also common noun phrases (NP) such as *the button*, freely in discourse, relying on the addressee's ability to understand which button or person is being referred to. This preference for certain entities, given a prior context, will be called *salience* in this paper. Salience corresponds to a prediction or expectation that a certain entity will be the topic of an utterance. Estimating the relative salience of each entity in the universe of discourse is an important task in computational models of referring behavior - both in producing felicitous noun phrases and also in interpreting connected discourse. The long-term objective of our research program is to create robust, accurate algorithms for reference interpretation in automated agents. This task is impossible without a firm understanding of contextual effects on referring behavior.

It has been well-established in the computational linguistics literature that discourse history can be interrogated to estimate the salience of entities in a sentence one is trying to interpret. However, recent technology improvements create opportunities for human-computer conversations in which several contextual factors in addition to the discourse history are in play at the same time, each impacting entity salience in different ways. The goal of our project is to create conversational software agents that can carry on a *situated* conversation with a human partner. For the purposes of this paper, situated language will be defined as language having these properties:

Immersion. The conversation takes place within a 3D setting that is perceptually available to the conversational partners. The partners can speak to each other face to face within the setting.

Mobility. Both conversational partners are at liberty to move about in the world, independently of each other, to gather information or change their perceptual perspective of the world.

These characteristics distinguish situated language from other interaction paradigms. The bulk of reference processing algorithms in computational linguistics have been developed using data collected in traditional experimental settings where conversational partners are explicitly prevented from exploiting extra-linguistic contextual clues, such as gesture and gaze [13]. However, in the current study, we examine situated language between two human partners, using new data generated in our lab. This allows us to investigate the interplay between the discourse context and the conversational setting and its effect on the interpretation of referring expressions in a visually-rich domain.

The primary focus of the present work is to develop a model of visual attention that can be used to interpret *exophors*, references to items in the discourse setting. Similar to the way that an anaphor constitutes a repeated mention of an item introduced into the context by the linguistic history, an exophor is a repeated mention of an item already introduced into the context by the physical world, in other words, a cross-modal coreference. Our hypothesis is that the world that is visually perceptible to the conversational partners will be likely to shape the content of their discussion, especially when they are performing a task involving objects in that world. Moreover, a likely source of denotations for exophors are items that the speaker's attention is directed toward as the utterance is produced. Given these two factors influencing the dialog, our aim is to test a method of tracking one speaker's view of the world over the course of a dialog, and use that information as input in a reference resolution algorithm to interpret ambiguous referring expressions. Our eventual goal is to construct a model that fuses attentional information provided in the visual channel with that provided by the discourse history. In the present work, we perform pencil-and-paper analyses and offline simulations of our model, as a first step in developing the algorithms that will eventually be implemented.

2 Motivation and Background

2.1 Overview

Our work is motivated by the goal of building automated agents or interactive characters that cohabit a virtual space with a human partner. Such agents will be able to discuss the world they are in, as well as collaborate, reason, plan, and perceive the virtual world. In order to design this type of agent, we first need to learn how human beings behave in similar environments.

The data used to inform algorithm development in this study was collected by placing pairs of human partners in a first-person graphical world, rendered by the QuakeII game engine[1]. In the virtual world, the partners collaborate on a treasure hunt task. One person in each pair of players, "the leader", was given the list of tasks. The other player, "the follower", had no prior knowledge of these tasks. This setting forced the players to converse in order to solve their task. The partners communicated through headset-mounted microphones, and an audio recording of their dialog was collected and transcribed. In addition, each player's movement and activity in the virtual world was recorded to video tape. The QuakeII game engine allows the two partners to move about in the world independently and manipulate objects. As he moves about, each player sees a first-person view of the virtual world. We trapped separate recordings from each person's viewpoint.

It is obvious that understanding natural language is important in order to collaborate in this domain, and that the two partners discuss not only items that they see but also items that have been discussed or seen in the past. There is evidence in the literature showing that visual context influences how people organize and interpret the meaning of spoken language [25, 21]. Figure 1 shows a sample dialog fragment from our study[2], which contains instances of both cross-modal and linguistic coreference over a chain of references to the helmet. Before utterance 1, speaker F sees the helmet in the room. The expression *it* in utterance 1 denotes the physical helmet in the world[3]. After this mention, the helmet is repeated several times. This discourse fragment also shows the high concentration of referring expressions in this domain. The partners are unlikely to successfully complete their task without correctly interpreting these phrases.

F: I see it {vn:AH} I see *the helmet*
L: yeah
F: yeah {vn:doo} {vn:ack} and to pick *it* up I do control right
L: yes

Fig. 1. A sample dialog demonstrating both linguistic and cross-modal coreference

---

[1] www.id.com
[2] The notation {vn:} signifies non-word vocal noise.
[3] Forty-six utterances prior to this point in the dialog, the partners had discussed finding the helmet, so the speaker's use of *it* in this example is partially anaphoric.

The visually-perceived Quake world is not only the source of semantics for the conversational partners, but also impacts their focus of attention. For example, when the partners walk through a door and suddenly have a view of a set of new objects in a room, their attention is fixated on the new objects they have just discovered. Therefore, our computational model must attempt to calculate what items of interest might be discussed next, given a 2D plane of pixels which represents the field of view of one partner. There are a variety of issues that must be addressed in utilizing this visual information as input to language processing.

Video Segmentation and Alignment with Language. One of the most interesting challenges in this domain is that the field of view is an ongoing data stream. This stream must be broken into units in order to compute which items are within the speaker's field of view at each point in the stream. We will call these units *visual context frames*. The frequency at which these frames are captured will affect the sensitivity of any algorithm that uses the resulting data. If the sample frequency is too low, many visual events might be missed. The highest available sample frequency is the frame rate of the video (30Hz).

Gaze Direction. In each video context frame, our system should discern which items the viewer is looking at. An object's proximity to the center of the field of view [18] turned out to be a poor indicator of its visual salience. This is primarily because many of the subjects had difficulty making fine-grained movements using their keyboard controls, so they would sometimes pan just until the object came

Fig. 2. The speaker's view when he said "will you punch *that little button* over there?"

Fig. 3. The view at the word "*cabinets*"

Fig. 4. The view at the word "*them*"

into view and then stop. Figure 2 shows an example. Although the speaker is talking about the button, it is not in the center of his field of view. Also, an item may move out of view by the time the speaker refers to it. For example, Figures 3 and 4 show the speaker panning the scene while saying "There's a couple of cabinets here. While I'm here, let me see if I can open the cabinets and not fall into *them.*"

Foreground vs. Background Objects. Each visual context frame contains not only items of interest in the task, but also walls, floors, ceilings, etc. These background items are in view in most frames, therefore a model that simply favors items that have been seen frequently/recently will over-weight such items.

2.2 Computational Linguistics Background

Computational Models of Referring. An automated agent that can collaborate in rich domains such as ours will need sophisticated reference understanding software. For collaborative agents, reference resolution is the module that provides a mapping from the noun phrases spoken by the user to the objects the user intend to denote. For example, "Let's see what's in that room" might be a command to explore a particular room, and the reference resolution module determines which room the system thinks the user meant. For a given referring expression, there are many possible places to search for the referent: the physical context, items previously mentioned in the discourse, mutually known objects from the *ambient context* such as 'the president', etc. Although individual algorithms vary in their details, resolution systems primarily rely on linguistic information such as syntax or semantics or the combination of these two. With the development of multi-modal systems, researchers have begun to incorporate visual information into the resolution process [19, 20, 22]. Most of this work is still very preliminary. For example, Campana et al. [5] propose incorporating eye-tracking into a reference resolution module to take advantage of gaze information, but the idea is yet to be evaluated.

The process of reference resolution is normally modeled as two separate steps. First, the *context management* step prepares a set of possible referents that might be referred to in subsequent discourse. To contain the full list of available referents, a system interacting in situated discourse will need:

- A *Linguistic Context* (LC) that contains a list of possible referents that are introduced by the verbal interaction. This is used to track the attentional state as it is portrayed by the discourse. Generally, the entities added are only those that were mentioned as nominal constituents, however more recent algorithms also add high-order referents such as events and propositions [7, 4]. In systems with a visual interaction component, GUI items are added to the LC [17, 3, 15], because they are considered to be part of the communication process.
- A *Mutual knowledge Context* (MC) that contains a list of the objects assumed to be known to both parties before the conversation begins. For example,

an airline reservation system might initialize its MC with a list of airline companies and cities.
- A *Visual Context* (VC) to track items in the world that the partners have interacted with and might discuss. In systems such as [9, 18] that allow the user to move through a virtual world, items encountered by the users are also added to the context. The VC represents the attentional state of the conversational participants based on their extra-linguistic perception in the world.

Taken together, these lists are meant to represent all of the entities which might be mentioned in subsequent discourse.

Each time the context is updated with new entities, the relative salience of all the items is adjusted. A small subset of LC entities comprise the current linguistic focus of attention. A large variety of techniques exist for calculating the focus or salience ranking from linguistic cues [28, 24, 27, 1]. For example, items encountered or discussed recently carry more salience than items that have not been mentioned recently. The salience update process might also take into account attributes of the visual world. For example, Kaiser et al [16] developed a model of visual salience for an augmented reality application using four factors: *time, stability, visibility*, and *center-proximity*. Time represents persistence: the portion of frames over a certain window in which the object appears. Stability results in a penalty for objects that enter and leave the region multiple times. Visibility represents the amount to which the user's gesture overlaps with the object's visible projection, and center-proximity gives items at the center of the user's gaze or gesture a higher salience ranking.

The second step, *interpreting referring expressions*, is triggered as each referring expression is encountered and the context is searched for a semantically compatible referent. The search may integrates a number of different properties of the expression itself, the local linguistic context in which it appeared, and the context as it existed when the expression was spoken. Information about the expression itself can include lexical semantics, such as the lexical head and agreement features, and also the form of the expression, i.e. whether it was a pronoun, description, or locative adverb, etc. Different NP forms indicate different relationships with the context, and therefore different search procedures are invoked for each form. For example, to interpret a pronoun, the search begins with entities that are judged to be most salient [12, 2, 11].

Information about the local context might include the predication context, for example the expression might have been used to describe the PATIENT of a PUSH() action. This information, combined with a semantic resource that defines basic categories and the semantic restrictions on particular argument positions, can provide a powerful level of discrimination for resolving ambiguous phrases. A wide assortment of search methods for anaphora resolution have been developed, some exploiting syntactic structure or semantic features, others using statistical preferences (see [26] for a recent survey).

2.3 Visual Perception and Context

We aim to determine the topic of an utterance using the visual context of the speaker. Our definition of VC is the set of all objects that have ever been within the speaker's field of view and the timing information associated with their appearances and disappearances. In our study, visual salience is defined as that property of an entity in the VC that makes it the most likely topic of a person's utterance. We believe that a speaker's allocation of visual attention provides a reliable indicator of the relative visual salience of objects in the scene. Hence, factors influencing visual attention can provide valuable information about an object's visual salience, which could then be used to determine the topic of utterance.

Research in visual perception has shown that several factors influence the focus of our attention when presented with complex scenes. A well established theory in visual attention research is that the deployment of visual attention can be "guided" by the result of preattentive visual processing [8, 14]. The preattentive visual processing stage is of interest since it is known to be sensitive to certain features, such as color, orientation, curvature, and size, which form a feature-space in which the objects of our Quake world vary greatly. An object's novelty within this feature-space causes it to "pop-out", making it the focus of attention in later stages of visual processing [23].

Based on the role of the preattentive visual processing stage, we provide a simple measure that quantifies the novelty of an object, and how it changes over time, using a *Uniqueness* (U) parameter. Our definition of visual salience does not require an object to be in the current field view for it to be selected as the topic of the utterance. When an object falls out of view, its saliency is determined not only by its Uniqueness, but also by the amount of time since it was last seen ("recall delay"), and its exposure time before it dropped from view. The positive and negative effects of recall delay and exposure (or presentation) time on visual memory performance [10] is well known, and form the basis of the *Recency* (R) and *Persistence* (P) terms respectively.

3 Visual Salience Algorithm

As a person moves through the virtual world, different entities enter the visual context, and the factors affecting their visual salience need to be updated periodically.

Uniqueness (U): As mentioned earlier, the novelty of an object influences its pop-out, and hence influences the deployment of attention over the scene. The Uniqueness term models an object's novelty based purely on how frequently the object appears within the field of view. However, this model can be enhanced using computer vision based approaches that utilize object features pertinent to the preattentive visual processing stage (Sect. 2.3) to determine novelty. In our current formulation, an entity such as the floor, by virtue of being almost constantly visible within a given period, should be assigned a small U value, while an uncommon object (e.g., the button in the Quake world (Fig.2)) should get assigned a relatively larger value. Initially all objects are assigned a maximum U value of 1, signifying equal Uniqueness. At each instant, the U value of an entity is penalized by a quantity proportional to the frequency of its occurrence in the field of view over a time window called the Uniqueness Window (T_u),

$$U_{i,j} = U_{i,j-1} - \delta$$
$$\delta = k \times \left(\frac{n_{i,j}}{T_u}\right) \quad (1)$$

where the subscripts i and j represent the object i.d., and the current time instant, respectively. $n_{i,j}$ is the number of times object i was seen between $j - T_u$ and j. The constant of proportionality between the penalizing factor δ and the frequency of occurrence is denoted by k.

An object seen often in the recent past would have a large δ value when computed over a small T_u, and its Uniqueness would thus be heavily penalized. This seems consistent with the phenomenon of object-based Inhibition-of-Return (IoR), which states that "people are slower to return their attention to a recently attended object" [6]. It should be noted that while in our model the U values of all visible objects are penalized, a more faithful model of object-based IoR would penalize only objects that were attended to in the scene.

Recency (R): Once an object drops out of the field of view, we assume that the probability of it being the target of a referring expression decays with time. This relation is analogous to the well known decay of visual memory with increase in recall time [10]. A zero-centered Gaussian is chosen to model R. This profile represents a slow decay in R immediately after an object disappears, followed by a period of rapid decay that leads to an almost constant near-zero value,

$$R_{i,j} = e^{-\left(\frac{t_{i,j}}{\sqrt{2}\sigma}\right)^2}, \quad (t \geq 0) \quad (2)$$

where σ stands for the standard deviation (describe later), and $t_{i,j}$ is the length of the time interval measured from j since object i was last seen. Note that all objects currently visible have a maximum R value of 1.

Persistence (P): Analogous to the effect of presentation time on visual memory, the R values of different objects at each time instant should not only depend on $t_{i,j}$, but also on how long they were visible before disappearing. Persistence is a simple measure of an object's exposure time, computed as the frequency of occurrence of an object within a time interval, called the Persistence Window (T_p),

$$P_{i,j} = \frac{m_{i,j}}{T_p} \quad (3)$$

where $m_{i,j}$ is the number of times an object i was seen between $j - T_p$ and j. Intuitively, T_p should be large enough to allow objects to acquire significant P values, and at the same time small enough to ensure that P values of temporally distant objects fade with time. The dependence of R on P is established by making the value σ in Eq.2 proportional to P, such that

$$\sigma = c \times P_{i,j}$$

where c is a constant.

Visual Salience (S): At any instant, the combination of an object's Uniqueness and Recency determines its visual salience (S). Since neither a novel object seen a long time ago, nor a common one that is currently visible, have high visual salience, we define S such that only objects with large values of U and R would get assigned a high visual salience value.

$$S_{i,j} = U_{i,j} \times R_{i,j} \qquad (4)$$

Since the range of U and R is [0,1], an $S_{i,j} = 1$ corresponds to maximum visual salience.

4 Our Study

We trained our model using a randomly selected five minute discourse segment involving one pair of participants. We chose a small segment of data in this pilot study due to the time-consuming process of manually annotating the video frames. In order to quantify the relationship between the visual and linguistic contexts without the possible interference of factors such as mutual knowledge, we modeled the visual context of a single participant. Frames of the selected video segment were manually annotated at a fixed time interval with a list of objects visible in that frame. The manual annotation assumed perfect object segmentation, that is, each visible object was given a unique label[4], however locative entities such as rooms were not considered in the frame annotations. The algorithm takes a stream of frames in sequential order to create the VC. For each sample frame, the objects in the visual context were ordered by their visual salience as computed by the algorithm described in the previous section.

4.1 Our Baseline Algorithm

The closest comparable approach to our work presented in the literature is [18], where "centrality" and "size" are used to determine visual salience in a simplistic simulated 3D world. However the assumptions that form the basis of their approach do not hold in our domain, thus preventing a direct comparison. As described in Sect. 2.1, salient objects may not necessarily be in the centre of the field of view. Further, since *all* entities in the perceptible world are potential referents, the size feature is also inappropriate, since background objects like walls, floors, and ceilings will always be assigned high salience. We hence resort to a discourse-based salience model to provide a comparison to the proposed approach.

---

[4] Background entities such as walls, floors and ceilings were determined to be possible targets of reference and were, therefore, identified and annotated by color and texture.

The discourse-based reference resolution method worked as follows, and was hand-simulated rather than automatically produced.

- The transcripts were manually segmented into units. Each independent clause and each full constituent that was not part of a complete clause was considered an independent unit.
- The LC was updated after each unit by forming a new context frame containing an id for the referent of each base noun phrase in the unit, and adding this frame to the beginning of the LC list.
- The salience of items in each frame was determined using a left-to-right, breadth-first ordering on NP-based arguments in the utterance [26].
- Given an expression to match to a referent, each update frame in the context was searched in order of recency starting from the utterance containing the RE. Items in each context frame were compared to the RE and semantically incompatible items were discarded. For example, the plural pronoun *them* would not match a singular item, and a description such as *the button* only matches buttons.

4.2 Experiments

Table 1 shows the count of test items in the development dataset used to tune our visual attention algorithm, and the agreement between the item the speaker referred to and the most salient item in the visual context, as computed by our model. Items in the VC are rank-ordered using the visual attention model described above in Section 3. The number of test items in this table is small because we eliminated referring expressions that referred to items that were never seen, such as generic entities and propositions, in order not to penalize the algorithm for items it cannot track.

As the table demonstrates, using visual salience alone (Column (a)), we were able to identify which entity the user is speaking about 41.5% of the time. Column (a), titled "Absolute Highest Rank", shows whether the most-salient item, as calculated by our visual salience algorithm, was the correct referent. The next set of columns show the effect of adding lexical semantics as a filter on

Table 1. Performance on different referring forms in the development corpus

RE Form	Count	Highest Ranked Absolute VC (a)	Highest Ranked Semantic Match VC (b)	LC (c)
A/An N	2	1	2	1
The N	19	7	16	11
This/These N	1	0	0	1
That/Those N	4	3	3	2
This/That/These/Those	5	2	2	2
It/Them/They	10	4	4	6
Total	41	17 (41.5%)	27 (65.9%)	23 (56.1%)

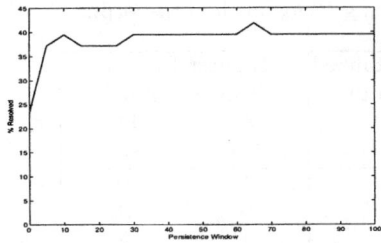
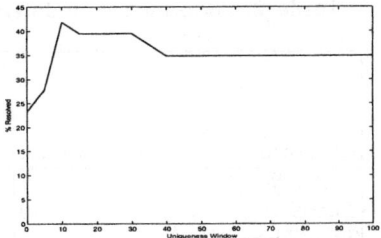

Fig. 5. Accuracy using varying Persistence Windows

Fig. 6. Accuracy using varying Uniqueness Windows

items in the context. For example, if the speaker says *a button*, the algorithm was considered correct if the most salient button was the correct referent. Column (b) shows the performance of our algorithm with the addition of lexical semantic filtering. As a comparison, Column (c) shows the performance of the discourse-only baseline algorithm, also with lexical semantic filtering.

Effect of Persistence Window and Uniqueness Window. Figures 5 and 6 show the effect of varying the length of the Persistence and Uniqueness Windows on the performance of reference resolution for the development dataset. To account for the possibility of dependence on both parameters, Persistence Window and Uniqueness Window were varied simultaneously. The maximum resolution performance was then determined for each parameter to obtain the optimum lengths of the Persistence and Uniqueness windows in the final algorithm. Figure 5 shows that, for the training dataset, the optimum Persistence window is of length 65 seconds and the optimum Uniqueness window is 10 seconds long. We use these same values for the testing dataset.

Effect of Sampling Frequency. The frequency of update of the visual context can affect not just the performance of reference resolution but the space and time complexity of the resolution algorithm as well. Keeping in mind the goal of enabling real time reference resolution, it is important to minimize the complexity of the algorithm. The performance of the reference resolution model was observed while varying the frequency of visual context updates. The accuracy of reference resolution increases with the frequency of visual context update (0.5Hz=31.70%, 1Hz=41.46%).

We used the development dataset to train these parameters, then tested the algorithm again on two new portions of video, totalling approximately 9 minutes, which used different speakers. The performance is shown in Table 2. The results from our algorithm in this segment are slightly lower compared to that obtained on the development set.

Two examples of correctly resolved noun phrases are shown in Figures 7 and 8. In Figure 7, several objects are in the scene. The Quake logo was correctly chosen for the pronoun *that* in spite of the presence of the table and background objects such as the walls, ceilings and floors. In Figure 8, the correct referent (the button

Table 2. Performance on different referring forms in the test corpus

RE Form	Count	Highest Ranked Absolute VC (a)	Highest Ranked Semantic Match VC (b)	Highest Ranked Semantic Match LC (c)
A/An N	1	1	1	0
The N	19	1	9	10
This/These N	8	1	4	2
That/Those N	6	4	4	4
This/That/These/Those	13	6	7	5
It/Them/They	20	8	10	18
Total	67	21 (31.3%)	35 (52.2%)	39 (58.2%)

Fig. 7. Speaker's view when he said "yeah so *that* needs to go there"

Fig. 8. Speaker's view at "is there *a second button* there"

on the right) associated with the phrase *a second button* was identified by our system. Since this ambiguous expression is the first mention of this object in the discourse, it will never be resolved by the baseline system.

5 Conclusions and Future Work

In this preliminary study we have created a model that assigns a salience measure to each object in the visual context, and automatically updates this value as the speaker moves around in the world and interacts with his surroundings. We use only this salience assignment (along with some minimal assistance from other linguistic analyses, namely semantics) to determine the referents produced by the speaker in two different discourse segments, with promising results. Several aspects of the model were based on findings from related research on visual attention and memory. The results obtained encourage us to believe that the assumptions made were well-founded. Because our visual salience model performs strongly compared to a baseline using linguistically-determined salience, we expect that a fused model, using both sources of evidence, will perform better than currently available methods for tracking the attentional state of a dialog in service of reference resolution.

We intend to develop an automatic process to identify the different objects in the field of view of a user by employing geometric constraints and knowledge of the Quake world. Another stage of the pipeline that needs automation is the segmentation of the transcript into discourse units. With these improvements, we would be able to test our model on larger data sets, and also explore the effect of faster update rates (> 1 Hz). Further, we also intend to incorporate visual characteristics of objects into our model to enable us to better discriminate between objects in the visual context based on their salience.

References

1. Saliha Azzam. Resolving anaphors in embedded sentences. In *Proceedings of the 34th Annual Meeting of the Association for Computational Linguistics (ACL '96)*, pages 263–269, 1996.
2. Susan E. Brennan, Marilyn W. Friedman, and Carl J. Pollard. A centering approach to pronouns. In *Proceedings of ACL '87*, pages 155–162, 1987.
3. Donna K. Byron. Improving discourse management in TRIPS-98. In *Proceedings of the 6th European Conference on Speech Communication and Technology (Eurospeech-99)*, 1999.
4. Donna K. Byron. Resolving pronominal reference to abstract entities. In *Proceedings of the 40th Annual Meeting of the Association for Computational Linguistics (ACL '02)*, pages 80–87, 2002.
5. E. Campana, J. Baldridge, J. Dowding, B. A. Hockey, R. W. Remington, and L. S. Stone. Using eye movements to determine referents in a spoken dialogue system. In *Workshop on Perceptive User Interfaces*.
6. S. E. Christ, C. S. McCrae, and R. A. Abrams. Inhibition of return in static and dynamic displays. *Psychonomic Bulletin and Review*, 9(1):80–85, 2002.
7. Miriam Eckert and Michael Strube. Dialogue acts, synchronising units and anaphora resolution. *Journal of Semantics*, 17(1):51–89, 2000.
8. H. E. Egeth, R. A. Virzi, and H. Garbart. Searching for conjunctively defined targets. *J. Exp. Psychol:Human Perception and Performance*, 10:32–39, 1984.
9. Malte Gabsdil, Alexander Koller, and Kristina Striegnitz. Natural Language and Inference in a Computer Game. In *Proceedings of the 19th International Conference on Computational Linguistics (COLING02)*, Taipei, 2002.
10. G. Gillund and R. M. Shiffrin. A retrieval model for both recognition and recall. *Psychological Review*, 91:1–67, 1984.
11. Barbara J. Grosz, Aravind K. Joshi, and Scott Weinstein. Centering: A framework for modeling the local coherence of discourse. *Computational Linguistics*, 21(2):203–226, 1995.
12. R. Guindon. Anaphora resolution: short term memory and focusing. In *Proceedings of the 23rd Annual Meeting of the Association for Computational Linguistics (ACL '85)*, pages 218–227, 1985.
13. P. Heeman and J. Allen. The Trains spoken dialog corpus. CD-ROM, Linguistics Data Consortium, 1995.
14. J. E. Hoffman. A two-stage model for visual search. *Perception and Psychophysics*, 25:319–327, 1979.

15. M. Johnston, S. Bangalore, G. Vasireddy, A. Stent, P. Ehlen, M. Walker, S. Whittaker, and P. Maloor. Match: An architecture for multimodal dialogue systems. In *Proceedings of the 40th Annual Meeting of the Association for Computational Linguistics (ACL '02)*, pages 376–383, 2002.
16. Ed Kaiser, Alex Olwal, David McGee, Hrvoje Benko, Andrea Corradini, Xiaoguang Li, Phil Cohen, and Steven Feiner. Mutual disambiguation of 3d multimodal interaction in augmented and virtual reality. In *Proceedings of the 5th international conference on Multimodal interfaces (ICMI 2003)*, Vancouver, B.C., Canada, November 2003.
17. Andrew Kehler. Cognitive status and form of reference in multimodal human-computer interaction. In *Proceedings of the 17th National Conference on Artificial Intelligence (AAAI-2000)*, Austin, Texas, July 2000.
18. J. Kelleher and J. van Genabith. Exploiting visual salience for the generation of referring expressions. In *Proceedings of the 17th International FLAIRS conference*, 2004.
19. John Kelleher and Josef van Genabith. Dynamically updating and interrelating representations of visual and linguistic discourse. *submitted to Artificial Intelligence*.
20. John Kelleher and Josef van Genabith. Visual salience and reference resolution in simulated 3-d environments. *Artificial Intelligence Review*, 21(3):253–267, 2004.
21. Pia Knoeferle, Matthew W. Crocker, Christoph Scheepers, and Martin J. Pickering. The influence of the immediate visual context on incremental thematic role-assignment: Evidence from eye-movements in depicted events. *Cognition*, 2004.
22. Scheutz Matthias, Eberhard Kathleen, and Andronache Virgil. A real-time robotic model of human reference resolution using visual constraints. *Connection Science*, 2004.
23. U. Neisser. *Cognitive Psychology*. Appleton, Century, Crofts, New York, 1967.
24. Candace L. Sidner. Focusing in the comprehension of definite anaphora. In M. Brady and R. Berwick, editors, *Computational Models of Discourse*, pages 363–394. 1983.
25. M.K. Tanenhaus, M.J. Spivey-Knowlton, K.M. Eberhard, and J.E. Sedivy. Integration of visual and linguistic information in spoken language comprehension. *Science*, 268:1632–1634, 1995.
26. Joel R. Tetreault. *Empirical Evaluations of Pronoun Resolution*. PhD thesis, University of Rochester, 2004.
27. Marilyn A. Walker. Limited attention and discourse structure. *Computational Linguistics*, 22(2):255–264, 1996.
28. Terry Winograd. *Understanding natural language*. New York: Academic Press, 1972.

Meaning in Context

Henning Christiansen[1] and Veronica Dahl[2]

[1] Roskilde University, Computer Science Dept.,
P.O.Box 260, DK-4000 Roskilde, Denmark
henning@ruc.dk
[2] Dept. of Computer Science, Simon Fraser University
Burnaby, B.C., Canada
veronica@cs.sfu.ca

Abstract. A model for context-dependent natural language semantics is proposed and formalized in terms of possible worlds. The meaning of a sentence depends on context and at the same time affects that context representing the knowledge about the world collected from a discourse. The model fits well with a "flat" semantic representation as first proposed by Hobbs (1985), consisting basically of a conjunction of atomic predications in which all variables are existentially quantified with the widest possible scope; in our framework, this provides very concise semantic terms as compared with other representations. There is a natural correspondence between the possible worlds semantics and a constraint solver, and it is shown how such a semantics can be defined using the programming language of Constraint Handling Rules (Frühwirth, 1995). Discourse analysis is clearly a process of abduction in this framework, and it is shown that the mentioned constraint solvers serve as effective and efficient abductive engines for the purpose.

1 Introduction

Natural language semantics, or the relation between language and the world, has been one of the main concerns of linguists and computational linguists over the past few decades. Most efforts have gone into devising representation schemes for the world knowledge a sentence expresses. A plethora of such formalisms has resulted from different ways of viewing the world and how to represent it.

Compositional styles of semantics in the Montague tradition [17] typically result in interesting but very large and complex semantic expressions, involving functions, disjunctions, nested expressions, etc. It has recently been argued [18] that most approaches to semantics so far do not constitute semantics at all, but an alternative kind of syntax, that is: instead of being interpretations in the logical sense, they are simply elaborate paraphrases of the natural language utterance they purport to be the "meaning" of.

We propose instead a model of context-dependent natural language semantics in which sentence meanings are expressed in a context with which they interact as opposed to being functions parameterized by context. Details of a sentence are interpreted in context and may in turn contribute to this context, so that the terms representing a particular sentence comes closer to a purified form of the "intention" of that sentence. Depending on application, we may even go so far as to say that the meaning of a sentence *is* its contribution to context.

This model is formulated in terms of possible worlds so that a context, here taken to mean the knowledge about the world collected so far, designates a class of worlds which are compatible with that knowledge. Our representations are true interpretations, in the sense that utterances analyze into a dynamically growing knowledge base which models the world so far, and provides as well the context within which future utterances will be analyzed. Individuals in the world do relate to names (constants in the knowledge base), relationships in the world to predicates in the knowledge base, and so on, as in any interpretation in the logical sense. Successive utterances weed out other possible worlds by further materializing the emerging world being described, so that there is a dynamic interaction between context and interpretation.

Example 1. Consider an everyday discourse that refers to different individuals, one of them usually called "Peter", and let us abstractly identify the individual as p17. In that discourse we take the meaning of "Peter" as a reference to this particular individual. In the same discourse, the occurrence of the pronoun "he" may refer to the same individual, and in our model we consider the meaning of this "he" to be the reference to p17. That occurrence of "he" could have been replaced by "Peter" without changing the overall meaning of the sentence or discourse, the pronoun is used for convenience only. (This as opposed to considering the meaning of "he" as an occurrence of a generic reference device radically different from a proper noun.)

Another speaker may refer to "the tall, red-haired man carrying a laptop"; the meaning in context of this expression may be exactly the same as the mentioned occurrences of "Peter" and "he", namely the reference to p17. The speaker may have chosen the longer expression for different reasons: he may not know the name, or may want to communicate to other people whom he (the speaker) believes do not know the name. In our Meaning in Context model, we see the meaning of the sentence (1) below as a wish to indicate (or: as having the purpose of indicating) that a particular individual has won a particular piece of hardware, and not as a wish to indicate other (and in a certain sense irrelevant properties of the individual) which abstractly might be represented by a logical fact won(p17,f450).

The tall, red-haired man carrying a laptop won a brand new Ferrari. (1)

However, this sentence *presupposes* that the individual in question possesses certain properties such as being tall, etc. While the concise (contextual, or pragmatic) meaning of (1) may be represented by the expression won(p17,f450), this can also be seen as just another contribution the meaning of the discourse.

From the viewpoint of discourse analysis, (1) provides the following little knowledge base,

```
tall(X), read_haired(X), carries(X,Y), laptop(Y),          (2)
won(X,Z), ferrari(Z), brand_new(Z)
```

The use of absolute references such as p17 and f450 may be problematic from a philosophical point view and may also make the implementation difficult, so we have replaced them by logical variables.

In general, our methodology is applicable to "flat" meaning representations (e.g., those involving a conjunction of atomic predications in which all variables are existentially quantified with the widest possible scope). Of course, this leaves open the question of fine-tuned treatments of quantification. Previous approaches (e.g., [11]) maintain existential quantifiers only by tricks such as reifying universal variables as typical elements of sets and by a sort of skolemization of dependent existentially quantified variables. Future work would have to incorporate a subtler treatment that considers as well the many nuances in NL quantifiers. However, flattening everything into predications does have the advantage of allowing direct knowledge base creation — that is, direct construction of a true interpretation — as a result of an utterance's analysis, and one that takes context into account at that.

Discourse analysis based on this Meaning-in-Context model related to interpretation by abduction [12] as it consists basically of "guessing" those real-world hypotheses that are necessary for the discourse to reflect the world in a truthful way. The possible worlds semantics underlying our approach fits particularly well with recently developed techniques for language analysis and abduction based on the paradigm of Constraint Handling Rules [10] which is a declarative extension to the Prolog programming language for writing constraint solvers; we can, in fact, demonstrate a natural correspondence between the possible worlds semantics and such constraint solvers. We refer to implementations using two paradigms, CHR Grammars [4,6] which provide a bottom-up analysis system in which parsing itself is treated as a constraint solving process and A^2LP [7] which incorporates abduction and other facilities in a more traditional Prolog and DCG setting. These technologies can handle contexts consisting of a variety of hypotheses: atomic hypotheses as shown above, so-called explicit negation of hypotheses, assumptions in the sense of Assumption Grammars [9], and also implications with existentially (globally) as well as universally (locally) quantified variables.

We can mention other advantages of the approach that are rather obvious, but not shown in details here due to lack of space. Lexical ambiguities that are difficult to resolve otherwise can be handled be taking the current context into account, or perhaps being delayed until more context information have been collected in the subsequent discourse. We can also provide rules that activate pre-existing contexts when sufficient amount of indication is found, thus making available new vocabulary and ontology.

The present paper is inspired by earlier work presented at CONTEXT'99 [3] but here given simpler and more general form and accompanied by relevant implementation techniques developed in the meantime.

2 General Presentation of the Meaning in Context Model

A discourse is grounded on some kind of real world. We use *real world* as a metaphor for some imagined or historically placed setting, in which events can take place and different properties may or may not hold, perhaps limited in time and space. It may, for example, be the planet Earth in the 20th century or the setting for an adventure which includes true unicorns and witches who really fly by means of brooms. A professionally constructed set of lies also reflects a "real world" in this sense, i.e., the illusion of a reality that someone wants to convince the public exists.[1] Real worlds are unwieldy entities that cannot be characterized fully by any set of facts, be it finite or infinite.

The word "context" is used by different authors and communities for different but often interrelated and dependent notions. Linguists often refer to the context of a phrase or word as the text that surrounds it. Another everyday usage of "context" refers to a section of the real world in which some events or a discourse takes place, and is often intertwined and confused with another meaning, namely knowledge about the same thing.[2] Our usage complies with the last view: to us, a *context* is a collected body of represented knowledge, typically what has been learned by attending a given discourse up to a certain point. In this respect, we agree with Stalnaker's work, and the possible worlds, Kripke-style semantics for contexts and the terminology we introduce is essentially a formalization of his ideas summarized in [21] .

Let $\mathcal{W}$ denote the set of all worlds, one of which may be designated the *real* world (the existence of a particular real world is, however, not essential for our purpose). No specific representation of context needs to be fixed, but we will, however, assume that it is decomposable in the sense that it can be separated into statements or propositions that we call *context facts* whose truth can be evaluated separately. Let, thus, $\mathcal{C}$ refer the set of all context or parts thereof and assume an operation $\cup : \mathcal{C} \times \mathcal{C} \to \mathcal{C}$ being associative, commutative, and idempotent; intuitively, no distinction is made between a proposition and the set consisting of it.

The notion of truth is represented by an entailment relation $\models: \mathcal{W} \times \mathcal{C}$ with the intuitive meaning that $w \models c$ if and only if c is a correct property in w. Entailment is *monotonic* in the sense that

$$w \models c_1 \cup c_2 \quad \text{implies that} \quad w \models c_1 \text{ and } w \models c_2.$$

This property should not be confused with the distinction between monotonic and non-monotonic reasoning as our model by no means excludes non-monotonic logics for reasoning with context.

---

[1] In the present paper, we disregard the consequences of possible disagreement between such worlds and the real, real world.

[2] The definition (one out of two) given by Webster's dictionary [22] indicates this duality "... 2. the set of circumstances or facts that surround a particular event, situation, etc."

We define the *semantic function* W which maps any element of $\mathcal{C}$ into a set of its possible worlds, i.e.,

$$W(c) =_{\text{def}} \{w \in \mathcal{W} \mid w \models c\}.$$

The following monotonicity property follows from the definition:

$$W(c_1 \cup c_2) \subseteq W(c_1) \cap W(c_2)$$

Whenever, for contexts c and c', we have that $W(c) \subseteq W(c')$ we say that c *logically implies* c', written $c \models c'$. In case $c \models c'$ and $c' \models c$, we say that the two contexts c and c' are *equivalent* which is indicated symbolically $c \equiv c'$.

In general, it may be expected that $W(c)$ is either infinite or empty; a context c is *inconsistent* whenever $W(c) = \emptyset$ and *consistent* otherwise. Let $\top$ denote a prototypical inconsistent context and $\bot$ a prototypical context without information content at all, i.e., with $W(\bot) = \mathcal{W}$; a context with one of these properties is identified with the relevant of $\top$ and $\bot$.[3]

The notion of a discourse is not restricted particularly to text or speech, and applies also to general sequences of sensor signals or other sorts of meaning-bearing events, or a perhaps a mixture of all kinds.[4]

An alphabet Σ is assumed, and we define a *sentence* as an element of some designated subset $\mathcal{S}$ of Σ^+.[5] A *discourse* is any finite concatenation of sentences, i.e., we let

$$\mathcal{D} =_{\text{def}} \{s_1 \cdot \ldots \cdot s_n \mid n \geq 0, s_i \in \mathcal{S}\}$$

refer to the set of all discourses; ϵ denotes the empty discourse, and concatenation is indicated by juxtaposition or made explicit (as above) by a dot. For simplicity of definition, it is assumed that the decomposition of a discourse into sentences is unique, but otherwise a discourse may be as ambiguous as ever.

A *(contextual) meaning function* is a function $M : \mathcal{D} \to \mathcal{C}$ which satisfies the following *prefix properties*:

$$M(\epsilon) = \bot$$
$$M(d \cdot s) \models M(d), \text{ for any discourse } d \text{ and sentence } s$$

In other words, a meaning function M extracts the content of a discourse and puts in into a represented form, and the composition of M and W provides a possible worlds semantics for discourses. The second prefix property indicates that the longer discourse, the more context knowledge is learned and the

---

[3] The symbols $\top$ and $\bot$ correspond to top and bottom elements in the algebraic lattice of $\mathcal{C}$ with $\cup$ induced by W.

[4] Whether or not a discourse is considered an element in the real world is not important for us at this level; this may or may not be assumed as convenient.

[5] The term "sentence" should be taken as a convenient usage for any syntactic entity that directly contributes to context and do not necessarily comply directly with a syntactic notion of a sentence.

fewer worlds possible. We may classify as *rubbish* (*boring*), any discourse d with $M(d) = \top\ (= \bot)$.

We assume also the following about M, which we may call *syntax independence*. For any discourses d, d', and sentence s we have that

if $M(d) \equiv M(d')$, then for any sentence s, we have $M(d \cdot s) \equiv M(d' \cdot s)$

This means that the interpretation of a sentence depends on what has been said and not the way it has been said. This property indicates that any piece of information that can affect the interpretation of future sentences needs to be passed through the context, so one way to handle coordination of parallel sentences is to include (the memory of) the already spoken words as part of the context.

Finally, we introduce the notion of the *accomodation function* $A : \mathcal{C} \times \mathcal{S} \to \mathcal{C}$ given by a meaning function M which is intended as a formal equivalent to Stalnaker's notion of accomodation [21]:

$$A(M(d), s) = M(d \cdot s), \text{ for any discourse } d \text{ and sentence } s$$

We notice that a meaning function gives rise to an accommodation function and vice versa.

Example 2. The very first sentence in Nikos Kazantzakis' novel about Zorba the Greek [15] consists of the following four words in Greek.

$$\begin{array}{llll} Tov & \pi\rho\omega\tau o\gamma\nu\acute\omega\rho\iota\sigma\alpha & \sigma\tau ov & \Pi\epsilon\iota\rho\alpha\acute\iota\alpha \\ \text{him} & \text{first-meet/know-1stSingPast} & \text{in-the} & \text{Pireus} \end{array} \quad (3)$$

Greek grammar is very compact, for example the $-\alpha$ inflection of the verb indicates first person, singular, and past tense, so the subject is implicit in the sentence. It translates into "I met him for the first time in Pireus". The author has a specific real world in mind, and even the opening sentence (3) of the novel provides us quite a lot of important information about this world, and actually information that is central to the very end of the story. In the terms defined above, an accomodation function may be involved which produces a first context telling that at least two persons are involved in the story, one of whom is the novel's "I", and something about the relationship between the two, etc. In a formal, flat representation this context might be something like this:

```
person(X), male(X), person(Y), i(Y), X≠Y, event(E),
    past(E), place_of(E,P), name_of(P, Pireus),           (4)
 meeting_event(E,X,Y), knows(X,Y), first_of_its_kind(E)
```

A meaning function assigns a set of possible worlds to this context, worlds that all includes Pireus and two persons who made each others acquaintance there, etc. As the novel's discourse goes on we learn more and more about X and Y and other events in which they and other characters are involved, thus narrowing down to fewer and fewer worlds. For the indicated representation and its meaning function, it may be assumed that any context is inconsistent if it contains event(E1), event(E1), before(E1,E2), first_of_its_kind(E2).

3   Discourse Analysis as Constraint Solving

The analysis of a given discourse will be considered in a sequential manner, reflecting the time span in which the discourse takes place and the fact that a discourse may continue even if we thought that it was finished.

In this section we refer to constraints in the sense of constraint logic programming (CLP) [10, 13]. A *constraint* is a formula of (typically) first-order logic, often restricted to atomic formulas, but this needs no be the case; a *constraint store* is a finite set of constraints. A semantics is assumed for constraints and constraint stores, e.g., in terms of logical axioms.

A *normal form* is assumed among semantically equivalent constraint stores, which can be briefly characterized as having a minimal textually representation that is an acceptable answer to a user. We may as an example expect that $\{p(x), q(y), x = y\}$ has the normal form $\{p(x), q(x)\}$ (in case p and q do not depends on each other and the equality sign has the usual meaning). Special normal forms are *false* (failure) and *true* (the empty constraint store) which, in our applications to discourse analysis are equivalent to $\top$, resp., $\bot$.

A *constraint solver* keeps the constraint store normalized in an incremental fashion, i.e., when a new constraint arrives to a normalized store, it re-establishes normalization while respecting semantics. In practice, however, a constraint solver often only approximates this property by, for example, returning a store that is equivalent to *false* but given as a set of other constraints. This may be due to inherent, undecidable properties or simply as a shortcut made for reasons of efficiency. For ease of definitions, we ignore the possible imperfections of practical constraint solvers and accept *the* stores produced by the constraints solvers as "normalized".

A CLP system consists of a *driver* that produces constraints during the process of solving some problem, and the constraint solver maintains normalization. Typically, the driver is a Prolog engine running some program and the constraint solver reports back successes, failures, and implied unifications.

For discourse analysis, we identify context with constraint store and, thus, context facts with constraints. A constraint solver is thus, a function $C : \mathcal{C}_{\text{norm}} \times \mathcal{C} \to \mathcal{C}_{\text{norm}}$ where $\mathcal{C}_{\text{norm}}$ refers to the set of constraint stores (contexts) in normal form. The fact that a constraint solver must respect the semantics of the constraints can now be expressed as $C(c, c') \equiv c \cup c'$.

A driver for discourse analysis can be a parser written in Prolog or any other kind of syntax analyzer working in a sequential manner. A syntax analyzer can be depicted as a function S from sentences to constraints; in general the analyzer may utilize the current constraint store so we can write $S : \mathcal{S} \times \mathcal{C}_{\text{norm}} \to \mathcal{C}$.[6] So when talking about syntax analysis, we have abstracted away the possible construction of a syntax tree or similar representation, and indicate only the "semantic tokens" produced in the shape new constraints (i.e., new context facts).

---

[6] In many cases, it is sufficient with a context independent analyzer $S : \mathcal{S} \to \mathcal{C}$ as the constraint solver will adapt the constraints produced to the current context.

A syntax analyzer S is correct with respect to accomodation function A whenever

$$A(c,s) = C(c, S(c,s)) \text{ for any normalized context } c \text{ and sentence } s.$$

The other way around, if S is the given entity, we can say that it defines an accommodation function and in turn a meaning function by the condition above.

Disjunctions and context splitting

It may be the case that new context facts indicate a choice between two or more incompatible hypotheses, for example, that the pronoun "he" refers to either Peter or Paul, but with no indication of which is the right one. We can assume the context c_0 being extended with the new piece of information $x = peter \vee x = paul$.

In principle, a constraint solver could treat disjunctions as constraints, i.e., consider the formula $x = peter \vee x = paul$ as *one* constraint. This does not fit with standard CLP technology that eliminates equations by means of unification, i.e., replacing any occurrence of a variable by the indicated value. Instead the constraint store is made subject of a *splitting*, which means that constraint store

$$c_0 \cup \{x = peter \vee x = paul\}$$

gives rise to two new constraint stores

$$c_0 \cup \{x = peter\} \quad \text{and} \quad c_0 \cup \{x = paul\}$$

and the analysis of the remainder of the discourse needs to be performed twice, once for each store and, of course, recursively in case the store splits again later.

This notion of splitting in a constraint solver has been formalized in the language of $CHR^{\vee}$ [2]. A CLP system based on Prolog implements splitting by means of backtracking. Another strategy is to maintain the different constraint stores in parallel, so that when a new context fact comes in from the discourse, a copy is added to all different stores each of which is normalized before next sentence is analyzed; this is applied in the CHR Grammar system [6] which is briefly mentioned below. We leave out the straightforward formalization in terms of possible worlds of systems including splitting (based on the union of possible worlds for the alternative contexts).

Constraint Handling Rules

The language of Constraint Handling Rules, CHR, is an extension to Prolog intended as a declarative language for writing constraint solvers for CLP systems; here we give a very compact introduction and refer to [10] for details.

Constraints are first-order atoms whose predicate are designated constraint predicates and a constraint store is a set of such constraints, possible including variables, that are understood existentially quantified at the outermost level. A constraint solver is defined in terms of rules which can be of the following two kinds.

$$\text{Simplification rules: } c_1,\ldots c_n \;\texttt{<=>}\; Guard \mid c_{n+1},\ldots,c_m$$
$$\text{Propagation rules: } c_1,\ldots c_n \;\texttt{==>}\; Guard \mid c_{n+1},\ldots,c_m$$

The c's are atoms that represent constraints, possible with variables, and a simplification rule works by replacing in the constraint store, a possible set of constraints that matches the pattern given by the *head* $c_1, \ldots c_n$ by those corresponding constraints given by the *body* $c_{n+1}, \ldots, c_m$, however only if the condition given by *Guard* holds. A propagation rule executes in a similar way but without removing the head constraints from the store. In addition, rule bodies and guards may include equalities and other standard relations having their usual meaning. The declarative semantics is hinted by the applied arrow symbols (bi-implication, resp., implication formulas, with variables assumed to be universally quantified) and it can be shown that the indicated procedural semantics agrees with this. This is CHR explained in a nutshell.

The declarative semantics provides a possible worlds semantics for constraint stores (alias contexts). To see this, define for a set of rules R, a *model* of R as any set of ground (i.e., variable-free) constraints with the property that any rule in R evaluates to true in M in the usual way (see, e.g., [10] for precise definitions or use any standard textbook of mathematical logic). In general, there may be an infinity of models of a set of rules R, which can be understood as a collection of all possible worlds that we refer to as $\mathcal{W}_R$. In general, we may use $\mathcal{W}_S$ to refer to the set of models of any formula or set of formulas S.

A constraint store c represents the knowledge collected up to a certain point and its possible worlds should be exactly all those models of R in which the information of c is true, i.e., those models of c that are also models of R. We define, thus,[7]

$$\mathcal{W}_R(c) =_{\text{def}} \mathcal{W}_R \cap \mathcal{W}_c = \mathcal{W}_{R \cup c}. \qquad (5)$$

The condition (5) indicates also that the R can be thought of as defining an implicit initial context whose meaning function restricts to those worlds in which R holds.

Example 3. Documentaries assume implicitly the laws of physics whereas this is typically not the case for space movies.

As a consequence of the properties noticed earlier, we see that a CHR program together with a syntax analyzer induces an accommodation function and meaning function in the strict sense defined above.

Example 4. In example 2, we indicated a flat discourse representation which is well suited for CHR. A constraint solver for this representation may include the following rules.

```
X=/=X ==> fail.
before(E1,E2), first_of_its_kind(E1) ==> fail.
meeting_event(E,X,Y) ==> knows(X,Y).
```

---

[7] Strictly speaking, this does not comply with our assumption that a world cannot be characterized by any, even infinite, set of facts. It is easy to repair this flaw by the introduction of a "true" possible worlds semantics for logical models.

```
place_of(E,X) \ place_of(E,Y) <=> X=Y.
i(X) \ i(Y) <=> X=Y.
```

The first two rules identify inconsistent states, the next one adds a new hypothesis implied by another one. The rule about `place_of` tells that the place of an event is unique, and when it is applied during a discourse analysis, it may identify inconsistency if two different locations for the same event are asserted or help resolve an anaphoric reference. In the same way, the last rule indicates that the narrator "I" of a story is unique. Notice that the two last rules are so-called simpagation rules of CHR which indicate that the constraints following the backslash are removed from the constraint store and the others stay (thus they abbreviate special kinds of simplification rules).

In our own work, we have extended CHR so that the constraint store may contain dynamically created CHR rules as well, which is useful in many application for language processing. So, for example, the sentence *"all green objects are on the table"* can be modelled by adding `green(X) ==> on(X,the_table)` to the constraint store.

4 On the Relation to Abductive Reasoning

For reasons of space, we cover this topic in a very compact manner; detailed arguments can be found in [4, 6, 7].

Generation of a discourse D is an inherently deductive process based on known premises of a grammar G and the speaker's known context C. The relationship between the components is that

$$G \wedge C \to D. \tag{6}$$

For this discussion, let us instantiate this pattern by assuming that G is a DCG [19], C a set of Prolog facts, and D an answer produced by a Prolog interpreter.

For discourse *analysis*, grammar and discourse are known but the premise C of (6) is the unknown to be found. This is by definition an abductive problem for which methods developed for Abductive Logic Programming [14] (ALP) apply. In general, a set of *integrity constraints* (ICs) are needed: ICs are logical conditions that must be satisfied by any context C for it to be consistent in the sense defined above. An ALP interpreter enforces the ICs and, in case of splitting, discards inconsistent branches. (In the case of generation, the ICs are not necessary, but they are implicit in the tacit assumption that the *given* C represents some real world.)

It has been explained above how a CHR program P defines a meaning function which especially can identify inconsistent contexts. Now comes an important point: When such a P is combined with a syntax analyzer written in Prolog (as a DCG, for example), the overall functioning is exactly that of an abductive interpreter with P as integrity constraints. When a discourse D is given as query, the corresponding context is produced as the final constraint store or, if splitting occurs, as the disjunction of the alternative constraint store produced.

5 Discourse Analysis in A²LP and CHRG

A²LP is an extension to Prolog with abduction and assumptions defined by means of a few CHR rules explained in [7]. Such assumptions, known from Assumption Grammars [9], are related to abduction but provide scoping principles that are useful for modeling many linguistic phenomena; here we show only the abduction part of A²LP. Prolog's built-in grammar notation can be used with A²LP, and we show a grammar for simple discourses on still-life pictures.

Context representation is given as facts about the immediate physical relationship between objects, so, e.g., i_on(a,b) denotes that a is situated directly upon b, similarly for the predicate i_in. A constraint solver defines a meaning function or, alternatively in ALP terminology, provides the integrity constraints of an abductive logic program.

```
i_on(X,Y), i_on(Y,X) ==> fail.
i_in(X,X) ==> fail.
...
container(C) ==> thing(C).
i_in(the_box,the_vase) ==> fail.
i_in(_,C) ==> container(C).
```

It identifies a number of impossible situations and indicates properties and classes of some known objects, for example that the_box cannot be inside the the_vase. The following rules apply semicolon as in Prolog, i.e., disjunction by splitting, to restrict to a universe with exactly four objects.

```
thing(X) ==> X=the_flower ; X=the_box ; X=the_vase ; X=the_table.
container(X) ==> X=the_box ; X=the_vase.
container(the_flower) ==> fail.   container(the_table) ==> fail.
```

Finally, let us introduce a rule that defines the everyday notion of one thing being on another thing, which may or may not involve an intermediate object.

```
on(X,Y) ==> i_on(X,Y) ; i_on(X,Z), i_on(Z,Y) ; i_in(X,Z), i_on(Z,Y)
```

The following grammar rule defines a little syntax analyzer for simple sentences.

```
sentence --> [A,is,on,B], {thing(A), thing(B), on(A,B)}.
```

Analyzing *"the flower is on the table"* gives rise to a context constructed as the disjunction of different final states, which allows the placement of the flower in a number of different positions. Continuing the discourse with *"the flower is in the vase"* and assuming an analogous rule for sentences about *"in"*, we cut down to possible worlds in which the flower is placed in a vase on accordance with the traditional picture of the ideal home.

There is no reason to include more sophisticated examples as it is well-known that Prolog's DCGs can model a large variety of linguistic phenomena, and it is well-known that a flat representation in the sense of [11] is very general. In addition, the newest version of A²LP also can handle of context facts in the shape of rules, as shown in the end of section 3.

It also clear that a constraint solver written in CHR can work together with other and more sophisticated syntax analyzer. Due to lack of space, we leave out a presentation of the CHR Grammar system [5, 6], but it must be mentioned as it provides a very flexible kind of grammars that are compiled into CHR rules that run as a robust bottom-up parser that can interact with abduction and constraint solvers written in CHR as shown above.

6 Extensions and Application

We have only showed very simple examples but it should be emphasized that CHR is a very powerful language for expressing different properties; it is Turing-complete so every computable function can be used as meaning function. We indicate a few relevant applications.

Resolving Lexical Ambiguity. A classical example is "bank" which can be a river bank or a financial institution, but in most cases it is "clear from context" for a human what is meant. The following two rules provides a self-explanatory example of how this can be expressed.

```
nautical_event(E,X,Y) \ place(E,bank) <=> place(E,river_bank).
financial_event(E,X,Y) \ place(E,bank) <=> place(E,financial_bank).
```

They give a dynamic interaction between context and analysis. In case a relevant `nautical_event` is known at the point when `place(···,bank)` arrives, the rule applies immediately; otherwise the resolution of `place(···,bank)` may be delayed while analysis continues and done when the relevant context indicators are encountered.

Activation of predefined context elements. "... tire ... gearbox ... break ... 200 kmh ...". There is no doubt these people are talking about cars, so "boot" is likely not a piece of footwear but a place for your luggage. A systematic way of treating this phenomenon is to write rules that dynamically sum up a number that reflects the number of indicators, and when this is sufficiently high, call a constraint that activates the context.

```
indication(carContext,K) ==> K>0.8 | activate_context(cars).
```

Here `activate_context` can be a premise in other rules (as above) but also be a predicate programmed in Prolog that installs a sub-lexicon.

Non-monotonic Reasoning. The monotonicity assumptions in our general definitions does not exclude nonmonotonic reasoning as would be needed to handle liars in a more clever way that just noticing inconsistency. Here standard reification technique applies: add an extra attribute to each predicate in question, intitially uninstantiated, and set to `notTrusted` if it is learned that that the speaker is a liar. Intuitively, this may seem better than a traditional non-monotonic system that would *remove* the problematic facts: The human listener would still remember that these lies were told by a certain speaker and may have this in mind (i.e., in context) when analyzing what follows.

7 Related Work

Seeing the meaning of linguistic expressions as context-dependent is not a new idea, and the principle has been applied in papers which are too numerous to mention here, including at the recent Context conferences. A precursor from 1975 is [20], presenting a context-dependent semantic model for natural language whose definition of interpretation function looks very much, at the surface, like our accommodation functions, but the remaining part of that model is difficult to compare to ours (which clearly has benefitted from insight gained in logic programming, especially variants with abduction and constraints).

A search on the web based on the title of the present paper gave an interesting hit, a paper entitled "Meaning in Context: Is there any other Kind?" [16] from 1979 that argues for a shift in attitude in phenomenology, sociolinguistics, and ethnomethodology based on very much the same intuition that we have based our work on.

The view of discourse analysis as abduction appears in many works in the last twenty years or more, with [12] usually considered a central reference; [4] gives more references. These methods have never become widespread in practice, and we may hope that the application of constraint logic programming techniques as we advocate may result in, or inspire to, efficient and useful systems.

Abduction in CHR was proposed by [1] and refined in other publications already mentioned; basically it can be explained as a transformation of abduction into deduction, and it is interesting to compare this with a paper from 1991 [8] (at a time when CHR did not exist) that pointed out an isorphism between a class of abductive problems and deduction.

8 Conclusion and Perspectives

We have argued for a Meaning-in-Context approach to natural language semantics, grounded on overall conceptual considerations, given a possible worlds formalization of it, and in a straightforward way explained effective implementations based on recent advances in constraint logic programming. CHR is the subject of intensive research and development these years, including on adding priorities and weightings to the language — which will be an essential addition for the paradigm presented here.

Acknowledgements. This work is supported by the CONTROL project, funded by Danish Natural Science Research Council, the Velux Visiting Professor Programme, and Canada's NSERC Discovery Grant program.

References

1. Abdennadher, S., and Christiansen, H., An Experimental CLP Platform for Integrity Constraints and Abduction. Proceedings of FQAS2000, Flexible Query Answering Systems. Larsen, H.L., Kacprzyk, J., Zadrozny, S., (Eds.) *Advances in Soft Computing series,* Physica-Verlag (Springer), pp. 141–152, 2000.

2. Abdennadher, S., and Schütz, H. CHR^V: A flexible query language. Proceedings of FQAS'98, Flexible Query Answering Systems. Andreasen, T., Christiansen, H., Larsen, H.L. (Eds.) *Lecture Notes in Artificial Intelligence* 1495, pp. 1–14, Springer, 1998.
3. Christiansen, H., Open theories and abduction for context and accommodation. 2nd International and Interdisciplinary Conference on Modeling and Using Context (CONTEXT'99) Bouquet, P., Brezillon, P., Serafini, L. (eds.) *Lecture Notes in Artificial Intelligence* 1688. Springer-Verlag, pp. 455–458, 1999.
4. Christiansen, H., Abductive Language Interpretation as Bottom-up Deduction. In: Natural Language Understanding and Logic Programming, Proceedings of the 2002 workshop, ed. Wintner, S., *Datalogiske Skrifter* vol. 92, Roskilde University, Comp. Sci. Dept., pp. 33–47, 2002.
5. Christiansen, H., CHR Grammars web site. Source text, manual, and examples, http://www.ruc.dk/~henning/chrg, 2002.
6. Christiansen, H., CHR grammars. *International Journal on Theory and Practice of Logic Programming, special issue on Constraint Handling Rules*, to appear 2005.
7. Christiansen, H., Dahl, V., Assumptions and Abduction in Prolog *Proceedings of WLPE 2004: 14th Workshop on Logic Programming Environments and Multi-CPL 2004: Third Workshop on Multiparadigm Constraint Programming Languages Workshop Proceedings*, September 2004, Saint-Malo, France. Susana Muñoz-Hernández, José Manuel Gómez-Perez, and Petra Hofstedt (Eds.) pp. 87-101.
8. Console, L., Theseider Dupré, D., Torasso, P., On the Relationship between Abduction and Deduction. *Journal of Logic and Computation* 1(5), pp. 661–690, 1991.
9. Dahl, V., and Tarau, P. From Assumptions to Meaning. *Canadian Artificial Intelligence* 42, Spring 1998.
10. Frühwirth, T.W., Theory and Practice of Constraint Handling Rules, *Journal of Logic Programming*, Vol. 37(1–3), pp. 95–138, 1998.
11. Hobbs, J., Ontological Promiscuity. *Proceedings of the 23rd conference on Association for Computational Linguistics*, 8-12 July 1985, University of Chicago, Chicago, Illinois, USA, Proceedings, ACL, pp. 61–69, 1985.
12. Hobbs, J.R., Stickel, M.E., Appelt D.E., and Martin, P., Interpretation as abduction. *Artificial Intelligence* 63, pp. 69-142, 1993.
13. Jaffar, J., Maher, M.J., Constraint logic programming: A survey. *Journal of logic programming*, vol. 19,20, pp. 503–581, 1994.
14. Kakas, A.C., Kowalski, R.A., and Toni, F. The role of abduction in logic programming, *Handbook of Logic in Artificial Intelligence and Logic Programming*, vol. 5, Gabbay, D.M, Hogger, C.J., Robinson, J.A., (eds.), Oxford University Press, pp. 235–324, 1998.
15. Kazantzakis, N., Βίος και Πολιτεία του Αλέξη Ζορμπά. [Eng.: Life and career of Alexis Zorbas.] 1946.
16. Mishler, E.G., Meaning in Context: Is there any other Kind? *Harward Educational Review* Vol. 49, No. 1, pp. 1–19, 1979.
17. Montague, R., *Formal philosophy* Yale University Press: New Haven.
18. Penn, J.R., The Other Syntax: Approaching Natural Language Semantics through Logical Form Composition To appear in "Proceedings of first International Workshop on Language Processing and Constraint Solving, Roskilde, Sept. 1–3, 2004", eds. Christiansen, H., Skadhauge, P.R., Villadsen, J., *Lecture Notes in Artificial Intelligence* 3834, 2005.
19. Pereira, F.C.N., and Warren, D.H.D., Definite clause grammars for language analysis. A survey of the formalism and a comparison with augmented transition grammars. *Artificial Intelligence* 10, no. 3–4, pp. 165–176, 1980.

20. Rieger, C., Conceptual overlays: A mechanism for the interpretation of sentence meaning in context. *Proc. IJCAI-75* pp. 143–150.
21. Stalnaker, R., On the representation of context. *Journal of Logic, Language, and Information*, vol. 7, pp. 3–19, 1998.
22. *Webster's Encyclopedic Unabridged Dictionary of the English Language.* 1989 Edition, Gramercy Books, dilithium Press, 1989.

Descriptive Naming of Context Data Providers

Norman H. Cohen, Paul Castro, and Archan Misra

IBM T. J. Watson Research Center,
19 Skyline Drive, Hawthorne, New York 10532, USA
{ncohen, castrop, archan}@us.ibm.com

Abstract. Much context data comes from mobile, transient, and unreliable sources. Such resources are best specified by descriptive names identifying what data is needed rather than which source is to provide it. The design of descriptive names has important consequences, but until now little attention has been focused on this problem. We propose a descriptive naming system for providers of context data that provides more flexibility and power than previous naming systems by classifying data providers into "provider kinds" that are organized in an evolving hierarchy of subkinds and superkinds. New provider kinds can be inserted in the hierarchy not only as subkinds, but also as superkinds, of existing provider kinds. Our names can specify arbitrary boolean combinations of arbitrary tests on data-source attributes, yielding expressive power not found in naming schemes based on attribute matching.

1 Introduction

A number of systems obtain services from network resources such as sensors, cameras, printers, cellular phones, and web services. These resources may be mobile, they may be ephemeral, and their quality of service may fluctuate. It has become widely accepted that such systems should not require users to name a specific resource from which they wish to obtain services, but rather, to describe what the resource is expected to provide. This approach, known as descriptive [1], data-centric [2], or intentional [3] naming, allows a name resolver to discover an appropriate resource at runtime.

Descriptively named resources are important in context-aware computing, because a context-aware application often requires a particular type of context information rather than information from a particular source. For example, the application may require the location of a particular individual; it should be possible for the application to ask for this location without considering whether it is deduced from the location of the individual's cell phone, laptop computer, or car. Descriptive naming allows a name resolver to select the best available provider of data, based on current conditions, and to select a new provider when those conditions change. It makes an application robust against the failure of any one device or service. It facilitates the frequent addition or removal of context-data providers without modification of applications that use them. It allows an application to be ported easily to an environment with a different set of context-data providers.

To accommodate a wide variety of applications, the scheme for writing a data-provider query must be flexible enough to describe any provider of context data. Different applications may need to query, for example, for providers of Fahrenheit temperatures at a given latitude and longitude, Celsius temperatures of the patient in a given hospital bed, current prices of IBM stock in U.S. dollars, the number of the room where a given active badge was last sensed, and the identification numbers of all vehicles in a specified zone with excessive engine temperatures. Clearly, it is untenable to establish a fixed vocabulary of concepts and data types to be used in provider queries.

The rest of this paper is organized as follows. Sect. 2 summarizes our approach, explaining what is unique about it and why it is preferable to previous approaches. Sect. 3 describes the programming model that we assume for context applications. Sect. 4 explains the nature of our descriptive names and the underlying model. Sect. 5 describes a prototype implementation of our naming system. Sect. 6 discusses related work and Sect. 7 presents our conclusions. The focus of the paper is the proposed naming model, not the prototype implementation. We believe that there must be fundamental improvements to the structure and semantics of names for providers of context data before a scalable and continually upgradeable context infrastructure can be practically deployed.

2 Our Contribution

The design of descriptive names for providers of context data has profound consequences for the expressiveness of queries, the efficiency of name resolution, and the ability of the naming system to grow to accommodate previously unforeseen kinds of resources. Nonetheless, until now, little attention has been focused on this issue, or on the consequences of various design decisions. We have explored the issues to be considered in designing a descriptive naming system for sources of context data, developed a descriptive naming scheme, and implemented a prototype system that resolves our descriptive names.

To name a data provider, we must be able to specify the kind of data we need—for example, Celsius temperature readings or the price of IBM stock—and the constraints we place on providers of that kind of data—for example, that the temperature readings be taken within a specified region, or that the stock price reflect a ticker delay of no more than twenty minutes. To identify the kinds of data we need, we propose a vocabulary of *provider-kind names*. To express constraints, we propose a combination of parameters, predicates on attributes of individual data providers, and filters for selecting from among eligible data providers according to application-specified metrics.

One challenge in obtaining a useful vocabulary of provider-kind names is to allow queries with varying degrees of specificity. For example, it should be possible for one application to request data providers giving at least the latitude and longitude of a vehicle with a given identifier, and for another to request data providers giving at least the latitude, longitude, and elevation of such a vehicle. Any query for providers of latitude and longitude should be satisfied by providers of latitude, longitude, and elevation. Another challenge is to enable the vocabulary to evolve in a disciplined way—

to incorporate new kinds of data providers unrelated to previously existing kinds, to add a specialization of an existing provider kind, or to add a generalization of two or more existing provider kinds (so that a query for providers of the new kind can be satisfied by providers of any of the existing provider kinds). We address these challenges by organizing provider kinds in a multiple-inheritance hierarchy of subkinds and superkinds, analogous to the hierarchy of subclasses and superclasses in an object-oriented programming language. Our hierarchy is novel in that a new provider kind can be stipulated to be both a subkind of certain existing provider kinds and a superkind of certain other existing provider kinds; rather than just adding new provider kinds below existing provider kinds in the hierarchy, we can sandwich a new provider kind between existing provider kinds.

As in other descriptive naming schemes, each data provider is assumed to have a set of named properties that can be used for expressing constraints. In contrast to schemes in which a descriptive name consists of a set of attribute-value pairs, tested for equality with provider property values, we allow a descriptive name to include any boolean combination of arbitrary tests on property values. For example, to test whether x and y properties of a data provider specify coordinates within a certain rectangle, we test whether that value of each property is *less than* one specific value and *greater than* another; this constraint cannot be expressed by equality tests. To test whether the x and y properties specify any point within a *pair* of rectangles, we test whether the point lies within the first rectangle, *or* within the second rectangle; this constraint cannot be expressed by listing a set of tests on individual properties, all of which must be satisfied simultaneously. We supplement properties that a boolean constraint may refer to with named *activation parameters* that are required be specified in a descriptive name for a provider of a particular kind. These specific values may be needed to establish a connection with a data provider.

Some descriptive naming systems have a fixed metric associated with named entities for determining which entity a name should resolve to when there are several eligible candidates. We improve upon this approach in two ways. First, we enable an application to specify any numeric combination of property values to be used as a metric. Different descriptive names can specify different metrics. Second, rather than simply using the metric to determine that one data provider will be selected over another, our scheme allows descriptive names for sets of multiple data providers, such as those with the 10 highest metric values or all those with a metric value greater than 50.

Our property tests, application-defined metrics, and activation parameters are closely integrated with our system of provider kinds. The properties that may be named in a boolean constraint or a metric are determined by the provider kind; any property defined for a particular provider kind must be defined for subkinds of that provider kind, thus ensuring that a boolean constraint or metric that can be applied to a provider of a given kind can also be applied to a provider of any subkind of that kind. Similarly, the activation parameters for which values must be supplied in a descriptive name are determined by the provider kind. Any activation parameter required to be specified for providers of a given kind is also required to be specified for providers of all superkinds of that kind.

3 A Programming Model for Context Applications

In our model, context-aware applications obtain data from *data providers*. A data provider has a *current value* that may change from time to time. All data providers respond to requests for their current values. Some data providers are also active, taking the initiative to report that they have generated new values. A web service that responds to requests for the current price of a given stock is a passive data provider. A sensor that issues a report whenever it detects motion is an active data provider.

An application obtains context data from a service with a straightforward interface: The application issues a descriptive name called a *provider query* to the service, and the service responds with a list of one or more handles for context-data providers, registered with the service, that satisfy the query. Each handle has a descriptor reporting distinguishing properties of its data provider. Given a data-provider handle, an application can request the current value of the data provider, or subscribe to receive a notification each time the data provider generates a new value. Thus, the subject of a provider query is not a value, but the continuously evolving stream of values associated with a data provider.

4 The Nature of a Provider Query

We discuss underlying concepts related to provider queries in Sects. 4.1 and 4.2, and turn to provider queries themselves in Sect. 4.3.

4.1 Provider Kinds

Every data provider is registered with the name-resolution service as belonging to some provider kind. The definition of a provider kind specifies the type of the values returned by the provider, the names and types of its activation parameters, and a set of attributes describing properties of the provider. Activation parameters provide the information needed to initialize a data-provider handle. Activation parameters might include, for example, the unique identifier of a particular real-world entity, or an authentication token.

Fig. 1 defines a provider kind for providers that return the location of a vehicle with a specified vehicle identification number (VIN). A provider of kind `VINTo-Location` provides values of type `LatLongType` and is activated with a parameter `vehicleID` of type `VINType`. The definition in Fig. 1 says nothing about the semantic

```
Provider kind VINToLocation:
    Type of provided values: LatLongType
    Activation parameters:
        vehicleID: VINType
    Properties:
        radiusOfErrorInMeters: float
        freshnessInSeconds: int
```

Fig. 1. Definition of provider kind `VINToLocation`

relationship between the `VINType` value used for activation and the `LatLongType` value provided—for example, that the value provided is the location of the vehicle with the specified VIN. The definition could apply just as easily to a kind for providers that give the location of *the registered owner of* the vehicle with the specified VIN. We expect a provider kind to reflect a particular semantic relationship; providers with different semantics belong to different provider kinds, say `VINToVehicle-Location` and `VINToOwnerLocation`, that may happen to have the same provided type, activation parameters, and properties.

We do not attempt to formalize the semantics of a provider kind. Rather, we rely on the humans who name provider kinds in queries to be familiar with the intended semantics of those provider kinds, just as users of a relational database are expected to be familiar with the semantics of the tables and columns they name in SQL queries.

4.1.1 Subkinds and Superkinds

Provider kinds can be organized into hierarchies of superkinds and subkinds, such that a query for a provider of kind k can be satisfied by a provider of any subkind of k. To formalize this hierarchy, we assume that the types to which provided values and activation-parameter values belong are themselves organized in a supertype-subtype hierarchy. A provider kind p is allowed to be the direct parent of a child provider kind c only if each of the following conditions holds:

- The type of value provided by c is a subtype of the type of value provided by p.
- For each activation parameter of kind c, kind p has an identically named activation parameter, and the type of each parameter of c is a supertype of the type of the corresponding parameter of p. (Thus the set of parameter values understood by a provider of kind c includes *at least* every parameter value understood by a provider of kind p; p may have "extra" parameters that have no counterpart in c, which can be ignored when activating a data provider of kind c as if it were of kind p.)
- The set of properties of c is a superset of the set of properties of p.

To these formal conditions, we add an informal one:

- The semantics of c (as understood informally by a human) are consistent with the semantics of p.

(The formal conditions determine when it is *legal* for p to be a direct parent of c, and the informal condition determines when it is *appropriate*.) The *superkinds* of a provider kind k consist of k and the superkinds of all direct parents of k; if x is a superkind of y, then y is a *subkind* of x. (Every provider kind is a subkind and a superkind of itself.)

For example, suppose the type `LatLongType`, giving a two-dimensional location in terms of latitude and longitude, has a subtype `LatLongElevType`, giving a three-dimensional location that also includes elevation. Fig. 2(b) defines a kind for providers of three-dimensional locations of vehicles with a given VIN. Because `LatLongElevType` is a subtype of `LatLongType`, `VINTo3DLocation` can be a subkind of `VINToLocation`. Then a query for a provider of kind `VINToLocation` could be satisfied by a provider of kind `VINTo3DLocation`; an application would use the `LatLongElevType` values it receives from the provider as if they were

Descriptive Naming of Context Data Providers   117

(a)
```
Provider kind VINToLocation:
  Type of provided values: LatLongType
  Activation parameters:
    vehicleID: VINType
  Properties:
    radiusOfErrorInMeters: float
    freshnessInSeconds: int
```

(b)
```
Provider kind VINTo3DLocation:
  Type of provided values: LatLongElevType
  Activation parameters:
    vehicleID: VINType
  Properties:
    radiusOfErrorInMeters: float
    freshnessInSeconds: int
```

(c)
```
Provider kind VINToGPSLocation:
  Type of provided values: LatLongType
  Activation parameters:
    vehicleID: VINType
  Properties:
    radiusOfErrorInMeters: float
    freshnessInSeconds: int
    satellites: int
```

(d)
```
Provider kind VINToGPS3DLocation:
  Type of provided values: LatLongElevType
  Activation parameters:
    vehicleID: VINType
  Properties:
    radiusOfErrorInMeters: float
    freshnessInSeconds: int
    satellites: int
```

Fig. 2. Definitions of provider kinds (a) VINToLocation (as in Fig. 1), (b) VINTo 3D-Location, (c) VINToGPSLocation, and (d) VINToGPS3DLocation

LatLongType values. Some providers of vehicle-location information might use GPS receivers, and for those providers it is meaningful to define an additional property, the number of GPS satellites contributing to the reading. Fig. 2(c) defines a provider kind for these GPS-based providers. VINToGPSLocation is also eligible to be a subkind of VINToLocation, since its properties include all the VINToLocation properties. Any query for a provider of kind VINToLocation can be satisfied by a provider of kind VINToGPSLocation. We can also define a provider kind for GPS-based providers of three-dimensional location, as shown in Fig. 2(d). VINToLocation, VINTo3DLocation, and VINToGPSLocation all qualify to be direct parents of VINToGPS3DLocation. If we define VINTo3DLocation and VINToGPSLocation to be direct parents of VINToGPS3Dlocation, we obtain the hierarchy shown in Fig. 3.

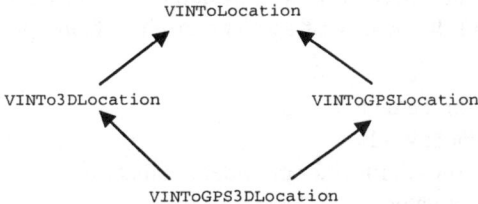

Fig. 3. A multiple-inheritance subkind hierarchy. A query for a provider of a given kind can be satisfied by a provider of that kind or of any kind directly or indirectly below it in the hierarchy

4.1.2 Bottom-Up Definition of Superkinds

Traditionally, subtype hierarchies are built from the top down; that is, the definition of a type names its direct parents, which must have been defined earlier. In contrast, the definition of a new provider kind names both direct parents and direct children. Thus the new provider kind can be installed as a superkind of existing kinds, as a subkind of existing kinds, or both.

Just as top-down growth of a hierarchy allows for specialization, bottom-up growth allows for *generalization*. Generalization allows the vocabulary of provider queries to evolve without disruption as new provider kinds are devised. We give two examples.

First, suppose a standard type TelematicsData has been extended independently by company X to type XTelematicsData and by company Y to type YTelematicsData. Each company markets a device that reports a value of its own extended type, given a VIN. The companies define corresponding provider kinds VINToXTelematicsData and VINToYTelematicsData. We are managing a fleet that had been using X's device to obtain standard TelematicsData values (treating XTelematicsData values as TelematicsData values, ignoring X's extensions). We have now added vehicles with Y's devices to the fleet. So that we can write a query that will find *all* providers of TelematicsData values, we define a new provider kind, VINToTelematicsData, as a superkind of VINToXTelematicsData and VINToYTelematicsData.

A second form of generalization involves activation parameters. Suppose we have data providers of kind VINToLocation, which provide the location of a vehicle given its VIN, and data providers of kind PlateToLocation, which provide the location of a vehicle given its plate number. Suppose we have both the VIN and plate number of all vehicles of interest. Rather than issue one query for VINToLocation and, if that fails, a second query for PlateToLocation, we can define a new provider kind VINAndPlateToLocation, which takes *both* a VIN and a plate number as activation parameters. Then VINAndPlateToLocation can be defined as a superkind of VINToLocation and PlateToLocation. We can issue a single query for VINAndPlateToLocation data providers, which will be satisfied by both VINToLocation and PlateToLocation data providers.

4.2 Provider Descriptors

Every data provider has a *provider descriptor* conveying the identity of the provider and information about its state. This may include static information about the nature and capabilities of the data provider as well as dynamic information such as the provider's current value and the quality of service currently being provided. A provider descriptor includes:

- a unique identifier for the data provider
- the name of its provider kind
- values for the properties defined for providers of that kind
- the provider's current value

If a provider kind s is a subkind of a provider kind k, a descriptor for a provider of kind s includes at least the properties found in a descriptor for a provider of kind k.

4.3 Provider Queries

A provider query is a test that a provider descriptor either passes or fails. It includes:
- the name of a provider kind, indicating that a provider of that kind or one of its subkinds is desired
- values for the activation parameters associated with that provider kind
- a predicate, possibly referring to the values of properties associated with the provider kind, to be applied to the property values in a given provider descriptor, indicating whether the descriptor should be considered to satisfy the query
- a *selection mechanism* for determining which provider descriptors, among those determined to satisfy the query, should be returned in the query result

A selection mechanism can have one of the following forms:

- *all*, indicating that the result should contain all provider descriptors satisfying the query
- *first(k)*, indicating that the result should contain at most the first k provider descriptors found that satisfy the query
- *top(k,expression)*, where *expression* is a numeric expression involving properties in the provider descriptor for the specified provider kind, indicating that the result should contain up to k provider descriptors for which *expression* has a maximal value
- *ge(expression,n)*, where *expression* is a numeric expression involving properties in the provider descriptor for the specified provider kind, indicating that the result should contain all descriptors for which *expression* has a value greater than or equal to n

4.3.1 Predicates Versus Activation Parameters

Predicates and activation parameters play distinct roles. A predicate tests whether properties of a data provider satisfy certain conditions, but need not constrain a property to hold one specific value. Activation parameters supply specific values needed to establish a connection to a data provider.

Sometimes, the same information must be supplied as an activation parameter and in a predicate. Consider a query for providers of IBM stock prices: Some of these providers might be general stock-quote services, which require a stock symbol to be passed as an activation parameter; others might be dedicated specifically to providing the price of IBM stock. Such providers belong, respectively, to provider kind PriceBySymbol and its subkind IBMPrice, defined in Fig. 4. A query for PriceBySymbol can be satisfied by a provider of either kind. However, such a query would also be matched by providers belonging to other subkinds of PriceBySymbol, such as IntelPrice and MicrosoftPrice. To filter out these other data providers, we write a query that not only specifies a value of "IBM" for symbolParameter (as required for providers of kind PriceBySymbol) but also specifies the predicate symbolProperty="IBM".

With descriptive naming schemes that test attributes only for equality with specific values, there is no need for both parameters and properties: The string "symbol=IBM"

```
┌─────────────────────────────────────────┐  ┌─────────────────────────────────────────┐
│ Provider kind PriceBySymbol:            │  │ Provider kind IBMPrice:                 │
│   Type of provided values: USDollars    │  │   Type of provided values: USDollars    │
│   Activation parameters:                │  │   Activation parameters: (none)         │
│     symbolParameter: string             │  │   Properties:                           │
│   Properties:                           │  │     symbolProperty: string              │
│     symbolProperty: string              │  │     tickerDelayInMinutes: int           │
│     tickerDelayInMinutes: int           │  │                                         │
└─────────────────────────────────────────┘  └─────────────────────────────────────────┘
```

Fig. 4. Definition of provider kind `PriceBySymbol` and its subkind `IBMPrice`

acts both as a specification of the value to be used for `symbol` (when activating a data provider requiring a specific value) and as a test to be performed on the value of the property `symbol` (when filtering provider descriptors that contain a `symbol` property). However, by restricting a query to be, in essence, a conjunction of equalities, such a scheme precludes queries for, say, a stock-price provider with a ticker delay *less than* 20 minutes.

4.3.2 Semantics of a Provider Query

We define the semantics of a provider query operationally: A provider query specifying a provider kind *pk*, activation parameters $ap_1, \ldots, ap_n$, predicate *p*, and selection mechanism *sm* is resolved as if by the following steps:

1. Attempt to activate every data provider registered as belonging to some subkind of *pk*, using the activation-parameter values $ap_1, \ldots, ap_n$.
2. For each successfully activated provider, construct a provider descriptor appropriate for kind *pk* with the properties and current value of that provider.
3. Apply the predicate *p* to each provider descriptor and include all those for which the result is *true* in a set of candidates.
4. Apply the selection mechanism *sm* to select a result set from the set of candidates.

The effect of these steps can often be achieved more efficiently: If the predicate does not refer to dynamic properties of a data provider, it can be applied to an *approximate descriptor*, containing only static properties, created without actually activating the provider. (This approach is reminiscent of the approximate caches of [4].) If the predicate refers to dynamic properties other than the provider's value, it is necessary to activate the provider, but not to retrieve its current value. For a selection mechanism of the form *first*(*k*), the query processing can be stopped after *k* descriptors have been obtained. Traditional indexing and query-optimization techniques can be applied to static properties to avoid the retrieval of provider descriptors that cannot possibly satisfy the predicate in a query.

A predicate referring to a data provider's value is potentially costly: In the worst case, resolving a name entails activating every provider that has a suitable provider kind, and requesting its current value. However, this feature is also very powerful: We can query for all vehicles currently located in a specified region, or all sensors currently sensing out-of-range temperatures. Fortunately, a name resolver can be designed so that the cost is borne only by queries that explicitly refer to a provider's value.

5 Prototype Implementation

We have implemented a resolution service for our provider queries. The service is part of the CxS [5] middleware for context-aware applications. CxS collects and combines data from a wide variety of data providers. Context-aware applications running on CxS include one using active-badge data to issue context-aware reminders, one setting priorities for hospital nurses based on data from simulated monitors, and one estimating the availability of individuals based on a variety of context data.

All runtime values in CxS have XML representations, and belong to XML Schema [6] types. Each provider kind is registered as providing values of a particular XML Schema type, and as having activation parameters of particular XML Schema types. XML Schema types can have subtypes; the subtype hierarchy is used to determine whether one given provider kind is allowed to be a direct parent of another.

A CxS provider descriptor is an XML document. XQuery [7], a language that can specify computations on the contents of an XML document, is a natural medium for specifying computations on the values in a provider descriptor. The predicate in a CxS provider query is a boolean XQuery expression. The selection mechanisms *top(k,expression)* and *ge(expression,n)* contain numeric XQuery expressions.

CxS is implemented in Java. Registrations of CxS data providers and provider kinds are stored in a relational database accessed through JDBC. Data-provider registrations are indexed by provider kind. We select from the database all provider registrations whose provider kinds are subkinds of the kind specified in the provider query. Our prototype implementation then naively processes provider queries in accordance with the operational definition given in Sect. 4.3.2: An attempt is made to activate each data provider whose registration was retrieved, using the activation-parameter values found in the query. If this attempt is successful, a completely filled-in XML provider descriptor is obtained, and the XQuery predicate in the provider query is applied to this descriptor. If the predicate is true, the provider descriptor is added to a list of candidates. After each retrieved provider registration has been processed in this way, the selection mechanism is applied to the list of candidates to obtain the result of the query. Despite the fact that we have not yet implemented any of the performance improvements envisioned in Sect. 4.3.2, we are able to process over 10 provider queries a second using the IBM Java 1.42 JVM running on a 3GHz Pentium 4 processor with 500MB RAM.

While CxS provides a compelling proof-of-concept for the descriptive naming system proposed in Sect. 4, the focus of this paper is on the naming system itself rather than on any particular implementation of it. Therefore, further details about CxS (such as mechanisms for continually rebinding to the best available providers over the lifetime of a provider query) are beyond the scope of this paper.

6 Related Work

As explained in Sect. 2, two limitations characterize previous approaches to descriptive naming. Some approaches lack expressiveness, effectively restricting the conditions that can be tested to a conjunction of equalities. Some approaches do not support a hierachical classification of descriptively named entities akin to our provider-kind hierarchy.

The Lightweight Directory Access Protocol (LDAPv3) [8] allows directory-entry attributes to be tested by an arbitrary boolean search filter. However, an LDAP directory hierarchy does not reflect superkind-subkind relationships: A query for an entry of type t is not satisfied by an entry whose type is any subtype of t. In the Service Location Protocol (SLPv2) [9], queries include an LDAPv3 search filter. Every SLP service belongs to some *concrete service type*, and concrete service types may be grouped into *abstract service types*. However, only a two-level hierarchy is supported.

In the Ninja project's Service Discovery Service [10], a service has an XML *service description*, analogous to our provider descriptors. A query is a service description with some of the elements removed, and specifies a conjunction of equalities between corresponding values in the query and the service description. The Intentional Naming System [1] and its follow-on Twine [11] take a similar approach: Both queries and resource descriptions are *attribute-value trees*. A query matches a resource description if and only if each path from the root of the query tree has a corresponding path in the resource description, so a query is effectively a conjunction of equality tests. While [1] contemplates adding ordering comparisons to queries, [11] exploits the fact that a path in a query matches a path in a resource description only if both paths can be hashed to the same value.

In the *directed diffusion* paradigm [2,12], queries are called *interests*. In [2], an interest is expressed as a set of attribute-value pairs, interpreted as a conjunction of equality tests. Every interest includes an event code naming the subject of the query. Event codes roughly correspond to provider-kind names, but with no inherent relationships among the notions they denote. An enhancement using attribute-comparator-value triples instead of attribute-value pairs is presented in [12]; an interest is still equivalent to a conjunction of simple tests, but the simple tests may include ordering comparisons as well as equalities.

The Jini [13] Lookup Service discovers Java objects representing services. A service object is described by a *service item* that includes a *service identifier* and *attribute sets*. A query is a *service template* that may include a service identifier, Java types, and *attribute-set templates*. A service template matches a service item if its service identifier matches the service identifier in the service item, the service object in the service item is an instance of each Java type in the service template, and each attribute-set template in the service template matches an attribute set in the service item. The hierarchy of Java service-object subclasses can play a role analogous to our subkind hierarchy, but the attribute-set template specifies a conjunction of exact matches with specified values.

Universal Description, Discovery, and Integration (UDDI) [14] is a framework for issuing queries to discover web services. The UDDI registry was originally conceived of as a "yellow pages" directory, in which services are looked up by locating a business in a given industry and then examining the services offered by that business. Businesses are categorized by industry according to a hierarchy, but this hierarchy has no semantic role in the processing of queries. Services are not looked up based on their semantic properties.

In contrast to these approaches that are less flexible than ours, ontology-based systems aspire to provide greater flexibility. The OWL Web Ontology Language [15], based on the Resource Description Framework [16], is one notation for defining ontologies. Ontology-based systems aim to support unstructured (e.g., natural-language) queries, and to apply common-sense reasoning to deduce facts that are not explicitly represented. By defining relationships between terms in different vocabularies, an ontology also provides a bridge between the vocabularies of a query and a provider descriptor, allowing providers of "thermometer reading" to be discovered in response to a query for "temperature." The goals of ontology-based systems are ambitious, but their promises are unproven. Development of ontologies is labor-intensive, so few exist yet, and it is not clear that resources will exist in the long run to maintain them. Our hierarchy of provider kinds can be viewed as a kind of primordial ontology, with less ambitious and therefore more attainable goals. We seek to classify only data providers rather than arbitrary knowledge, and we do so in a highly constrained manner.

7 Summary

We have proposed a powerful approach for naming context data providers. Our names *describe* the desired properties of value streams rather than identifying particular data providers. Descriptive naming allows a name resolver to select the best available provider dynamically, isolates client applications from dependence on one particular provider, allows providers to be added to or removed without modifying client applications, and makes applications portable to environments with different sets of resources.

Data providers registered with the name resolver are classified according to a multiple-inheritance hierarchy of provider kinds. New provider kinds can be inserted in this hierarchy not only below, but also above specified existing provider kinds, facilitating the introduction of a new provider kind that generalizes previously existing provider kinds. The current state of a registered data provider is described by a provider descriptor with content that depends on its provider kind. A provider query specifies the name of a provider kind, a set of activation-parameter values meaningful for that provider kind, a predicate applicable to descriptors for providers of that kind, and a selection mechanism specifying how data providers are to be selected from among those that are eligible. The descriptor includes the current value of a data provider, enabling queries for all data providers currently providing values that satisfy a particular condition.

Our descriptive names, or provider queries, are applicable to arbitrary domains, and to sorts of context-data providers not yet conceived of. At the same time, they are precise and unambiguous. A provider query is amenable to consistency checks to ensure that it refers to only attributes that are meaningful for the kind of data provider it describes. It is possible to perform general tests on attributes, including range tests and disjunctions.

The naming system we have proposed has been implemented in the CxS middleware. However, the true test of our approach can only come over the course of years. The approach should be deemed successful if it supports the incorporation of new, unanticipated provider kinds, and if it supports the precise expression of queries by new, unanticipated applications.

Acknowledgements

The CxS system and the context-aware applications described in Sect. 5 were developed by a team of IBM researchers. At various times, this team has included, in addition to the authors, James Black, Marion Blount, John Davis, Maria Ebling, Qi Han, Srikant Jalan, Hui Lei, Barry Leiba, Apratim Purakayastha, Wolfgang Segmuller, and Daby Sow.

References

1. Mic Bowman, Saumya K. Debray, and Larry L. Peterson. Reasoning about naming systems. ACM Transactions on Programming Languages and Systems 15, No. 5 (November 1993), 795–825
2. Chalermek Intanagonwiwat, Ramesh Govindan, and Deborah Estrin. Directed diffusion: a scalable and robust communication paradigm for sensor networks. Proceedings of the Sixth Annual International Conference on Mobile Computing and Networking (MobiCom 2000), Boston, Massachusetts, August 6–11, 2000, 56–67
3. William Adjie-Winoto, Elliot Schwartz, Hari Balakrishnan, and Jeremy Lilley. The design and implementation of an intentional naming system. Proceedings of the 17th ACM Symposium on Operating Systems Principles (SOSP '99), December 12-15, 1999, Kiawah Island Resort, South Carolina, published as Operating Systems Review 33, No. 5 (December 1999), 186–201
4. Chris Olston, Boon Thau Loo, and Jennifer Widom. Adaptive precision setting for cached approximate values. Proceedings of the ACM SIGMOD International Conference on Management of Data, Santa Barbara, California, May 21–24, 2001, 355–366
5. Norman H. Cohen, James Black, Paul Castro, Maria Ebling, Barry Leiba, Archan Misra, and Wolfgang Segmuller. Building context-aware applications with Context Weaver. IBM Research Report RC 23388, October 22, 2004
6. David C. Fallside, ed. XML Schema Part 0: Primer. W3C Recommendation, May 2, 2001 <URL: http://www.w3.org/TR/xmlschema-0/>
7. Scott Boag, Don Chamberlin, Mary F. Fernandez, Daniela Florescu, Jonathan Robie, Jérôme Siméon. XQuery 1.0: An XML Query Language. W3C Working Draft, May 2, 2003 <URL: http://www.w3.org/TR/xquery/>
8. M. Wahl, T. Howes, and S. Kille. Lightweight Directory Access Protocol (v3). IETF RFC 2251, December 1997 <URL: http://www.ietf.org/rfc/rfc2251.txt >
9. E. Guttman, C. Perkins, J. Veizades, M. Day. Service Location Protocol, Version 2. IETF RFC 2608, June 1999 <URL: http://www.ietf.org/rfc/rfc2608.txt>
10. Steven E. Czerwinski, Ben Y. Zhao, Todd D. Hodes, Anthony D. Joseph, and Randy H. Katz. An architecture for a secure service discovery service. Proceedings of the Fifth Annual ACM/IEEE International Conference on Mobile Computing and Networking (MobiCom '99), Seattle, Washington, August 15–19, 1999, 24–35
11. Magdalena Balazinska, Hari Balakrishnan, and David Karger. INS/Twine: a scalable peer-to-peer architecture for intentional resource discovery. International Conference on Pervasive Computing (Pervasive 2002), Zurich, Switzerland, August 26–28, 2002, 195–210
12. John Heidemann, Fabio Silva, Chalermek Intanagonwiwat, Ramesh Govindan, Deborah Estrin, and Deepak Ganesan. Building efficient wireless sensor networks with low-level naming. Proceedings of the Eighteenth ACM Symposium on Operating Systems Principles (SOSP 2001), Banff, Alberta, October 21–24, 2001, 146–159

13. Sun Microsystems. Jini Technology Core Platform Specification. Version 2.0, June 2003 <URL: http://wwws.sun.com/software/jini/specs/>
14. Tom Bellwood, Luc Clément, Claus von Riegen, eds. UDDI version 3.0.1. UDDI Spec Technical Committee Specification, October 14, 2003 <URL: http://uddi.org/pubs/uddi_v3.htm>
15. Deborah L. McGuinness and Frank van Harmelen, eds. OWL Web Ontology Language overview. W3C Candidate Recommendation, August 18, 2003 <URL: http://www.w3.org/TR/ owl-features/ >
16. Frank Manola and Eric Miller, eds. RDF Primer. W3C Working Draft, October 10, 2003 <URL: http://www.w3.org/TR/rdf-primer/>

Quotations and the Intrusion of Non-linguistic Communication into Utterances

Philippe De Brabanter

Institut Jean Nicod (CNRS-EHESS-ENS), 1bis avenue de Lowendal,
75007 Paris, France
phdebrab@yahoo.co.uk

Abstract. When linguists or philosophers of language study communication, they are naturally biased towards linguistic communication. This has resulted in a situation in which very little attention is being paid to the fact that many of our daily utterances are actually a mixture between linguistic and other expressive means, such as noises, gestures and facial expressions. It is precisely on this phenomenon that I wish to focus. I show how Clark & Gerrig's notion of demonstration can be usefully applied to intrusions of non-linguistic material into spoken or written utterances. Although I agree with relevance theorists that the linguistic bias should be abandoned, I am led to propose an account of non-linguistic demonstrations as linguistic constituents, at least in the particular category of data that I examine, namely utterances in which some gesturing appears to stand in for a missing linguistic constituent. In those cases, I contend, non-linguistic demonstrations are recruited to fulfil various syntactic functions. I justify this somewhat paradoxical stance by exploring the essential similarities between quotations and non-linguistic demonstrations.

1 Introduction

The present study needs to be situated against the backdrop of a long-standing dispute in language scholarship. This dispute, which arose in connection with theories of quotation, concerns the status of non-linguistic 'things' in sentences and utterances. For ease of exposition, I distinguish four positions:

1. <u>Radical anti-thing position</u>: proponents of this view ([1], [2], [3]) held, as a matter of principle, that sentences or utterances never contain non-linguistic things. The principle can be summed up as follows: the entities that an utterance talks about do not occur in it, only designations of these entities do.
2. <u>Qualified anti-thing position</u>: people like Carnap and Reichenbach granted that "it is quite possible to take as a name for the thing, the thing itself, or, as a name for a kind of thing, the things of this kind" [4]. However, Reichenbach remarked, "such a practice […] would often lead to serious difficulties, for instance if we wanted to use this method for denoting lions and tigers" [5]. In other words, though theoretically acceptable, the method is quite unworkable in practice.
3. <u>Pro-thing position</u>: people like Christensen [6], Searle [7], Recanati [8] rejected position 1 as unfounded. To them, it was clear for instance that the quoted words occurring in an utterance were the very thing denoted by the quotation.

The debate centred on actual objects or substances being inserted into linguistic utterances. It was not concerned with non-linguistic *communication*: everything was envisaged from the vantage point of verbal communication, taken as the paradigm form. Yet, the match or handful of sand placed on a sheet of paper (as a stand-in for the words "match" or "sand") do not just happen to be there by accident. Neither does the speaker's rendition of a bird's song in Searle's example:

[...] an ornithologist might say 'The sound made by the California Jay is ...' And what completes the sentence is a sound. [7]

The 'things' occurring in these utterances are introduced intentionally by someone trying to achieve particular communicative intentions. In other words, it seems realistic to regard the placing of a match or sand on paper and the insertion of bird song into an utterance as stimuli provided in order to communicate something. This is the sort of characterisation that would be advocated by proponents of a possible fourth position in the debate:

4. The 'rejection of the linguistic bias' position: these scholars (Sperber and Wilson [9], Carston [10]) take the general phenomenon of human communication as their starting-point. One of their central assumptions is "that most verbal utterances are a complex of linguistic, paralinguistic, facial and vocal gestures, which appear to function as a single signal receiving a unified interpretation" [10]. What these diverse gestures have in common is that they serve the purposes of *ostensive* communication. Communicators produce stimuli in order to enrich the audience's *cognitive environment* by making certain assumptions (more) manifest to them. For instance, I can sniff ostensively (= the stimulus) in order to make it possible for my audience to notice the smell of gas I have just become aware of (cf. [9]). I could also seek to achieve that effect by saying "There's a smell of gas". This stimulus too would enable my audience to add to their cognitive environment the assumption that the room in question smells of gas.

One's general conception of communication has direct implications on how the context is to be represented. On the traditional language-centred view, every component of an act of utterance that is *not* linguistic is part of the context. This means that whatever pertains to non-linguistic communication is regarded as 'just another' aspect of the context that can be useful in interpreting the utterance. Thus, Lyons [11] lists among the components of the context of utterance "the appearance, bearing and attitude of the various participants in the language-event [...] of which the utterance in question is a constitutive part". By contrast, in the relevance-theoretic framework, when a sentence is uttered as part of the performance of an act of communication, this sentence is usually just one of the stimuli produced. Other stimuli are, for instance, paralinguistic features, gestures and facial expressions. As for the context, it comprises all the assumptions that are accessible to the audience at the time of utterance. At that moment, the non-verbal stimuli produced simultaneously with the utterance being processed are not part of the context. However, communication is a dynamic process, and once the stimuli have been processed, they become part of the context against which a subsequent act of communication will be interpreted.

2 Why Opt for a Linguistic Approach?

I think the relevance-theoretic picture is essentially correct. It makes a whole lot more sense to set communicative stimuli apart from the set of assumptions against which the stimuli can generate contextual effects than to separate linguistic stimuli from everything else (thereby turning the context into a mixed bag of non-verbal communicative stimuli and background assumptions).

However, though my sympathies are entirely with the relevance theorists, this whole paper consists in an attempt to offer a *linguistic* treatment of certain instances of non-linguistic communication. What are the motivations behind this apparent inconsistency? In this section, I offer some basic considerations towards answering that question. More arguments will be brought up towards the end of the paper.

My data consist of utterances in which a piece of non-linguistic communication appears to be called upon to stand in for a linguistic constituent which remains covert. Here is an example:

(1) I got out of the car, and I just [DEMONSTRATION OF TURNING AROUND AND BUMPING HIS HEAD ON AN INVISIBLE TELEPHONE POLE]

One has to imagine (1) being produced in spoken conversation. The square-bracketed string in small capitals is a description of the sort of gesturing that takes place in the conversational setting. Use of this expedient is rendered necessary by the impossibility to depict most non-linguistic actions in writing. The demonstration (a term I explain below) plays a similar role to a Verb Phrase. It is a more striking (more entertaining, more lively) way of conveying information that could also have been communicated by uttering "turned around and hit my head against this pole".

What is special, and intriguing, about (1) is that the gesturing in it seems to perform a linguistic function. Of course, there are many other cases in which gesturing of this type does not appear to stand in for an unrealized linguistic constituent. Think of Sperber & Wilson's example of the person sniffing ostensively to signal a smell of gas. Or take this other example from Clark and Gerrig [12], in which a child draws the attention of his adult friend:

(2) Herb! [POINTS TO EVE] + [PUTS AN IMAGINARY CAMERA TO HIS EYES AND CLICKS THE SHUTTER]

The child means to inform Herb that Eve is taking a photograph. This, he achieves by producing various types of stimuli, linguistic and non-linguistic. But the latter are not embedded or incorporated into a linguistic structure, as was the case in (1). Hence, it would be absurd to analyze the gestures in (2) in linguistic terms.

In this paper, I concentrate on hybrid occurrences like (1) and suggest that they are liable to a linguistic analysis. The main arguments supporting this option are based on a fundamental similarity between non-verbal gesturing and quotations. But before arguing that quotations are demonstrations, I must make it clearer what a demonstration is.

3 Demonstrations

Clark & Gerrig explain that they understand the verb "demonstrate" as meaning "illustrate by exemplification" [12]. Demonstrations, then, are essentially iconic. Thus, I can demonstrate my brother's manner of scoring goals at football by shooting the ball (real or imaginary) with, say, my left foot, and in a certain characteristic position. I can demonstrate Jimi Hendrix's soloing by pretending to move my fingers smoothly and swiftly along a guitar fingerboard (real or imaginary) (see [12] for more examples). In all such instances, my demonstration involves the production of an example of that which I seek to demonstrate. However, demonstrations are selective, in this sense that only some of their aspects are *depictive*, i.e. meant to be iconically related with that which I wish to demonstrate. For instance, the fact that, say, I am wearing my work clothes while demonstrating my brother's soccer skills is not depictive. And neither is, say, my using a tennis racket instead of a guitar in the second case.

Example (2) is useful in that it allows us to distinguish between two kinds of non-linguistic gesturing. The first, the indexical kind, is illustrated by the child's pointing to Eve. Such gestures, Clark & Gerrig call *indications*. The second, iconic, kind is illustrated by the child's depicting Eve's photographic behavior. This is demonstration proper. This distinction shows that, notwithstanding the label, *demonstrative* pronouns do *not* demonstrate, because they are devoid of the required iconic dimension.

4 Quotations as Demonstrations

The idea that quotations are a species of demonstration goes back to Clark & Gerrig [12], and has recently been revived and developed by Recanati [13], [14]. It is easiest to show in what respects quotations can be considered to be demonstrations by looking at direct speech. When I quote, say Keith's words, I can obviously not produce those very tokens that were uttered by Keith. What I do instead is produce an utterance-token that instantiates the same utterance-type as the original words. In other words, I exemplify that type. Take:

(3) Keith replied, "Righ', I admi' my elimination in the firs' round was a bi' of a disappoin'men' ".

In (3), the words in quotation marks are displayed; i.e. the addressee's attention is drawn to them, but only inasmuch as they exemplify (the same type as) Keith's earlier utterance.

I have used the word "type" in the singular, but that is a simplification. In reality, the displayed token in (3) instantiates many different types, only one of which is the *utterance*-type. Among the instantiated types, only some are targets of the depiction, and these depicted targets need not even include the utterance-type. Assuming that I have uttered (3) aloud, my omission of final "t's" may indicate that I intended to demonstrate other types as well (other properties exemplified by the words in quotes). For instance, I may have wished to demonstrate Keith's way of talking (his accent), or more broadly, the pronunciation characteristic of the dialect spoken by Keith, say Cockney. Or perhaps I wanted to demonstrate Keith's stupidity or the amazing candor tennis players. All it takes for a successful demonstration is that, in the context of utterance, some link be recoverable between the quoted words and the type demonstrated. In other words, all it takes is that, contextually, the quoted words be an exemplification of that type.

A similar analysis also applies to cases of plain metalinguistic quotation. Although, in such cases, the *demonstratum* often boils down to the expression-type, that is not always the case. Take:

(4) "About" is preposition.

When (4) is uttered (i.e. written down) by a grammarian, the demonstrated type is the expression-type "about", because the grammarian's point is to talk about a word, full stop. But different utterances of (4) can be imagined. For instance, spoken utterances in which "about" is produced (i) with a second vowel as in "coo" rather than "cow" or (ii) very slowly. Because of the added mimicry, these displayed tokens are exemplifications of additional types, say, (i) the Scottish accent (in general) or some particular (contextually salient) individual's Scottish accent; and (ii) some salient agent's slow-wittedness. The interpreter who identifies these demonstrata makes inferences on the context of utterance in order to select those aspects of the displayed tokens that are 'depictive'; i.e. those to which the utterer deliberately seeks to draw attention. What is most important, however, is that the core mechanism is always the same: each demonstrated type is made accessible via the displayed token.

5 Closed and Open Quotations

Examples (3) and (4) share an important extra feature: they contain quotations that function as nominal constituents. These nominal constituents are singular terms that refer to a linguistic entity, a piece of discourse in (3), an expression-type in (4). But not all quotations behave like that. Not all quotations are referential NPs. Consider:

(5) Gerald said he would "consider running for the Presidency".
(6) The girl showed the soldier "her" house in Jaffa.

Using the terminology first put forward by Quine [1], we can say that the quoted sequence in each of the above is *used* at the same time as it is *mentioned*. It is mentioned because it comes between quotation marks (and is therefore envisaged *as a linguistic entity*), and it is used because it plays its ordinary syntactic and semantic role in the embedding sentence. This means, notably, that the sentence does not break down grammatically and semantically if the demonstration is removed. To verify this, just leave out the the quotation marks:[1] (5) still means that Gerald said he would

---

[1] This operation is no more than a convenient means of suggesting what the utterance is like without the quotation. Unlike many authors (e.g. [15]), I do not take it that removing the quotes eo ipso amounts to removing the quotation.

consider running for the Presidency, and (6) that a girl showed some soldier a house in Jaffa to which she had some connection.

Now try doing the same with (3) and (4). What you get is:

(3_1) Keith replied, Righ', I admi' my elimination in the firs' round was a bi' of a disappoin'men'.
(4_1) About is a preposition.

If (3_1) is grammatical and interpretable at all, it means something very different from (3). For instance, in the absence of the quotation, "I" and "my" no longer refer to Keith but to the speaker. As for (4_1), it is simply not grammatical: a preposition, as such, cannot fill the subject position in an English sentence.

Recanati calls (3) and (4) *closed* quotations, while he calls (5) and (6) *open* quotations.[2] The difference between closed and open quotations is a matter of *linguistic recruitment*: in (3) and (4), says Recanati, the demonstration (i.e. the *act* of illustrating by exemplification) is recruited to fulfil a syntactic function in the matrix sentence, object in (3), subject in (4). This idea, that the act itself is recruited, is crucial. Previous accounts all dealt with referential quotations very differently, ascribing reference to (i) the complex made up of quotes and enclosed expression (so-called *Name Theory*), (ii) the enclosed expression (*Identity Theory*), or (iii) the quote marks themselves (*Demonstrative Theory*) (See [16] for more details). This was because the scholars in question all took quotation to be an essentially linguistic phenomenon. That is precisely what Clark & Gerrig and Recanati disagree with. To them, quotation is "at bottom, a *paralinguistic* phenomenon, like gesturing or intonation" [13]. The notion that the act, not the words, is recruited (already prefigured in [12]) is an original feature of Recanati's framework which will play a major role later in this paper.

One consequence of linguistic recruitment is that the words in quotes are, in Quinean terms, only mentioned, not used in the matrix sentence. This means that they play no syntactic or semantic role in it. Only the recruited demonstration does. As Davidson [17] puts it, the quoted words are semantically and syntactically *inert*. This stands in stark contrast with the quoted segments in (5) and (6), which each fulfil a syntactic function at the level of the matrix sentence. An analysis of (5), for instance, will produce a tree with separate terminal nodes for "consider", "running", "for", etc. (this would not be the case for "righ' ", "I", "admi' ", "my", "elimination", etc. in (3)).

This inertness of the quoted words is an essential feature of closed cases. It is what explains that we are basically free to quote whatever we like. Thus, closed quoting in English is by no means restricted to well-formed English expressions:

(7) Galileo said, "E pur si muove".
(8) "Tropstical" doesn't mean a thing.
(9) And the Martian guy shouted, "fgey cvjrv=§:/b"ç"çà"=+:/ à)à)à)à))àva".

---

[2] (5) and (6) actually belong to a sub-class of open quotations. I do not discuss other open quotations in this paper.

Although the above sentences seemingly contain foreign words (7), pseudo-words (8) and strings never encountered in any known language (9), they are perfectly well-formed English sentences. That is simply because all we find in the syntactic trees for these sentences are demonstrations recruited as NPs (in positions that require the presence of NPs).

6 Closed Non-linguistic Demonstrations

I now turn to what I think are uncontroversial cases of non-linguistic demonstration doing duty *as a linguistic constituent* of an utterance:

(10) Piano student plays passage in manner μ
    Teacher: It's not [PLAYS PASSAGE IN MANNER μ] —it's [PLAYS PASSAGE IN MANNER μ']. (from [18])
(11) John went [IMITATION OF A CAMEL BELCHING].

Though both examples explicitly involve a non-verbal demonstration, a minor difference can be observed. In (10), there is no difference in kind between the displayed token and the demonstratum: both are musical passages and it is possible, for once, that *all* the aspects of the demonstration are depictive. The demonstration in (11) is a more usual instance of mimicry. The sound produced by John in not a camel belch, though it is meant to be close enough for the intended demonstratum to be identifiable. Most cases of non-linguistic demonstrations occurring in discourse would actually tend to be like (11): demonstrations of bird songs, doors closing, water dripping from the ceiling, objects crashing into each other, etc.

The obvious step to take is to propose an account of (10) and (11) in terms of linguistic recruitment. This is easily done for (10): the playing of the piano passage is turned into an NP. In (11), things are slightly more complicated. After the reporting verb "to go", we have an adjunct rather than an NP. This poses a problem for Recanati's original account, which limits syntactic recruitment to nominal positions. The question is important, but I cannot address it in detail here. I can only make the following remarks: "to go" is used indifferently to report speech or non-linguistic sounds. It is easy to accept, I believe, that, in "And then Sheila went 'I think Mike is a better guitarist then Rob' ", the complement of "to go" is a direct speech report (as it would be after "to say").[3] Moreover, there is no good reason to contend that the sequence in quotation marks is used at the same time as being quoted: it does no other job than demonstrating (some aspects of) a previous utterance of Sheila's. Both this unambiguous semantic function and the grammatical similarities with "to say" point to the fact that the quotation must have been recruited syntactically, albeit as an adjunct. If that is granted, then there is every reason to extend this analysis to the imitation of the camel belching: this bit of mimicry too is a demonstration, intended as a depiction of a genuine camel belch. Therefore, it must also be a recruited demonstration.

---

[3] A reviewer claims that my argument is problematic because "go" can actually only report non-linguistic sounds. This is simply not true. Examples of "go" introducing direct speech are so common that this use is widely recorded even in basic dictionaries (e.g. *Concise Oxford*, 8th ed. (1990); *Robert & Collins* 6th ed. (2002), etc.). See also [12].

7 Non-English Words in Open Quotations

Having dealt with non-linguistic demonstrations in closed cases, I now turn to the quotation of non-English sequences in open positions. Consider the following examples:[4]

(12) If you were a French academic, you might say that the parrot was *un symbole du Logos*. (Julian Barnes, *Flaubert's Parrot*, 1985: 18)
(13) Each tablet in the war cemetery would commemorate Monsieur Un Tel, *lâchement assassiné par les Allemands*, or *tué*, or *fusillé*, and then an insulting modern date: 1943, 1944, 1945. (Julian Barnes, *Cross Channel*, 1996: 105)

The main difference between these utterances and the "Gerald" example mentioned earlier is the presence of cross-linguistic quotations. But, just like "consider running for the Presidency" in example (5), neither "*un symbole du Logos*" nor "*lâchement assassiné par les Allemands*" occurs as a referential term denoting a linguistic object. On the contrary, these sequences each appear to be used with their ordinary semantics (i.e. French semantics). This makes it tempting to say simply that they are mentioned *and* used just as in the "Gerald" example. But there is a problem with use: what does it mean for a French expression to be used in an English sentence?

Let me make the point more strikingly. The "Gerald" example can be paraphrased as a pair of sentences:

(5_1) Gerald said that he would consider running for the Presidency.
(5_2) Gerald said (the words) "consider running for the Presidency".

This paraphrase basically separates a 'use-line' from a 'mention-line'. Attempting to do the same with (12) and especially with (13) poses problems:

(12_1) ?? You might say that the parrot was un symbole du Logos.
(12_2) You might say "un symbole du Logos".

(13_1) ?? Each tablet in the war cemetery would commemorate Monsieur Un Tel, lâchement assassiné par les Allemands, or tué, or fusillé.
(13_2) French people use the expression "lâchement assassiné par les Allemands" to mean "killed in a cowardly way by the Germans", etc.

Some comments: the problem with the use-line (12_1) appears to result from the removal of the quotation device (the italics). I have put two question marks rather than a star because I do not want to commit myself from the outset to a judgment of ungrammaticality. As regards (13_1), the difficulty is the same as in (12_1), only made more striking by the recurrence of French expressions.

An in-depth discussion of the grammaticality status of (12_1) and (13_1) would take us too far afield. Let me simply remark that the well-formedness of (12) and (13) –

---

[4] The quotations in these and in (7) to (9), I call *cross-linguistic quotations*. (7) to (9) contain *closed* cross-linguistic quotations; (12) and (13) contain *open* cross-linguistic quotations.

the utterances *with* the italics – is not entirely guaranteed either: (12) and (13) differ from (12_1) and (13_1) only in terms of the presence vs. absence of markers of quotation. It would seem reckless to ground the grammaticality of (12) and (13) on the mere presence, in this case, of a typographical device.[5] Any proper substantiation of the well-formedness of (12)-(13) – and perhaps of (12_1)-(13_1) – would have to appeal to other arguments. I briefly sketch two that could be of help.

The first, the idea of shifts in the context (especially *translinguistic context-shifts*, or *language-shifts* for short), has been developed by Recanati [13], [14]. Roughly, Recanati assumes that the context involved in the interpretation of an utterance is made up of three components: a situation of utterance, a language, a circumstance of evaluation.[6] Each of these components is liable to shift in certain forms of open quotation. In my data, it is the language component that may to shift. Here is an example discussed by Recanati himself:

(14) My three-year-old son believes that I am a 'philtosopher'. (originally in [15])

This example is like (12) and (13) in all relevant respects. The only apparent difference – the fact that 'the other language' is not French but a little boy's idiolect — has no impact on the present discussion. Recanati's account goes like this: language-shifts are identified *pre-semantically*, which means that the addressee must have recognised them if a semantic interpretation of the utterance is to be possible. Like disambiguation, language selection is a necessary step toward the identification of *which sentence* has been uttered. Only when the addressee has determined which sentence has been uttered is s/he in a position to complete its semantic interpretation. In (14), the addressee must tag "philtosopher" as belonging to somebody else's language (the young son's idiolect), failing which, says Recanati [14], the utterance would be ungrammatical, since it would include a string that does not exist as a word in the speaker's context (read "language" in this case).

I believe that Recanati's account neatly captures what goes on in cases like (14), but also (12) and (13). Besides, it goes some way towards explaining the grammaticality of these utterances: if the various parts of the sentence are tagged for a language, then grammaticality judgments for each tagged segment will be made in terms of the grammar of the relevant language (possibly an idiolect). Still, there is one thing that Recanati's language-shifts do not do, and that is to enable us to predict at which positions in an utterance a shift is licensed. In other words, language-shifts are not enough to ground the grammaticality of open cross-linguistic quotations.

The second element likely to support an account of the acceptability of (12) and (13) is the idea that linguistic competence may be other than monolingual. This idea has been defended by many specialists of the phenomenon commonly known as *code-switching*. Code-switching can be defined as a "a speech style in which fluent bilinguals move in and out of two (or conceivably more) languages" [19], as in "This morning mi hermano y yo fuimos a comprar some milk" or "Sometimes I start a

---

[5] This appears all the more illegitimate because there are also *spoken* utterances that are in all relevant respects similar to (12) and (13).

[6] Although this is a different conception of the context from that of relevance theorists, I believe that it could be translated into a set of assumptions available to the audience.

sentence in English y termino en Español". Many linguists regard such utterances as well-formed sentences.[7] My contention — once again it is one that requires considerable extra substantiation — is that the grammar (the theory) that explains these code-switched utterances is the same as (or at least very similar to) the grammar that underlies examples (12) and (13). If that is accepted, then

(i) examples (12) and (13) (and possibly their counterparts with no marking), although they contain non-English words that are *used* as part of the matrix sentence, can be regarded as grammatically well-formed;
(ii) there are two principles that can be used to assess grammaticality: the 'grammar of code-switching' determines permissible switch sites, and the notion of language-shift provides a justification for assessing the grammaticality of the quoted sequence in terms of another language than that of the matrix sentence.

8 Intrusion of Non-linguistic Communication in Non-nominal Positions

I now return to non-linguistic communication intruding into utterances, because it appears that demonstrations of that kind can enter discourse in other capacities than as NPs or Ns, namely those positions which, according to Recanati, license recruitment. At the beginning of the previous section, I pointed out a difficulty with respect to the idea that non-English words were *used* in an English sentence. After that, I sketched a tentative solution to the problem, by appealing to Recanati's language-shifts and the grammar of code-switching. Obviously, a similar problem arises here... but with more severe implications. By definition, non-linguistic material can simply not be used linguistically. This would be a contradiction in terms. Hence, no possibility of simultaneous use and mention as in the previous instances of open quotation that we looked at. I repeat (1) and offer two more examples:

(1) I got out of the car, and I just [DEMONSTRATION OF TURNING AROUND AND BUMPING HIS HEAD ON AN INVISIBLE TELEPHONE POLE]
(15) And then she [IMITATION OF SOMEONE RUNNING AWAY]
(16) Of course he made a point of looking very [IMITATION OF HUMILITY]

We saw above that the demonstration in (1) does the same job as a conjunction of VPs. Similarly, in (15), the demonstration stands in for, say, the VP "ran away" (it

---

[7] One reviewer contends that many generative linguists would deny that there is a grammar of code-switching. This may be so. But there is a consensus among *some* syntacticians that code-switching behaviour is underlain by a linguistic competence, in the same sense as monolingual performance is (See, f.i., [20], [21], [22])). Shana Poplack [23], a leading authority on code switching, writes: "Researchers first dismissed intra-sentential code-switching as random and deviant [...] but are now unanimous in the conviction that it is grammatically constrained". Note, finally, that Jeff MacSwan ([19], [23]) has developed a Chomskyan minimalist account of code-switching that appeals only to mechanisms already required by the grammar of monolingual sentences.

could also substitute for the V "ran" if, for instance, it was followed by the words "right into the street"). In (16), it stands in for, say, the adjective "humble" (it could also substitute for an adjective phrase if, for instance, "very" was left out).

Now the fact that we cannot say that the demonstrations in (3) are *used* seems to constrain us to judge these utterances ungrammatical. Indeed, if the demonstrated action is not used in (1) and (15), the right syntactic analysis seems to be that we have a verbless main clause whose absent VP cannot be said to be elliptical (as it could perhaps be if the "And then she" of (15) followed after "First *he* was there": there the gap could be explained in terms of ellipsis). Hence, there is no justification that I can see for the construction. A similar reasoning applies to (16).

We seem to have arrived at a negative conclusion: (1), (15) and (16) may be instances of communicative behavior, but they are not sentences in any conceivable sense. This entails too that, on a traditional understanding of semantics, they cannot be used to *say* anything in the technical sense of the term; they have no complete semantic interpretation. All the same, it is easy to acknowledge that they are successful communicative stimuli: they can easily be given an interpretation *in context*. With (1) the communicator conveys that he turned around and knocked his head against a pole at some time t. By producing (15), the communicator means that the referent of "she" ran away from a contextually determined place p at some time t. By producing (16), the communicator means that the referent of "he" was keen to look very humble at time t.

9 Recruitment in Non-nominal Positions?

Though the above line of reasoning squares with the relevance-theoretic rejection of the primacy of linguistic communication, I none the less do not find it altogether satisfactory. The possibility I would like to consider briefly is that (1), (15) and (16) *are* well-formed sentences. The essential motivation for this hypothesis is, once again, the analogy with quotation. My first point is that, as Clark & Gerrig put it, "the border between quotations and nonquotations isn't sharp". Thus, I have treated (9) [the Martian's quote] as a quotation and (11) [the camel belch] as a non-verbal demonstration. But this looks like an arbitrary decision. 'Quotations' of noises and sounds suggest that there may be a continuum from 100% linguistic to 100% non-linguistic demonstrations. My second point is that, as is again noted by Clark & Gerrig, speakers tend to treat examples like (1) as quotations. This is illustrated, notably, by the use of "to go" and "to say" for both linguistic and non-linguistic demonstrations (though in differing proportions). Besides, the very fact that, just like linguistic quotations, 'nonlinguistic quotations' "can be embedded as constituents other than NPs" [12] is further evidence, in Clark & Gerrig's eyes, that speakers treat them alike. My third point is the observation that the insertion of non-linguistic communication in non-nominal positions appears to be rule-governed: it is not true that just any sort of intrusion is possible at any spot in an utterance. Moreover, the positions at which a shift from linguistic to non-linguistic is allowed seem to largely overlap with the permissible switch sites for code switches and quotational language-shifts.[8] My fourth point is that if "nonlinguistic demonstrations [...] aren't generally considered part of 'the discourse' [...] only because they aren't linguistic, the argument is circular" [12].

My tentative proposal would consist in saying that a non-linguistic demonstration can be recruited linguistically in *other capacities* than as an N or NP. For example, the imitation of someone hitting their head against a pole in (1) and someone running away in (15) would be recruited as a VP or V, and that of humility in (16) would be recruited as an ADJ or ADJ Phrase.

The recruitment hypothesis for (1), (15) and (16) suggests an explanation for the following fact: remember that I initially introduced (1), (15) and (16) as the non-linguistic counterparts of open quotations. I single out (15) for convenience. There is an apparent discrepancy between (15) and its putative fully linguistic counterpart:

(15_1) And then she 'ran away',

where " 'ran away' " is an open quotation. The 'fully linguistic counterpart' triggers pragmatic effects that cannot be secured by the utterance involving gestural mimicry. In particular, the former implies that the utterer distances herself from the choice of "ran away", whereas the latter does not. However, once we accept that non-linguistic demonstrations in non-nominal positions do not do 'double-duty', then (15_1) can no longer be regarded as the fully linguistic counterpart of (15), because the quotation in (15_1) *does* double-duty (it is used and mentioned). It is therefore quite predicatble, on this account, that discrepancies should arise between (15) and (15_1).

I conclude this section by mentioning an alternative account, albeit a more adventurous one: the suggestion that hybrid utterances like (15) and (16) involve a 'language-shift' in a manner similar to (12), (13) and (14). Only, this time, the shift takes us from natural language to another system of communication (a pictorial or gestural system). On that assumption, (15) and (16) might be well-formed utterances, but only in a different sense. The idea would be that they comply with the rules of a hybrid 'grammar' mixing linguistic rules with the constraints of another system. This, of course, is less economical than the previous solution (or than the dismissal of (15) and (16) as ill-formed utterances), but it may be more compatible with psychological facts (provided we settle on a 'watered down' conception of grammar). In particular, the hybrid 'grammar' might amount to a number of cognitive constraints. For instance, it might be the case that some positions in a sentence are not available for a shift into another system because that would make interpretation too costly from a cognitive point of view. Note that, if this explanation turned out to be correct, it would shed some interesting light on syntacticians' theories of code-switching. Perhaps the assumed grammar for code-switching boils down to some general cognitive constraints too. But of course, at this stage, these are just conjectures.

---

[8] This is still no more than a hypothesis, albeit one that is empirically testable. But consider these examples, which I deem unacceptable:
* That's [?? MIME OF A CAUSAL RELATION (equivalent to "because")] he hadn't arrived yet.
* The cat climbed up [?? IMITATION OF DEFINITENESS (equivalent to "the")] tree and then he fell.

10 Conclusion

The present study brings to light a strong kinship between quotations and non-linguistic demonstrations. One implication of this is that, in future, any account of quotation should also be able to be integrated within an account of demonstration and, in turn, of human communication as a general phenomenon. Another virtue of the paper is that it highlights the importance of various kinds of stimuli in linguistic or 'mixed' communication. Conventional symbols have always been in the centre of attention. There has been a lot of interest in indexes as well: many linguistic signs have been shown to be not just symbols but indexes too. Icons, by contrast, have always tended to remain in the background. A study like this one should show that icons matter in (or together with) linguistic communication. Not only do all quotations have an iconic dimension, but it appears that, provided they are *demonstrated*, all sorts of non-linguistic 'things' can enter into utterances as well.

As hinted above, I have been led to take a slightly uncomfortable stance on non-linguistic communication. On the one hand, I have advocated discarding the linguistic bias present in too many studies by philosophers and linguists. On the other, I have ended up defending an account of non-linguistic gesturing as playing a *linguistic* role (though only in the sort of data I have chosen to examine). Some might think that my insistence on linguistic recruitment has caused me to relapse into the traditional linguistically biased view according to which everything that is not linguistic (including gestures and so on) belongs to the context. The traditionalist could claim that what happens in recruitment is precisely that *an element of the context* gets 'raised' to linguistic status and thereby ends up playing a role in a linguistic utterance. However, I still believe that the account I have offered is compatible with a notion of communication and of the context such as that advocated by relevance theory. I stand by the thesis that non-linguistic communicative stimuli are part and parcel of utterances. My only point was to show that, in *particular circumstances*, those communicative behaviors can be made to fulfil a linguistic function. In other circumstances, they belong with communicative stimuli, not with the background elements that make up the context.

References

1. Quine, W.V.O.: Mathematical Logic. Harvard University Press, Cambridge, Mass. (1940)
2. Tarski, A.: The Semantic Conception of Truth and the Foundations of Semantics. Journal of Philosophy and Phenomenological Research 4 (1944) 341-75
3. Rey-Debove, J.: Le Métalangage. Etude linguistique du discours sur le langage. Le Robert, Paris (1978)
4. Carnap, R.: The Logical Syntax of Language. Transl. by A. Smeaton. Kegan Paul, Trench, Trübner & Co, London (1937)
5. Reichenbach, H.: Elements of Symbolic Logic. Macmillan, New York (1947)
6. Christensen, N.E. The Alleged Distinction between Use and Mention. Philosophical Review 76 (1967) 358-67
7. Searle, J.R.: Speech Acts: An Essay in the Philosophy of Language. Cambridge U. P., Cambridge New York (1969)

8. Recanati, F.: La transparence et l'énonciation. Pour introduire à la pragmatique. Seuil, Paris (1979)
9. Sperber, D., Wilson, D.: Relevance. Communication and Cognition. Blackwell, Oxford Cambridge, Mass. (1995)
10. Carston, R.: Explicature and Semantics. In: Davis, S., Gillon, B. (eds.): Semantics: A Reader. O.U.P., Oxford (2004) 817-845
11. Lyons, J.: Semantics. Cambridge University Press, Cambridge (1977)
12. Clark, H.H., Gerrig, R.J.: Quotations as Demonstrations. Language 66 (1990) 764-805
13. Recanati, F.: Open Quotation. Mind 110 (2001) 637-87
14. Recanati, F.: Oratio Obliqua, Oratio Recta: An Essay on Metarepresentation. MIT Press, Bradford Books, Cambridge, Mass. (2000)
15. Cappelen, H., Lepore, E.: Varieties of Quotation. Mind 106 (1997) 429-50
16. Washington, C.: Identity Theory of Quotation. Journal of Philosophy 89 (1992) 582-605
17. Davidson, D.: Quotation. Theory and Decision 11 (1979) 27-40
18. Horn, L.R.: A Natural History of Negation. The University of Chicago Press, Chicago London (1989)
19. MacSwan, J.: A Minimalist Approach to Intrasentential Code Switching. Spanish-Nahuatl Bilingualism in Central Mexico. Unpublished PhD. Online at: http://www.public.asu.edu/~macswan/front.pdf (1997)
20. Joshi, A.: Processing of sentences with intrasentential code switching. In: Dowty, D.R., Karttunen, L., Zwicky, A.M. (eds.): Natural language parsing. Psychological, computational, and theoretical perspectives, Cambridge U.P., Cambridge (1985) 190-205
21. 21 Hudson, R.: Syntax and Sociolinguistics. In: Jacobs, J. et al. (eds.): Syntax. Ein internationales Handbuch zeitgenössischer Forschung. An International Handbook of Contemporary Research. Walter de Gruyter, Berlin New York (1995) 1514-1528
22. Bullock, B. E., Toribio A. J.: Phonetic evidence of syntactic constraints in Spanish-English bilingual code-switching. Paper presented at the Seventh Conference of the European Society for the Study of English, Zaragossa, September 2004.
23. Poplack, S.: Code-switching (linguistic). In: Smelser, N., Baltes, P. (eds.): International Encyclopedia of the Social and Behavioral Sciences. Elsevier Science Ltd. (2001) 2062-2065
24. MacSwan, J.: The Architecture of the Bilingual Language Faculty: Evidence from Intrasentential Code Switching. Bilingualism: Language and Cognition 3-1 (2000) 37-54

Mobile Phone Talk in Context

Mattias Esbjörnsson and Alexandra Weilenmann[*]

Mobility, Interactive Institute, P.O. Box 240 81,
SE-104 50 Stockholm, Sweden
mattias.esbjornsson@tii.se
alexandra.weilenmann@tii.se

Abstract. In light of recent attempts to design context-aware mobile phones, this paper contributes by providing findings from a study of mobile phone talk in context. We argue the benefits of investigating empirically the ways in which a place is interactionally constituted as appropriate, or not, for a mobile phone conversation. Based on a study of naturally occurring mobile phone talk, we show how people handle calls in potentially difficult situations. Availability is negotiated, and it is not always agreed on whether a situation is appropriate. These findings pose challenges to the design of context-aware telephony.

1 Introduction

Design providing remote awareness is a well-investigated issue in HCI (Human Computer Interaction) and CSCW (Computer Supported Cooperative Work), and has been studied empirically and evaluated over the years [13, 8, 5]. Recently, with the advent of mobile collaborative technologies, mobile awareness has become a topic of interest. There are a number of attempts to design context-aware applications, i.e. technology providing users with the possibility to see remote participants' location and activity, in order to determine whether or not to initiate communication [20, 25, 19]. However, it has been argued that many of these systems apply a simplified view of context [9]. Further, context-aware systems often derive from a strong technological focus, underbuilt by few, if any, empirical findings.

In this paper, we aim to contribute to the field of context-aware mobile telephony by adding empirical findings from actual use of this mobile collaborative technology. We explore how the participants themselves provide awareness and show availability, by investigating the talk over mobile phones. This is done by the detailed analysis of recordings of mobile phone conversations. Data has been collected by audio-recordings of the conversations, as well as video-recordings providing additional contextual information. A few excerpts are used in the paper to illustrate our findings on how the appropriateness of mobile phone conversations is established in respect to place and activity. We show how the appropriateness of having a conversation in the situation where the answerer or the caller is located, is negotiated and discussed in a more or less explicit fashion by the participants. It becomes clear that places and activities,

---

[*] Authors are listed alphabetically.

which might seem inappropriate for conducting mobile phone conversations, are not always treated as such. Furthermore, the participants do not always agree on whether a place and an activity is appropriate or not, leading to negotiations. We discuss what these findings might imply for design of context-aware mobile telephony.

2 Related Work

This work originates in two fields; that of context-aware applications, particularly mobile phones, and that of studies of mobile phone use. These two fields are outlined, below, along with a brief introduction to Conversation Analysis (CA), to ground the empirical approach used to understand mobile phone talk in context.

2.1 Context-Aware Mobile Phones

As a crucial aspect in cooperative work, awareness has been much investigated in CSCW and HCI, as part of co-located work [8, 1, 12] as well as remote and mobile collaborative work [3, 18]. There have been attempts to design applications that provide awareness to mobile users. In this paper we focus on attempts to design mobile phones which to some extent provide users with information about context, and in specific, systems which aim at settling the appropriateness for conducting calls.

Several attempts to design context-aware mobile phones rely on the use of calendars. Milewski and Smith [20] present an application, the live address book, which aims at helping people make more informed telephone calls. The users are supposed to set availability status, refer to which telephone number on which she can be reached, and there is also room for a personal message. All this information has to be kept updated by the user. From the technical trial Milewski and Smith learnt about the effort in updating, the reliability of the information provided, ways of negotiating availability, etc.

Another example is Comcenter, [4], a system which shows awareness information about the recipient before starting a communication act. For instance, if an organization uses open calendars, the current calendar entry can be shown or a user specified message could be displayed, which enable some sort of rudimentary one-way "negotiation" before communication is initiated.

There are also attempts to combine calendar data with automatically retrieved contextual information. One example of this is SenSay, a context-aware mobile phone [25]. With a starting point in the 'troublesome' activity to keep the phone in its correct state, i.e. the ringer on or off, determining call priority, etc., a mobile phone is introduced which modifies its behavior based on its user's state and surroundings. SenSay uses data from a number of sources, including appointments scheduled in a calendar, if the user is in a conversation or speaking aloud, if physical activity is high, and level of ambient noise, to determine whether the user is "uninterruptible". SenSay provides the remote callers with the ability to communicate the urgency of their calls, makes call suggestions to users when they are idle, and provides the caller with feedback on the current status of the SenSay user.

The problem with many of these systems is that they rely on the idea that it is possible to build into the design an understanding of what is 'an uninterruptible activity'. Context is seen as a more or less stable entity. This approach has been criticized because it is based on a view of context as a representational problem [9]. Dourish introduces an interactional model of context, where he argues that context is an *occasioned* property, relevant to particular settings, particular instances of action, and particular parties to that action. Further, rather than taking context and content to be two separable entities, he instead argues that context arises from the activity. Drawing upon CA and ethnomethodological work, Dourish argues that context is not just there; rather it is actively produced, maintained and enacted in the course of activity at hand.

Based on Dourish's critique, along with the observation that there is a lack of studies based on empirical findings, we believe that it is necessary to examine today's actual use of the mobile phone, in order to design supportive tools.

2.2 Studies of Mobile Phone Use in Context

The mobile phone is increasingly receiving attention as a collaborative technology. There is work on the adoption of mobile phones [21], text messaging among teenagers [11] and mobile phone use among young people in general [26]. However, there is still a lack of studies using naturalistic data to understand the actual interaction with technology, as is common in CSCW studies with a stronger ethnographic focus. Most studies rely on accounts of use [e.g. 22, 17], rather than on analysis of the actual interaction. Although a useful method for some inquiries, we want to stress the benefit of understanding mobile phone use in context.

One example of relevance for the issues dealt with in this paper shows the sort of arguments that can be made using accounts or other forms of data. Many authors claim that the mobile phone privatizes public space, as it enables people to have private conversations in public places. For instance, in discussing mobile phone culture in Finland, Puro maintains that: "as someone talks on the phone, one is in her or his own private space. Talking on the mobile phone in the presence of others lends itself to a certain social absence where there is little room for other social contacts. The speaker may be physically present, but his or her mental orientation is towards someone who is unseen" [23, p. 23].

However, previous ethnographic field studies of mobile phone use in natural settings, display how in some situations, conversationalists include co-located others in the mobile phone communication, rather than withdrawing to have private conversations. One example of this is related in Weilenmann and Larsson's [28] fieldwork on public mobile phone use among teenagers in Sweden. In one illustrating excerpt from the field study, four girls all take part of a mobile phone call received by one of them, and they relate to the caller what is going on in the group at their end. From this field study it seems clear that the young people studied do not exclude their co-present friends when talking on the mobile phone, they remain attentive to the ongoing event as well as that on the phone. Studying mobile phones in context revealed that their usage can be a shared collaborative activity.

A similar lack of focus on naturally occurring interaction is present in most existing studies on mobile phone use in traffic (e.g. [19, 27]). Most of these studies take the starting point in experimental or laboratory settings to be able to "control" the variables, in order to investigate how the use of mobile phones influences driving, a highly complex activity per se. The methodological benefits of focusing on mobile phone use as part of natural traffic situations are discussed in [10].

2.3 Conversation Analysis

From the very beginning, CA has been closely linked to the analysis of phone conversations. One practical reason was that telephone calls were particularly suitable for CA methods. By making audio recordings of both ends of phone conversations the researcher would get access to much of the same interactional resources as the participants, since they also are only connected through audio. Most important to this is that, on the phone, participants have no visual access to each other.

For the purpose of this paper, it can be valuable to point out a few CA findings relating to the participants' availability for having a conversation, and show how CA can be useful when understanding the situated nature of mobile phone talk.

Schegloff [24] identifies a number of ways in which the second turns in the phone call (the caller's first turn) are constructed. Of specific relevance for the present study is the case where the second turn formulated as a "question or noticing concerning answerer's state". For instance, this can look like the following [24]:

| A: | Hello |
| C: | Hi can you talk |

Or

| A: | Hello |
| C: | Hello. You're home |

This deals with issues of availability for having a conversation, as well as recognizing where the answerer is located. Of course, in the second case, the fact that the caller knows that he is calling to a residence home, a landline phone, is obvious. If someone answers this call, the caller can be certain that the called is home where the phone is located. This is obviously different in the case with mobile phone calls.

Button and Casey [7] report on a phenomenon relevant when considering how availability is established in phone conversations. They show how questions about what the co-participants are doing, thus an "inquiry into immediately current events", what they call topic initial elicitors, occur after the identification and recognition section. They argue that these topic initial elicitors "make a display of availability for further talk but without, themselves, introducing topic material provides the opportunity for, as a preferred next activity, a newsworthy event reported in a next turn" [7, p. 172].

Taking a yet larger perspective on the telephone conversation, another study by Button deals with how a conversation is organized as part of a series of conversations. He found that arrangements may be oriented to as a "special status topic", which is specifically used to place the conversation on a closing track" [6, p. 251]). One way of doing this is through "projecting future activities", for instance, talking about whom should call a third person and make arrangements, etc.

2.3.1 Conversation Analytic Work on Mobile Phone Talk

There are so far few CA or CA inspired approaches to mobile phone conversations. Apart from the newness of mobile phone technology, one of the reasons is likely to be because it is relatively difficult to get recordings of mobile phone conversations.

In one of the first available studies of mobile phone use based on recordings, Laurier [15] investigates the ways in which mobile office workers talk about location when traveling by car. He seeks to explore "why people say where they are during mobile phone calls". Laurier's argument is that this is a question of location used to establish a mutual context in communication, between participants who are dislocated. The formulation of location in mobile phone conversation is tied to the business that needs to be done between the two people, and the place descriptions are thus doing a lot more than just formulating place. Laurier's study is ethnographic and uses video to capture the interaction in the car, where the researcher is present. This means that a rich description can be provided of the setting in which the driver/mobile phone user is located. Having this data becomes even more useful in another of Laurier's papers, where he expands upon the issue of how mobile workers handle their daily work and talk alongside the task of driving and maneuvering the car [16]. Other studies of mobile phone talk are more focused on the particulars of the conversations [2, 14].

Methodologically, it is noteworthy that Arminen [2] points out that ethnographic data of the local circumstances and constraints of the answerer could sometimes be used to shed light on what is happening in the opening of the conversations. In some situations it is difficult to answer the phone, but for some reasons it is done anyway. For instance, Arminen shows an example where a caller answers while being in the toilet of a train, and what interactional difficulties this entails in the conversation.

3 Data Collection

The empirical data presented in this paper derive from two separate projects. They both rely on recordings of naturally occurring conversations, but they are different in a number of ways.

The first piece of data was originally collected for a study looking at the impact of mobile phone talk on driving [10]. In order to understand how drivers combine mobile phone conversations and driving, seven drivers were studied. The person we follow here is called Eric. He is a salesman who travels over a vast geographical area. Each year he drives roughly 100.000 kilometers, to visit customers on a regular basis. He uses his car both as a means of transportation, and as a mobile office where he conducts paper work and mobile phone calls.

The analysis required the conversations, a comprehensive view of the traffic-situation, as well as a view of how the driver handled the vehicle. Accordingly the researcher sitting in the passenger seat collected the data by video recording the activities taking place in the car. By using a single video camera we were able to alter the perspective between activities in the car, and the traffic-situation. The video recordings show only some of the visual details that occupy the drivers' attention. Thus, the video camera is not a way of collecting complete observable data. It is rather a tool for the researcher that provides contextual data from one side of the conversation.

The data presented in Excerpt 1 comes from a driver who used a car-mounted phone; i.e. the mobile phone was put in a holder on the dashboard, connected to a speaker and a microphone. The participants in the study agreed on being recorded, and are presented in a way that protects their identity. They were also requested to inform us if any conversations were not appropriate to record, and should be deleted.

In the second study [29], the conversations of a teenage girl were recorded using a special recording device, which was built in order to collect mobile phone talk data. One person was recruited to have her calls recorded, an 18-year-old girl here called Nicky, living in a small suburb to Göteborg, Sweden's second city.

Also for this study it was made sure that the informant would feel that she was in control over what was recorded. The informant had the possibility of deciding which phone conversations to give to the researcher. After having recorded a conversation, she herself could delete it if she did not feel she wanted it to be used for the study. She was told to let her friends know that she would be part of this study, so that those who did not want to be recorded could say so. A few of her friends then chose not to be recorded. All names of persons appearing in the conversations have been changed.

For the latter study, we did not gather any ethnographic data. As discussed previously, leisure activities can be more difficult to get access to. This clearly led to some lack of insight into the context of the calls. On the other hand, the calls presented here, took place in situations which it is doubtful if any ethnographer with a video camera would have been given access to.

4 Mobile Phone Talk in Context

In the following we illustrate the various ways in which a place is interactionally constituted as appropriate for having a mobile phone conversation.

4.1 The Car as an Appropriate Place to Talk

The participants in the study on the impact of mobile phone talk on driving [10], noticeable favored a car in motion for phone conversations. They adapted their activities to make calls when driving, e.g. initiating conversations immediately after entering the car, and terminating conversations when reaching their destinations.

In the following example the explicit choice of the driving situation, consequently the car, as a suitable place for conversation is evident. This is observable in how they express themselves in the conversation, and in how Eric uses his phone.

Having had lunch at a hotel in the outskirts of Smalltown, Eric switches his handheld on and puts it back in the holder on the dashboard, immediately after entering the car. He receives a text message telling him that there are four messages in his voicemail. The amount of messages is caused by the fact that he turned off his phone during a lunch break, and during a recent visit at a customer's site. As he exits the parking lot, Eric returns the call from Fredrik:

	Conversation	Inside the car
Fredrik:	The Sport Shop, Sandstad, Fredrik	
Eric:	Hi Fredrik! Eric Sport Products	Eric keeps one hand on the steering wheel. Looks straight out on the road.
Fredrik:	Hi:::	
Eric:	Ho:w are you	Eric looks down from the road, probably adjusts the heating with his right hand.
Fredric:	I'm fine:	
Eric:	Sounds good:	Lowers his right hand. Only one hand on the steering wheel.
Fredrik:	Will you be in the car for a while?	
Eric:	if I will be in the car for a while? >yes< you can give me a call	
Fredrik:	>Yes<	
Eric:	Yes	
Fredrik:	I'll call you	Eric looks at the phone.
Eric:	Sounds great	Eric looks out on the left.
Fredrik:	>Bye<	
Eric:	Bye:	Eric looks down on the phone, and uses his right hand to end the call.

Excerpt 1. During this short conversation, Eric and Fredrik mutually agree on the car as an appropriate place for mobile phone conversations

The answerer gets out of the conversation before a topic is initiated. The way of doing this is to make arrangements for calling back later. Fredrik asks whether Eric will be in the car for a while. This might seem like an odd question, why would he want to know that? However, Eric does not take the question to be odd, rather he takes is as a question of whether he will be available for conversation a while later. He displays his understanding by saying "Yes you can give me a call".

Thus, their conversation clearly shows how they consider the time spent in the car as time available for incoming as well as outgoing calls, i.e. telephone hours. The excerpt nicely shows how both the driver, and the non-present conversationalist, orients themselves to the car as an appropriate place to talk. The time for transportation is time 'forced' to be spent in the car. Consequently, the time can be used for conversational work. Additionally, the ethnographic study revealed that Eric switches off the mobile phone during lunch breaks and visits at customer sites, and immediately switches it back on when entering the car. This behavior reinforces the view of the car as an appropriate place to talk.

The transcript displays how the place and the activities, e.g. the contextual factors, are non-static. Despite the fact that the environment within the car is a fairly stationary setting, several simultaneous activities takes place both within and outside the car, e.g. changing gears, adjusting the heat, doing maneuvers, adjusting the speed to other cars. The traffic situation is dynamic, and the driver has to adapt to these contingencies. Even if he focuses on several simultaneous tasks, the driving situation is appropriate for him to conduct other activities, in this case a mobile phone conversation.

Second, the transcript illustrates the difficulties in deciding, and defining in advance, the appropriateness of a call. The complexity in setting the appropriateness in time and space for a conversation, due to contextual factors, is visible in how they postpone the conversation to a later occasion.

Third, and this is important when considering design of context-aware applications, despite the inappropriateness in having a conversation, Fredrik answers the phone. For some reason he cannot talk at the moment but still takes the call, thus being able to reach an agreement on when to call back. This can also be a way to check whether the call is very urgent and should take precedence for the current activity. In any case, the conversation is held in a situation which might have been considered as "uninterruptible" by a system, and the fact that the conversation can take place despite this, allows the participants to reach an agreement on when to continue talking.

4.2 Making Place for Mobile Phone Talk in an Inappropriate Place

Moving on to a setting different from the car, we will consider an example where a call is received, and answered, in a classroom.

Nicky:	Hi![1]
Oscar:	Hi::
	(.)
Nicky:	What are you doing
Oscar:	I'm having a class: but it's no problem hhh
Nicky:	Okay h:
	(.)
Oscar:	Well::
Nicky:	You
Oscar:	Yes:
Nicky:	Tonight
Oscar:	Yes:
Nicky:	Whe:::n eh:: blublub do we get anything to eat?
Oscar:	No
Nicky:	We don't?
Oscar:	No
Nicky:	Okay then (.) then I'll have to eat now then

Excerpt 2. In this conversation, the answerer and caller adjust to the fact that the answerer is located in a classroom during class

Here we can see how the caller is informed that she has called someone who is in class at the moment. However, the answerer claims that his being in class is not a problem, thus displaying availability. He does this in one turn "I'm in class but it's no

---

[1] It is unclear here why the caller utters the first turn, the initial "hi". We are not sure whether there had been some interaction prior to this; this is all that is available on the recording.

problem". It is interesting that on the question "what are you doing?" he does not just answer that he is in class; he also says that it is not a problem. Probably this is because many of us would actually see this as an activity where one is (or should be) unavailable for talking on the phone, and he therefore needs to state that he is not one of these people. Presumably, it could be a problem for other people in his immediate surroundings, e.g. the teacher. Perhaps this is also a way then to "be cool", to show that he can do as he pleases.

So, in the next lines we can see how the caller and the answerer together are making place for mobile phone conversations to continue in this setting. The way that the conversation unrolls after he has said where he is, is peculiar. There is a long sequence with one word turns before she initiates the topic. This can be because she does not really have a topic, and comes up with one as the call develops. It could also be because she is orienting to him being in class, she might find this more problematic that he pretends to do. Therefore in initiating the topic step by step in short turns, she gives him the possibility of saying that he cannot talk. This argument is supported by the fact that she is hesitating and rephrasing her question about whether they will get anything to eat. She is perhaps searching for a way to formulate herself so that he does not have to give a lengthy answer, given the presumed inappropriateness of having a mobile phone conversation in class.

In this conversation, Nicky and the person in the classroom struggle with what it means to be in a classroom, and what sort of activities are appropriate in such a place. On the one hand, the answerer says that it is not problem for him to have a conversation during class, on the other hand, the conversation proceeds in a way that seems sensitive to place and situation.

4.3 An Inappropriate Place to Talk

In the previous excerpt, we saw how a mobile phone conversation was conducted in a potentially inappropriate place, and how the participants cooperated to make place for the conversation in this setting, using a set of conversational strategies. In the next excerpt, the answerer is in a place that also may seem inappropriate, namely a fitting room. It shows how the caller and answerer negotiate about the appropriateness of a place in terms of conducting a mobile phone conversation.

The answerer, Nicky, is in a fitting room talking to someone co-present, when she receives and answers the call. The answerer's first turn seems to have multiple recipients; she seems to orient herself to more than one listener. The utterance "I think that was nice yeah hi", has two parts. The first ("I think that was nice") presumably is meant for the other(s) present with her in the fitting room or in the shop. The second part ("yeah hi") is presumably meant for the caller. However, it might be more complex than this. The fact that the caller has access to the entire first turn makes it possible for the caller to use this as a resource. In hearing "I think that was nice", the caller can draw some conclusions about the location and activity in which the called is engaged. It might also be that the utterance is designed to give the caller this back ground information. This could then be a way of showing that she is already engageds in a conversation with someone co-present, meaning that she is busy. Also, if she

Nicky:	I think that was nice yeah hi
	((sound of door closing))
Richard:	What?
Nicky:	Hi
Richard:	Hi hi
Nicky:	But I can't talk now (.) cause [I'm:: in a fitting+
Richard:	[oh yeah what should I do about that then
Nicky:	E:hehehe:
Richard:	What are you doing then?
	(0.3)
Nicky:	I'm si+ I'm standing in a fitting room and trying on clothes
Richard:	Oh yeah:::
Nicky:	M
Richard:	Oh yeah:::
Nicky:	Yes!
Richard:	Yes (.) AND?
Nicky:	Heh I'm calling you later
Richard:	No you don't at all I'm not home h
Nicky:	Oh yeah oh well I'm [calling you tomorrow then
Richard:	[you'll have to call the mobile then
Nicky:	Yes
Richard:	Yes
Nicky:	Yes
Richard:	Hi
Nicky:	Hi

Excerpt 3. The caller and answerer negotiate the appropriateness of a fitting room as a place for conducting a mobile phone conversation

wants to get the conversation on a closing track from the beginning, letting the caller hear this piece of talk could be a strategy of displaying her unavailability. So, the access to background noise might play an important role in settling the suitability to continue a phone call. In this conversation, the caller could immediately in the opening of the conversation, potentially get some clues that the caller was busy, which he then, in a fashion that increasingly annoyed the answerer, chose to ignore.

After the greeting sequence, the answerer's first thing to say is that she "can't talk now". She thus tries to initiate a closing of the conversation in the beginning of the conversation. In line with the argument in Button, she is trying to place the conversation on a closing track [6] by saying that she will call him later, thus making arrangements for the future. Button identified this specific topic as being one used to begin the closing of a conversation. However, the caller is not cooperative in this matter. It takes Nicky quite a few turns after having initiated the closing, before she can actually get out of the conversation, and end the call. She says explicitly that she is unavailable for having a conversation - "I can't talk now" and begins her explanation to why she cannot do this "I'm in a fitting room". The caller does not seem to hear her explanation; just that she cannot talk right now. The question "what are you doing then?" seems to imply that he wants a good explanation for why she cannot talk to him right then. The second time she explains why she cannot talk; she does this by giving both location ("I'm standing in a fitting room"), and activity ("and trying on clothes").

In the beginning of the phone call, Nicky seems amused by the fact that she is answering while being in a fitting room, but as the conversation develops and she has difficulties ending the conversation, she seems more and more annoyed. Although this caller might have been unusually unwilling to cooperate, it is interesting to see how the called tries to get out of the conversation by saying what she is doing, and how this is treated by the caller.

The main point with this excerpt in relation to the topic of the paper, is that it shows how the caller and answerer negotiate the appropriateness of a place in terms of conducting a mobile phone conversation. In line with the argument of the case of the conversation in the classroom, this excerpt gives insights into the notion of what type of activities belongs in a certain place. Nicky shows quite vividly that she does not consider a fitting room an appropriate place to talk. However, it is clear that the caller does not agree, by his reaction when she says where she is. The conclusion that they do not agree on whether this situation is appropriate or not for having a conversation, clearly adds to the complexity of designing technology which provides the caller with contextual information about the called party.

5 Discussion

The excerpts from naturally occurring mobile phone conversations illustrate how the conversations have been dealt with, despite the difficulties and constraints set by the places and ongoing activities. The conversations have taken place in: a car, a class room, and a fitting room, all places where it is clear that other activities than talking on the mobile phone are normally going on.

5.1 Mobile Phone Talk in Context

The possible variation in places and activities attached to mobile phone use, contribute to the complexity of understanding mobile phone talk. Place and activity are important for understanding the appropriateness to conduct mobile phone conversations.

We have shown how the car is treated as an appropriate place for having a mobile phone conversation. There can be several reasons why this is so. First, the car is a place where talk can be carried out without interruption. The car in this sense is a private, secluded area. On the other hand, the car moves within a public domain, the road area, and the driver can be held accountable for actions within this place. The conversations however, are private within the car.

As a contrast to the car-example, this excerpt also shows us that a restricted secluded area is not necessarily always suitable for a conversation. The fitting room was not taken as appropriate for having a conversation by the answerer. This can be because of the activity taking place in here is private (changing clothes, grooming) as opposed to the more public activity of driving a car. Further, another explanation is that trying on clothes is difficult to do while talking on the mobile phone, simply because there is a need for holding the phone or keeping the headset adjusted, whereas it has been shown how drivers can handle the car while talking on the phone with ease. Also, fitting rooms ordinarily do not provide people with soundproof walls, so that the conversation taking place there is not necessarily private; other people can listen in. On the other hand, many people have conversations in public.

The study has provided us with insights on how to reach a mutual agreement on the appropriateness to continue the conversation over the mobile phone, despite the situation at hand. In the example from the study on the salesman in the car, we see how the remote part clearly understands that Eric is sitting in his car, probably driving, and how they despite or maybe even because of this, agree on the car as appropriate for talking. However, in the example of the call to the person in the classroom, the answerer states that he is in the classroom, but this is not a problem. Nevertheless, the caller clearly hesitates after hearing this, and the conversation continues a bit stumbling. The situation is a bit different in the fitting-room example. The answerer tries to end the conversation, without immediate success. The remote caller continues to talk, and seems to think it is okay to keep the conversation going. The lesson is that it is not always the case that people reach an immediate understanding on the appropriateness of conducting a mobile phone conversation, and people can have different opinions.

5.2 Challenges for the Design of Context-Aware Mobile Phones

The context-aware systems we focus on in this paper attempt to provide remote users with contextual information. In specific, these systems aim at settling the appropriateness for conducting the calls. Our findings can be used to pose some further challenges for the design of context-aware mobile phones.

First, it becomes clear that places and activities, which might seem inappropriate for conducting mobile phone conversations, are not always treated as such. Excerpt 1 and 3 illustrate how one party of the conversation cannot talk at the moment but still takes the call, thus being able to reach an agreement on when to return the call. This can also be a way to check the urgency of the call, and whether it should take precedence for the current activity. In any case, the calls are answered in situations that might have been considered as "uninterruptible" by a system, and the fact that the conversations can take place despite this, allows the participants to reach an agreement on when to continue talking.

Second, as we have seen, it is not only up to one single participant to decide upon the appropriateness for a conversation. The conversationalists do not always come to joint agreement on whether a place and activity is appropriate or not. Rather, it is an ongoing negotiation-work between the conversationalists. The data illustrate the difficulties in deciding, and defining in advance, the factors that influence the appropriateness of a call.

In general we believe that the design of context-aware mobile phones would gain from studies on everyday mobile phone use. The main problem with systems providing remote awareness is how they rely on the idea that it is possible to build into the design an understanding of what is an 'uninterruptible activity'. Our study illustrates how the participants reach a mutual understanding of a situation as appropriate for having a mobile phone conversation. This is not something which easily can be set in advance. The empirical data reinforces Dourish [9] arguments on context as not just being there; rather it is actively produced, maintained and enacted in the course of the activity at hand.

6 Conclusion

In this study, we have taken a closer look on the ways in which a place is interactionally constituted as appropriate for having a mobile phone conversation. The empirical data illustrate a number of instances on how mobile phone conversations are treated as part of the activities in which the answerer, or caller, is involved.

First, we have shown examples of how the appropriateness of having a conversation in the place where the answerer or caller is, is negotiated and discussed, in more or less explicit fashion, by the participants. Places and situations that might seem inappropriate for conducting mobile phone conversations are not always treated as such. Furthermore, the participants do not always agree on whether a place is appropriate or not, leading to difficulties.

Second, we have demonstrated how a mobile worker uses the car as his main place for taking and making mobile phone calls, and how his contacts orient to the car as a place for calling this particular person. The conventional view is that the car is not an accepted place for mobile phone use, seeing that it is banned in several countries (however not in Sweden). Our findings support previous studies [10, 16] in showing that drivers adjust the mobile phone conversations to the traffic situation.

Third, we have presented a discussion on the challenges involved when studying mobile phone use as part of mobile activities. We have argued the benefit of combining audio-recordings with video-recordings and ethnographic observations, thus making it possible to study how the called party treats the incoming call and deals with the local constraints of the setting. The additional contextual data provided using these methods contribute to the understanding of the conversations.

Fourth, in relation to the design of context-aware mobile phones, the empirical data point to the challenges in deciding and defining in advance the factors that influence the appropriateness of a call. Calls are answered in situations which might have been considered as "uninterruptible" by a context-aware mobile phone, and conversations take place despite this. The challenge therefore, becomes to design systems which allow for negotiations, and support the context work taking place in mobile phone talk.

References

1. Ackerman, M.S., D. Hindus, S.D. Mainwaring, & B. Starr (1997) Hanging on the 'Wire: A Field Study of an Audio-Only Media Space. In ACM Transactions on Computer-Human Interaction, vol. 4, no. 1, 1997, (39-66).
2. Arminen, I. & Leinonen, M. (2004) Mobile phone call openings – tailoring answers to personalized summons. Manuscript.
3. Bellotti, V. & S. Bly (1996) Walking Away from the Desktop Computer: Distributed Collaboration and Mobility in a Product Design Team, in Proc. of CSCW'96, ACM Press.

4. Bergqvist, J. & Ljungberg, F. (2000). ComCenter: A person oriented approach to mobile communication. In Proc. of CHI'00, ACM Press. (123-124).
5. Brave, S., Ishii, H. & Dahley, A. (1998). Tangible Interfaces for Remote Collaboration and Communication. In Proc. of CSCW'98, ACM Press. (169-178).
6. Button, G. (1991). Conversation-in-a-Series. In Talk and Social Structure: Studies in Ethnomethodology and Conversation Analysis, (Eds.) D. Boden, D.H. Zimmerman. Polity Press, Cambridge. (251-277).
7. Button, G. & Casey, N. (1984). Generating topic: the use of topic initial elicitors. In Structures of Social Action: Studies in Conversation Analysis. (Eds.) J.M. Atkinson, J. Heritage. Cambridge: Cambridge University Press. (167-190).
8. Dourish, P. & Bellotti, V. (1992). Awareness and Coordination in Shared Work Spaces. In Proc. of CSCW'92, ACM Press. (107-114).
9. Dourish, P. (2004). What We Talk About When We Talk About Context. In Journal of Personal and Ubiquitous Computing. 8(1). (19-30).
10. Esbjörnsson, M. & Juhlin, O. (2003). Combining Mobile Phone Conversations and Driving: Studying a Mundane Activity in its Naturalistic Setting. In Proc. of ITS'03, Ertico.
11. Grinter, R. E. & Palen, L. (2002). Instant Messaging in Teen Life. In Proc. of CSCW'02, ACM Press. (21-30).
12. Heath, C. & P. Luff (1991) Disembodied Conduct: Communication through Video in a Multi-Media Office Environment. In Proc. of CHI'91, ACM Press, (99-103).
13. Hill, J. & Gutwin, C. (2003). Awareness support in a groupware widget toolkit. In Proc. of Group'03, ACM Press. (258-267).
14. Hutchby, I. & Barnett, S. (forthcoming). Aspects of the Sequential Organisation of Mobile Phone Conversations. To appear in Discourse and Society. Vol 15, 2005.
15. Laurier, E. (2001). Why people say where they are during mobile phone calls. In Environment and Planning D: Society & Space. (485-504).
16. Laurier, E. (2002). Notes on dividing the attention of a car driver. Team Ethno Online.
17. Ling, R. (2000). "We will be reached": The use of mobile telephony among Norwegian youth. In Information technology and people 13(2). (102-120).
18. Luff, P. & C. Heath (1998) Mobility in Collaboration. In Proc. CSCW'98, ACM Press.
19. Manalavan, P., Samar, A., Schneider, M., Kiesler, S. & Siewiorek, D. (2002). In-Car Cell Phone Use: Mitigating Risk by Signaling Remote Callers. In Proc. of CHI'02, ACM Press. (790-791).
20. Milewski, A.E. & Smith, T.M. (2000). Providing Presence Cues to Telephone Users. In Proc. of CSCW'00, ACM Press. (89-96).
21. Palen, L., Salzman, M. & Youngs, E. (2000). Going Wireless: Behavior and Practice of New Mobile Phone Users. In Proc. of the CSCW'00, ACM Press. (201-210).
22. Palen, L. & Salzman, M. (2002) Voice-Mail Diary Studies for Naturalistic Data Capture under Mobile Conditions. In Proc. of CSCW'02, ACM Press. (87-95).
23. Puro, J.P. (2002). "Finland: A mobile culture" in Perpetual contact: Mobile communication, private talk, public performance, (Eds.) Katz, J. & M. Aakhus. Cambridge University Press, Cambridge. (19-29).
24. Schegloff, E., (1979). "Identification and Recognition in Telephone, Conversation Openings" In Everyday Language: Studies in Ethnomethodology (Ed) G. Psathas. New York: Irvington Publishers Inc. (23-78).
25. Siewiorek, D., Smailagic, A., Furukawa, J., Krause, A., Moraveji, N., Reiger, K., Shaffer, J. & Wong, F. (2003). SenSay: A Context-Aware Mobile Phone. In Proc. of ISWC'03.
26. Taylor, A. & Harper, R. (2002). Age-old practices in the 'New World': A study of gift-giving between teenage mobile phone users. In Proc. of CHI'02, ACM Press. (439-446).

27. Trbovich, P. & Harbluk, J. L. (2003). Cell Phone Communication and Driver Visual Behavior: The Impact of Cognitive Distractions. In Proc. of CHI'03, ACM Press. (728-729).
28. Weilenmann, A. & Larsson, C. (2001). Local Use and Sharing of Mobile Phones. In B. Brown, N. Green, and R. Harper (Eds.) Wireless World: Social and Interactional Aspects of the Mobile Age. Springer-Verlag. (99-115).
29. Weilenmann, A. (2003). "I can't talk now, I'm in a fitting room": Availability and Location in Mobile Phone Conversations. In Environment and Planning A, vol. 35, (9), (1589–1605).

Unsupervised Clustering of Context Data and Learning User Requirements for a Mobile Device

John A. Flanagan

Nokia Research Center, PO Box 407,
FIN-00045 Nokia Group, Finland
Tel: +358 50 483 6225, Fax: +358 7180 36855
adrian.flanagan@nokia.com

Abstract. The K-SCM is an unsupervised learning algorithm, designed to cluster symbol string data in an on-line manner. Unlike many other learning algorithms there are no time dependent gain factors. Context recognition based on the fusion of information sources is formulated as the clustering of symbol string data. Applied to real measured context data it is shown how the clusters can be associated with higher level contexts. This unsupervised learning approach is fundamentally different to the approach based, for example, on ontologies or supervised learning. Unsupervised learning requires no intervention from an outside expert. Using the example of menu adaptation in a mobile device, and a second learning stage, it is shown how user requirements in a given context can be associated with the learned contexts. This approach can be used to facilitate user interaction with the device.

1 Introduction

As the technology of mobile devices, such as mobile phones and personal digital assistants (PDAs), advances they have increased functionality and a capacity to receive and transmit large amounts of data. This increased functionality and data needs to be managed on such devices by the user, based on a constrained and slowly evolving User Interface (UI) (i.e. small screen and limited keypad) which can be quite onerous. Context awareness on a mobile terminal can be used to facilitate the interaction of the user with the terminal by recognizing the user context and determining and reacting to the user's needs in a given context, thus minimizing the user's interaction with the device through the UI.

One approach to context recognition is for an "expert" to define contexts and user needs in those contexts. Context awareness based on ontologies could be considered as one such approach [1], [2], [3], [4]. Learning is another possibility for defining contexts, however the approaches used in [5], [6], [7] are ultimately based on supervised learning which requires the intervention of an expert, or the user, at some point to label contexts or define the user needs in a given context. Furthermore there is still the problem that the learning approach cannot generalize beyond the training data which means either a very large collection of training data or the problem of adapting the learning, through a time dependent gain function, over a long period of time.

In what follows, as a guiding principle, it is assumed that if context awareness is to be useful then it must be possible to personalize it for each user and their personal mobile device. In both the ontology and supervised learning approaches, personalization most likely requires intervention by the user which in the case of the mobile devices discussed above serves to increase rather than decrease the functional complexity of the device. An unsupervised, continuous learning, approach to context recognition is now described. Furthermore a second learning stage is used to learn associations between contexts and user needs in those contexts, without explicit user intervention.

The Symbol String Clustering Map (SCM) [8] was presented as an algorithm for the unsupervised clustering of symbol string data based on adaptive learning. The analysis of sensor data for context recognition using the SCM has been described in [9]. The SCM was based on the classical learning paradigm mentioned above which more explicitly includes the assumption of a fixed learning period and a random sampling of the input data. Both of these assumptions are potentially limiting in real situations as it is difficult to predict how long the learning period for an unsupervised learning algorithm should be in order to sample a large number of potential inputs and also to have a random sampling of the input data. In what follows a modified version of the SCM algorithm is described and analyzed, now referred to as the K-SCM. The K-SCM does not require a random sampling of data and does not have any defined learning period or learning parameters that need to vary over time in any predefined manner.

In order to describe the K-SCM algorithm and understand what type of context learning it performs it is important to understand the input to the K-SCM and how it is generated. In Sec. 2 the model of the input to the K-SCM is described, in brief the K-SCM algorithm performs information fusion of different sources. The basic model of the K-SCM algorithm for performing information fusion in a real application is described in Sec. 3. In Sec. 4 the application of the K-SCM to context recognition, which is equated here with cluster discovery, using real measured data is described. The data [10], [11] has been recorded in a mobile environment and the processing of the data and its exploitation corresponds to what would happen in a real mobile device. In this example learning in the K-SCM is initiated by user interaction with the device, such as starting applications, accessing WEB pages etc., which means the learned contexts are those when the user interacts with the device. Using a second learning algorithm described in Sec. 5 the applications etc. that initiated learning are associated with those learned contexts. In brief, contexts where the user interacts with the device are learned and those interactions are associated with the learned contexts. The interactions can then be presented to the user, for example, in context specific short cut menus, each menu having a list of applications, WEB pages etc. useful to the user in that context. This approach indicates one means of exploiting the results of an unsupervised clustering algorithm such as the K-SCM. Section 6 has the conclusion.

2 Information Fusion and Context

In this section the context recognition problem is addressed as a problem in information fusion. In the basic assumption there are M information sources $\Psi_i, i = 1, \ldots, M$, with each source i having the possibility of being in one of the states

$$\Psi_i(t) \in \{\psi_{i,1}, \psi_{i,2}, \ldots, \psi_{i,m_i}\}, \quad m_i < \infty, \tag{1}$$

at time t. For example one of the sources Ψ_i could be the location information given by the Cell Identification (ID) code of the GSM phone network, where each state $\psi_{i,j}$ represents a distinct Cell ID. Another source Ψ_k could be the time, with each $\psi_{k,j}$ representing the hour.

At a given time it is possible to observe all the states of the information sources $\Psi(t)$ as,

$$\Psi(t) = \Big(\Psi_1(t), \Psi_2(t), \ldots, \Psi_M(t)\Big). \tag{2}$$

We have two possible interpretations of the *symbol string* $\Psi(t)$; first it defines a fusion of the information sources at a given time, second in terms of context we say it is a high level context, relative to the context represented by the states of the sources $\Psi_i(t)$. The K-SCM algorithm presented in the next section processes this type of symbol string data. The K-SCM essentially identifies clusters in the "noisy" symbol string data which are then interpreted as higher level contexts. In terms of modelling the probability distribution $P_\Psi(\Psi)$ of the Ψ, a classical mixture model of the symbol strings is used [12], with

$$P_\Psi(\Psi) = \sum_{j=1}^{K} \pi_j\, p_j(\Psi|\theta_j), \tag{3}$$

where $p_j(\Psi|\theta)$ is a uni-modal probability distribution with parameters θ_j. $P_\Psi(\Psi)$ is assumed to be a K-modal probability distribution which in this case implies there are K clusters or contexts in the symbol string data. The parameters π_j are the mixture probabilities such that $\sum_{j=1}^{K} \pi_j = 1$. However this is a static model and does not necessarily reflect the time evolution of a user's context as might be expected. For example if the user is in context i at time t, it is likely that $\pi_i \gg \pi_j, j \neq i$ and when the user is in context $k \neq i$ then $\pi_k \gg \pi_j, j \neq k$. This requires the parameters π_i to vary with time, resulting in a new form for the mixture model of (3),

$$P_\Psi(\Psi, t) = \sum_{j=1}^{K} \pi_j(t)\, p_j(\Psi|\theta_j), \tag{4}$$

where it is assumed here that the $p(\Psi|\theta_j)$ do not vary with time. The exact variation of π_i with time is not discussed any further, but the assumption is the user remains in a given context for a period of time, for example if the context is based on the fusion of time and location information. Figure 1 shows an illustration of how the variation of the $\pi_i(t)$ might be modelled for context

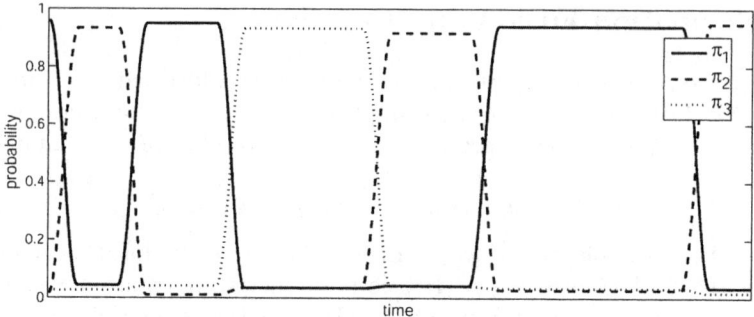

Fig. 1. Variation of the mixture probabilities $\pi_i(t), i = 1, \ldots, 3$

data in the case of $K = 3$. This dynamic model of the variation of $P_\Psi(\Psi, t)$ better reflects the type and nature of the context data used later in section 4.

3 The K-SCM Algorithm

The structure of the K-SCM algorithm described here is similar to the structure of the SCM algorithm described in [8], [13]. The operation of the SCM and K-SCM is similar in that for an input symbol string $\Psi(t)$ there are two basic operations. The first is to determine the node in the K-SCM which best matches $\Psi(t)$. The best matching node is referred to in Artificial Neural Network (ANN) terminology [14], as the Winner Take All (WTA) node or the winner. The second operation involves updating the winning node in such a manner that it better matches $\Psi(t)$. However there is also a fundamental difference in the way the learning is performed in the K-SCM. In the SCM [13] it was necessary to perform a segmentation of the recorded context data using the segmentation algorithm described in [15] and choose a fixed number of random samples of the context in the identified segments, independent of the length of the segment. The reason for this is related to the underlying assumption in typical unsupervised learning algorithms, such as the Self-Organizing Map (SOM) [16], that the input data to the learning is randomly sampled over the input space of the data. This assumption can be interpreted as a requirement that the probability distribution of the data be static, as described in (3), whereas in the current case the probability distribution of the data is assumed dynamic and given by (4). The determination of the winner node for an input $\Psi(t)$ and the update of the node are described in what follows, but first a description of the K-SCM structure and the terminology used.

In its most general form the K-SCM consists of N nodes and associated with each node is an entity representing some feature of the input data set. This entity for node i consists of a symbol string S_i and an associated weight vector W_i. Formally S_i is defined as,

$$S_i = [s_i^1, s_i^2, \ldots, s_i^M] , \quad (5)$$

where each s_i^j is in turn a symbol string associated with information source j. Each s_i^j is given by,

$$s_i^j = (s_{i,1}^j, s_{i,2}^j, \ldots, s_{i,m_i^j}^j),\qquad(6)$$

where $s_{i,k}^j \in \Psi_j$ one of the states of information source Ψ_j. It is also true that $s_{i,w}^j \neq s_{i,v}^j$, $v \neq w$. Similarly the weight vector W_i is defined as,

$$W_i = [w_i^1, w_i^2, \ldots, w_i^M].\qquad(7)$$

Each w_i^j in turn is given by,

$$w_i^j = (w_{i,1}^j, w_{i,2}^j, \ldots, w_{i,m_i^j}^j),\qquad(8)$$

where $w_{i,k}^j \in [0,1]$ a scalar. Each weight $w_{i,k}^j$ is associated with symbol $s_{i,k}^j$ in the symbol string. Associated with each node i are the state variables V_i^p, $V_i^T \in [0,1]$ and $\beta_i \in (0,1)$ defined and discussed later on. The method of determining the winning node is significantly different from the method used in [8], and indeed in many other ANN algorithms, where the similarity measure is static, unlike the measure in this case which is dynamic.

For a given input symbol string $\Psi(t)$ the first step is to determine the winning node. This requires calculating the intermediate parameters G_i^+ and G_i^- for each node i where,

$$G_i^+ = \sum_{j=1}^{M} \sum_{k=1}^{m_i^j} \delta_{i,k}^j w_{i,k}^j,\qquad(9)$$

and

$$G_i^- = \sum_{j=1}^{M} \sum_{k=1}^{m_i^j} (1 - \delta_{i,k}^j) w_{i,k}^j,\qquad(10)$$

with $\delta_{i,k}^j$ defined as, $\delta_{i,k}^j = 1$ if $s_{i,k}^j = \Psi_j(t)$ and $\delta_{i,k}^j = 0$ if $s_{i,k}^j \neq \Psi_j(t)$. In brief, G_i^+ corresponds to the sum of the weights associated with node i, whose corresponding symbols appears in the input $\Psi(t)$. G_i^- corresponds to the sum of the weights associated with node i, whose corresponding symbols do not appear in the input $\Psi(t)$. The variable V_i^p is initialized as $V_i^p = \phi \in [0.05, 0.08]$, a random number and initially $\beta_i = 1.0$. A series of iterations through all the nodes is made, at each iteration τ the value of $V_i^p(\tau)$ for each node is evaluated as,

$$V_i^p(\tau) = V_i^p(\tau-1) + 0.05\Big(G_i^+ - V_i^p(\tau-1).(G_i^- + G_i^+)\Big).\qquad(11)$$

If $V_i^p(\tau) < 0.05$ then $V_i^p(\tau) = \phi$. At each iteration $V_i^p(\tau)$ is compared with $\beta_i.V_i^T$, if $V_i^p(\tau) > \beta_i.V_i^T$ then node i is the winner node. If node i is not the winner node and $G_i^+ - V_i^p(\tau-1).(G_i^- + G_i^+) < \epsilon \approx 0.05$ then $\beta_i = \beta_i.\gamma$ with $\gamma \in (0,1)$ (e.g. $\gamma = 0.97$). The iteration is repeated over all nodes until a winner node is found at iteration τ_0. While not demonstrated here, it can be shown with the conditions described, a winner node is always found in a finite time, independent of the input and the state of the nodes.

Assume node v is the winning node at iteration τ_0 then it is updated as follows. For each $\Psi_j(t) \in \Psi(t)$, that is the symbol associated with the current state of source j, the following steps are carried out:

- If $\Psi_j(t) \in s_v^j$ then k is the index of $\Psi_j(t)$ in s_v^j.
- If $\Psi_j(t) \notin s_v^j$ then insert $\Psi_j(t)$ into s_v^j for example at position k and insert a small weight (i.e. ≈ 0.04) at position k of w_v^j. This also implies $m_v^j \to m_v^j + 1$.
- In both cases update the weights $w_{v,k}^j$ of w_v^j as follows, with $\alpha_1 \in [0,1]$ and $0 < \alpha_0 \ll \alpha_1$ constant gain factors,

$$w_{v,k}^j = w_{v,k}^j + \alpha_1(1.0 - w_{v,k}^j).(1.0 - w_{v,k}^j), \qquad (12)$$

$$w_{v,l}^j = w_{v,l}^j + \alpha_0(1.0 - w_{v,l}^j).(0.0 - w_{v,l}^j), \quad l = 1,\ldots,m_v^j, l \neq k. \quad (13)$$

- For $l = 1,\ldots,m_v^j$ if $w_{v,l}^j < \rho$ then $s_{v,l}^j$ is removed from s_v^j and $w_{v,l}^j$ is removed from w_v^j. ρ is defined as a very small fixed threshold (i.e. $\rho = 0.002$), the same for all nodes.
- The threshold V_v^T is adapted such that for $V_v^p(\tau_0) > V_v^T$

$$V_v^T = V_v^T + \gamma_1\left(V_v^p(\tau_0) - V_v^T\right) \qquad (14)$$

or else for $V_v^p(\tau_0) < V_v^T$ and $V = \max[0.2, V_v^p(\tau_0)]$ then

$$V_v^T = V_v^T + \gamma_0\left(V - V_v^T\right), \qquad (15)$$

with $\gamma_1 \approx 0.9$, $\gamma_0 \approx 0.001$. The most notable feature of the learning in the K-SCM is there is no learning rate that needs to be adapted in a predetermined fashion over time, all gain factors used in the learning are constant. This has very important advantages when applying such an algorithm in a real situation with real data. Based on the description of the algorithm it is not obvious that it can extract clusters from symbol string data. In the next section a real context data set, recorded in a mobile environment, is analyzed using the K-SCM and from the results it is more obvious what is happening.

4 Analyzing Time and Location with the K-SCM

In this section the data collection described in [10] and publicly available at [11] is processed using the K-SCM. A complete description of the data used is available in [11], but in brief, the data set consists of a set of feature files for 43 different recording sessions. In each recording session the same user carried a mobile phone, sensor box and laptop PC, going from home to the workplace or vice-versa. During the journey the user walks, takes a bus and Metro and sometimes uses a car. On occasion slightly different or very different routes/modes of transport are taken. During the session, sensors recorded 3-axis acceleration, atmospheric pressure, temperature, humidity etc. The ambient audio was recorded on the laptop using a microphone and sound card. On the mobile phone, changes in the user's location were recorded as Cell ID and Location Area Code (LAC)[1] from the GSM network. Furthermore all user interaction with the mobile phone such as calls, SMS, WEB pages accessed, applications started etc. were logged and time stamped. After each recording session all recorded signals were processed and different features extracted. For example the audio signal energy was averaged over a 10 second interval and the average sampled every second. The accelerometer signal magnitude was averaged and sampled every second to give the user activity. The dynamic range of all features was divided into different levels and each level numbered. For example the temperature range was divided into 10 levels, the user activity into 6 levels. At every second of the session every feature was quantized into its corresponding current level and denoted by an integer which is in fact the symbol. Note the numerical value representing the quantized state of the feature is not related in any way to the feature itself, it is purely a representation. Figure 2 shows an extract from one of the session files with the state of the features of each source on the same line, one line for each second. Each user interaction with the terminal was time stamped and recorded in a separate interaction file.

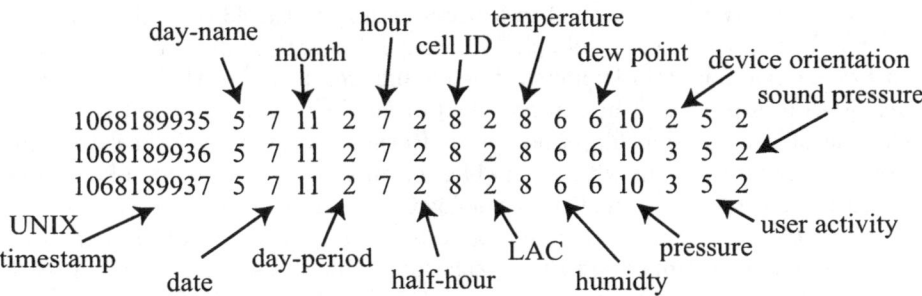

Fig. 2. Extract from a session feature file indicating the feature represented in each column

In the example discussed here the input to the K-SCM consists of a subset of these features, with $M = 4$ sources relating to the time and location information which are available in currently available mobile devices. The m_i of (1) are given by $m_1 = 4$ (i.e. morning, afternoon, evening, night), $m_2 = 24$ (i.e every hour), $m_3 = 2$ (i.e. hour:15-hour:45), $m_4 = 6$ (i.e. different LAC's), corresponding to a symbol string of the form

$$("day\ period", "hour", "half\ hour", "LAC").$$

---

[1] In an urban area a cell may have a radius of several 100 meters with a LAC consisting of several cells.

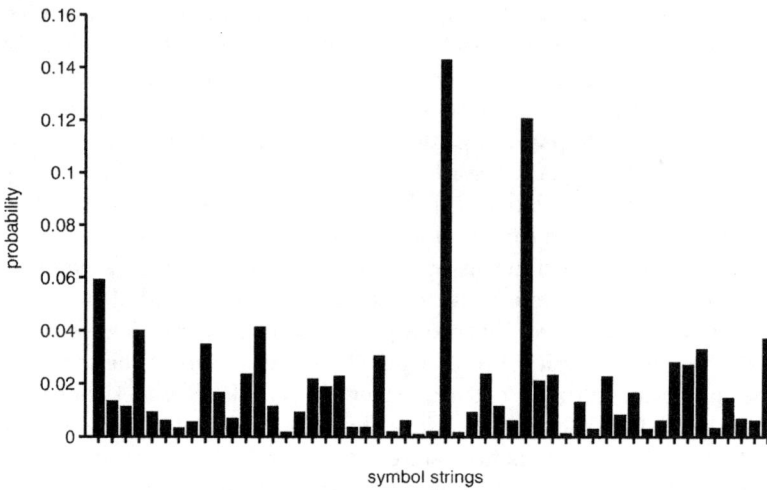

Fig. 3. Histogram, with each bar representing a distinct symbol string of time and location features, of all the inputs used in training the K-SCM

A K-SCM with $N = 25$ was used to process data from the 43 recording sessions. This number of nodes was used mainly because it is easy to illustrate the results and for the example data it offers a fine enough resolution of the data clusters. Each session is processed in chronological order. For reasons related to Sec. 5, each line of each session file is not used. Rather the learning is based on the recorded user interactions with the mobile terminal. The time stamp of each line of each session file is scanned in chronological order. If at a given time stamp there is a user interaction with the mobile terminal, recorded in the interaction file, with the same time stamp then this line from the session file along with the next 20 lines (i.e. 20 seconds) are used as input to the K-SCM algorithm. The reasoning behind this approach is that all user contexts are not learned, rather only contexts where the user interacts with the terminal in some way and hence only the contexts relevant to the interaction of the user with the terminal. This also serves to reduce the memory and computational requirements of the application.

Figure 3 shows a normalized histogram of the distribution of all the inputs used during the training, where each bar represents the probability of a distinct symbol string being used in the training. There are a total of 50 distinct input symbol strings. Figure 4 shows an illustration of the symbol strings of the K-SCM after two passes through the set of 43 recording sessions, where the initial starting condition of the K-SCM was randomly initialized symbols and weights. The reason for basing this statistic on the second run is to allow the SCM to converge to some stable state from its random initialization and it makes the presentation easier as the initial, unnecessary symbols are pruned from the symbol strings. For the sake of presentation the K-SCM is presented as a 5×5

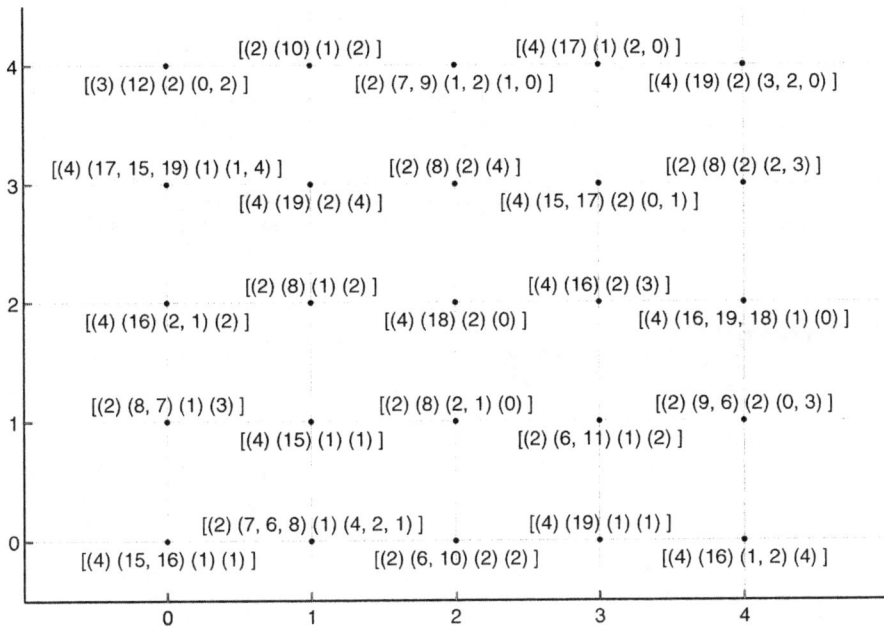

Fig. 4. Illustration, in grid form, of the symbol strings associated with the K-SCM after learning with time/location data

grid, with each point of the grid representing a node. The symbol string S_i, of (5), associated with each node i is enclosed in '[]'s. Inside these brackets are the source symbol strings s_i^j, of (6), enclosed in '()'s. Similarly Fig. 5 shows the associated weight vectors of each node with the W_i of each node enclosed in '[]'s and the w_i^j of each source j enclosed in '()'s.

Figure 6 shows the same distribution of the symbol strings as in Fig. 3 but only for symbol string inputs with a probability ≥ 0.025. On top of each bar, in '[]' brackets, are the indices of the node in the lattice of Fig. 4 that was most often the winner node when this symbol string was input during the second run through the training set. Above each node number, in () brackets, is the fraction of times this principle node was the winner node for this input. For example, symbol string $(4, 15, 1, 1)$ is the most dominant input and node $[1, 1]$ was the winner for this input every time during the second run. The next most dominant input is symbol string $(4, 16, 1, 4)$ with node $[4, 0]$ being the winner 0.99 of the time. The other 0.01 of the time which corresponds to 3 separate times it was assigned to other nodes, probably at the beginning of the second run. Similarly for all the other most dominant inputs shown in Fig. 6. By comparing the resulting symbol strings, weight vectors and histograms an insight into the functioning of the K-SCM can be obtained. Obviously the most dominant input $(4, 15, 1, 1)$ is uniquely represented at node $[1, 1]$ and all the corresponding weights are close to 1. On the other hand inputs $(4, 19, 1, 0)$ and $(4, 16, 1, 0)$ are represented by the same node $[4, 2]$ and while the weights associated with the common symbols

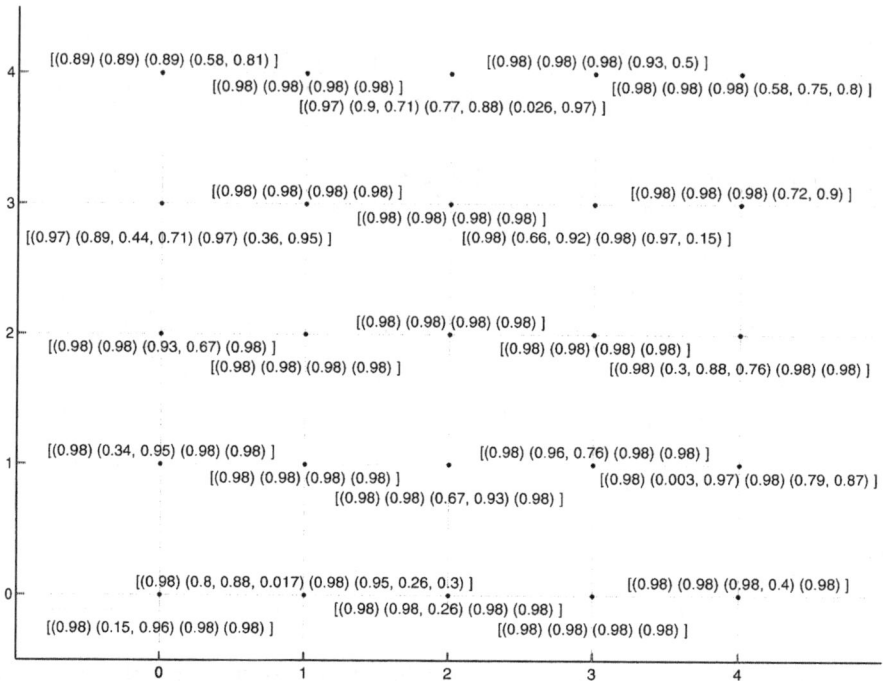

Fig. 5. Illustration, in grid form, of the weight vectors associated with the K-SCM after learning with time/location data

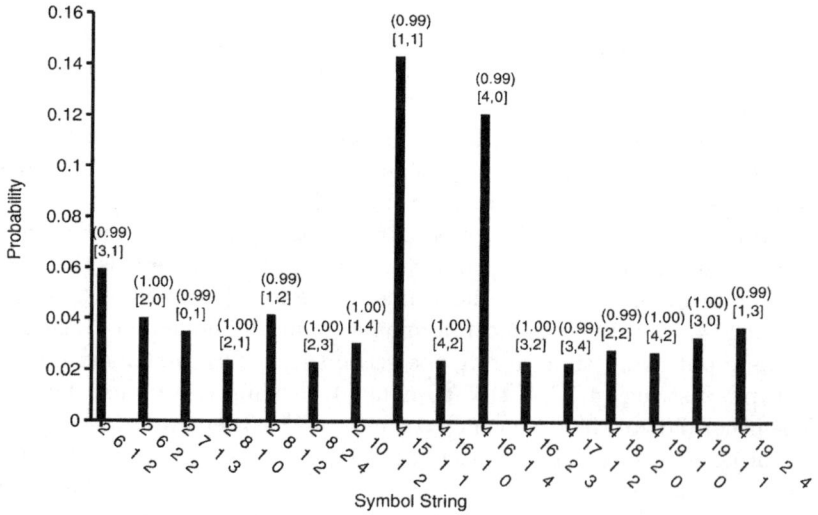

Fig. 6. Histogram of the symbol strings with probability > 0.025 used in the training of the K-SCM

4, 1, 0 are close to 1 those associated with 16, 19 are significantly < 1. The overall probability of the (4, 19, 1, 0) and (4, 16, 1, 0) strings are smaller relative to the probabilities of (4, 15, 1, 1) which may explain why (4, 15, 1, 1) and (4, 19, 1, 1), despite being very similar, are not represented by the same node, (4, 19, 1, 1) being uniquely represented at [3, 0]. On the other hand the input (3, 12, 2, 2), despite not even appearing in the histogram of Fig. 6, is represented at node [0, 4], probably due to the fact that it is quite different from the other symbol strings in the data set because of the 3, 12 symbols. This type of result indicates the K-SCM is performing a clustering of the input data as opposed to finding the most probable input symbol strings.

In terms of generating higher level context, based on Fig. 4, for example node [1, 1] represents a time/location context interpreted as being in location 1 at time 4, 15, 1 which corresponds to between 15:15–15:45 hours in the afternoon. On the other hand the context represented at node [4, 2] corresponds to being in location 0 in the evening around the times 16:00–19:00. While both higher level contexts have the same time/location information they are different in the time frame that they represent.

Using the K-SCM starting from different initial conditions, it shows a very robust behavior and while not converging to the exact same result each time, on analysis, it does consistently and robustly extract the principle clusters in the data. During learning one observed behavior is the splitting of clusters. For example a single node may initially represent several inputs, but during learning based on the number of repeated similar type inputs, the cluster can "split" which means the inputs represented by a single node are separated so that, for example, the inputs are split into two sub-groups and each sub-group is represented by two different nodes. Varying the number of nodes in the K-SCM, as expected varies the clusters identified. As the number of nodes decreases, in general, each node represents more inputs and vice-versa when the number of nodes increases.

Similar formation of clusters has been achieved using the data from the sensors such as the temperature, user activity etc. which are of a statistically very different nature from the time/location data. However such sensor data is typically quite noisy and not suitable for directly driving an adaptive menu, unlike the time and location information used here.

5 Context Sensitive Adaptive UI for a Mobile Device

In the previous sections it has been shown the K-SCM can be used to extract clusters from different forms of context information represented as symbol strings. Each cluster is defined as representing a distinct user context when the user interacted with the device, starting an application or accessing a WEB page etc. The principle of the K-SCM is unsupervised learning which means that no external supervision, for example from the user, is required during the learning. However this also means that it is impossible to know what the resulting K-SCM looks like or what exactly is represented where in the K-SCM. The obvious ques-

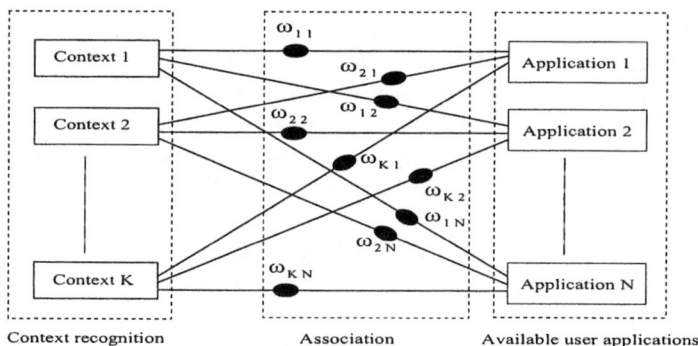

Fig. 7. An illustration of how to associate a user's use of different applications in different, recognized, user contexts

tion is how can the results of the K-SCM's context learning/recognition ability be exploited, for example on a mobile device, for the benefit of the user, without requiring a detailed post learning analysis of the resulting K-SCM.

One approach is now proposed that implicitly uses the user's interaction with the mobile device. Note, while this approach is applied to the K-SCM in this work it does not explicitly require the K-SCM to be the context recognition algorithm. Figure 7 shows an illustration of one means of implementing a method for relating user interaction with a mobile terminal with the user's context. It is assumed there are a set of K recognizable contexts. At any given time the recognition algorithm recognizes the user is in context j, if the user is in context j for a continuous period of time $[0, T]$ and the user uses applications $a_1, a_2, \ldots, a_l$ in context j then there is a weight $\omega_{ja_i} \in [0, 1]$ associated with each of these applications. At time T the weights are updated as,

$$\omega_{ja_i} = \omega_{ja_i} + \alpha \left(1.0 - \omega_{ja_i}\right) \quad i = 1, \ldots, l , \qquad (16)$$

where $0 < \alpha \ll 1$ is an adaptation parameter, assumed constant here. Also assume there are a set of weights $\omega_{jb_1}, \omega_{jb_2}, \ldots, \omega_{jb_m}$ for applications $b_1, b_2, \ldots, b_m$ which are not used by the user in the period $[0, T]$, then these weights are updated as,

$$\omega_{jb_i} = \omega_{jb_i} + \beta \left(0.0 - \omega_{jb_i}\right) \quad i = 1, \ldots, m , \qquad (17)$$

with $\beta \ll \alpha$. With this adaptation it is very simple to show that $\omega_{jk} \to a$, where $a \propto p_{jk}$, and p_{jk} is the probability that the user uses application k when in the recognized context j [8]. This leads to the possibility of an adaptive user interface on the mobile terminal that uses shortcut menus to different applications. Hence when the user is recognized as being in a context j, the shortcut menu presented to the user is determined by the weights ω_{jk} and the order of the applications in the shortcut menu is determined by the magnitude of the weights, with the most likely ones being top of the list in the menu. Hence if the user wants to use an application in a given context, the most likely application, based on previous experience, is the application at the top of the list that the user can very easily start up with few keypad presses.

Table 1. Table of ordered according to rank in ()'s of applications, WEB pages and communication types for context [1, 1] in Fig. 4

Application		WEB Page		Comm. Type	
Opera	(0.7845)	file://localhost/	(0.8473)	PacketData	(0.9575)
www	(0.4540)	http://news.bbc.co.uk/	(0.8430)	Shortmessage	(0.4464)
Messaging	(0.4531)	http://www.newscientist.com/	(0.5972)		
Locus	(0.2635)	http://my.opera.com/	(0.5920)		
SnakeEx	(0.2570)				

Such a learning mechanism was implemented for the K-SCM resulting from the learning in Sec. 4. The result of learning the user interaction is shown in Table 1 for context/node [1,1] in Fig. 4. The applications, WEB pages and communication types are listed in menus and ranked according to the weights ω_{jk}. The benefit of such an application is more difficult to evaluate, what is true however is that the the algorithm as described here does always provide the most likely applications at the top of the lists. In order to evaluate this adaptive menu application it would need to be implemented in a real mobile terminal and probably used by many different users. The basic principle could also be implemented to benefit the user in a more hidden mode, where the user is not presented with a shortcut menu of applications, but rather for example if in a given context the user is very likely to use the WEB browser, when this context is recognized, the browser could be initiated, to speed up the connection and initialization etc which would enhance the "feel" and "use" of the terminal for the user. This examples illustrates how it is possible to exploit the unsupervised learning of the K-SCM in a very simple and useful manner.

6 Conclusion

The K-SCM algorithm is presented as a means for on-line, unsupervised, clustering of symbol string data. On-line in the sense that there is no need to adapt gain factors over time, or perform other functions in order to achieve a random sampling of the data. Applied to the context recognition problem for information fusion the clusters generated by the K-SCM are associated with higher level contexts. In the specific example shown, the higher level context is based on the fusion of time and location information. As the learning in the K-SCM is unsupervised, in order for it to be exploited it must be interpreted in some way. The easiest means of interpretation is to learn associations between the learned contexts and different requirements in those contexts. For example in a context aware mobile terminal the applications used or WEB pages accessed by a user in a given context are associated with that context using a simple learning mechanism. Hence when the user is again in that given context the mobile terminal could offer a shortcut option to those applications/WEB pages. This means the unsupervised context recognition can be exploited, through a second phase of quasi-supervised learning, in a meaningful manner for the user without requiring any explicit user input.

Acknowledgements

This work has been performed in the framework of the IST project IST-2004-511607 MobiLife, which is partly funded by the European Union. The author would like to acknowledge the contributions of his colleagues, although the views expressed are those of the author and do not necessarily represent the project.

References

1. Chen, H., Finin, T., Joshi, A.: An ontology for context-aware pervasive computing environments. Knowledge Engineering Review **18** (2004) 197–207 Special Issue on Ontologies for Distributed Systems.
2. Wang, X., Zhang, D.Q., Gu, T., Pung, H.K.: Ontology-based context modeling and reasoning using OWL. In: Workshop on Context Modeling and Reasoning at IEEE International Conference on Pervasive Computing and Communication (PerCom'04), Orlando, Florida, US (2004)
3. Khushraj, D., Lassila, O., Finin, T.: sTuples: semantic tuple spaces. In: Mobile and Ubiquitous Systems: Networking and Services. (2004) 268–277
4. Korpipää, P., Mäntyjärvi, J.: An ontology for a mobile device sensor-based context awareness. In: Proc. Context03. LNAI no. 2680, Springer-Verlag (2003) 451–459
5. Schmidt, A., Aido, K.A., Takaluoma, A., Tuomela, U., van Laerhoven, K., van de Velde, W.: Advanced interaction in context. In: Proc. Intl Symp. Handheld and Ubiquitous Computing, LNCS 1707, Springer-Verlag (1999) 89–101
6. van Laerhoven, K., Cakmakci, O.: What shall we teach our pants? In: Proc. 4th Intl Symp. Wearable Computers, IEEE CS Press (2000) 77–83
7. Korpipää, P., Mäntyjärvi, J., Kela, J., Keränen, H., Malm, E.J.: Managing context information in mobile devices. Pervasive Computing, IEEE **2** (2003) 42–51
8. Flanagan, J.A.: Unsupervised clustering of symbol strings. In: Intl' Joint Conference on Neural Networks, IJCNN'03, Portland Oregon, USA (2003) 3250–3255
9. Himberg, J., Flanagan, J.A., Mäntyjärvi, J.: Towards context awareness using symbol clustering map. In: WSOM 03 (Intelligent systems and innovational computing), Kitakyushu, Japan (2003) 249–254
10. Flanagan, J., Murphy, D., Kaasinen, J.: A Nokia context recording database with synchronized user interaction. In: Benchmarks and a Database for Context Recognition: Workshop Proceedings Pervasive 2004, ETHZ, Switzerland (2004)
11. Mayrhofer, R.: (Context database) Available at, http://www.soft.uni-linz.ac.at/-Research/Context_Database/.
12. Hand, D., Mannila, H., Smyth, P.: Principles of Data Mining. MIT Press, Cambridge, US (2001)
13. Flanagan, J.A., Himberg, J., Mäntyjärvi, J.: Unsupervised clustering of symbol strings and context recognition. In: IEEE Intl' Conf. on Data Mining (ICDM02), Maebashi City, Japan (2002) 171–178
14. Haykin, S.: Neural Networks, A Comprehensive Foundation. 2nd edn. Prentice-Hall (1999)
15. Himberg, J., Korpiaho, K., Mannila, H., Tikanmäki, J., Toivonen, H.: Time series segmentation for context recognition in mobile devices. In: IEEE Intl' Conf. on Data Mining (ICDM2001). (2001) 203–210
16. Kohonen, T.: Self-Organizing Maps. 3rd edn. Springer, Berlin (2001)

Identification of Textual Contexts

Ovidiu Fortu and Dan Moldovan

Human Language Technology Research Institute,
University of Texas At Dallas,
Department of Computer Science University of Texas at Dallas,
Richardson, TX 75083-0688, (972) 883-4625
{fovidiu, moldovan}@hlt.utdallas.edu

Abstract. Contextual information plays a key role in the automatic interpretation of text. This paper is concerned with the identification of textual contexts. A context taxonomy is introduced first, followed by an algorithm for detecting context boundaries. Experiments on the detection of subjective contexts using a machine learning model were performed using a set of syntactic features.

1 Introduction

The identification of contexts in text documents becomes increasingly important due to recent interest in question answering, text inferences and some other natural language applications. Consider for example the question "Who was Kerry's running mate in the last US Presidential election?" Here one can identify a temporal context, i.e. last year, and a domain context namely US Presidential election. Proper identification of such contexts may help a Question Answering system locate the right documents, by using contextual indexing for example, and provide correct answers by using contextual reasoning.

Although contexts in natural language were studied for some time, not much work has been done on context taxonomy, boundary detection, and how to use them in language understanding. However, considerable work was done on contexts in AI in general. John McCarthy's work on formalizing the contexts ([1], [2]) provides a general framework for a context-based representation of information. His theory is general and flexible, and as a consequence it can be adapted to formalize some complex aspects of textual contexts.

Consider for example the sentence:

Mary said the dog ate the cake, but I don't believe it. The pronoun "it" refers to an entire sentence, and thus the representation of the phrase above in first order logic is complicated. The use of contexts can simplify the representation:

Ist(C_M, "the dog ate the cake") $\wedge \neg$ Ist(C_S, "the dog ate the cake"),

where C_M is the context of Mary's perspective, while C_S is the context of the speaker's perspective. We used McCarthy's notation, i.e. Ist(c, p) means that proposition p is true in context c.

This paper proposes to represent textual contexts as objects, provides a concept taxonomy, and a method for the identification of contexts in text. Experiments were performed on the detection of subjective contexts using a machine learning model.

2 Textual Contexts

There is an important distinction between the concepts in McCarhty's theory and the ones discussed in this paper. While McCarthy's objective was to create a general and flexible theory for Artificial Intelligence applications, the objective of this paper is a lot more narrower. We deal with the problem of identifying contexts and determining the correct boundaries of these contexts. For example:

"John got a new job on Monday. He got up early, shaved, put on his best suit and went to the interview."

The word "Monday" appears only in the first sentence. Although the second sentence has absolutely no reference to any date, a human interprets the text considering that all actions in the second sentence happened on Monday as well. Thus, the temporal information given in the first sentence also applies to the second sentence. To cover both intra-sentence and inter-sentence contexts we refer in this paper to *textual contexts* as a set of clauses or sentences that have some implicit information in common. The commonality is the context we are trying to identify.

For the purpose of text understanding, we need this common information, because it helps in interpreting the sentences, like in the example above. The complexity of the common information shared by the sentences that belong to the same textual context makes the context detection a hard problem. Our solution for coping with this difficulty is to avoid direct detection of complex contexts. Instead, we define very simple classes of contexts (see section 3) and focus on them. Some of the different contexts that we find will overlap (contain

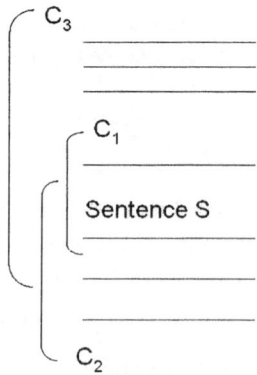

Fig. 1. Example of contexts layout

common sentences). For each sentence, we can use information from all contexts that contain it. The combination of all these pieces of information can be viewed as a more complex context. Thus, even if we do not detect complex contexts, we can reconstruct approximations from the simple ones. Figure 1 shows how a sentence can be contained in several textual contexts. The brackets show the boundaries of the contexts C_1, C_2 and C_3, all of which contain sentence S. For understanding S, we can use simultaneously the information of all the three contexts.

3 Context Taxonomy

As seen above, context structures in discourse can be used for text understanding. In the following, we give a taxonomy of contexts that defines the goals of the context detection task.

Let us consider a very simple abstract example. Let a, and b be predicates and the proposition: $a \wedge b$. One can take a context A such that $Ist(A, a)$, another one B such that $Ist(B, b)$, or AB such that $Ist(AB, a \wedge b)$.

By definition, the logic formulas above are valid, but there may be some other contexts in which a, b and $a \wedge b$ hold. In fact, we can take all combinations of predicates and define contexts - a very large number of contexts (exponential with the number of predicates). We can easily do the same with discourse elements, and thus obtain different, but valid representations of the same thing. The point is that contexts are not an inherent property of a text, like for example the subject of a sentence which is uniquely determined, but rather a way of representing the information in the text.

This is why the detection of arbitrary contexts, even if accurate, may not prove useful. Thus, the need to establish types of contexts which have useful meaning and properties. Once the context types are established, one can explore possibilities for representing information in a text using contexts. For the types of contexts that have been chosen, one can also find properties that allow the detection, and last but not least axioms that allow inferences with contexts.

Our purpose is to identify contexts that are useful for the analysis of text, and thus the types of contexts that we present here are intended for general use. Our criterion for the classification is the nature of the information they contain.

General Objective (Factual) Contexts. These contexts contain statements that are generally accepted as true (such as scientific facts, for example). One can identify them in various types of documents, from scientific papers to news articles. Example:

"Citrate ions form salts called citrates with many metal ions. An important one is calcium citrate or "sour salt", which is commonly used in the preservation and flavoring of food."

Note that even the sentences are general facts, they can not be taken separately without solving the coreferences (pronoun "one", which plays the role of subject in the second sentence, for example).

The sentences associated to these contexts are generally expressed using the present continuous tense. They are also marked by the presence of quantifiers (like "all", or "commonly" in our example). The presence of nouns with high generality (placed close to top in WordNet hierarchy) is also a fact that indicates objective contexts.

Subjective contexts. We introduce this type of contexts to handle feelings, beliefs, opinions, and so on. The truth value of sentences that express such facts is relative to the speaker's state of mind. From the point of view of the real world, the truth value of such sentences may vary from speaker to speaker and from situation to situation. One can identify some subclasses of this type of contexts:

1. Statement contexts, containing statements that people make.
2. Belief-type contexts, representing personal beliefs of the speaker
3. Fictive contexts, dealing with facts that are not real, like dreams and products of imagination.

It is necessary to treat the statements as contexts, because it is possible that different people say contradictory things about the same subject (in fact this is very common in debates or news articles about arguable things). By simply taking the stated facts without considering the contexts, one easily obtains contradictions, which puts an end to inferences.

Belief contexts are very similar to statement contexts, but in general it is less likely that facts that are true in a belief context are also true in the real world. In fact, in many cases people explicitly state that what they say is a belief in order to place a doubt over the truth of what is said. Example:

"John, when is this homework due?
I believe it's on Monday."

John thinks the homework is due on Monday, but he is not certain. He does not refuse to answer the question, but he avoids assuming the responsibility implied by a clear answer. In general, it is implied that people believe what they say (according to Grice's principles). This is why the explicit statement of belief is redundant, and thus an indication of possible falsity.

The main purpose of this classification is to offer support to establishing the relation between what is stated in the text and the reality. From this point of view, the three subclasses can be viewed as a finer grain ordering; the content of a statement is in general more likely to be true than a belief, which in turn is more likely to be true than a fiction.

Probability/Possibility/Uncertainty Contexts. Probability and probable or possible events are handled naturally in human inference and as a consequence, are present in human languages. This is why one can consider the manipulation of information related to possible events using contexts. Example:

"Perhaps Special Projects necessarily thinks along documentary lines. If so, it might be worth while to assign a future jazz show to a different department - one with enough confidence in the musical material to cut down on the number of performers and give them a little room to display their talents.

Time/Space(dimensional) Contexts. Many aspects of our lives are related to time and space constraints. This makes temporal and spatial information frequently present in human reasoning and language. Since propositions that are true at a certain time or in a certain place are not necessarily true outside their time frame or place, it is useful to define time and space related contexts. From the point of view of the limitations they impose, the dimensional contexts can be further divided in:

- Time contexts - mainly temporal restrictions
- Space contexts - location/space restrictions
- Event contexts - combine time and space restrictions

Domain Contexts. These contexts deal with restrictions regarding the domain of applicability of the statement. Example:

John is very good at math, but he is slow at English.

Although these contexts are not as frequent as the time/space based contexts, they are still important for the correct interpretation of the text. The beginning of such a context is generally marked by the presence of constructions with prepositions like "in" and "at" and nouns that denote domains (like "math" above).

Necessity Contexts. These contexts deal with necessary conditions for something to happen. Example:

In order to achieve victory, the team must cooperate. The strikers must come back to help the defenders.

These contexts are marked by the presence of the modal verb "must" or some substitutes like "have to", "need to". Adverbs like "necessary" and nominalizations like "requirement" can also do the same job.

Planning/Wish Contexts. This type of context contains information about someone's plans (or wishes). Example:

After I cash my paycheck, I go to the casino and win a lot of money. I buy a nice car and a big house. Then I quit my job and go on vacation.

The taxonomy presented above covers a wide variety of examples. It is a starting point for solving specific problems. The classes presented here are not necessarily disjoint. In fact, it is possible to have a context that has characteristics from several classes. The issue here is not to accurately classify a given context; our purpose is to identify the textual contexts of types presented above and exploit the information they contain for text understanding. If several contexts of different types overlap (or even coincide), it is not a conflict, but richness in information.

4 Textual Context Identification

4.1 Approach

The atomic unit of text for our problem is a verb (or place-holder like a nominalization) together with their arguments. In general, this unit is a clause. For simplicity, we will use the term "clause" instead of "atomic unit" from now on. A textual context will then be a set of such atomic units (clauses).

The main property of discourse exploited by the context detection mechanism proposed in this paper is the simple fact: utterances that belong to the same context tend to "stick together" in clusters. A *cluster* in a textual context is a maximal set of adjacent clauses that belong to the same textual context. A textual context is composed of one or several clusters. Thus, a natural approach to the context identification problem is first to identify the clusters and then group them to construct full contexts.

The identification of such a cluster is equivalent to the identification of its boundaries. Each clause can be a border for a context, and the problem of identification of contexts is reduced to the problem of deciding whether a given clause is boundary or not. Let us consider the following example:

"We're getting more 'pro' letters than 'con' on horse race betting", said Ratcliff. "But I believe if people were better informed on this question, most of them would oppose it also. *I'm willing to stake my political career on it"*.

All clauses of this text belong to the context of Ratcliff's statements, except for the one in non-italics, of course. Note that both the first clause (the border that marks the beginning of the textual context) and the last one have nothing special on their own. They could appear anywhere in text in the same form, without marking the beginning or the end of a context. Thus, the identification of context borders must depend mostly on the characteristics of discourse. However, the phrase in non-italics clearly indicates the presence of a subjective context, regardless of the surrounding clauses.

The flexibility of the natural language makes it possible for contexts to be introduced by countless syntactic patterns. Moreover, different types of contexts are introduced differently (more details in 4.2). Thus, for each type of context we must have a specific procedure for identifying the clues that indicate the presence of the context. For simplicity, these clues will be called *seeds* from now on.

Once we have identified the seeds, we can proceed to the identification of the clusters. The general method for this is to take each seed and try to add to the cluster associated to it the adjacent clauses through an iterative process. During this process, the text span that is associated to the context is growing, starting from the seed; this is the intuitive justification for the use of the term *seed* for designating the clue phrase that signals the existence of a context. Note that we have to consider both clauses that appear before the seed and after the seed to make sure we don't miss anything.

Of course, the process of growing a seeds is not the end. As mentioned above, it is possible that several text spans belonging to the same context are not adjacent. The simplest example is the dialogue; each participant says something, and the utterances of one of the speakers participating in the dialogue are scattered. For solving the problem, we need to compare the clusters obtained and group them as necessary.

In a nutshell, our algorithm for identification of contexts has three stages:

- Stage 1: Identify the seeds.
- Stage 2: Grow the seeds into clusters.
- Stage 3: Group the clusters obtained at the previous step to obtain the full contexts.

Next we provide a detailed description of these steps.

4.2 Seed Identification

The identification of the seeds is the first step towards the detection of contexts. As mentioned above, it is important to point out that different types of contexts are introduced by different seeds; for example, the fact that a certain clause contains the verb "say" has little if any relevance for the detection of a dimensional context. However, if we are trying to detect subjective contexts, the same clause becomes a very powerful indicator of a context presence. Therefore, we take the types of contexts one at a time, and adapt the search procedure for each one. We can not cast the problem as a multi-classification problem, having a class for each type of context and a null class for clauses that do not play the role of seeds, as it is possible that a single clause functions as seed for two different types of contexts simultaneously:

The Fulton County Grand Jury said Friday an investigation of Atlanta's recent primary election produced "no evidence" that any irregularities took place.

In this example, the clause in italics simultaneously introduces two contexts, namely a temporal context (the context of the last Friday before the text was written) and the context of the statements of the Fulton Jury.

Table 1 contains just a few examples of words that can signal a seed. Note that the property of being a seed applies to the whole clause, not to a single word. Thus, if a clause contains one of these words, it is not necessarily a seed. In general, we should use either a rule-based system or machine learning algorithm to decide if a given clause is a seed or not. These words are in general polysemous; obviously, not all their senses signal the presence of a seed. Thus, a word sense disambiguation system is needed for a better identification of seeds.

4.3 From Seeds to Clusters

The detection of all types of contexts at once is unlikely to produce good results, since the detection of each type of context boils down to solving of a more specific problem, and the problems for each type of context are somehow different in nature. For each type of context, there are different procedures for finding seeds and it is natural to have different procedures for boundary detection.

Table 1. Examples of words that may indicate a seed

	verb-seeds	noun-seeds	adverb-seeds	adjective-seeds	other
objective	NA	NA	commonly	abstract	in general
subjective	say, tell	statement, declaration	reportedly	aforesaid	according to
probability	can, might	chance, possibility	perhaps	likely	if
temporal	NA	day, year	now, yesterday	NA	during, before
domain	NA	NA	scientifically	NA	at, in
necessity	have to, need to	requirement	necessarily	compulsory	NA
planning	want, plan	plan	hopefully	desired	after

Markers. Since it is a set of adjacent clauses, a cluster is completely determined by its boundaries; all we need to do is to mark the first and the last clause of the cluster. For this purpose, we use a B (beginning) marker for the first clause and an E (end) marker for the last one.

Initially, we set a B and an E marker on each seed-clause. Thus, after the initialization, each cluster has only one clause, namely the seed. After that, the markers are moved, enlarging the cluster in both directions. This growth process is somehow similar to the growth of a seed, and this is the reason why the term of seed has been adopted for naming the clause that indicates the presence of a context.

The decision to move a marker (expand the boundary) is taken using a machine learning algorithm. The markers can move when there is cohesion between the cluster and the next clause. The general principle underlying the method of deciding the movement of the markers is that the text must be coherent; however, since coherence is very hard to detect, we must rely on cohesion, which is closely associated to coherence. The problem of text cohesiveness is also unsolved, but we can introduce features that relate to cohesiveness so that the learner may exploit them (more about that in 5.2).

A marker is considered stuck when it meets a marker that comes from the opposite direction (due to the assumption that contexts do not overlap). It is also stuck when a decision "not move" is taken, since the features are the same, and thus any new evaluation would yield the same result.

For example, let us see the steps of the algorithm applied to the example from 4.1:

Table 2 shows the evolution of the cluster boundaries (assuming that the decision of moving the markers is perfect). At the end of the algorithm, the clause that has the B marker is the beginning clause, and the one that has the E marker is the last one in the cluster.

Marker Movement. The decision to move a marker can easily be cast as a binary classification problem. There are two possible decisions, namely to move

Table 2. Marker movement

No.	clauses	markers' positions					
		initial	step 1	step 2	step 3	step 4	step 5
1	"We're getting more 'pro' letters than 'con' on horse race betting",		B	B	B	B	B
2	said Ratcliff	B,E					
3	"But I believe			E			
4	if people were better informed on this question,				E		
5	most of them would oppose it also.					E	
6	I'm willing					E	
7	to stake my political career on it".						E

or to stop. This decision can be taken using a classifier automatically constructed by a machine learning algorithm.

In a nutshell, the second stage of context detection is as follows:

1. Initialize markers placing one B (from beginning) marker and one E (ending) marked at each seed
2. While we ca still move any marker, do the following:
   - for each beginning marker B, if decision(cluster(B), previous clause) = yes, move B upwards
   - for each ending marker E, if decision(cluster(E), next clause) move E downwards
3. Output the clusters obtained.

where decision(cluster(marker), clause) is the procedure that tells if the marker must be moved, machine learning or manually constructed. (Cluster(marker) is the cluster that is bounded by the marker).

4.4 Cluster Grouping

The final stage in context detection is the grouping of the clusters obtained at stage 2. This process will also depend on the type of context. It is a very simple task for subjective contexts, because all we need to do is group all statements, beliefs, etc. by the speaker to whom they belong.

For a dimensional context, the problem is also simple if the context is associated to a fixed, known temporal interval, like a day or an year. However, when we deal with spans of several days, for example, things get more complicated.

In general, the contexts considered in this paper are rich in information, like the situations in situation calculus. The need for efficiency in communication has made the natural language very compressed; definitely, a text is not built on the principle "what you see is what you get". What we find in the text is merely a projection of the real context, which resides in the speaker's mind. As a consequence, it may become difficult to determine to which context a certain cluster belongs. Note that failure to add some cluster to a context is preferable to addition of a cluster to the wrong context. If a cluster that belongs to a context is wrongfully declared a new context, the impact on interpretation of sentences is low, provided that the information defining the context is correctly extracted. Incorrect introduction of a cluster in a context is, however, leading to falsity, because the information of the context will by assumed to be true for the cluster as well.

5 Experimental Implementation: Detection of Subjective Contexts

Since different procedures need to be developed for each class of contexts, a context detection system will be very large. The lack of a corpus compelled us to test the strategy of detection on a single class, namely the subjective contexts. The detection of subjective contexts follows the general algorithm described above. To complete the description, we only need to specify two aspects:

1. The procedure for the detection of the seeds
2. The way the decision for marker movement is taken

The most difficult of the two is the decision for marker movement. The rest of this section describes the details of this second problem and the experimental results.

5.1 Subjective Seed Detection

The class of subjective contexts has three subclasses, and for each of them there are different types of seeds. The presence of the statement contexts is indicated by the semantic relation "topic", which appears with communication verbs or their place-holders:

- communication verbs: say, state, tell, reply, answer, inform, declare, announce, etc.
- non-communication verbs used with a sense that implies communication: add, estimate, notify, etc. Example:
  John said he had seen Paul at school. In the detention room, *he added*.
- expressions that replace the communication verbs: *to have an answer for, one's explanation is*, etc.

The situation is similar for the belief contexts, the only difference coming from the fact that we have another set of verbs:

- verbs that express beliefs: believe, think, consider, etc.
- expressions that introduce the idea of belief: from one's point of view, for him, etc.

At last, fictive contexts are generally marked by the presence of nouns/verbs that express fictiveness:

- dream, novel, play, story, tale
- to imagine, to picture, to fancy

Example: *H.E.Bates has scribbled a farce called "Hark, Hark, the Lark"! It is one of the most entertaining and irresponsible novels of the season. If there is a moral lurking among the shenanigans, it is hard to find. Perhaps the lesson we should take from these pages is that the welfare state in England still allows wild scope for all kinds of rugged eccentrics . Anyway, a number of them meet here in devastating collisions. One is an imperial London stockbroker called Jerebohm. Another is a wily countryman called Larkin, [...]*

The author of this article is introducing us to the world of a literary work. The seed clause in this case is not the noun "farce" itself, but a reference to it (the adverb "here").

Practically, we need to detect the idea of communication (belief, fiction) in text; this can be achieved using a semantic relation detector that searches for the corresponding semantic relations, or even with the help of an word-sense disambiguation system, by selecting the right senses of the verbs. This is because, as seen in 4.2, not all senses of these words are signaling seeds; for example, from the 8 senses of verb "tell" in WordNet, only the first 7 imply communication.

For subjective contexts, the communication verbs are by far the most frequent manner of introducing a context (manual evaluation showed that more than 90% of the seed clauses contain such verbs). Moreover, for these verbs most of the senses (and also the most frequently used) imply communication.

For our experiments, we used gold standard seeds (we did this part of the task manually).

5.2 Machine Learning for Marker Movement

The decision for marker movement is taken using a classifier automatically constructed using Support Vector Machines. The features that we considered for the encoding of the training examples are listed below:

1. first connector
2. second connector - the first two features refer to the punctuation signs or conjuncts that make the connection between the current clause and the next one; if there is no such connector, we use the "null" value
3. direction of movement (up, down)
4. distance (number of clauses) from the seed
5. seed verb (the WordNet synset of the verb of the seed clause)
6. connector path from the seed - the sequence of connectors from the seed to the current clause; in this sequence, we eliminate the duplicate null values
7. verb tense of previous clause
8. verb tense of current clause
9. end of sentence encountered (the ".", "?" etc)
10. number of coreferences - this feature is computed by counting the coreference chains between words in the current context cluster (as much as it has grown at this point) and the next two clauses; the number three was arbitrary chosen; the reason we don't take it to be 1 is that strong cohesion between

the clauses of the context and the ones following the current clause is an indication that we should include the current clause in the context
11. marker clash (true if another marker of the same type, coming from opposite direction is encountered)

Example:

1. The Fulton County Grand Jury said Friday
2. an investigation of Atlanta's recent primary election produced "no evidence"
3. that any irregularities took place .
4. The jury further said in term end presentments [...]

Clause number 1 is the seed, and the three following clauses have the feature vectors given below:

Table 3. Example of feature codification

1	2	3	4	5	6	7	8	9	10	11	target
null	null	down	1	say	null	past	past	f	0	f	y
null	that	down	2	say	null+that	past	past	f	0	f	y
par	further	down	3	say	null+that+par+further	past	past	t	1	t	n

The value "par" means that a paragraph end has been encountered. Note that the connectors are not necessarily placed at the beginning or ending of the clauses; in clause 4 of the example, the word "further" acts as connector, as it establishes the relation between the text segments. In general, all discourse markers (cue phrases - see [3]) must be considered.

Features 1, 2, 3, 4, and 6 are designed to characterize the structure of the text. The following example shows the rationale behind these features:

Mary said the dog ate the cake and *I don't believe it.* In this situation we have only one connector. The phrase has a low degree of ambiguity, but the addition of a comma before the conjunction "and" would make it even better. The interpretation is that the second phrase does not belong to the context of Mary's statements. However, this changes if we add another connector:

Mary said the dog ate the cake and that *I don't believe it.* Now, the second clause is clearly a part of the context of Mary's beliefs. The example considered here is as simple as can be, since it spans over a single short, clearly structured sentence. In spite of this simplicity, the detection of its boundaries is not straight forward, since we have to be able to correctly interpret the relation between the two clauses. Semantic information is also needed, for example:

Mary said the dog ate the cake and John was laughing. By replacing the second clause we have obtained an ambiguous phrase; it is not clear whether Mary said that John was laughing or if John was laughing while Mary was speaking. Syntactic information alone is not enough to solve such a problem.

We introduced features 7 and 8 to take advantage of the fact that there is a correlation between the change of the verbal tense and the structure of discourse. Feature number 9, though different in nature, has the same rationale as 7 and 8.

Finally, the number of coreferences (feature 10) was introduced as an additional measure of the cohesion of the text.

Our model is based on syntactic features only; we expect that the addition of semantic information will improve the results.

5.3 Results

We used a set of news articles collected from SemCor corpus for our experiments. We generated feature vectors (like in table 3) for the subjective contexts in these articles. We have obtained roughly 1000 training instances that we used for the evaluation of the model. The machine learning we used was Support Vector Machines ([4]), and the method of evaluation was 20 fold cross validation. Since the features are discrete, we encoded them in numerical vectors. The encoding procedure is widely used in N.L.P. applications: assign each feature a number of dimensions equal to the number of values; this way each value for each feature has a unique entry in the vector, which takes value 1 if the corresponding feature valued is encountered, and 0 otherwise. The kernel that we used was the RBF kernel. The accuracy of the decision to move or stop markers was 90.4%.

Table 4. Experimental results

	1	2	3	4	5	6	7	8	9	10
1	90.30									
2	88.69	88.83								
3	88.54	85.16	87.95							
4	90.01	88.69	88.25	89.86						
5	90.60	88.10	87.51	89.72	90.45					
6	89.42	84.28	88.10	90.01	89.72	90.01				
7	89.72	88.10	88.39	90.30	90.60	90.45	90.30			
8	90.89	87.81	87.95	89.86	89.28	90.01	90.60	90.30		
9	88.69	81.93	87.66	88.54	88.69	88.83	89.13	89.28	89.28	
10	90.45	88.39	88.69	90.16	89.72	89.72	90.01	90.01	89.57	89.72

In table 4, the field (i,j) contains the accuracy obtained if we remove the features i and j; (i, i) is the accuracy obtained by removing only feature number i. In the evaluation of the performance of the learning, we dropped feature 11 (marker clash). We did so because in the annotated data a marker clash can only happen at the boundaries of the clusters, while in real life this is obviously not true. By using this feature, the accuracy grows to nearly 95%. This proves that the learner deduces the rule that markers coming from opposite directions must stop when they meet.

We notice that the contribution of the features is rather uniform (none of the features has a major impact on its own); the coreference seems to have the lowest contribution.

6 Discussion

Table 4 shows that the feature no. 6 (connector path) does not have the impact that we expected. Intuitively, this feature carries a considerable informational baggage. In many instances, the human annotator can correctly guess the decision (if the marker must move) solely based on this feature. The conclusion is that the learner, as we use it now, can not fully exploit this feature.

The cost of an error in the decision of moving a marker is not constant; it depends on the distance between the clause where the marker stops and the one where it should stop. The further the marker stops, the worse the error is. Currently, our system does not take this fact into account.

References

1. John McCarthy. Notes on formalizing contexts. In Tom Kehler and Stan Rosenschein, editors, *Proceedings of the Fifth National Conference on Artificial Intelligence*, pages 555–560, Los Altos, California, 1986. Morgan Kaufmann.
2. Douglas B. Lenat and R. V. Guha. *Building Large Knowledge-Based Systems: Representation and Inference in the CYC Project*. Addison-Wesley, Reading, Massachusetts, 1990.
3. Diane J. Litman. Classifying cue phrases in text and speech using machine learning. In *National Conference on Artificial Intelligence*, pages 806–813, 1994.
4. Chih-Chung Chang and Chih-Jen Lin. *LIBSVM: a library for support vector machines*, 2001. Software available at http://www.csie.ntu.edu.tw/~cjlin/libsvm.
5. Diane Litman and Julia Hirschberg. Disambiguating cue phrases in text and speech. In *13th. International Conference on Computational Linguistics (COLING-90)*, Helsinki, Finland, 1990.
6. Barbara J. Grosz and Candace L. Sidner. Attention, intentions, and the structure of discourse. *Comput. Linguist.*, 12(3):175–204, 1986.
7. William C. Mann. Discourse structures for text generation. In *Proceedings of the 22nd conference on Association for Computational Linguistics*, pages 367–375. Association for Computational Linguistics, 1984.
8. Douglas B. Lenat. CYC: A large-scale investment in knowledge infrastructure. *Communications of the ACM*, 38(11):33–38, 1995.

Investigation of Context Effects in Iterated Prisoner's Dilemma Game

Evgenia Hristova and Maurice Grinberg

Central and Eastern European Center for Cognitive Science,
New Bulgarian University,
21, Montevideo street, Sofia 1618, Bulgaria
ehristova@cogs.nbu.bg, mgrinberg@nbu.bg

Abstract. Context effects during Prisoner's Dilemma (PD) game playing are investigated. The Cooperation Index (CI) – a quantity computed as a relation between payoffs – defines a cooperativeness scale, along which PD games can be distributed. Context is selected by manipulating the CI range of the games played. It is found that the level of cooperation depends not only on the CI of the current game but also on the CI of the other games in the sequence. The influence of context on the full CI scale is investigated by introducing probe games covering the whole CI range. The results are compared with the prediction of a model that takes into account the current game, the previous play history, and the predicted opponent's move. The model is found to be sensitive to both the CI and to the CI of the other games form the game set and a very good agreement between the model and the experimental data was found.

1 Introduction

The Prisoner's Dilemma (PD) game is used as a model for exploring the determinants of cooperative behavior in social situations. The dilemma for the players lies in the fact that the best possible outcome for all results from behavior when each participant refrains from trying to maximize her own self interest in the short run. Here we study the players' behaviour in the dynamic context of the iterated PD game. So the results may be applied to situations in which players build long term relationships. The PD game has attracted much attention in the scientific literature and considerable numbers of papers have appeared. In order to explain human behavior in iterated PD game playing, stochastic models ([14], [15]) and reinforcement learning models (e.g. [4], [7], [11]) were put forward. Several different simple strategies (e.g. tit-for-tat, Pavlov etc.) have also been investigated in multi-agent simulations and have shed light on their evolutionary stability and how they can lead to interesting emergent behavior (e.g. [3]).

All other factors put aside, cooperation in PD games is of course related to the game payoffs. The cooperation index (CI), computed as a ratio between the payoffs in PD game, received a considerable theoretical and empirical support as a reliable predictor of the cooperation rate in PD game [9], [13], [14]. In the same time, the question of how people estimate CI remains an open question. Findings from psychophysics and experiments involving choices between risky options sustain the interest in the topic. Lockhead [10] summarizes a lot of psychophysical experiments that demonstrate that people have very poor representation of absolute magnitude of stimuli. Stewart et al. in [18] demonstrated that the utility of a prospect depends on the options in the choice set. Another research work [17] shows that there are strong sequence effects in categorization of simple perceptual stimuli. All these findings point to the fact that magnitudes of stimulus properties are identified with respect to the properties of the other stimuli in the set. Recently, in a couple of experiments [8], [19], it was demonstrated that the influence of CI is context dependent and can be modified by the distribution of the games CI's. Vlaev [19], for instance, found context effects in a series of one-shot PD games. In ref. [8], the influence of different contexts, defined by different ranges of the game CI's, on subsequent cooperativeness judgments about game was established. In the latter experiment, however, the contexts effects were established by a subsequent game evaluation phase and not during actual playing. So, only indirect evidence about dynamic context effects was obtained and no theoretical account could be given.

In the present paper, we propose a new experimental procedure that allows investigating the context influence of different CI distributions during game playing on the full CI scale (see Section 3). In the same time, a model with dynamically updated parameters is formulated (see Section 5) that accounts very well for the experimental results and especially for the context effects observed. This model gives information on other factors that also influence player's decisions like the behavior of the opponent and the outcomes of previous games.

2 The Prisoner's Dilemma Game

2.1 The Payoff Matrix

The Prisoner's dilemma is a two-person game. The payoff table for this game is presented in Figure 1. The players simultaneously choose their move – C (cooperate) or D (defect), without knowing their opponent's choice.

		Player II	
		C	D
Player I	C	R, R	S, T
	D	T, S	P, P

Fig. 1. Payoff table for the PD game. In each cell the comma separated payoffs are the Player I's and Player II's payoffs, respectively

R is the payoff if both cooperate (play C), P is the payoff if both "defect" (play D), T is the payoff if one defects and the other cooperates, S is the payoff if one cooperates by playing C and the other defects by playing D.

Furthermore, the payoffs satisfy the inequalities $T > R > P > S$ and $2R > T + S$. This structure of the payoff matrix of that game offers a dilemma to the players: there is no obvious best move. As well known, D is the strongly dominant strategy for both players because each player receives a higher payoff by choosing D rather than C whatever the other player might do. So the Nash equilibrium of the game or each player is always to defect and play D. However, if both players play D, the payoffs (P, P) are lower than the payoffs (R, R) they would have received if both had cooperated (had chosen the dominated C strategy). By choosing to cooperate however, they have to trust that their opponent will also cooperate and take the risk of getting the lowest payoff – S (taken to be 0 in the present experiment).

2.2 The Cooperation Index

Personality factors put aside, it is reasonable to assume that cooperation in PD games depends on the relative magnitudes of the payoffs. Rapoport and Chammah [14, 13] have proposed the quantity $CI = (R-P)/(T-S)$, called *cooperation index,* as a predictor of the probability of C choices and it has been experimentally proven [14, 12] that cooperation increases monotonously with CI. In Figure 2, two examples of PD games with different CI are presented – a highly cooperative PD game with CI equal to 0.9 and a non-cooperative PD game with CI equal to 0.1.

		Player II	
		C	D
Player I	C	56, 56	0, 60
	D	60, 0	2, 2

		Player II	
		C	D
Player I	C	56, 56	0, 60
	D	60, 0	50, 50

Fig. 2. Examples of PD games with different CI. The first game has a CI=0.9, the second one has CI=0.1

Further in the paper we use the CI to define a scale (a cooperativeness scale), along which the PD games can be distributed. Thus we can explore the influence of different game distribution with respect to cooperativeness (measured by CI) and investigate the existence of context phenomena in judgment and decision-making similar to those found in psychophysics (see ref. [19]).

3 Experimental Design

The present work can be best understood in the light of the results obtained in a previous experiment [8].The experiment had two phases. In the first, three groups of subjects played games with different CI ranges. In the second, all the participants were asked to make judgments about the cooperativeness of games, covering the full CI scale.

It was found that the judgment scale in the second phase is influenced by the playing experience during phase one. The change is the largest for subjects who took CI into account during their playing. Thus, subjects that played games with low CI's tended to give lower estimates for the cooperativeness of the game and vice versa. No context effect was found for the subjects who were less influenced by CI in phase one. Their judgments, however, depended on CI despite the fact that the latter was not used during actual playing.

These results support the conclusions that previous experience influences the judgments of the cooperativeness of PD games. Furthermore, they suggest that contexts determined by different CI distributions are influencing actual game playing (not only subsequent judgment). The present paper is aimed at the investigation of this influence.

The goal of the present experiment is to find a way to directly evaluate changes in the CI scale during playing of iterated PD games when the context, determined by different CI distributions, is manipulated.

3.1 Choice of PD Payoff Matrices

The payoff matrices were randomly generated with magnitude of payoffs held in certain limits. This is done in order to avoid memory effects or big differences in the payoffs for games that could favor different strategies (e.g. subjects could pay more attention to games with higher payoffs than to games with the same CI but with much smaller payoffs).

The CI is a quantity invariant with respect to the possible linear transformations of the payoffs. However, Oskamp and Perlman [12] claimed that the average payoff per trial $((T+R+P+S)/4$ (see Figure 1) is a very important factor with significant effect on the level of cooperation. Taking this into account, we generated the games so that T is between 22 and 78 points (mean 53), R was between 15 and 77 points (mean 45), P was between 4 and 47 points (mean 17). For simplicity we set $S = 0$.

3.2 Design and Procedure

Each subject played 270 PD games against the computer. On the interface, the moves were labeled in a neutral manner as '1' and '2'. For convenience, further in the paper, we will continue to use *cooperation* (C) instead of move '1' and *defection* (D) instead of move '2'.

The computer used a probabilistic version of tit-for-tat that takes into account the two previous moves of the player. For instance, in the case when the subject makes one and the same move during the last two games, the computer plays the same move with probability 0.8.If the case of two different moves the probability for any move is 0.5. This was done to allow the subject to choose her own strategy (followed by the computer) without easily become aware of the computer strategy.

The payoffs were presented as points, which were transformed into real money and paid at the end of the experiment. After each game the subjects got feedback about their and the computer's choice and could monitor permanently the total number of points they have won and its money equivalent.

In order to make the subjects concentrate exclusively on the payoff table when making their choice, they were instructed to try to maximize their payoffs and not for instance to compete with the computer.

The game was presented to them in a formal and a neutral formulation to avoid other factors and contexts as much as possible. The terms 'cooperation' or 'defection' were not mentioned in the instructions to further avoid influences other than the payoff matrix. Subjects were not informed about the existence of CI.

The subjects received information about the computer's payoff only for the current game and had no information about the computer's total score. This was made to prevent a shift of subjects' goal – from trying to maximize the number of points to trying to outperform the computer. In this way, the subjects were stimulated to pay more attention to the payoffs and their relative magnitude and thus indirectly to CI.

As the main goal of the present experiment was to find context effects due to the different game distribution with respect to CI, three CI ranges of games were constructed:

- Full CI range – PD games covering the full CI scale (CI's equal to 0.1, 0.3, 0.5, 0.7 and 0.9);
- High CI range – PD games taken from the higher range of the CI scale (CI's equal to 0.7 and 0.9);
- Low CI range PD – games only from the lower range of the CI scale (CI's equal to 0.1 and 0.3).

In order to obtain information about the full CI scale 25 predefined *probe games* – 5 with each of the CI's - 0.1, 0.3, 0.5, 0.7 and 0.9 – were incorporated. They were the same for all subjects in all groups and were positioned evenly. Every tenth game played was a probe game. The purpose of using probe games was to measure the cooperation levels for the full CI scale during play while changing the context as slightly as possible (especially for the High-CI and the Low-CI ranges).

The experiment consists of three experimental conditions that differed in the CI distributions of the games played (see Figure 3):

- *Full-CI-range condition* – each subject played games from the Full CI range, intermixed with the probe games.
- *High-CI-range condition* – each subject played games from the High CI range, intermixed with the probe games.
- *Low-CI-range condition* – each subject played games from the Low CI range, intermixed with the probe games.

In each experimental condition all the subjects played one and the same set of PD games (the sets are the same within condition and different between conditions). Each participant played 270 PD games among which were the probe games. They were the same in all three context conditions and were on the same places in the game sequences – each tenth game, starting form game 20 and ending at game 260. We expected subjects participating in different context condition to show different playing behavior in the probe games.

There were in total 72 participants (34 males and 38 females). In the Full-CI-range condition participated 24 subjects, in High-CI-range condition – 26 subjects and 22 subjects took part in the Low-CI-range condition. All were university students with an average age of 23 (ranging from 18 to 35). Each subject was randomly assigned to one of the three experimental conditions and after being instructed played 5 training games. All subjects were paid according to the number of points they have obtained.

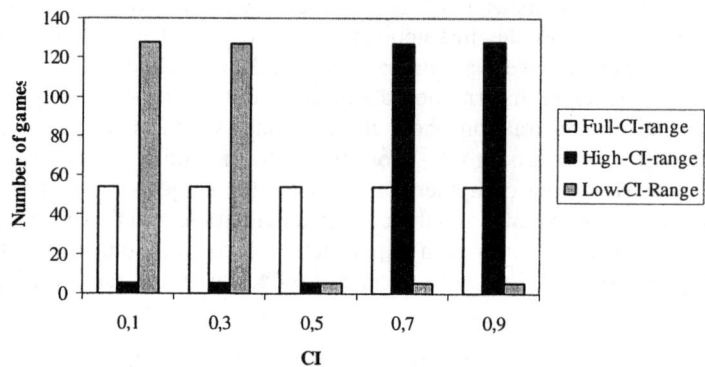

Fig. 3. Distribution of PD games played in different experimental conditions: Full-CI-range (*white bars*), High-CI-range (*black bars*), and Low-CI-range (*grey bars*)

4 Experimental Results

4.1 Mean Cooperation Rates in Different Experimental Conditions

The cooperation rates in different experimental conditions are significantly different ($F(2, 69) = 7.7$, $p = 0.001$). Participants cooperated in 44% of the games in the High-range condition, in 34% of the games in the Full-range condition, and in 23% of the games in the Low-range condition (see Figure 4).

This result is expected, as participants in different groups played PD games with different CI's. And as mentioned before, it is well known that subjects cooperate more in PD games with higher CI's. It is interesting to check, however, if there are differences in cooperation for games with equal CI in the different experimental groups.

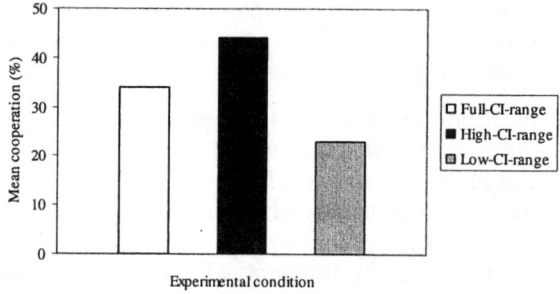

Fig. 4. Mean cooperation in the Full-CI-range (*white bar*), High-CI-range (*black bar*), and Low-CI-range (*grey bar*) conditions

4.2 Mean Cooperation for Games with the Same CI in Different Experimental Conditions

If cooperation rate during playing were not context sensitive, one would have obtained no difference between the cooperation in games with the same CI in the different experimental conditions.

The mean cooperation for games with given CI is presented in Figure 5. Participants in the Low-CI-range condition cooperated significantly less than participants in the Full-CI-range condition for PD games with CI=0.1 (21% and 28 % respectively, χ^2=25.9, p<0.001) and CI=0.3 (24% and 30% respectively, χ^2=10.5, p=0.001). Participants in the High-CI-range condition cooperated significantly more than participants in the Full-CI-range condition for PD games with CI=0.7 (43% vs. 38%, χ^2=7.42, p=0.006) and CI=0.9 (47% and 40 %, χ^2=16.6, p<0.001).

The latter results are a clear indication that there are context effects present. Players cooperate more when they play only highly cooperative games (PD games with CI=0.7 and 0.9 in the High-CI-range condition) than when they play the same games mixed with other less cooperative games (PD games with CI=0.7 and 0.9 in the Full-CI-range condition). On the other hand, players cooperate less when they play only low- cooperative games (PD games with CI=0.1 and 0.3 in the Low-CI-range condition) than when they play the same games mixed with other, more cooperative games (PD games with CI=0.1 and 0.3 in the Full-CI-range condition).

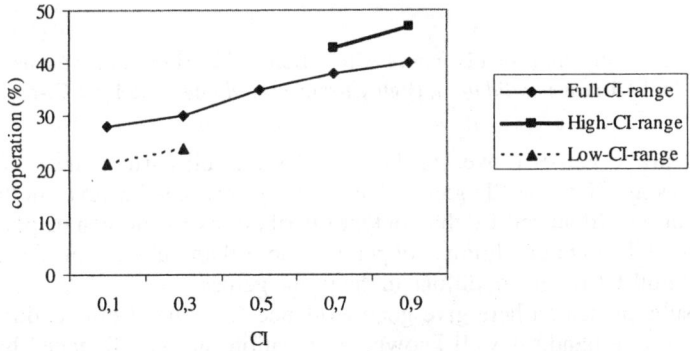

Fig. 5. Mean cooperation for PD games with different CI. Three experimental conditions are presented: Full-CI-range (*solid line*), High-CI-range (*bold line*), and Low-CI-range (*dotted line*)

4.3 Cooperation Rates for the Probe Games

In each context condition (Full-CI-range, High-CI-range and Low-CI-range conditions) PD games with CI's covering the full CI-scale were included – the probe games. They were intermixed with the other games as explained in section 3.2.

The number of probe games in which subjects cooperated was analyzed in a repeated-measures analysis of variance with CI as a within-subjects factor and the experimental condition (Full-CI-range vs. Low-CI-range vs. High-CI-range) as a between-subjects factor. Averaged data are presented in Figure 6 (although analysis was performed using the actual number of games the results are presented in percents for simplicity).

The main effect of CI is significant ($F(4, 276)=8.7$, $p<0.001$). The interaction between CI and the experimental condition is not significant. Participants in all experimental conditions cooperated more with increasing CI. The effect of the experimental condition is significant at $p=0.058$ ($F(2, 69)=2.96$). Post-hoc comparisons show that the mean difference in cooperation in probe games between High-CI-range condition and Low-CI-range condition is significant at 0.032 level and the mean difference between High-CI-range and Full-CI-range conditions is significant at 0.054 level.

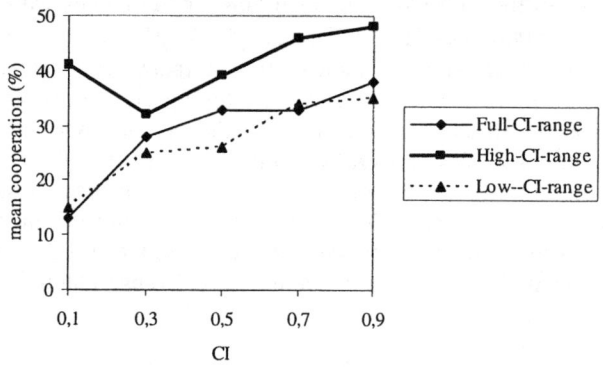

Fig. 6. Mean cooperation for PD games with different CI. Three experimental conditions are presented: Full-CI-range (*solid line*), High-CI-range (*bold line*), and Low-CI-range (*dotted line*)

Playing probe games (covering the full CI-scale) all participants cooperated more with increasing CI of the PD games. There are differences however in the mean level of cooperation influenced by the context distributions of the games played. Subjects in the High-CI-range conditions cooperated more than subjects in the Low-CI-range and in the Full-CI-range conditions in the probe games.

The results presented here give good evidence for context effects during PD game playing. On one hand, as well known, cooperation rate is influenced by the CI of a given game; but on the other hand it also depends on the context created by the CI's of the other games played in a particular game session.

5 The Model

A model that in principle could describe the phenomena investigated in the previous sections must have mechanisms for accounting for the game CI and in the same time be sensitive to the other games in the game set. The logic of the model, presented here, is as follows. Subjective values of the C and D moves ($V(C)$ and $V(D)$) are used to determine the probability of move C. In order to establish the value of a particular move it is not enough to use the values of the possible payoffs. The player should also take into the account the expectations about the opponent's move. This is an application of the subjective expected utility model in the domain of game playing. What we add are weights associated with each possible game outcome (CC, CD, DC, and DD), computed on the basis of previous payoffs (see the discussion below).

A similar in spirit model has been used in a different context in ref. [2]. Our model can be viewed as based on the general framework of the subjective utility theory [16] but with dynamic determination of the utilities and expectations about the probabilities of other player's move.

The model we have used has the following form:

$$V(C) = w_{CC} \text{Pff}(CC) P_{op}(C) + w_{CD} \text{Pff}(CD)(1-P_{op}(C)), \quad (1)$$

$$V(D) = w_{DC} \text{Pff}(DC) P_{op}(C) + w_{DD} \text{Pff}(DD)(1-P_{op}(C)), \quad (2)$$

and

$$P(C) = V(C)/(V(C)+V(D)), \quad (3)$$

where:

P(C) is the probability of move C;
V(C) and V(D) are the values of moves C and D.
Pff (CC), Pff (CD), Pff (DC), and Pff (DD) are the current payoffs R, S, T and P, respectively.
$P_{op}(C)$ is the predicted probability for the opponent to play C.

w_{CC}, w_{CD}, w_{DC}, and w_{DD} are weights that stand for the importance of the specific game outcome (CC, CD, DC or DD). These weights are computed as running averages of the payoffs received in the games with respective outcome and thus depend on previous payoffs:

$$[w_{xy}]_{new} = (1-\alpha)[w_{xy}]_{old} + \alpha \text{Pff}(xy), \quad (4)$$

where x, y = C, D; Pff(xy) is the received payoff; and $0<\alpha<1$.

Pop(C) is also calculated as a running average over the past opponents moves:

$$[P_{op}(C)]_{new} = (1-\beta)[P_{op}(C)]_{old} + \beta M_{op}, \quad (5)$$

where $0<\beta<1$ and M_{op} is the opponent move.

Because of the way the w's and the $P_{op}(C)$ are calculated (see eqs. (4) and (5)) they are responsible for the context sensitivity of the model. In that sense the model resembles the reinforcement learning model. However, the use of the current game payoffs in eqs. (1) and (2) ensures that the move will depend also on the game at hand and on its CI. This property is not available in typical reinforcement based models ([4], [7], [11]) in which the probability for cooperation is based only on past games.

The only parameters of the model are the averaging parameters (α and β) and the initial cooperation probability P(C). All these parameters are fixed to 0.5 in the simulations presented below. The model played against the computer player used in the experiment. The model was run 30 times in each context condition (Full-CI-range,

High-CI-range, and Low-CI-range) with the respective game set from the experiment. The model played five training games and then the running averages in eqs. (4) and (5) started to be evaluated for the remaining part of the simulation.

6 Comparisons Between Model Predictions and Experimental Data

Model predictions were compared with the data form the experiment with real subjects. Several measurements are chosen in order to estimate how well the model describes the experimental data.

The model results are compared to the experimental data on the basis of the overall cooperation, the cooperation with respect to CI (see Section 6.1), and the cooperation in the probe games (see Section 6.2). As a second criterion of goodness of fit, we use the number of different game outcomes during the game sessions (see Section 6.3). The game outcome cumulative statistics is a relatively independent indicator of the behavior of both players (different game outcomes are possible for the same number of C and D moves of both players).

6.1 Cooperation Rates

Theoretical cooperation rates are presented in Figure 7. They are very similar to the analogous data from the experiment (for comparison see Figure 5) - cooperation rates are highest in the High-CI-range condition and lowest in the Low-CI-range condition. We have also obtained different cooperation rates in games with the same CI in the different experimental conditions (p<0.001).

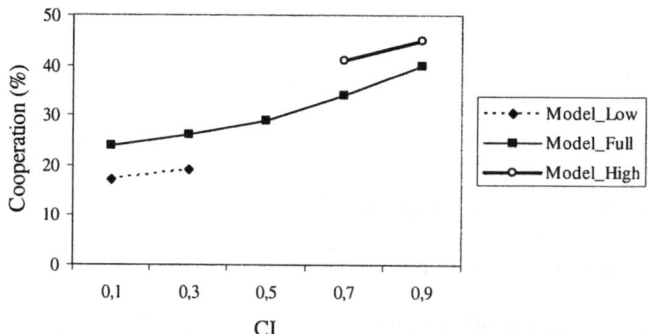

Fig. 7. Theoretical results for the mean cooperation in PD games with different CI in each context condition: Full-CI-range (*solid line*), High-CI-range (*bold line*), and Low-CI-range (*dotted line*)

The games in which the subjects or the model cooperated is compared with a repeated measures analysis of variance with CI as a within-subjects factor and the treatment condition (experiment vs. model) as a between subjects factor. The analysis is performed for each context condition separately. For all of them, the only statistically significant factor is CI (p<0.001) (see Figure 8). The influence of the treatment condition (experiment vs. model) is not statistically significant, nor is the interaction between CI and the treatment condition. The model predictions and the experimental data are statistically undistinguishable. This is an evidence of the good matching of model and experiment.

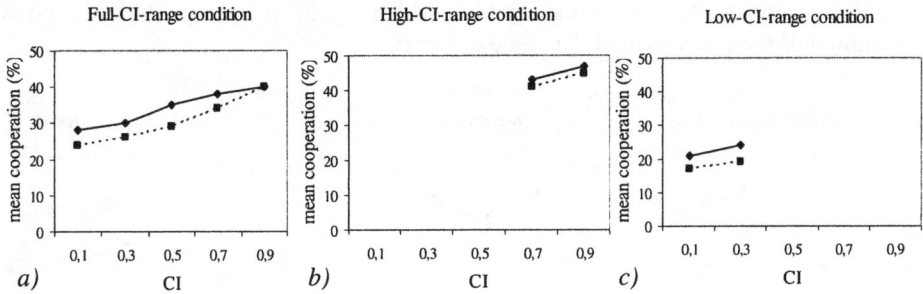

Fig. 8. Comparison for the cooperation rates between the model predictions (*dotted lines*) and the experimental data (*solid lines*) in each context condition: a) Full-CI-range; b) High-CI-range; c) Low-CI-range

6.2 Cooperation in the Probe Games

The model cooperation rates for the probe games, which cover the full CI scale, are presented in Figure 9. Again the model results are very similar to the experimental ones (for comparison see Figure 6). The context effect observed in the experimental data is also present in the model predictions - there are differences in the mean level of cooperation influenced by the context distributions of the games played (F(2, 87) = 33.06, p<0.001).

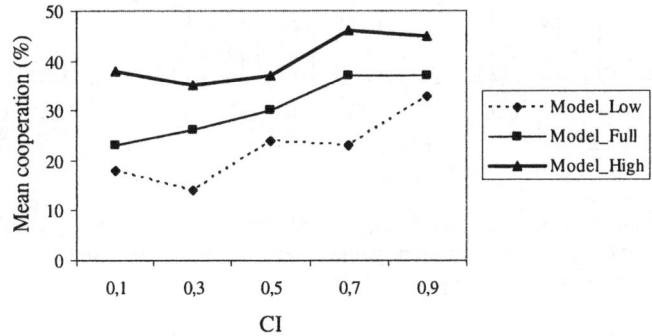

Fig. 9. Theoretical mean cooperation for PD games with different CI in each context condition:. Full-CI-range (*solid line*), High-CI-range (*bold line*), and Low-CI-range (*dotted line*)

The number of games in which the subjects or the model cooperated is compared with a repeated measures analysis of variance with CI as a within-subjects factor and the treatment condition (experiment vs. model) as a between-subjects factor. The analysis is performed for each context condition. In all context conditions the only statistically significant factor is CI ($p<0.001$) (see Figure 10). The influence of the treatment condition (experiment vs. model) is not statistically significant, nor is the interaction between CI and the treatment condition. The experimental and theoretical results are statistically undistinguishable, which again is a sign for the good description of the experimental data by the model.

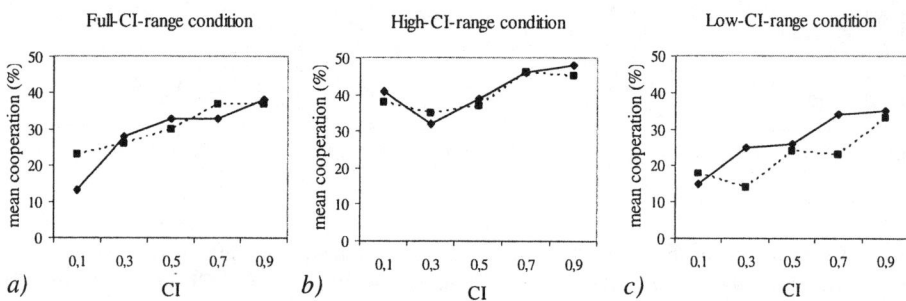

Fig. 10. Comparison between the model predictions (*dotted lines*) and the experimental data (*solid lines*) for the mean cooperation in the probe games for each context condition: a) Full-CI-range; b) High-CI-range; c) Low-CI-range

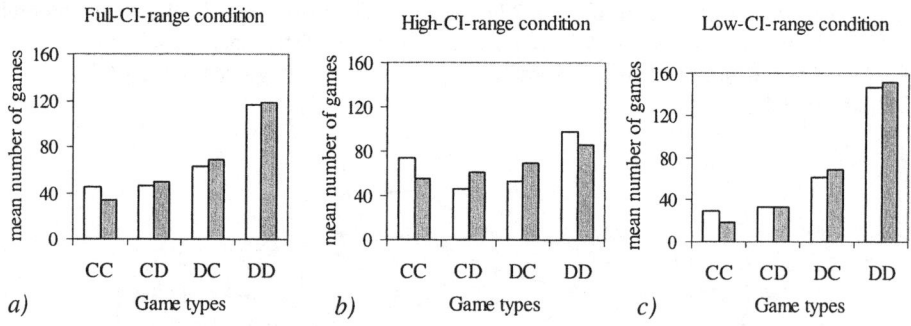

Fig. 11. Comparison between the model predictions (*grey bars*) and the experimental data (*white bars*) for the number of different game types in each context condition: a) Full-CI-range; b) High-CI-range; c) Low-CI-range

6.3 Number of Different Game Types

There are 4 possible game types – CC, CD, DC and DD. Each game type is bound to certain payoff (e.g. R, 0, T and P, respectively for Player I, see Figure 1). The same number of C and D moves, can produce different distributions of game types. Thus the number of different game types is a relatively independent characteristic of the game dynamic and must be accounted by the theoretical model.

The means for the number of different game types in the simulation and in the experiment were compared for each context condition (Full-CI-range, High-CI-range, and Low-CI-range). A chi-square test was used to evaluate the fit of the model to experimental data (see Figure 11). The distribution of different game types is undistinguishable statistically between the experimental data and the model data in the Full-CI-range condition ($\chi^2=3.86$, p=0.277) and in the Low-CI-range condition ($\chi^2=5.54$, p=0.1.36). In the High-CI-range condition the chi-square test is significant ($\chi^2=15.9$, p=0.001), still it is seen in Figure 11c that the model results approximate well the experimental data.

7 Discussion and Conclusions

In the present paper, context effects, related to different CI distributions of PD games during game playing, were investigated. In order to compare the cooperation rate in different context conditions, a set of probe games was incorporated in the game sequence of each experimental condition. Thus, it became possible to determine the influence of the context on the whole CI scale. The experimental results showed that the cooperation rate depends not only on the current game payoffs and CI but also on the CI's of the other games in the game set. If a PD game with a given CI is presented in the context of games with higher CI values, people tended to cooperate significantly more than in the other experimental conditions. The results obtained in the paper give good evidence for context effects during PD game playing.

In order to explore the context sensitivity observed, a model was put forward. Inspired by the subjective utility theory, it takes into account previous payoffs that determine dynamically the utility of the game outcomes; the prediction about the probability for cooperation of the opponent, based again on previous experience; and the current game payoffs. This form of the model turned out to be very sensitive both to the CI of the given game and to the context related to the other game CI's. The agreement between the theoretical results and the experiment is very good despite the fact that only three parameters define the model and they were kept fixed and the same for all the three conditions. The model reproduced quite accurately the dependence of cooperation on CI and context and the cumulative number of different game types. This result gives evidence that the quantities included in the model are essential for the explanation of the phenomena observed. Further investigations of the characteristics of the model are being performed in order to clarify the specific contribution of each term in the success of the simulation.

References

1. Andreoni, J., Miller, J.: Rational cooperation in finitely repeated Prisoner's Dilemma: Experimental evidence. Econ. J. 103 (1993) 570-585
2. Antonides, G.: Mental accounting in a Sequential Prisoner's Dilemma game. J. Econ. Psychol.15 (1994) 351-374

3. Axelrod, R: The Evolution of Cooperation. Basic Books (1984)
4. Camerer, C., Ho, T.-H., Chong, J.: Sophisticated EWA Learning and Strategic Teaching in Repeated Games. J. Econ. Theory 104 (2002) 137-88
5. Colman, A.: Game theory and its applications in the social and biological sciences. Oxford: Butterworth-Heinemann Ltd. (1995)
6. Colman, A.: Cooperation, psychological game theory, and limitations of rationality in social interaction. Behav. Brain Sci. 26 (2003)139-153
7. Erev, I., Roth, A.: Simple reinforcement learning models and reciprocation in the prisoner's dilemma game. In: Gigerenzer, G., Selten, R. (eds.) Bounded rationality: the adaptive toolbox, Cambridge, Mass. MIT Press (2001)
8. Hristova, E., Grinberg, M.: Context Effects on Judgment Scales in the Prisoner's Dilemma Game. Proceedings of the 1st European Conference on Cognitive Economics (2004)
9. Jones, B., Steele, M., Gahagan, J., Tedeschi, J.: Matrix values and cooperative behavior in the Prisoner's Dilemma Game. J. Pers. Soc. Psychol. 8 (1968) 148-153.
10. Lockhead, G.: Psychophysical scaling: Judgments of attributes or objects? Behav. Brain Sci. 15 (1992) 543-601
11. Macy, M., Flache, A.: Learning dynamics in social dilemmas. PNAS 99 (2002) 7229-7236
12. Oskamp, S., Perlman, D.: Factors affecting cooperation in a Prisoner's Dilemma game. J. Conflict Resolut. 9 (1965) 359-374
13. Rapoport, A.: A note on the 'index of cooperation' for Prisoner's Dilemma. J. Conflict Resolut. 11 (1967) 100-103
14. Rapoport, A., Chammah, A. M.: Prisoner's Dilemma: A study in conflict and cooperation. Univ. of Michigan Press (1965)
15. Rapoport, A., Mowshowitz, A: Experimental Studies of Stochastic Models for the Prisoner's Dilemma. Behav. Sci. 11 (1966) 444-458
16. Schoemaker, P.: The expected utility model: Its variants, purposes, evidence and limitations. J. Econ. Lit. 20 (1982) 529-563
17. Stewart, N., Brown, G. D. A., & Chater, N.: Sequence effects in categorization of simple perceptual stimuli. J. Exp. Psychol. Learn. 28 (2002) 3-11
18. Stewart, N., Chater, N., Stott, H., Reimers, S.: Prospect relativity: How choice options influence decisions under risk. J. Exp. Psychol. Gen.132 (2003) 23-46
19. Vlaev, I. & Chater, N.: Effects of Sequential Context on Judgments and Decisions in Prisoner's Dilemma Game. Proceedings of the 25th Annual Conference of the Cognitive Science Society. Erlbaum, Hillsdale, NJ. (2003)

Context-Aware Configuration: A Study on Improving Cell Phone Awareness

Ashraf Khalil and Kay Connelly

Department of Computer Science, Indiana University,
150 S. Woodlawn Ave. Bloomington, IN, 47405
{akhalil, connelly}@cs.indiana.edu

Abstract. As the number of mobile devices we carry grows, the job of managing those devices throughout the day becomes cumbersome. This is especially true for cell phones. Despite the many benefits they provide, cell phones create problems that arise from a mismatch between the user's context and the cell phone's behavior. In large part, the mismatch occurs because owners do not remember to frequently update their cell phone configuration according to the current context. It is desirable for mobile devices to automatically configure themselves based on the context of the environment and user preferences.

Given the personal attachment between people and their mobile devices such as cell phones, context-aware automatic configuration may not be the preferred solution for users. We have conducted an in situ experiment to examine the feasibility, effectiveness and people's reactions to such a solution. Our results show that people prefer automatic configuration over configuring their devices by hand, and they are willing to adopt it in real life. The results also suggest that a hybrid, passive-active, context-aware configuration approach is preferred over a purely passive or active one.

1 Introduction

Cell phones are currently the most ubiquitous communication device the world over [1]. The mobile nature of cell phones is changing the way people have traditionally mapped activities to places [2]. Places usually dictate the structure of the activities that take place within, but cell phones are, to a large extent, loosening or dissolving that relationship. Wellman has described this as a shift from Place-to-Place communication to Person-to-Person communication [3]. This new social order has created both opportunities and problems for mobile phone owners and for society in general.

Mobile phones offer great accessibility and flexibility. No longer do people have to remain in a fixed location to carry on conversations over the phone. Having the ability to remain in constant contact with people via the phone also gives people an additional sense of security [4]. However, most people already consider cell phone use in public places to be annoying [5-7]. Wei and Leung [8] have conducted a large study that shows that when people are asked about the contexts in which they find cell phone use irritating, 81% responded restaurants or cafes, 80% answered classes or libraries and 79% cited airport or train stations. Bautsch, et. al [9] found that most people think there should be etiquette guidelines created for public mobile phone use. Many rough attempts can be found in newspaper and magazine articles by authors fed up with rude users. Wireless World gets biblical with the "Ten Commandments" of Mobile phone Etiquette [10]. It is not uncommon anymore to see a sign saying "No Cell Phones Allowed" in some public places. An increasing number of places, such as churches, commuter trains and even parliaments, as in India, are using cell-phone jammers to restrict cell phone usage. Despite the fact that jammers are illegal in most countries, more and more countries, such as Japan and France and Mexico are approving their use in public [11, 12].

In addition to annoying surrounding people, inappropriate calls can cause inconvenience, disruption and embarrassment for the owner. Such calls can also lead to an increased level of stress and errors, or even put the owner in a dangerous situation, as in the case of receiving calls while driving [13, 14]. The effect of interruptions has been shown to be disruptive to task performance even when the interruption is ignored [15]. Mobile phones create new dilemmas for users: Do they really want to be reached anywhere and anytime? What is the appropriate state for their cell phone in different places? And to whom should users give their number?

Most of the problems mentioned above can be reduced or eliminated by reducing the mismatch between the cell phone state and the context of the owner and the surrounding space. In other words, it is important to make cell phones more aware of their context and surroundings.

We predicted that context-aware configuration may contribute greatly in decreasing the mismatch and provide more socially acceptable cell phones. One approach is to identify a set of daily activities that have a consistent mapping between different activities and different configurations. We conducted an experimental study to examine the viability of such a solution. We report the findings of our study examining people's reactions to cell phones that are more aware and can change their configuration automatically.

Section 2 examines the existing solutions to this problem and describes how they fail. We briefly describe a preliminary study we performed on workday context in Section 3, and give the details of our in-situ study in Section 4. Section 5 goes over our results, including the accuracy of the context input we used, the consistency of the mappings between context and configuration in our user population, and overall user feedback. We conclude in Section 6.

2 Existing Solutions

Many solutions have been proposed to address the social disturbance caused by cell phone interruptions. We categorize these solutions as follows:

2.1 Human Intervention

One proposed solution is to have the government and policy makers formulate regulations concerning cell phone use in public places [8]. Another one is by educating the public about cell phone use etiquette. This may reduce the problem but will not likely resolve it since cell phone interruptions often occur simply because the owner has forgotten to switch his cell phone to the correct configuration. Another proposed solution is to provide callers with contextual information about the receiver to help them make more educated decisions about the appropriateness of the call before making it [16-18]. The contextual information can be entered explicitly by the phone owner or automatically by other means such as sensors. This type of solution, however, leaves many questions unanswered:

- What type of information should the receiver publish?
- How often should information be published/updated?
- Will people adopt a manual solutions that requires extra effort and time?
- Would context information improve the match between the caller's and receiver's expectations?
- In terms of privacy, will receiver's be willing to publish information, and to whom?

We believe the caller-based approach is somewhat orthogonal to our work, in that the context information could be published automatically. However, there are so many privacy issues involved with publishing one's context that it could interfere with our test results, so we decided not to pursue this approach at this time.

Another proposed solution provides cell phone owners with more configuration options and thus more control over their devices. Quiet Calls is a system that uses pre-recorded messages to carry on a conversation discreetly without the need to talk [19]. Additionally, Calls.calm uses web pages to provide information about the receiver's context and enables the caller and the receiver to interact to determine whether or not to continue with the call [16]. Such a system enables the recipient to determine the context to send to the caller in real-time. This kind of solution may decrease the level of the interruption for people around the receiver, but does not eliminate it for the receiver, as she is still expected to receive the call and act upon it. Additionally, it is expected that the caller will respond appropriately to the receiver's preferences.

2.2 Automatic Intervention

It is clear that the solutions cited above solve part of the problem. One main drawback is the need for the receiver to frequently update his published information to reflect the current context. This overhead has been stated as one of the main drawbacks in most collaborative systems [20]. The other main drawback of publishing one's context or information is the privacy violation. As a result, most systems have decreased the details of published personal information to overcome the privacy concerns, but this has not allowed callers sufficient information to make accurate judgments. We propose instead to empower cell phones with more capabilities and options so that they are more autonomous and flexible to adapt their state (i.e. on, off, loud, vibrate, quiet) dynamically according to the owner's context. This, together with having more informed callers, can provide a better solution to the problem.

The awareness of cell phones can be greatly improved by augmenting them with the capabilities of gathering information about the owner (receiver) and inferring the appropriate state or behavior. Contextual information can be gathered by sensors or from other cues such as the calendar book. As part of the SenSay project [21] researchers at CMU have augmented the cell phone with many different sensors to capture the context of the owner. The cell phone's behavior is changed dynamically depending on the owner's context. Schmidt et. al. introduced an adaptive cell phone that changes its profile automatically based on the recognized context [22]. The phone chooses to ring, vibrate, adjust the ring volume, or keep silent depending on whether or not the phone is on a table, in a suitcase, outdoors, or in hand.

However, given the personal connections people feel toward their cell phones, more aware and autonomous cell phones may not be a welcomed idea for many people. Moreover, people's sense of control decreases as a cell phone's autonomous capabilities increases [23]. This study is set to examine the validity of such a solution. We have designed an experiment that allows us to measure people's reaction to having more autonomous cell phones.

3 Preliminary Survey

The experiment was conducted in two stages. Preliminary data was collected in the first stage to help us better design the main part of the experiment. The benefit of a two-stage experimental approach in the context of Ubicomp was argued by Antifakos et al. [24]. The goal of the survey is to investigate how people categorize their daily activities as well as the variation of this categorization across different groups. The data was gathered by an online survey. We had a total of 72 participants divided among graduate students, undergraduates, professors and staff. The participants were distributed among 7 different majors or areas of study. The survey results show that the participants tend to do very similar activities irrespective of their major or occupation. However, we found that the frequencies of activities are different among different groups. Table 1 details the users' most frequent activities. This list of categories was used in the latter part of the experiment to ease the process for the users.

4 Experiment

The main goal of the experiment was to assess the likely value of the automatic device configuration approach. We examined whether automatic cell phone configuration, based on the user's context, improves the overall user experience. We also investigated the approach for automatic configuration and whether it should be passive, where users are aware of the change and have more control over it, or active, where the change is made without any notification and the user has less control over it.

Table 1. List of the most common activities

Activity Category	Different Labels (given by participants)
Meeting	Meeting with advisor, meeting students
Email	Email, checking email, read email
Food eating & Preparation	Lunch, cooking dinner, eat dinner, breakfast. Prepare dinner
Researching and Studying	Work on research, projects, reading papers, research, study, Homework, work on the lab
In transit	On road, drive to campus, return home, travel to work, moving from apart to school, take a bus home,
Classes	Seminar, class, attend classes,
Recreational Activity	recreational activity, work out, karate training,
Teaching	Teach,
Relaxing	Relax, watching TV, nap,
Sleeping	Sleep, nap,
Office Hour	Office hour,
Presentation	Presentation, seminar

4.1 Natural Setting

System evaluating in a natural setting is the best way to provide accurate data. This is especially true for Ubicomp systems because it is their inherent nature to interact with users in their natural environment. With that in mind, we chose to conduct our experiment in a college setting. College environments offer an ideal place for the development and testing of ubiquitous systems. They are very dynamic and active places with different groups interacting. Moreover, mobile devices and especially cell phones are extremely common on college campuses and students greatly depend on their cell phone to organize activities and keep in touch with their friends and family. In addition, campuses are highly connected environments with extensive support for mobile and wireless computing.

Many of the early ubiquitous technologies were deployed and tested in campus environments. Weiser [25] predicted that the compact nature of the campus environment will put it at the forefront of ubiquitous computing. The Active Campus project [26], designed for campus environments, is one of the largest ubiquitous computing projects in terms of its scale and the services it provides. The Aware Campus Guide [27] is another example of the early ubiquitous applications that have allowed users to annotate physical space with text notes. Several other ubiquitous applications have also been designed and deployed on campus environments to enrich students' classroom educational experiences.

4.2 Method

Design: The study consists of a context-aware cell phone configuration application. The application simulates a cell phone that changes its configuration (loud ring, quiet ring, vibrate, on, off) depending on the context of its owner. The context is derived from the calendar book. During the study, the participant carries a Palm PDA that runs the application, and during the day she receives simulated phone calls at random times (Figure1). According to the context of the participant, the application notifies the user differently about the received call. The cell phone configuration can be in any of four different states: Loud, Quiet, Off, and Normal. In the loud state, the phone rings loudly when a call is received, while it vibrates in the Quiet state. Normal state is the default state that takes whichever configuration has been set up by the owner. In the Off state, the phone is off, and if a call is received then a voice mail message will be generated the next time the phone is in any other state. Moreover, if the participant misses a phone call, he will be notified of that missed call the next time he answers a phone call. The four different states were identified from the online preliminary survey mentioned in the previous section.

After receiving the notification of either an incoming call, missed call or a voice mail message, the participant is asked whether the configuration of the cell phone, reflected by the notification mechanism, is appropriate or not. If the answer was inappropriate, then she is asked to select the most appropriate configuration. After that, the participant is asked to select his location and activity.

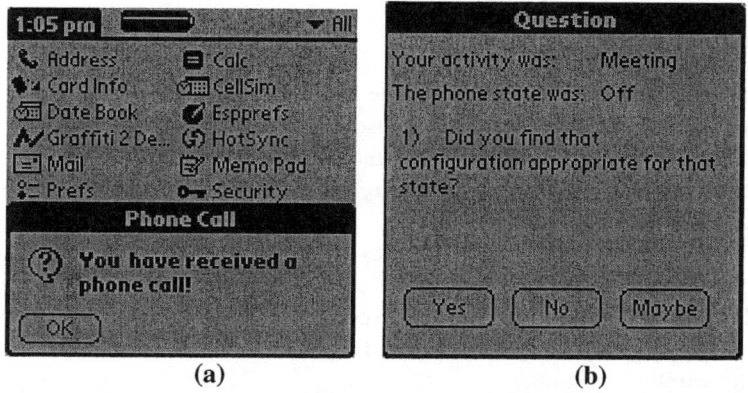

Fig. 1. Figure (a) shows the notification message that appears once a call is received. Figure (b) shows the question asked once the user press the "Ok" button on figure (a)

We chose to use a cell phone simulator that is running on a Palm PDA instead of using a real cell phone because the PDA provided us with more programming flexibility and with greater means to collect, store and manage the data in the field. At the same time, the selected PDAs had the same notification capabilities as cell phones, such as ringing, vibrating, LED, and volume control. We were only interested in measuring the appropriateness of the configurations in terms of social ramifications of them rather than the identity of the callers or any other factors. The simulation provided us with more control over the study, which enabled us to examine only the factor of interest while eliminating others such as caller identity.

Duration: The experiment duration was chosen to be 5 working days. This period was selected because most activities are repeated in either daily or weekly intervals. In addition, we conducted the experiment only during the week rather than on the weekends because we were mostly interested in the days when the participants are busy and interactive in a campus environment. In this case, the cost of interruption or misconfiguration is rather higher for both the user and the surroundings and thus the value of the application is highlighted.

Participants: 11 students both graduate and undergraduate from Indiana University participated in the study. Participants were aged 20-28 and 3 of them were males. All participants reported to have owned cell phones for more than a year and have busy daily schedules with many different activities throughout the day. 10 participants fully completed the study. One participant collected very little data due to a family emergency. This data was not considered in the evaluation process.

Equipments: The study was conducted using Tungsten T3 running Palm OS 5.2 and our cell phone simulator. The devices are equipped with ringing, vibration and volume control capabilities as well as a color display. Each participant was provided with a PDA for the duration of the study.

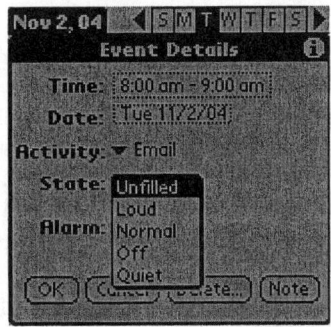

Fig. 2. After filling in the activity in the calendar book, the participant is asked to map it to the most appropriate configuration

Procedure: Participants were individually given a brief overview on how to use the PDA and then they were introduced to the cell phone simulator and how to use it. They were asked to fill in the calendar with their activities at the beginning of every day of the study with all the activities that last at least 15 minutes. Every activity is mapped by the participant to the cell phone state that best fits that activity as shown in figure 2. The participants were advised to think of the PDA as their own cell phone that is changing its configuration dynamically depending on the owner's context. After finishing with the experimental study, end-of-study interviews were conducted in one-on-one sessions that lasted approximately 40 minutes.

Design Tradeoffs: The fact that participants received simulated phone calls rather than real ones might have introduced some bias in their evaluation of the calls. In order to treat all calls with the same level of importance and factor out personal preferences, we asked participants to think of the calls as received from anonymous callers. With the simulated phone calls, participants still had to deal with social ramifications of receiving calls in public spaces and with inappropriate alerts that could have been caused by the calls. Also, in most cases, the mapping of activity to configuration should not be affected by the fact that the calls are simulated.

5 Results

During the study, a total of 340 calls were made, all generated by the simulator. Participants received an average of 30 calls and 4 voice mails. Even though participants missed 31% of the initial calls, they received reminders about many of the missed calls, and thus they had the chance to evaluate them. Overall participants evaluated 85% of all the calls; the rest were not evaluated due to the fact that the application only stored a partial list of missed calls and the participants were reminded only about the last three missed calls. In addition, in some cases, the Palm device had to be reset during the study, and thus a few stored reminders were lost.

5.1 Evaluating Context-Aware Cell Phone Configuration

Overall participants rated 87% of the evaluated calls as having the appropriate configuration and 9% as having an inappropriate configuration. The rest were evaluated as not having the exact configuration, as in the case where "Maybe" was selected to answer the question in Figure 1.b. Out of the missed calls that were later evaluated, 36% were missed unintentionally due to the fact that participants failed to notice the alert, usually due to low volume, and the rest were missed intentionally. Neither intentionally or unintentionally missed calls were directly interpreted as having an inappropriate configuration, and in most cases participants did not mind missing the calls because they did not want to be interrupted. Only 14% of the missed calls were evaluated as having an inappropriate configuration. One participant commented that it is worse to get embarrassed from having the phone ring in the middle of a meeting than to miss a phone a call. Most of the calls with inappropriate configuration were received when the participants were either in transition between activities or dealing with unplanned activities such as 'on the phone', 'taking a break' or 'having a conversation'.

During the end of the study interview, participants were asked to rate the usefulness of the application on a scale of 1 to 6, with 1 being the most useful and 6 being annoying. 40% of the participants rated it as very useful while the rest rated it as useful. These results are particularly interesting given the fact that 9%-13% of the calls were evaluated as having inappropriate or inexact configuration and that these were received in a real-life environment that could have caused frustration or embarrassment for the participants. The fact that evaluation occurs after notification makes the evaluation very accurate and reflects the real feeling of the participant that could not be obtained otherwise.

This shows that the positive value of the context-aware configuration is strong to the extent that people are willing to accept a certain level of inaccuracy in return. All participants were willing to use such an application in real life if their cell phones were equipped with it. All participants expressed a liking for the fact that configuration changed automatically depending on the context. Comments included:

> *I did not have to worry about forgetting to set the ringer to the appropriate setting when entering a certain situation.*

> *I like how it changes state without you having to tell it to. I always forget to turn my cell[off] in class and turn it on after.*

When the participants were asked about whether inaccuracies in the system frustrated them, all of them answered negatively. However, two of the participants said it was inconvenient at some times.

5.2 Consistency of Mapping Activities to Configurations

One important goal of the experiment was to examine how people map their activities to different configurations and to check for consistency in the mappings. In order for a context-aware configuration to be determined automatically by the cell phone and not as specifically directed by the owner, as in our experiment, there needs to be a predictable pattern of mapping from activities to configuration. Such consistency has ramifications in the broader field of context-awareness in the sense of whether a given context provides data reliable enough to generate the same behavior every time it occurs.

Our results showed that, for each individual, mapping from activities to configurations is consistent, i.e. each activity very frequently mapped to the same configuration. The results showed that 89% of activities have a predictable desired configuration for an individual. This consistency, however, did not hold across participants. This means that for the same activity people map configurations differently. This could be caused by the fact that activities are not the only factors that dictate the appropriateness of public interruption and other personal factors come to play. As a consequence, mapping can not be generalized across participants, and personalization is needed to reach the best daily configurations for each individual. A more detailed discussion can be found in [28].

5.3 Preferred Interactivity Level

Context-aware applications often provide for different levels of interactivity with the users. Chen and Kotz identify two different categories for context-aware computing based on their interactivity: passive context-awareness and active context-awareness [29]. Passive context-awareness offers context information but leaves the application's action or behavior to be determined by the user. On the other hand, active context-awareness autonomously changes the application's behavior without the user's explicit approval. Barkhuus and Day have since introduced *personalization* as a third level of interactivity [23]. Personalization in applications allows the user to specify the exact application behavior or settings for a given context. In our study we examined the preferred type of interactivity for context-aware configuration. All participants have owned cell phones for more than 6 months and thus have experienced personalized interaction, while our application provided them with active context-awareness throughout the experiment period.

As discussed earlier, all participants highly ranked the usefulness of the system, and all were willing to use automatic context-aware configuration in real life if their cell phones were equipped with it. However, participants differed on the level of interactivity with the application they were willing to accept. During the interviews, participants were asked whether they would like to be notified in the case of any automatic configuration change that is trigged by a context switch. All participants reported wanting to be notified but with a varying level of frequency. Two participants wanted to be notified before any configuration change, while the rest wanted to be notified only for certain kinds of dramatic configuration changes. For example, two participants wanted to be notified when the configuration is turned to "Loud" state while 3 others wanted to know when the configuration changes to "Off" state. Thus a hybrid context-aware configuration also is preferred over a stand-alone passive or active version. These responses show that both passive and active context-awareness are preferred over personalization. Barkhuus and Day obtained the same results when they used different context-aware services for mobile telephony [23].

5.4 Controls Versus Convenience

Naturally, any context-aware application takes some control from the user in exchange for the convenience and benefits of the services provided by the application. As a result, designers must constantly deal with the limit of control the users are willing to give up. This is directly related to the issue of interactivity level discussed in the previous section. The three levels of interactivity provide for varying levels of control and convenience; personalization offers the most control and the least convenience while the level of control decreases and that of convenience increases with passive and active context-awareness, respectively. As part of our experiment, we wanted to indirectly evaluate the willingness of participants to concede some of the control they have over their devices for future context-aware smart spaces or even smart devices that are equipped with sensors capable of providing context information. This inquiry is also relevant in the case of spaces initiating a particular device configuration [30].

Participants were asked whether they would be willing to use the service even if they could not be explicitly involved in context mapping or in deciding about the nature of the mapping from activities to configurations. 40% of the participants answered negatively while the rest answered positively. This shows the importance of the involvement of users in the decision making process of context-aware applications. One way of achieving that is to follow the accountability and intelligibility design principles proposed by Bellotti and Edwards [31].

6 Conclusion

We have presented a study to evaluate the feasibility and the effectiveness of context-aware configuration for cell phones. We first introduced the problems caused by the wide-spread usage of cell phones and then the different types of solutions that have been proposed to solve them. The main goal of all the solutions is to create more socially acceptable cell phones by decreasing their disruption, embarrassment, and annoyance for both the user and the surrounding environment. We tested for context-aware configuration as a viable solution for the problem. Though different projects [21] have explored the idea of context-aware configuration for cell phones, we are the first to study its viability and effectiveness from the user's perspective. Toward that, we conducted an experiment to examine the solution in a natural university setting.

Our results suggest that context-aware configuration provide a very desirable solution that is found to be preferred over the personalization approach. Further, our results suggest that for this solution to be adopted, it is very important for the users to be involved in the process of mapping context or activity to configuration. This is especially true because the results showed that there was a poor consistency of mapping activity to configuration across different participants. This has ramifications for the broader field of context-aware applications in which designers tend to generalize the application's behavior for a given context across different users and environments. Such generalization should not be assumed without rigorous examination of variation across different users and environments.

Moreover, our results suggest that a hybrid approach of both active and passive context-aware configuration is preferred over either individual solution, as a hybrid approach appears to provide the right balance of convenience and control. This was the case for our application, though further examination is needed to make more conclusive results.

The fact that the study uses simulated phone calls may have caused some bias in how people evaluated the appropriateness of some configurations. It is possible that people are less annoyed when they miss a simulated phone call as opposed to a real call, but this naturally depends on other factors such as the identity of the caller and the message of the call. Also, we expect people to be more accepting of interruptions made by friends or significant others as opposed to interruptions made by anonymous callers like those in our study. Even though we did not specify the identity of the callers in simulated phone calls, participants stated that they thought of the calls as having been made by anonymous callers. Still, this study serves as a starting point for evaluating the feasibility of the context-aware configuration approach. To achieve a more complete understanding of the approach, future studies with real cell phones are needed to account for roles played by other factors that could not be measured in our simulated cell phone study.

A recommended approach to future work is to integrate the approach of empowering cell phones to be more context-aware with the approach of empowering the caller to be aware of the receiver's context. We believe a promising solution is the one that offers the right balance between the two approaches. This is due to the fact that interruption appropriateness can only be determined in the context of both the initiator and the receiver [32]. For example even if the receiver is in a meeting, he still might be awaiting a call from somebody regarding updates related to the meeting. Interruption in such situations is appropriate even if it may seem otherwise from an outsider's perspective. It is also important, and it is in our plan, to extend our results to non-college settings and include weekend days in future studies.

References

1. REUTERS, *Mobile phone users double since 2000*. Dec, 2004.
2. Agre, P.E., *Changing places: Contexts of awareness in computing*. Human-Computer Interaction, 2001. **16**(2-4): p. 177-192.
3. Wellman, B., *Physical Place and Cyber Place: The Rise of Personalized Networking*. International Journal of Urban and Regional Research, 2001. **25**(2): p. 227-52.
4. Palen, L., M. Salzman, and E. Youngs. *Going Wireless: Behavior & Practice of New Mobile Phone Users*. In Proceedings of *ACM 2000 Conference on Computer supported cooperative work (CSCW '00)*. 2000. Philadelphia, PA.
5. Monk, A., et al., *Why are Mobile Phones Annoying?* Behaviour and Information Technolog, 2004. **31**(1): p. 33-41.
6. Wadler, J., *The nuisance of overheard calls: Cell phones are everywhere but good manners may not be*, in *New York Times*. 1998. p. A41.
7. Lasen, A., *A comparative Study of Mobile Phone Use in London, Madrid and Paris*. 2002
8. Wei, R. and L. Leung, *Blurring public and private behaviors in public space: policy challenges in the use and improper use of the cell phone*. Telematics and Informatics, 1999. **16**(1-2): p. 11-26.
9. Bautsch, H., et al., *An investigation of mobile phone use: a socio-technical approach*. 2001
10. Briody, D., *Thou shalt learn and abide by the ten commandments of cell phone etiquette*, in *InfoWorld*. 2000. p. p.59B.
11. Wylie, M., *Cell-phone jammers enforce quiet, but illegally*, in *Cleveland Plain Dealer*. 2000.
12. *Cell-phone jammers answer prayers of some*. 2004, Associated Press. p. Appeared on Nov. 2, 2004, editions of the Milwaukee Journal Sentinel.
13. Eyrolle, H. and J. Cellier, *The effects of interruptions in work activity: Field and laboratory results*. Applied Ergonomics, 2000. **31**: p. 537-543.
14. *Study: All cell phones distract drivers*. August 16, 2001, CNN.
15. Cutrell, E., M. Czerwinski, and E. Horvitz. *Notification, Disruption, and Memory: Effects of Messaging Interruptions on Memory and Performance*. In Proceedings of *Interact 2001*. 2001. Tokyo, Japan.
16. Pedersen, E.R. *Calls.calm: Enabling Caller and Callee to Collaborate*. In Proceedings of *CHI 2001*. 2001.
17. TANG, J.C., et al. *ConNexus to awarenex: extending awareness to mobile users*. In Proceedings of *CHI 2001*. 2001.
18. Milewshi, A.E. and T.M. Smith. *Providing Presence Cues to Telephone Users*. In Proceedings of *CSCW 2000*. 2000. Philadelphia, PA.
19. Nelson, L., S. Bly, and T. Sokoler. *Quiet Calls: Talking Silently on Mobile Phones*. In Proceedings of *CHI 2001 Conference on Human Factors in Computing Systems*. 2001.
20. Grudin, J., *Groupware and Social Dynamics: eight challenges for developers*. Communications of the ACM, 1994. **37**(1): p. 92-105.
21. Siewiorek, D., et al. *SenSay: A Context-Aware Mobile Phone*. In Proceedings of *IEEE International Symposium on Wearable Computers (ISWC)*. 2003. New York, NY.
22. Schmidt, A., et al. *Advanced interaction in context*. In Proceedings of *First International Symposium on Handheld and Ubiquitous Computing (HUC'99)*. 1999. Karlsruhe, Germany.
23. Barkhuus, L. and A.K. Dey. *Is context-aware computing taking control away from the user? Three levels of interactivity examined*. In Proceedings of *UBICOMP 2003, 5th International Symposium on Ubiquitous Computing*. 2003.

24. Antifakos, S., A. Schwaninger, and B. Schiele. *Evaluating the Effects of Displaying Uncertainty in Context-Aware Applications*. In Proceedings of *Ubicomp'04. 6th International Conference on Ubiquitous Computing*. 2004. Nottingham, UK.
25. Weiser, M., *The Future of Ubiquitous Computing on Campus*. Communications of ACM, 1996. **41**(1): p. 41-42.
26. Griswold, W.G., et al., *ActiveCampus - Experiments in Community-Oriented Ubiquitous Computing*. IEEE Computer, 2004. **37**(10): p. 73-81.
27. Burrell, J., et al. *Context-Aware Computing: a test case*. In Proceedings of *Ubiquitous Computing*. 2002. Gothenburg, Sweden.
28. Khalil, A. and K. Connelly. *Improving Cell Phone Awareness by Using Calendar Information*. In Proceedings of *INTERACT 05*. 2005. Rome, Italy. To Appear.
29. Chen, G. and D. Kotz, *A survey of context-aware mobile computing research*. 2000, Department of Computer Science, Darthmouth CollegeTR2000-381.
30. Connelly, K. and A. Khalil. *Towards Automatic Device Configuration in Smart Environments*. In Proceedings of *UbiSys Workshop*. 2003.
31. Bellotti, V. and W.K. Edwards, *Intelligibility and accountability: human considerations in context aware systems*. Human Computer Interaction, 2001. **16**(2-4): p. 193-212.
32. Dourish, P. and V. Bellotti. *Awareness and Coordination in Shared Workspaces*. In Proceedings of *ACM Conference on Computer Supported Cooperative Work (CSCW)*. 1992.

Context-Aware Adaptation in a Mobile Tour Guide

Ronny Kramer, Marko Modsching, Joerg Schulze,
and Klaus ten Hagen

Department of Computer Science,
University of Applied Sciences Zittau / Görlitz, Germany
K.tenHagen@HS-ZiGr.de

Abstract. Tourists who don't select standard guided tours have to rely on a map or on signs to explore an unknown city on their own. They often end up looking for information about an attraction they're at – but there's none available. They need a guidance leading them wherever they want, giving them information about whatever they find and furthermore care for their return in time. This is the main objective of the Dynamic Tour Guide (DTG). The DTG is a mobile agent enabling a personalized spontaneous guided tour. It selects attractions, plans an individual tour, provides navigational guidance during the execution and offers environmental information. This kind of ambient intelligence is based on the analysis of all available contextual information to support the tourist in any possible way with the help of mobile devices.

1 Motivation

Tourists sometimes stand in front of closed facilities. They would have needed current information in advance, since e.g. museums have different opening hours or might offer special expositions. On a summer weekend restaurants might be fully booked, whereas in November many restaurants will be closed. Because of this lack of information (availability), many tourists are unprepared and thus following signs, studying maps or attending a guided tour on the spot. As human tour guides generally serve groups of tourists they follow predetermined routes to the major sights. Therefore the majority of the tourists end-up on the beaten tracks. Interesting sights just a couple of hundred yards off the main tourist arteries are rarely visited.

The ideal is to have a local guide, who understands the individual interests and timeframe, knows the local situation, gives a personal tour, and fits into a pocket. This is the objective of the Dynamic Tour Guide (DTG). The purpose is to devise a tour, just like an expert guidance would do after getting to know a tourist's preferences, by means of new technologies like semantic matching and context aware computing.

Section 2 will deal with related projects. The next two sections will describe the concept of the DTG and its architecture. Following is a definition of the term context in particular. The main emphasis is placed on the sections 6 to 10 which describe the realisation of the DTG with focus on the tour adaptation due to contextual changes and the interaction with common navigation software available for Pocket PC's.

A conclusion shall summarize the most important innovations and point out possibilities for future research.

2 Related Work

Tour Guides have been an important topic for research activities for a long time. The following important projects aim at similar developments to the DTG:

- GUIDE [3] is a mobile tour guide which uses cell based positioning instead of GPS. The tourist can select sights by categories which he wants to visit during a tour. A route is computed. Considering closing-times, the order can be adapted. Reaching a point of interest (POI), context-sensitive information is provided.
  →The DTG creates a tour automatically using the tourists' personal interest profile. The concepts of tour adaptation and context-driven interpretation are mentioned but the DTG covers them much further.
- Cyberguide [1] was one of the first mobile tour guides. It works outdoor with GPS and indoor with infrared to determine context information like users' position and orientation. There is no tour computed, but the user can receive information but anything he/she sees, wherever he/she is. Requesting a route to a desired POI is possible too. In addition it shall provide the option to create a kind of diary about the whole tour.
  →The main difference is the computation of a whole tour based on the tourists' interests by the DTG, which demands solutions for additional challenges.
- The Crumpet project [16], [4] enables a mobile agent to find certain sights, to present them on a map and to calculate a route to a selected one.
  → The sights are found because of their locality. The user has to decide for her-/himself whether they are interesting for her/him and if she/he has enough time to visit them.
- The software developed by eNarro [8] provides predetermined tours presenting the most important sights in many big cities all over the world. The tourist needs a PDA with a special player and the content for the particular tour. She/he also has to have navigation software which will lead her/him to the different places. The attractions are then presented using audiovisual information.
  → The user can only select an existing tour for available cities. The sights being shown to the tourists are pre-selected to suit the interests of a broad demographic.

Predetermined or self-made tours are not the objectives of the DTG, it intends to generate an individual tour in real-time. Additionally it pays attention to the local situation like opening hours by always having up to date information via web services. All together it addresses the following challenges:

- Acquisition of a tourists interests in a mobile context to seed the profile
- Ranking of TBBs by semantic matching
- Computation of a tour in less than 5 seconds
- Context aware interpretation of the environment
- Tour tracking and adaptation

3 Concept

A tourist would like to spontaneously explore a destination. She/He has an interest profile, a start and end point and a given time period. This is the personal context of the tourist which needs to be mapped with the environmental context at the destination. Context means all available information at a certain location for a certain time, what will be defined more precisely in one of the next sections. The challenge is to compute an optimal tour given the personal and local context. The tourist can modify the proposed tour. During the execution of the tour the tourist will be guided to the next Tour Building Block (TBB), e.g. a sight or restaurant, using standard navigation software, like Mappoint or Navigon [14].

When the tourist starts walking the DTG is able to measure the actual walking speed of the tourist on this day given the conditions of the sidewalks and streets. This update of the personal context might make a recalculation of the entire tour necessary. As soon as a tourist approaches a point where a TBB becomes visible the DTG will provide introductory information suitable to the direction from which the tourist is approaching the TBB. Otherwise the tourist might soon get disorientated when approaching from the opposite side. As long as he is situated close by the TBB, the tourist will receive information about it via a headset. Some tourists will decide to explore the TBB further by e.g. walking into the court yard. In this case additional information appropriate to the current context is provided. As soon as he leaves the TBB, the DTG will stop providing further information and restart the navigational guidance towards the next TBB. In the case the tourist stays much longer than initially assumed the tour for the remaining amount of time is recalculated.

On the way to the next TBB some tourist might get distracted by another attraction be it another sight or simply a shop. Then the DTG will interrupt the navigational hints and provide information for the current context if available. In case of a spontaneous visit to a local store the DTG will simply wait for the tourist to leave the location to continue on a tour recalculated for the remaining amount of time. Despite the navigational guidance through audio hints some tourist might get on the wrong path. The navigation software will try to get the tourist back on a path towards the next TBB. However beyond a certain deviation it might be more meaningful to adapt the sequence of TBBs to the new location of the tourist.

4 Architecture

Each sight, as a possible component of the tour (TBB = Tour Building Block), is semantically modeled by a content provider using the DTG AuthoringTool. Each TBB will have its own web service (WS). A service provider like a restaurant will wrap the local restaurant management system by a WS. This WS will provide the semantic model, current information, e.g. opening hours, and a transactional interface to e.g. reserve a table. The WSs of the TBBs are registered at a UDDI registry.

The DTG server is executing a semantic match algorithm to rank the sights for a specific tourist. A computationally more demanding task of the DTG server is the computation of a tour as a sequence of TBBs.

Audio hints and a map for navigation are provided by standard navigation software to guide the tourist to the next TBB. The DTG provides information about a TBB as the tourist approaches it. Furthermore it adapts the higher-level plan for the remaining time to the actual walking speed and staying time at a TBB.

Expectedly most people will own a mobile device in the next couple of years, cities will be covered with WLAN access points or GPRS/UMTS coverage will be available and DGPS will provide localization with a precision of at least 1 m (EGNOS/SISNeT, 2005 [6]; DGPS, 2005 [5]). This enables the following features and interactions:

- Localisation:
  The mobile device is aware of its current position, either in a city via e.g. the Global Positioning System (GPS-WAAS) or inside buildings like museums via WLAN, Infrared grids or RFIDs.
- Service discovery:
  After arrival at a destination the DTG will determine the next DTG server in a UDDI registry. Based on the personal context like the maintained interest profile and the time period set by the tourist, the DTG will discover the local context like sights and services at this destination, interrogate the corresponding web services to update the current information and then compute potential tours.

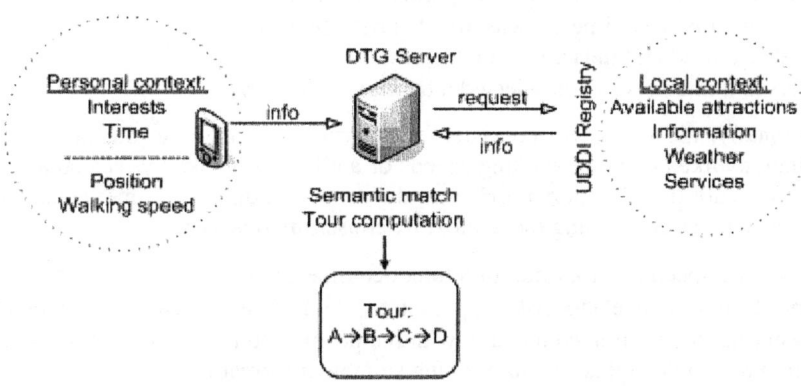

Fig. 1. Discovery of the environment

- Navigation:
  After the tourist has selected and optionally modified a tour, the local navigation software will visualize the tour on a map and guide the tourist via audio information. In the background the DTG will consistently track the execution of the ongoing tour for contextual changes, e.g. any deviations like changing walking speeds or additional breaks. Deviations beyond a certain threshold will trigger a recalculation of the tour to make sure that the tourist arrives at the desired endpoint in time.

5 Context Definition

Context spans the situational information. Any feature characterizing an entity and its environment determines its context. This context can be divided into different areas:

1. Personal context: The personal context includes ones personal information. It is defined by static elements like name or interests and dynamic elements like walking speed and current position
2. Local context: The local context consists of ones environmental information. These are for instance street and number of the actual position or the local weather.
3. Service context: The services context describes the available services. Static elements are information about a close-by sight, whereas data about a current exhibition or availability of a table in a restaurant are dynamic.

A context aware system is able to adapt its functionality because of filtered out contextual information [11]. This is called ambient intelligence; the personal context is mapped with the services context and the local one. The DTG does so by determining the sights nearby the current position and requesting their availability, rating them according to the personal interests and creating a tour limited to the specified duration. The computed tours strongly depend on the given contexts and thus are highly individual. Expectedly none of them is similar to another. Depending on:

- the available time period → the tours will differ in length.
- the current position → start- and endpoints are different
- the current time (daytime or season) → TBB's like restaurants or exhibitions are opened or closed and thus are available or not
- the personal interests → the selected attractions will vary

Additionally, the DTG will consistently supervise the ongoing tour and react on any deviations like changing walking speeds or additional breaks by recalculating the tour to make sure that the tourist arrives at the desired endpoint in time. Hence it has to react on changes concerning the context by constantly observing:

- The walking speed and tour duration to notice time problems
- The position to realize a tourist's approach to a sight or to get aware of distractions
- The walking direction in connection with the position to be able to call the tourist's attention to visible sights and start giving suitable information

6 Context Driven Tour Adaptation

Based on the tourist's personal context a tour starting at the current position, including sights according to the interest profile and ending in the given time frame is generated. Fig. 2 presents a concrete example of that scenario.

The first station of the tour is attraction A. The tourist is navigated to it starting from his/her original position. As the tourist arrives, appropriate audiovisual information is being presented on the mobile device as long as he/she keeps standing in front of it. As he/she leaves the attraction the presentation stops and the navigation to attraction B begins. On the way to attraction C the tourist suddenly leaves the path to go to another attraction (E) which seems interesting to him/her. Information about all sights in a specific area around the changing current position is being downloaded in advance. Thus information about attraction E is readily available too. As the tourist spends some time listening to this information and exploring the attraction, she/he looses time. The tour has to be recalculated, as C and D are not reachable both in the time left. The DTG alerts the tourist that it's time to go on with the tour in order to reach at least attraction C. After C the tourist is led to the desired end point.

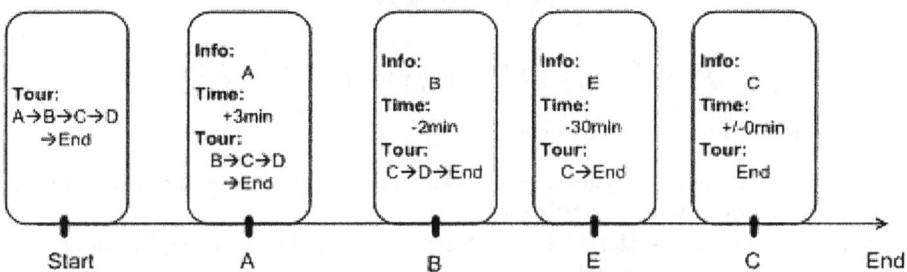

Fig. 2. Tour adaptation during execution

Instantaneous presentation of information applicable to the current context requires that all contextual information is available at the mobile device. Given the limited amount of memory of many mobile devices that is a challenge. However in a given time frame a pedestrian can only move within a very limited area. Therefore a preloader will be called at regular intervals, e.g. 5 minutes, to download all contextual information for the area the tourist can reach within the next 10 minutes. As a side effect this forward looking caching makes the DTG more robust in situation with incomplete mobile coverage.

7 Context Driven Information Provision

As mentioned above, the contexts are first used for the tour computation. The tourist's personal context, especially considering his actual position, interests and time frame, affects the selection of the sights. The contexts are secondly made use of for supervising the tour execution. The DTG is always aware of context changes which appear most clearly when approaching a TBB. The following figure shall demonstrate the interactions resulting from that scenario:

The relevant TBB is located near a certain street. It is modelled by storing all necessary data in an XML file. This also includes a separation of the area around the TBB into virtual rectangles. For every rectangle the upper left and the lower right coordinates are known. The coordinates of the tourist's current position are transmitted by a GPS receiver. To determine if the tourist is situated in one of the rectangles it needs to be checked whether his coordinates are smaller than the upper left and bigger than the lower right ones:

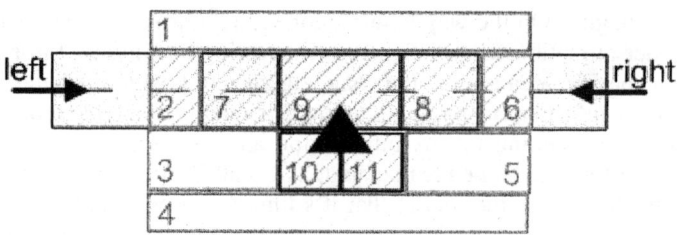

Fig. 3. TBB model

$$P^{up\_left}(x, y) \leq P^{tourist}(x, y) \leq P^{low\_right}(x, y)$$

If the tourist is situated in rectangle:

- 1-6 (green), he gets navigation hints towards the sight when approaching it or he is navigated back when departing too far
- 7 and 8 (red), he is alerted to the forthcoming sight when approaching it or navigated to the next one when leaving
- 9-11 (blue), he receives information about the sight via audio

Example:
The navigation software installed on the mobile device is performing the task of leading the tourist straight to the desired TBB. The tourist is approaching it coming from the left side. The required information derived from his personal context is his current position and his moving direction, which can be determined by comparing the actual position with the position before. While walking, his position steadily changes. The DTG permanently checks the position to detect entering a rectangle:

1. The tourist enters rectangle 2. As he hasn't been in nr 1, 3, 4 or 7 before, nothing happens.
2. The tourist enters rectangle 7. This is the position the sight can be seen from. As he comes from nr 2, he is alerted that he will soon reach the TBB on his right side.
3. The tourist enters rectangle 9. The DTG starts giving information about the sight like its architecture style, history and so on by playing an audio file.
4. The tourist moves around the TBB and enters rectangle 10. The DTG changes the audio file to provide information about the backside of the TBB.
5. The tourist departs from the sight and enters rectangle 4, maybe to take a photo. The DTG now stops giving information but alerts the tourist that he is leaving the right path and recommends going back.
6. The tourist walks back into rectangle 10. The DTG continues giving information to the tourist.
7. The tourist enters rectangle 11. The DTG keeps on giving information.
8. The tourist enters rectangle 9. As he comes from nr 11, the DTG still gives further information if available.

9. The tourist leaves entering rectangle 8. As he has been in nr 9 before, the DTG just stops giving information and lets the navigator do the navigation to the next station of the tour.
→ Please note the information presented in rectangles 7 and 8 strongly depends on the direction the tourist enters them from:
If she/he comes from the right side and reaches rectangle 8, she/he is alerted that she/he can see the TBB on her/his left hand. And if she/he then reaches nr 7 coming from 9, the information presentation will stop and switch to the navigation for the next TBB.

8 Interactive Navigation

The DTG relies on a standard navigation package for the task of guiding the tourist from one sight to another. This navigator is a separate program using his offline available geographical data for navigation and his stored map data for showing instructions and routes on a map. Furthermore the navigator gives instructions via audio to avoid the user holding the PDA in field of view all the time.

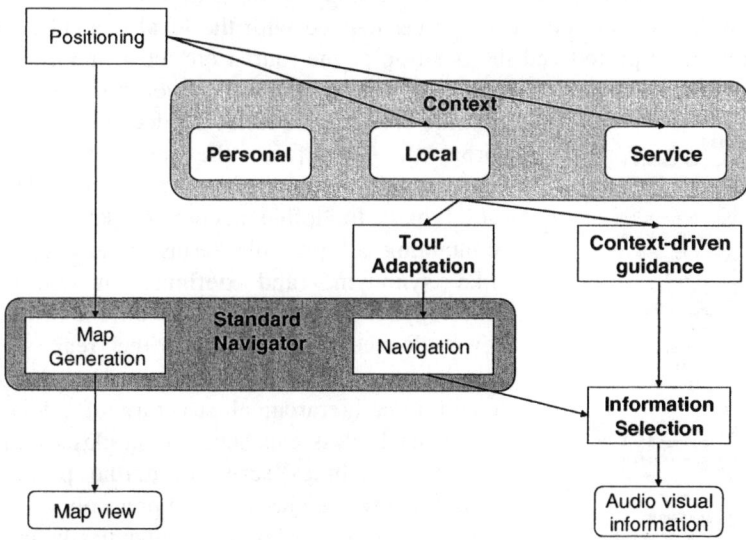

Fig. 4. Context-driven tour execution and information presentation

One of the main targets of the DTG is to ensure that the tourist gets back at the time he has established at the beginning. Therefore the continuation of the tour has to be observed constantly. The initial tour for example is calculated for a walking speed of 5 km/h. If the tourist walks slower through the streets, the tour has to be adapted by removing some TBBs from the tour. Another important issue is a sight the tourist discovers because it is located along the way but not a part of the tour. In this case the nav+igator permanently tries to bring the tourist back onto the regular route. Instead the DTG recognizes that the tourist is interested in this new sight and interrupts the navigator or adds the sight to the tour as a TBB. After that the rest of the tour has to be rearranged because the tourist hasn't the time to attend all other planned TBB's.

Also a significant point is the time spent at each TBB. The DTG will plan the tour based on average information. However the tourist might stay longer at each TBB and hence the estimates for the time at each TBB need to be increased. Again this might make a re-planning for the remaining tour necessary.

This kind of context-aware navigation can be compared with seafaring, where the navigator is the coxswain and the DTG is the captain who directs the whole trip.

Fig. 4 displays the complex functionalities of the DTG described above: the context driven tour adaptation, information provision and navigation by interacting with a standard navigator.

9 Semantic Matching

Selecting TBB's according to the personal interests of a tourist, semantic matching is applied. This task is different for any tourist as the contexts always differ. The personal context of the tourist has to be mapped with the local one. The interests, the available time period and the position of the tourist are most important. Based on this information a human expert can decide which tour would possibly fit best, but the challenge is to let the decision be made by a program. Therefore the computer needs to understand the meaning of certain data. The solution is to define a common knowledge base, containing all possible terms, arranging relations like synonyms and defining attributes – an ontology. It's a model of a specific area of reality. Every concept, existing in the real world, is displayed as a class. Relations between classes result in a hierarchical structure of all concepts, where each class can have parent classes and child classes. Attributes serve to define properties in order to describe classes more precisely.

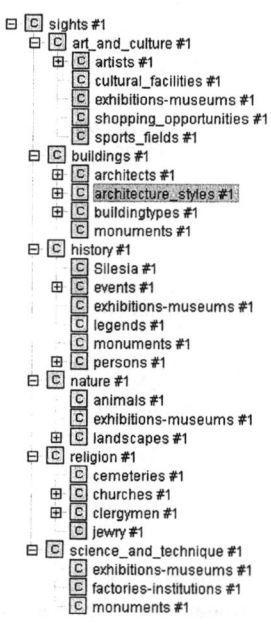

Fig. 5. Interest taxonomy

The ontology is used to semantically model the interests of the tourist and the TBB. At the beginning an ontology will be defined for a single destination. This ontology will have to be extended slightly in order to be used for other destinations in the same area. As the system is being applied to other regions it is important to maintain a hierarchical ontological system in order to enable reuse of the interest profiles. Otherwise a tourist would have to describe his interests from scratch whenever he enters a new region, which at best will lead to very shallow interests profiles.

All existing sights of a city (here: Goerlitz, 2003 [9]) are grouped into main categories of interests, which don't have anything in common. These are art & culture, buildings, history, nature, religion and science & technique.

Each category is subdivided. This allows a more precise modelling of interests. For example if a tourist is interested in buildings, he can either select a certain type of building like bridges, castles and so on, or he can opt architectural styles like baroque, art nouveau or others. The idea is that if a tourist's preference doesn't comply with a feature of a sight itself, but with a neighbour class (subclass or parent class) in the ontology, the tourist will probably be interested in that sight as well. That means that close-by classes are expected to be semantically similar so that relationships become visible easily. One example shall illustrate the way similarities are identified. If somebody is interested in animals (a subclass of nature), he's likely to be interested in nature in general. Thus also sights being described as landscape will satisfy his desires in some measure.

The TBBs are then sorted into this hierarchy by the content providers using an AuthoringTool to create the TBB models. Most TBBs will be listed in different branches of the hierarchy, e.g. a church might be listed under *Religion/Churches* and *Architecture styles/Middle ages/Gothic*. The sorting process results in the creation of an XML-profile that contains all chosen categories with all belonging superclasses.

The tourist is expressing her/his interests using the branches of the hierarchy. She/he will go through the exercise at one destination, and then rightfully expect that this investment will be reused at the next.

The ontology, the interest profile and the TBB models are used by the semantic match algorithm to compute the degree of similarity. As mentioned above, the ability to deal with several degrees of similarity is important, since if there aren't any sights available that cover the tourist's interests exactly, ones that meet related interests should be considered as well. These are the TBB's with a high amount of Interest Matching Points (IMP). Therefore the semantic match algorithm evaluates the hierarchical part of the ontology, which is a directed graph, with the given interest profile. The node presenting this interest is evaluated with 1. There are two functions the rest of the nodes can be evaluated with, whereas each node is restricted to have exactly one parent-node. Going up, the IMPs of the nodes are divided by two:

$$y \leftarrow f^u(x) = \frac{1}{2}x$$

Going down, the subnodes receive the same IMPs as their parent node:

$$y \leftarrow f^d(x) = x$$

Presumed node B was chosen as the starting point, an evaluated graph looks like this:

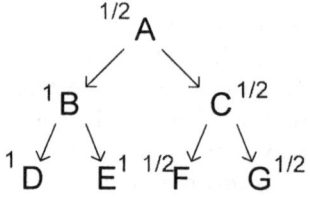

Fig. 6. Evaluated graph

For node D and E function $f^d(x)$ fits, so they also receive the IMPs of 1. Node A receives ½ because of function $f^u(x)$. C is rated with ½ by $f^d(x)$ starting in A, and then starting in C $f^d(x)$ rates F and G with ½ too.

Depending on the amount of interest fields in the profile the whole hierarchical structure is rated several times. Each time the instances receive points. At the end the points are summed up. The following example shall demonstrate that:

Shown in Fig. 7 is the ontological hierarchy including the TBB's. The interest profile of a tourist contains the following paths:

```
<interest>
        <class>baroque</class>
        <superclass>architecture</superclass>
        <superclass>building</superclass>
</interest>
<interest>
        <class>tower</class>
        <superclass>shape</superclass>
        <superclass>building</superclass>
</interest>
```

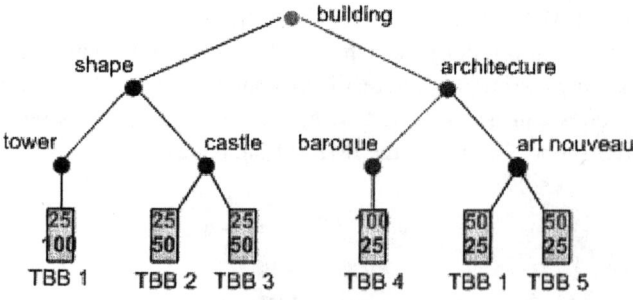

Fig. 7. Hierarchy with TBB's

All TBB's (rectangles marked with TBB 1 to TBB 5) receive points twice because of two different interest fields contained in the profile. TBB 4, a baroque building, meets the first interest of 'baroque' exactly and is rated with 1 which means a maximum of 100 points. For the second choice 'tower' the rate is ¼. All together it reaches 125 points. TBB 1 is a tower built in art nouveau style, therefore it belongs to two branches of the hierarchy. For the first valuation of the interest 'baroque' the node 'art nouveau' is rated with ½. Hence TBB 1 gets 50 points. The second valuation of the interest 'tower' results in a rate with value 1 for the node 'tower'. TBB 1 belonging to that node receives 100 points in addition. In total it gets 150 points, as only the maximal amount of points of each rating process is relevant:

```
foreach interest in profile
    foreach TBB
        TBB.IMP += MaxPoints(TBB, interest)
```

10 Tour Computation

After the semantic match algorithm has assigned IMPs to each TBB a tour can be computed. A valid tour is a sequence of TBBs that can be visited within the time allocated by the tourist. Each TBB has an average duration of visit. Since 20 TBBs with the same start and end point lead to $(20-1)!/2 = 6*10^{16}$ possible tours, valid tours can't be cached in advance and thus need to be computed online. The challenge is to compute a valid tour that maximises the IMP. When the tourist asks her/his mobile device to compute a tour she/he is most likely standing with the mobile device in her/his hands somewhere within the destination. Given that situation the tourist won't care too much if the tour presented to him after e.g. 5 seconds has a few less IMPs than the optimal tour. For most tourists the optimal tour is irrelevant – actually any tour – if the computation takes more than 5 seconds.

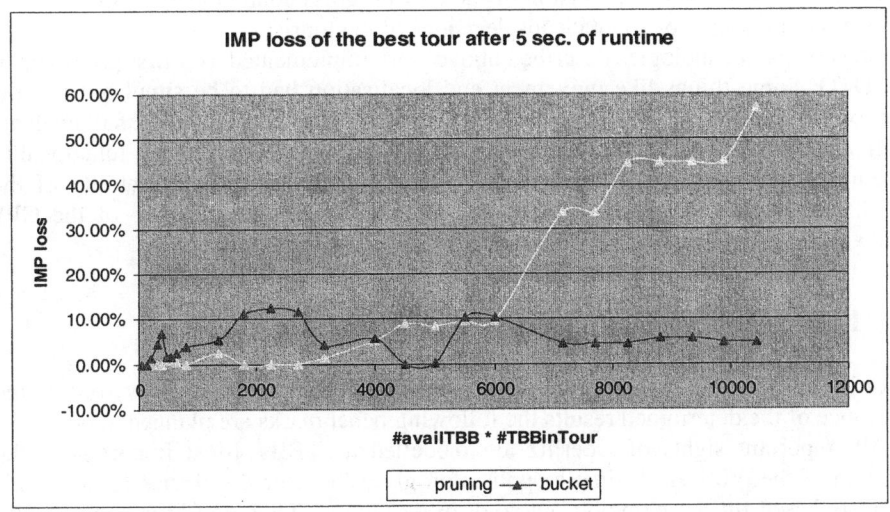

Fig. 8. IMP loss of the best tour after 5 sec of runtime versus the optimum

The used approximation algorithm is based on a depth first search. Fig. 8 compares two variants of the search algorithm for a suit of benchmarks. The y-axis is scaled by the product #availTBB * #TBBinTour. #availTBB gives the number of TBBs the algorithm can choose from. #TBBinTour is number of TBBs in a tour or the depth of recursion. This product is a measure of complexity. The y-axis is scaled by the reduction of IMPs compared to the optimal solution. The heuristics to select a TBB for the candidate list, sort the candidate list and to insert a new TBB are discussed in (ten Hagen et al, 2004 [17]).

The pruning algorithm removes a TBB from the list after it has been added to a tour. This is a deviation from the standard depth first, where a candidate is only removed from the candidate list for all nodes below the node of insertion. Given the insertion heuristics this removal avoids the computation of identical tours. The bucket algorithm divides the candidate list into chunks and processes them sequentially. The outcome of processing a partition of the candidate list is an approximation of the optimal tour, which is improved further using the candidates from the next bucket. As Fig. 8 clearly shows, the effect of working with a candidate list of finite length is extremely effective for a larger set of available TBBs. For a destination with 1000 available TBBs and 10 TBBs in a tour the complexity product would be 10,000 and the reduction of IMP still less than 7%.

11 Conclusion

The DTG uses innovative technologies in order to create individual, context-aware tours. Independent of location and time it determines the necessary information by detecting and interrogating available web services. It provides user guidance by giving navigation instructions and by offering the right information at the right time and place. A permanent supervision of the tour progress continuously adapts the tour to external influences or spontaneous decisions of the tourist.

Most of the technologies described above were implemented as a first prototype of the DTG. Some things like movement and localisation had to be simulated for this version but the system is working fine. A field study (see next chapter) shall evaluate both algorithms. Current developments tend to complete the whole functionality which will be evaluated in another field study concerning the applicability of the DTG. The main problem the functioning depends on is the precision of the GPS receivers and signals.

12 Future Work: Field Studies

In order to check the usability of the semantic matching technology as well as the relevance of the determined results the following benchmarks are planned to do:

All important sights of Goerlitz are modelled as TBBs. Most important is the creation of the profiles. A simple application gives the user the chance to define his interests based on the common ontology. Every single TBB profile is semantically matched with the interest profile to derive a list of TBB with IMP > 0. Different interest profiles should lead to different TBB lists. Thus a tourist who's interested in

Religion will see completely different sights than a tourist who likes the architecture of the Wilhelminian style. Depending on were the sights are situated both might walk completely different routes, leading them to completely different areas of a city. The differences might be calculated as a percentage.

To judge whether the algorithm ranks the TBBs correctly, the user having specified his interests before, has to rate some representative sights from different interest categories by text and picture. The results of the user are compared to those of the algorithm. The differentiation of the single rates should be low to prove that the algorithm works properly.

Acknowledgements

This project is part of VESUV in cooperation with Siemens AG, Microsoft's European Innovation Center (EMIC) and Fraunhofer Institute IGD. VESUV is supported by the Federal Ministry for Economics and Labour (BMWA).

References

1. Abowd, Gregory D.; Atkeson Christopher G.; Hong, Jason; Long, Sue; Kooper, Rob; Pinkerton, Mike (1997): Cyberguide – a mobile context-aware tour guide, Baltzer/ACM Wireless Networks.
2. Bussler, Christoph; Cardoso, Jorge; Fensel, Dieter; Sheth, Amit (2002): Semantic web services and processes: Semantic composition and quality of service, Irvine CA
3. Cheverst, Keith; Davies, Nigel; Mitchell, Keith; Friday, Adrian; Efstratiou, Christos (2000): Developing a Context-aware Electronic Tourist Guide: Some Issues and Experiences, Distributed Multimedia Research Group, Lancaster University, UK.
4. Crumpet (2004): http://www.ist-crumpet.org/
5. DGPS (2005): http://www.wsrcc.com/wolfgang/gps/dgps-ip.html; Wolfgang S. Rupprecht.
6. EGNOS/SISNeT (2005): http://esamultimedia.esa.int/docs/egnos/estb/sisnet/over6.htm; European Space Agency.
7. Godart, Jean-Marc (2003); Beyond the Trip Planning Problem for Effective Computer-Assisted Customization, *Information and Communication Technologies in Tourism 2003*, Andrew Frew et al. (eds.), Springer Computer Science
8. Enarro (2004): http://www.enarro.com
9. Goerlitz (2003): http://www.goerlitz.de/de/tourimus/kunst/kultur-denkmale2003.pdf
10. Horrocks, Ian; Li, Lei (2003): A software framework for matchmaking based on semantic web technology
11. Korkea-aho, Mari (2000): "Context aware applications survey", Helsinki University of Technology. http://www.hut.fi/~mkorkeaa/doc/context-aware.html#chap3.2
12. Lopez, Beatriz (2003): Holiday Scheduling for City Visitors, *Information and Communication Technologies in Tourism 2003*, Andrew Frew et al. (eds.), Springer Computer Science
13. Maedche, Alexander; (2003), Staab, Steffen; Services on the Move: Towards a P2P-Enabled Semantic Web Services, *Information and Communication Technologies in Tourism 2003*, Andrew Frew et al. (eds.), Springer Computer Science
14. Navigon (2004): http://www.navigon.de

15. Schmidt-Belz, Barbara; Posland, Stefan (2003a): "User Validation of a mobile Tourism Service"; Workshop "HCI mobile Guides", Udine (Italy)
16. Schmidt-Belz, Barbara; Laamanen, Heimo; Poslad, Stefan; Zipf, Alexander (2003b): Location-based Mobile Tourist Services – First User, *Information and Communication Technologies in Tourism 2003*, Andrew Frew et al. (eds.), Springer Computer Science
17. Ten Hagen, Klaus; Kramer, Ronny; Müller, Patrick; Schumann, Bjoern; Hermkes, Marcel (2004): Semantic Matching and Heuristic Search for a Dynamic Tour Guide, *Information and Communication Technologies in Tourism 2005*, Andrew Frew et al. (eds.), Springer Computer Science

Contextual Factors and Adaptative Multimodal Human-Computer Interaction: Multi-level Specification of Emotion and Expressivity in Embodied Conversational Agents

Myriam Lamolle[1], Maurizio Mancini[1], Catherine Pelachaud[1],
Sarkis Abrilian[2], Jean-Claude Martin[2], and Laurence Devillers[2]

[1] LINC, IUT de Montreuil, University Paris 8,
140, rue de la Nouvelle France, 93100 Montreuil
{m.lamolle, m.mancini, c.pelachaud}@iut.univ-paris8.fr
[2] LIMSI-CNRS, BP 133, 91403 Orsay
{sarkis, martin, devil}@limsi.fr

Abstract. In this paper we present an Embodied Conversational Agent (ECA) model able to display rich verbal and non-verbal behaviors. The selection of these behaviors should depend not only on factors related to her individuality such as her culture, her social and professional role, her personality, but also on a set of contextual variables (such as her interlocutor, the social conversation setting), and other dynamic variables (belief, goal, emotion). We describe the representation scheme and the computational model of behavior expressivity of the Expressive Agent System that we have developed. We explain how the multi-level annotation of a corpus of emotionally rich TV video interviews can provide context-dependent knowledge as input for the specification of the ECA (e.g. which contextual cues and levels of representation are required for enabling the proper recognition of the emotions).

1 Introduction

Multimodal Human-Computer Interfaces aim at enabling the combined use of several communication modalities between the user and the computer. Amongst them, Embodied Conversational Agents make use of a wide range of "natural" modalities such as speech, gesture, facial expressions. This rich set of modalities provides the user with different non-verbal behaviors depending on the current application context. Yet, the definition of the dynamics of these various modalities still remains to be done. For example, emotional behavior and expressivity of animated agents play a central role for the user, namely in Intelligent Tutoring Applications. But how to define the dynamics of each modality and their combination? at which level? how to select them by considering contextual factors?

We aim at creating an Embodied Conversational Agent (ECA) that would exhibit a consistent behavior with her personality and with contextual environment factors. The behavior of an agent depends not only on factors defining her individuality (such as her culture, her social and professional role, her personality and her experience), but also on a set of contextual (such as her interlocutor, the social conversation setting), and dynamic variables (belief, goal, emotion). These factors may act at different levels: they may act on what to say and when as well on how to say it and to express it. Thus they may act not only on the selection of a non-verbal behavior to convey a meaning (i.e. on the choice of the signals) but also on its expressivity (e.g. on their intensity level), in order to qualify it or to accentuate it.

To achieve such a goal we took a two-steps approach: 1) elaborate rules by analysis; 2) animate by copy synthesis. In the first phase we analyze and annotate a video corpus. We have elaborated an annotation scheme. Annotation of communicative behavior in social settings in extremely complex due to the large amount of variables acting in the communication process. Several annotation schemes of gesture [26] [30] [10], face [21], gaze [35], emotion [41] [22] exist. Each of these schemes are extremely rich in the data they encode and complex to use. When we have developed our annotation scheme, we had in mind the aim our study. Thus our annotation scheme encodes multimodal behaviors and complex emotions. Complex emotion may be defined as the combination of two affective states. Our annotation scheme encodes not only the signals being displayed but also their temporal evolution. Our second phase of study consists in animating an ECA. The ECA system takes as input the annotation made in the first phase and computes the face and gesture animation of the ECA.

Our expectation from this work is manifold. On one hand we aim at studying which perceptual cues are used to perceive a given emotion. The use of an ECA allows one to turn on and off given signals. By studying if subjects perceive from the synthesized animation, we can circumscribe which cues are the most salient to convey a given emotion. On the other hand, the copy synthesis method allows us to refine our animation model, in particular in relation to the modelling of gesture expressivity.

2   State of the Art

There has been a lot of psychological researches on emotion and nonverbal communication of facial and vocal expressions of acted basic emotions: anger, disgust, fear, joy, sadness, surprise [20], and also on expressive body movements [17] [32] [8] [9]. Indeed, research in non-verbal communication has already studied the relations between movements and emotions [18] [43] [31]. Yet, these studies were based mostly on acted basic emotions. Annotation of communicative multimodal behaviors in TV videos has also been addressed but without a focus on emotion [3] [27] or with the use of any annotation tool [5]. Thus, real-life multimodal corpora are indeed very few despite the general agreement that it is necessary to collect database that highlight naturalistic expressions of emotions [19]. These results from the literature in Psychology are very useful for the specification of Embodied Conversational Agents, but yet provide few details, nor do they study variations about the contextual factors of multimodal emotional behavior.

Several systems have been developed aiming at creating agents whose behaviors may be modulated by different factors: culture, emotion, social relationship, personality and so on. A first attempt was done by Barbara Hayes Roth [38] that developed a detailed and complex scheme to describe the characteristics of an ECA. Her model takes into consideration factors such as personality, habits, past memory, tastes. She elaborates a dialog and behavior model that uses this information to compute the animation of the agent. We are aware of very few other attempts. The role of social context in an agent's behavior have been considered. Poggi et al. [15] propose a model that decides whether an agent will display or not her emotion depending on several contextual and personality factors. Prendinger et al [36] integrate contextual variables, such as social distance, social power and threat, in their computation of the verbal and nonverbal behavior of an agent. They propose a statistical model to compute the intensity of each behavior. Rist and Schmitt [37] modelled how social relationship and attitudes toward others affect the dynamism of an interaction between several agents. Ruttkay and Noot [39] aim at creating agents with style. They developed a very complex representation language based on several dictionaries that reflect an aspect of the style (e.g. cultural or professional characteristics or personality) and that define the association between meanings and signals. The authors modelled explicitly how factors such as culture and personality affect behaviors.

But very few researchers have been using context specific multimodal corpora for the specification of an ECA [27]. In [11], the multimodal behaviors of subjects describing a house were annotated and used for informing the generation grammar of the Rea agent.

We distinguish our work from previously mentioned work in the sense that we do not model cultural and contextual factors per se, rather we modelled the different types of influences that may occur and how these ones may modulate an agent's behaviors at several levels.

3 Example Description

In this section we describe shortly an example for illustrating our approach. More details are provided in the following sections. The frame provided in figure 1 is from a video sample of a TV interview. The woman is reacting to a recent trial in which her father was kept in jail. As revealed by the manual annotation of such a video by 3 persons, the behavior displayed by this woman is perceived as a complex combination of anger and despair with temporal variation during the video clip. Furthermore, such emotional behavior is perceived in speech and in several visual modalities (head, eyes, torso, gestures).

Figure 2(b) shows a corresponding behavior displayed by an ECA thanks to a combination of manual specifications and automatic mapping between emotional tags and multimodal signs of emotion.

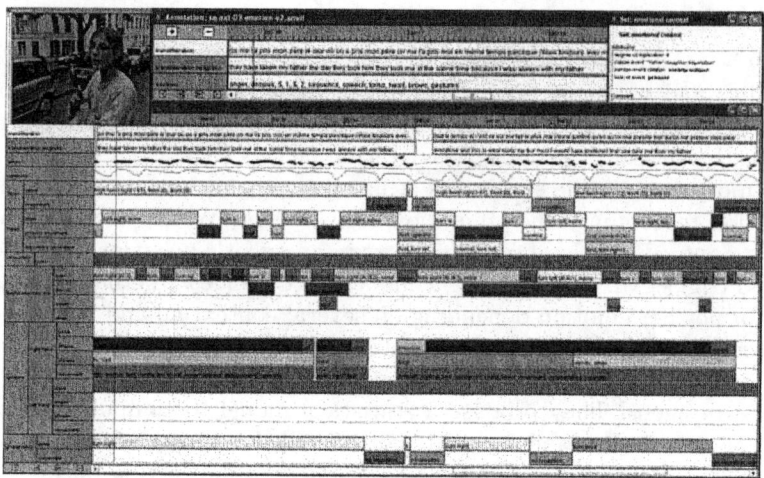

Fig. 1. Example of multi-level annotation with the Anvil tool: annotation of emotions, context, and multimodal behaviors

Fig. 2. (a) A real scene annotated by ANVIL displaying a blended emotional behavior combining sadness and anger. (b) A first simulation with the Greta system

4 Annotation and Modelling Emotional Behaviors

The annotation and modelling of emotional behaviors require representing multiple levels involved in the emotional process: the emotional context, the emotion itself and the corresponding multimodal behavior.

4.1 Emotion Labels

Three types of emotion annotations are generally used in research on emotion: appraisal dimensions, abstract dimensions and most commonly verbal categories. These verbal categories include both "primary" labels (anger, fear, joy, sadness, etc. [20]) and "secondary" labels for social emotions (e.g. love, submission).

Plutchik [34] also combined primary emotions to produce other labels for "intermediate" emotions. For example, love is a combination of joy and acceptance, whereas submission is a combination of acceptance and fear.

The number of labels required for annotating real-life emotions might be very high when compared to basic emotions. Actually, most of the emotion modelling studies have used a minimal set of labels to be tractable [7]. Instead of using these limited number of categories, some researchers define emotions using continuous abstract dimensions: Activation-Evaluation [19], Intensity-Evaluation [14]. But, these dimensions do not allow precise emotion representation as it is, for example, impossible to distinguish between Fear and Anger. Finally, the appraisal model is useful for describing the perception / production of emotion. The major advance in this theory is the detailed specification of appraisal dimensions that are assumed to be used in evaluating emotion-antecedent events (pleasantness, novelty, etc) [40].

4.2 Expressive Behavior

Conversation is made of action (the act of speaking) and perception (the act of listening). Speaker and hearer adapt each other behaviors as the interaction evolves. Interaction involves not only speech but also non-verbal behaviors. A speaker does not behave in the same way depending on several contextual factors: she adapts her speech content and her behavior depending on the evolution of the interaction, on her relation with her conversation partner, on how this partner reacts to her speech. Quantity of gesturing, smiling, gaze between speaker and listener are highly related [23]. The externalization of nonverbal behaviors does play an important role in the communication process. Their perception interacts with the judgement one made of the speaker. To model different agent's behavior we have decided to take such a stand point: to model what is visible; that is to model the signals and how they are produced. We do not model the processes that was made to arrive to the display of such and such signals, we simply model the externalization part.

We do not aim at modelling the different factors (such as culture, personality, profession) that characterize an agent. Our work does not intend either to model how different agents would differ in their emotional reaction to an event, what culture or personality mean in their emotional reaction or to model where does a certain type of behavior come from. We are interested in understanding and modelling how a given communicative act would be expressed quantitatively and qualitatively. We are aware that our work fully rely on the modelling of complex factors such as culture, role in a society and the like. But, for our synthetic agent, we have elaborated a computational model of emotional behavior and its expressivity, leaving on the side the modelling of the what, why, how and where does expressivity come from.

Having decided to approach the problem from the visible aspect of behaviors, we turn our attention to define a set of parameters to describe them. In previous work we have defined a taxonomy of communicative behaviors based on their communicative meaning [35]. The behaviors were defined as a (meaning, signal) pair. The pairs were elaborated based on video corpus analysis. To a given meaning may be associated different set of signals. For example the meaning 'emphasis' (emphasis of a word) may co-occur with a raise eyebrow, or a head nod, or a combination of both signals. Vice versa, a same signal may be used to convey different meanings; e.g. a raise eyebrow may be sign of surprise, of emphasis, or even of suggestion. The second element of the pair, the signal, was defined in a quite static manner: no notion of dynamic variation of, for e.g., intensity, temporal duration, strength of movement was built in. Since, now, we aim at creating expressive agents we had to overcome such a limitation. We have decided to define a signal not only by its static definition (such as facial expression, gesture shape) but also by other parameters. To define them we looked in the literature of perception studies to see which parameters were investigated [43, 24]. Six dimensions representing behavior expressivity are defined. They are described in the next section.

4.3 Expressivity Dimensions

The expressivity dimensions have been designed for communicative behaviors only. Each dimension acts differently for each modality. For the face the dimensions act mainly on the intensity of the muscular contraction and its temporal course (how fast a muscle contracts). On the other hand, for an arm gesture, expressivity works at the level of the phases of the gesture: for example the preparation phase, the stroke, the hold as well as on the way 2 gestures are coarticulated one in another. We follow the taxonomy proposed by D. McNeill [30] to characterize gesture phases.

- *Overall activation*: corresponds to the quantity of movement across several modalities during a conversational turn (passive/static or animated/ engaged). This parameter sets how many behaviors the agent displays while talking.
- *Spatial extent*: amplitude of movements. For the agent's face this parameter determines the quantity of physical displacement of the facial animation parameters involved in the expression. Then, spatial extent expressivity parameter will expand or condense the entire space in front of the agent that is used for gesturing.
- *Temporal*: duration of movements (e.g., quick versus sustained actions). The temporal parameter modifies starting and ending times of a facial expression. Gestures are synchronized with speech, but they may occur before the speech they accompany or after [30].
- *Fluidity*: smoothness and continuity of overall movement (e.g., smooth, graceful versus sudden, jerky). This parameter acts over several behaviors of a same modality. For two successive gestures, this dimension specifies how smoothly one gesture will map into the second one. While for the face it specifies the overall muscle contraction. Thus, as the movement gets more abrupt, there would be an increase of the muscles speed of contraction.

- *Power*: dynamic properties of the movement (e.g., weak versus strong). It corresponds to higher acceleration and deceleration magnitudes of the gesture. It also influences lip shape by controlling lip muscle tension.
- *Repetitivity*: tendency to rhythmic repeats of specific movements along specific modalities. This parameter aims to express how often a behavior is repeated. For gestures we refer to the technique of stroke expansion that we have previously introduced in [25] to capture coarticulation/superposition of beats onto other gestures. Stroke expansion repeats the meaning-carrying movement of a gesture so that successive stroke ends fall onto the stressed parts of speech following the original gesture affiliate.

5 Multi-level Representation for Naturalistic Corpus: Emotion, Multimodal Behaviors and Context

In order to model realistic emotional behavior, literature should be completed by the collection and annotation of context-specific audio-visual data. The EmoTV corpus features 50 videos samples of TV interviews with emotional behaviors [1]. A multilevel coding scheme has been defined after a first annotation phase. Emotion and multimodal annotations are annotated both at the global level of the video and at the level of individual emotional segments of the video. The contextual descriptors are also defined at the global level. The main difficult point of such a representation is to find the useful levels of description in term of granularity and temporality. The specificities of the multi-level coding scheme used for EmoTV are to enable annotation of both emotion labels and abstract dimensions, non-basic emotional patterns, two labels for labelling an emotion, the emotional context including some appraisal-based dimensions, a coarse temporal description of intensity variation in each segment and both a global description of perceived signs of emotion in the different modalities, and a more detailed description of multimodal behaviors in each segment [2].

Five sets of attributes represent the context namely *emotional context* including some appraisal dimensions (degree-of-implication, cause-event, person-event relation, time of event), *item interview context* (theme, place), *video-taped person* (age, gender, race), *overall communicative goal of the video-taped person* which combines consequence-event and communicative function, *recording context* (camera, character, acoustic quality, video quality).

Both verbal categories and abstract dimensions are used in order to study their redundancy and complementarity. In order to find an appropriate list of emotional labels, different strategies can be used [12] [14]. Two expert annotators labelled the emotion they perceived in each emotional segment, each time selecting one label of their choice (free choice). This resulted in 176 fine-grain labels (after a normalization phase) which were classified into the following set of 14 broader categories: anger, despair, disgust, doubt, exaltation, fear, irritation, joy, neutral, pain, sadness, serenity, surprise and worry. We have kept several levels of granularity. The coarse-grained level is composed of the 6 well-known Ekman classes [20] plus the "neutral" and "other" classes. The EmoTV coding scheme also features two classical abstract dimensions [13]: activation (passive, normal, active) and valence (negative, neutral, positive). The intensity (low, normal, high) and control dimensions (controlled, normal, uncontrolled) have also been added since they provide relevant information for the study of real-life emotion. Furthermore, for each segment coarse temporal descriptors for intensity variation are used.

The goal of the EmoTV corpus is to provide knowledge on the coordination between modalities during non-acted emotionally rich behaviors. It does not aim at providing detailed data on each individual modality.

The speech transliteration including non-verbal events markers was done using the Linguistic Data Consortium (LDC)[1] transliteration norm. Prosodic and spectral cues are automatically extracted.

In the videos only the upper body of people is visible. The coding scheme contains tracks for each visible modality: torso, head, shoulders, arms, facial expressions, gestures and global body. Torso, head and shoulders contain a description of the pose, and of the movement. Pose and movement annotations thus alternate. Head pose contains a primary position attribute (adapted from the FACS coding scheme): front, turned left / right, tilt left / right, upward / downward, forward / backward. A secondary position is available for representing combinations of positions (e.g. head to the right and down). Head primary movement observed between the start and the end pose is annotated with the same set of values as the primary position attribute. A secondary movement enables the combination of several movements. (e.g. head nod while turning the head). Tool-based annotation of gesture has already been studied [27]. We have kept some classical attributes and focused on repetitive and manipulator gestures which occur frequently in the EmoTV corpus.

The coding scheme enables the annotation of structural phases of gestures [30]: preparation (bringing arm and hand into stroke position), stroke (the most energetic part of the gesture), sequenceOfStroke (a number of successive strokes), hold (a phase of stillness just before or just after the stroke), retract (movement back to rest position). We have selected the following set of values for the gesture function (the gestures that are more frequent are listed first; representational gestures and emblems revealed to be very few after the annotation phase):

- *manipulator*: contact with body or object, movement which serve functions of drive reduction or other non-communicative functions, like scratching oneself; manipulator target (chest, hairs, eyebrows, nose, mouth); object that the video taped person is holding,
- *beat*: synchronized with the emphasis of the speech,
- *deictic*: arm or hand is used to point at an existing or imaginary object; deictic target (self, camera, other),
- *representational*: represents attributes, actions, relationships about objects and characters,
- *emblem*: movement with a precise, culturally defined meaning.

---

[1] http://www.ldc.upenn.edu/

Movement quality is annotated for torso, head, shoulders, gestures, global pose and movement. The attributes of movement quality that we selected as relevant in our corpus are: the number of repetitions, the fluidity (smooth, normal, jerky), the strength (soft, normal, hard), the speed (slow, normal, fast), the spatial expansion (contracted, normal, expanded).

6 Description of the GRETA ECA System

We have developed a system that generates the behaviors of a talking ECA. To determine speech-accompanying non-verbal behaviors the system relies on a taxonomy of communicative functions proposed by Isabella Poggi [35]. A communicative function is defined as a pair (meaning, signal) where meaning corresponds to the communicative value the agent wants to communicate and signal to the behavior used to convey this meaning. The former ones are represented as a set of goals and beliefs the speaker has the goal to communicate. In the taxonomy communicative functions are differentiated in information about speaker's beliefs, speaker's intentions, speaker's affective state and metacognitive information about speaker's mental state.

Our system, called Greta, takes as input the text the agent has to say and outputs the animation of the agent. The input text is augmented with information related to the ways the agent wants to say her text. Depending on the type of communicative acts that are specified in the input file, the agent will display different behaviors.

To control the agent we are using a representation language, called 'Affective Presentation Markup Language' (APML) where the tags of this language are the communicative functions [16].

Our system takes as input the text (tagged with APML) the agent has to say [33]. The system instantiates the communicative functions into the appropriate signals. The output of the system is the audio and the animation files that drive the facial model. The APML tags, corresponding to the meaning of a given communicative function, is converted into their corresponding facial signals. The conversion is done by looking up the definition of each tag into the library that contained the lexicon of the type (meaning, signals). Finally, we proceed with the animation generation for the agent. The animation is obtained by conversing each facial signal in their corresponding facial and body parameters.

7 A Representation Scheme for an Expressive Agent

We want to simulate that different agents may behave differently in a same situation and express their felt emotion differently. This representation allows us to define that an agent has a very expressive face or that she rarely uses wide arm movements, etc. For example, the simulation of one's nonverbal behavior, by two different agents to express anger produces two different perceivable animations. We do not aim at modelling what culture or personality mean, nor do we aim at simulating expressive animations. In this section, we detail the representation of the different levels of agent's expressivity [6] [28] in relation to modalities (face, gesture, gaze, posture, head) and we explain the computation of contextual factors effects.

7.1 Global Expressivity

In the input text, the tags are defined for the *default agent*. To allow for the generation of an expressive ECA, we associate to each agent a *behavioral profile* which specifies, on the one hand, the agent's expressivity, i.e. the agent's predispositions (which modalities the agent prefers to use) and the global expressivity (how the modalities are used), and on the other hand, the effects of the contextual factors.

The agent's predispositions represent the expressivity level of each modality. For example, an agent *Agent1* may be more expressive with the face and gestures than the default agent (i.e. her facial moves and her gestures are more visible than the *default agent's* one) but less than the posture. The predispositions, given as input, is constant during a dialog turn.

The own agent's expressivity is represented by her predispositions to display a communicative act in the different modalities. But, the agent can use a modality through different dimensions: spatial, temporal, fluidity, power, repetitivity and overallActivity (see section 4.3). These values lessen or accentuate the intensity, the velocity, the duration, the delay of the chosen signals for the corresponding modality in the animation engine to express the communicative acts specified in the input text. The *spatial* and *temporal* parameters are local to an communicative act and can be modulated by the *fluidity*, *power* and *overallActivity* parameters. For example, according to the agent's description factors, this agent gets a large fluidity in her movements but her gestures are close to her body (the spatial dimension is set to small). We also specify which modalities are more expressive than the others; i.e. which modalities display the most expressive behavior.

7.2 Modality Hierarchy

The predisposition behavioral profile, just explained, indicates the effects of the agent's expressivity for each modality. Another factor for distinguishing agents among each other, is the modality hierarchy [42]. This hierarchy represents the modalities over which the agent is the most expressive. She may mainly use her hands to communicate or her face will be very lively, almost grimacing. To each modality (face, gaze, gesture, posture, head), we associate a value which represents the preferential level in this hierarchy. In case several modalities have the same preferential level, we consider that agent's nonverbal behavior to express a communicative act is visible through several modalities [4].

8 System Overview

Given a tagged-input file, the system instantiates the tags into a set of signals. To do so, it looks in a library the signals that correspond to the given meanings.

Then it selects the signals that express the tags meaning, according to its attributes values and the agent's preferences. In the next sections, we describe the agent's contextual behavioral profile. We also detail the various selection stages of our system: the modality selection and the signals pre-selection. The first selection corresponds to determining which modality the agent uses; the second selection consists in ordering the set of possible behaviors having an equivalent meaning, from the most adequate solution to the least. This ordering takes into account the expressivity of the agent.

8.1 From Global to Local Non-verbal Behavior Specification

For each tag of the input text, the system has to decide the modality (face, gesture, gaze, posture, head one) to express the given meaning taking into account the weight of a communicative act and the global expressivity of the agent. This decision is based on the global agent's expressivity. Among the modalities that have at least one expression which allows the system to represent the meaning, the system chooses the one with the highest priority and that is not used yet, in order to prevent conflicts.

8.2 Pre-selection of Non-verbal Behavior

To obtain the local expressivity of each modality, the system selects a set of expressions from a library. The expression is selected if its range of values contains the wanted expressivity value [29]. Each local expression contains the signals (representing the non verbal behavior) to play by the animation engine.

Currently, if no expression is selected, the system chooses the nearest expression. So, the animation engine can display at least one non-verbal behavior for local expressivity.

Then, the system has to order the set of expressions based on the agent's definition. This ordering allows us to obtain a list of non-verbal behaviors in the order of the agent's preferential use. This pre-selection is sent to the animation engine of the Greta system that chooses the "most adequate" non-verbal behavior.

9 From Corpus Analysis to ECA Specification

In this section we briefly describe an example of generating the animation of an ECA from the annotation of a video. The image in figure 2(a) is from the EmoTV corpus. In section we have provided an example of the Anvil annotation for this video sequence.

In Greta we do not consider the complete annotation of the given video clip. As mentioned in section , we are concerned with the visible part of behaviors. So currently we leave aside all annotations regarding the description of the context. On the other hand we use the emotion labels as well as the description of the movement quality as input to our Greta system. We follow an analysis-synthesis loop approach to refine the animation of the ECA. The annotation of the video segment is re-written to follow the APML specification. In the example of figure

1 the annotated emotion is anger for the first half part of the segment and then it fades into despair for the rest of the segment. We have also used the annotation of the gesture strokes from the video segment to define emphasis tags in the corresponding APML text. This ensures that the gesture stroke of the ECA will happen with the emphasized words. Finally from the annotation of emotion and of multimodal behavior at the global level, we define the agent's behavioral profile. At this point, given the APML text and the agent's behavioral profile, the system automatically computes the expressivity parameters values (see 4.3) for each of the signals the agent has to produce. The animation engine considers both the signals and their expressivity to generate the agent's animation.

10 Conclusion and Perspectives

In this paper we have presented a methodology based on corpus analysis to create expressive ECAs. We have also proposed a representation scheme and a computational model for such an agent. We have explained how the annotation of expressivity in TV interviews is compatible with the specifications of our ECA. We will apply this protocol on a selection of video displaying basic and non basic emotional patterns. We will try to use the hybrid scheme used in the corpus for annotating each segment with two labels in order to consider non basic emotional patterns. The procedure will be validated via perceptual tests for evaluating how much the contextual cues, the emotion and the multimodal behaviors are perceptually equivalent in the original video and the simulation of the corresponding behavior by the ECA.

Acknowledgments

This work has been partially supported by the Network of Excellence Humaine (Human-Machine Interaction Network on Emotion) IST-2002-2.3.1.6 / Contract no. 507422 (http://emotion-research.net/). We are very grateful to Bjoern Hartmann for implementing the expressive behavior module and to Vincent Maya for his help in this project.

References

1. S. Abrilian, L. Devillers, S. Buisine, and J.-C. Martin. Emotv1: Annotation of real-life emotions for the specification of multimodal affective interfaces. In *HCI International 2005*, Las Vegas, USA, 2005.
2. S. Abrilian, J.-C. Martin, and L. Devillers. A corpus-based approach for the modeling of multimodal emotional behaviors for the specification of embodied agents. In *HCI International 2005*, Las Vegas, USA, 2005.
3. J. Allwood, L. Cerrato, L. Dybkær, and P. Paggio. The mumin multimodal coding scheme. In *Workshop on Multimodal Corpora and Annotation ,,* Stockholm, 2004.

4. Jens Allwood. Cooperation and flexibility in multimodal communication. In *CMC '98: Revised Papers from the Second International Conference on Cooperative Multimodal Communication*, pages 113–124, London, UK, 2001. Springer-Verlag.
5. H. Atifi and M. Marcoccia. L'expression et la mise en scène des émotions à la télévision : l'articulation des émotions vécues, racontées et attribuées. In *Oralité et gestualité (ORAGE 2001) : Interactions et comportements multimodaux dans la communication*, pages 179–182, Aix-en-Provence, 2001. L'Harmattan.
6. Venkata Rama Kiran Badam and Chaitanya P. Gharpure. A stochastic and multi-layered model for personality in computational agents. In *Personality in Computational Agents*, Logan, UT, USA, 2002. Department of Computer Science of Utah State University.
7. A. Batliner, K. Fisher, R. Huber, J. Spilker, and E. Noth. Desperately seeking emotions or: Actors, wizards, and human beings. In *SpeechEmotion-2000*, pages 195–200, 2000.
8. R. T. Boone and J. G. Cunningham. Children's understanding of emotional meaning in expressive body movement. In *Poster presented at Biennial Meeting of the Society for Research in Child Development*, Washington, DC, 1996.
9. R.T. Boone and J. G. Cunningham. Children's decoding of emotion in expressive body movement: The development of cue attunement. *Developmental Psychology*, 34(5):1007–1016, 1998.
10. G. Calbris. *The semiotics of French gestures*. University Press, Bloomington: Indiana, 1990.
11. J. Cassell, M. Stone, and Y. Hao. Coordination and context-dependence in the generation of embodied conversation. In *INLG*, pages 171–178, 2000.
12. R. Cowie. Emotion recognition in human-computer interaction. *IEEE Signal processing Magazine*, (18), 2001.
13. R. Cowie, E. Douglas-Cowie, S. Savvidou, E. McMahon, M. Sawey, and M. Schroeder. 'feeltrace': An instrument for recording perceived emotion in real time. In *ISCA Workshop on Speech & Emotion*, pages 19–24, Northern Ireland, 2000.
14. R. Craggs and M. M. Wood. A categorical annotation scheme for emotion in the linguistic content. In *Affective Dialogue Systems (ADS'2004)*, 2004.
15. B. DeCarolis, C. Pelachaud, I. Poggi, and F. de Rosis. Behavior planning for a reflexive agent. In *IJCAI'01*, Seattle, USA, August 2001.
16. B. DeCarolis, C. Pelachaud, I. Poggi, and M. Steedman. APML, a mark-up language for believable behavior generation. In H. Prendinger and M. Ishizuka, editors, *Life-like Characters. Tools, Affective Functions and Applications*, pages 65–85. Springer, 2004.
17. M. deMeijer. The contribution of general features of body movements to the attibution of emotions. *Journal of Nonverbal behavior*, 13:247–268, 1989.
18. M. deMeijer. The attribution of agression and grief to body movements : the effect of sex-stereotypes. *European Journal of Social Psychology*, 21:249–259, 1991.
19. E. Douglas-Cowie, N. Campbell, R. Cowie, and P. Roach. Emotional speech; towards a new generation of databases. *Speech Communication*, (40), 2003.
20. P. Ekman. Basic emotions. In T. Dalgleish and M. J. Power, editors, *Handbook of Cognition & Emotion*, pages 301–320. John Wiley, New York, 1999.
21. P. Ekman and W. Friesen. *Facial Action Coding System*. Consulting Psychologists Press, Inc., Palo Alto, CA, 1978.
22. P. Ekman and E. Rosenberg, editors. *What the Face Reveals : Basic and Applied Studies of Spontaneous Expression Using the Facial Action Coding System (Facs) (Series in Affective Science)*. Oxford Univ Press, 1998.

23. P. Feyereisen and J.D. de Lannoy. *Gestures and speech: Psychological investigations*. Cambridge University Press, 1991.
24. P.E. Gallaher. Individual differences in nonverbal behavior: Dimensions of style. *Journal of Personality and Social Psychology*, 63(1):133–145, 1992.
25. B. Hartmann, M. Mancini, and C. Pelachaud. Formational parameters and adaptive prototype instantiation for MPEG-4 compliant gesture synthesis. In *Computer Animation'02*, Geneva, Switzerland, 2002. IEEE Computer Society Press.
26. A. Kendon. Human gesture. In T. Ingold and K. Gibson, editors, *Tools, Language and Intelligence*. Cambridge University Press, Cambridge, 1993.
27. M. Kipp. *Gesture Generation by Imitation. From Human Behavior to Computer Character Animation*. PhD thesis, Universität des Saarlandes, 2004.
28. Sumedha Kshirsagar. A multilayer personality model. In *SMARTGRAPH '02: Proceedings of the 2nd international symposium on Smart graphics*, pages 107–115, New York, NY, USA, 2002. ACM Press.
29. V. Maya, M. Lamolle, and Pelachaud C. Influences on embodied conversational agent's expressivity: Toward an individualization of the ecas. In *Proceedings of AISB 2004*, 2004.
30. D. McNeill. *Hand and mind - what gestures reveal about thoughts*. University of Chicago Press, 1992.
31. J. Montepare, E. Koff, D. Zaitchik, and M. Albert. The use of body movements and gestures as cues to emotions in younger and older adults. *Journal of Nonverbal Behavior*, 23(2):133–152, 1999.
32. J. Newlove. *Laban for actors and dancers*. Routledge, New York, 1993.
33. C. Pelachaud, V. Carofiglio, B. De Carolis, and F. de Rosis. Embodied contextual agent in information delivering application. In *First International Joint Conference on Autonomous Agents & Multi-Agent Systems (AAMAS)*, Bologna, Italy, July 2002.
34. R. Plutchik. *The psychology and Biology of Emotion*. Harper Collins College, New York, 1994.
35. I. Poggi, C. Pelachaud, and F. de Rosis. Eye communication in a conversational 3D synthetic agent. *AI Communications*, 13(3):169–181, 2000.
36. H. Prendinger, S. Descamps, and M. Ishizuka. Scripting affective communication with life-like characters in web-based interaction systems. *Applied Artificial Intelligence*, 16(7-8):519–553, 2002.
37. T. Rist and M. Schmitt. Applying socio-psychological concepts of cognitive consistency to negotiation dialog scenarios with embodied conversational characters. In *Proc. of AISB'02 Symposium on Animated Expressive Characters for Social Interactions*, pages 79–84, 2003.
38. D. Rousseau and B. Hayes-Roth. Personality in synthetic agents. Technical report, Stanford Knowledge Systems Laboratory Report KSL-96-21, 1996.
39. Zs. Ruttkay, V. van Moppes, and H. Noot. The jovial, the reserved and the robot. In *proceedings of the AAMAS03 Ws on Embodied Conversational Characters as Individuals*, Melbourne, Australia, 2003.
40. K. R. Scherer. Emotion. In M. Hewstone & W. Stroebe, editor, *Introduction to Social Psychology: A European perspective*, pages 151–191. Oxford: Blackwell., 2000.
41. K.R. Scherer, A. Schorr, and T. (Eds.) Johnstone. *Appraisal processes in emotion: Theory, Methods, Research*. New York and Oxford: Oxford University Press, 2001.

42. M. Theune, D. Heylen, and A. Nijholt. Generating embodied information presentations. In O. Stock and M. Zancanaro, editors, *Multimodal Intelligent Information Presentation*, pages 47–69. Kluwer Academic Publishers, 2004. ISBN= 1-4020-3049-5.
43. H.G. Wallbott. Bodily expression of emotion. *European Journal of Social Psychology*, 28:879–896, 1998.

Modeling Context for Referring in Multimodal Dialogue Systems

Frédéric Landragin

Thales Research and Technology,
Domaine de Corbeville, F-91404 Orsay Cedex, France
Frederic.Landragin@thalesgroup.com

Abstract. The way we see the objects around us determines speech and gestures we use to refer to them. The gestures we produce structure our visual perception. The words we use have an influence on the way we see. In this manner, visual perception, language and gesture present multiple interactions between each other. The problem is global and has to be tackled as a whole in order to understand the complexity of reference phenomena and to deduce a formal model. This model may be useful for any kind of man-machine dialogue system that focuses on deep comprehension. We show how a referring act takes place into a contextual subset of objects, called 'reference domain,' and we present the 'multimodal reference domain' model that can be exploited in a dialogue system when interpreting.

1 Introduction

The understanding performance of natural language dialogue systems more and more relies on their pragmatic abilities. Indeed, modeling the context and modeling the interpretation process are particularly complex aspects of pragmatics for multimodal dialogue systems. For systems where a user interacts with a computer through a visual scene on a screen, the combination of visual perception, gesture and language involves interactions between the visual context, the linguistic context and the task context. There has already been several proposals related to the representation of the linguistic and the task contexts, considering components such as dialogue history, salience, focus of attention, focus spaces, topics, frames, plans and so on. Still, less attention has been put on how to deal with the visual context in such a framework: some works focus on structuring the visual scene into perceptual groups [24], others focus on the management of a visual focus of attention and on the relations between this notion and salience [1]. What we want to do here is to integrate all these perceptual, linguistic and cognitive aspects (see Figure 1), for the interpretation of reference to objects phenomena. To us, this has to be done by using an unified framework, in order to compare and to merge the various information from the various contextual aspects into homogeneous structures.

It is with this aim that we have been developed since several years the 'multimodal reference domain' model. As opposed to approaches like the DRT (Dis-

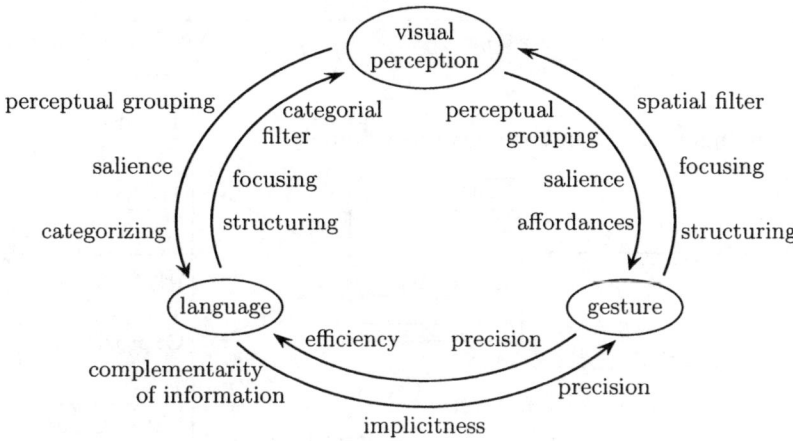

Fig. 1. Some interactions between visual perception, gesture, and language

course Representation Theory) [11], this model has been built with multimodal concerns from the early first phases of the design. As opposed to approaches based on domains of quantification [4], it takes into account the previous utterances when delimitating the context. Reference domains are then linked to each others. With these two strong points, the reference domain model appears to be useful when designing a multimodal dialogue system. The visual context as well as the linguistic context (dialogue history) can be represented by sets of reference domains, which can be easily compared. In this paper we want to show that multimodal dialogue systems need to take into account the visual and linguistic contexts in a same manner, in order to manage in a proper way all contextual information. We first present the main principles of our model. In the next section, we describe in details how we translate the visual context into visual reference domains. We then describe how a multimodal utterance (a verbal referring expression together with a pointing gesture) from the user can be interpreted with the help of reference domains.

2 Reference Domains

The basic idea of the 'multimodal reference domain' model is that when we interpret a multimodal referring expression, we take into account not the complete context (for instance all objects that are present in the communicative situation), but only a reduced part of it (for instance objects that are in the focus of attention of the participants). This part constitutes a 'reference domain.' Reference domains can come from visual perception, language or gesture, or can be linked to the dialogue history or the task constraints. Visual domains may come from perceptual grouping, for example to model focus spaces [1]. Some domains may come from the user's gesture, others from the task constraints. All of them are structured in the same way (see Figure 2). They include a grouping

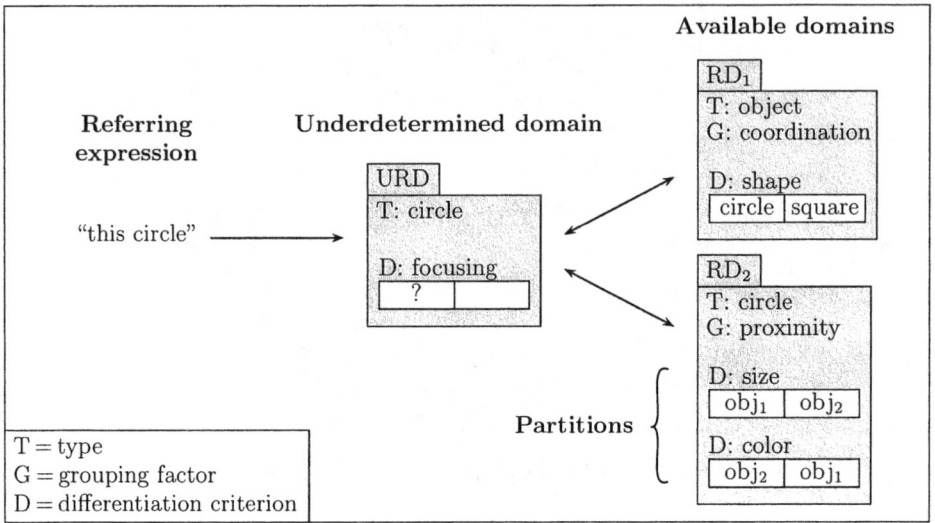

Fig. 2. Interpretation of a referring expression using reference domains

factor ('being in the same referring expression,' 'being in the same perceptual group'), and one or more partitions of elements. A partition gives information about possible decompositions of the domain [22]. Each partition is characterized by a differentiation criterion, which represents a particular point of view on the domain and therefore predicts a particular referential access to its elements ('red' compared to 'not-red,' 'focused' compared to 'not-focused'). With these formal aspects, reference domains consist of a way to represent data structures maintained in a dialogue system, with a cognitive inspiration.

One important point of the model is the creation of a new reference domain. The linguistic and contextual clues are sometimes not sufficient for the delimitation of such a domain. For this reason, we propose to manage underdetermined reference domains, as it is done in [22] and then in [16], and as it is showed in Figure 2. The linguistic and gestural information allow to build an underdetermined domain that groups all constraints. In Figure 2, the referring expression that is currently treated is "this circle." Such a demonstrative nominal phrase implies that a particular circle is focused. This interpretation constraint can be translated into an underdetermined reference domain, that consists of a partition where one element is focused. "This circle" is making a contrast between a particular circle and other circles. So the reference domain in which the interpretation occurs must include only circles. This is the role of the 'type' attribute.

Then, the reference resolution process consists of the unification of this underdetermined domain with the domains that appear in the context. In Figure 2, two reference domains are available in the context, RD_1 that comes from the dialogue history, and RD_2 that comes from the visual context. More precisely, RD_1 was built on at a previous stage of the interaction, when interpreting a referring expression such as "a circle and a square." RD_2 groups two objects because of their proximity. The domain with the best unification result is kept for the referent identification.

In the next sections we will first focus on perceptual phenomena that are at the early beginning of reference, including salience and grouping aspects. We then focus on referring phenomena, including multimodal aspects, and we conclude on the algorithm for multimodal reference resolution based on the management of reference domains. Such an algorithm has been developed in the framework of several European project: IST MIAMM (see http://www.miamm.org) and IST OZONE (see http://www.extra.research.philips.com/euprojects/ozone). We don't want here to describe the implementation of this algorithm, because it requires the presentation of a lot of technical problems that are not of importance here (such as the calculus of salience scores, or algorithms for the recognition of gesture trajectories). We want to focus on the exploitation of contextual and communicative clues for the interpretation of multimodal referring expressions, in order to emphasize the way the context can be modeled with reference domains.

3 Perceptual Phenomena

3.1 Focusing (Salience)

Since we consider that salience is at the origin of referring phenomena, we want here to clarify the way to take visual salience into account in multimodal dialogue systems. In the absence of information provided either by the dialogue history or the task history, an object can be considered as salient when it attracts the user's visual attention more than the other objects. Several classifications of the underlying characteristics that may make an object be perceived as salient have been proposed. For instance, Edmonds [5] has provided some specific criteria in direction-giving dialogues when the objects are not mutually known by the instructor and learner. However, such classifications are by far too dependent upon the task to be achieved (for example there is one specific classification for each type of object) and narrows down on the notion of salience to specific aspects. Merging them and adding to them the major results of pictural arts studies (Itten [9], Kandinsky, etc.) may lead us to contemplate a more generic model which in turn could be implemented for an application-driven system.

First, a salience model requires a user model of perception. Indeed, visual salience depends on visual familiarity. Some objects can be familiar to all users. It is the case for human beings: when a picture includes a human (or when a virtual environment contains an avatar), he will be salient and the user's gaze will be first attracted by his eyes, and then his mouth and nose, as well as his hands, when a specific effort has been made to simulate natural gestural behavior. For other objects, familiarity depends on the user. When a photographer enters a room, the pictures on the walls might be more salient than the computer on the table; whereas it might be the opposite for a computer scientist. Everyone acquires his own sensitivities, for example his own capacity in distinguishing colors. The choice of the right color term can show these sensitivities. Somebody may prefer to name 'red' a color that somebody else is used to naming 'pink.' No need to be color-blind for that.

Second, a salience model needs a task model. Visual salience depends on intentionality. When you invite colleagues in your office, you search chairs in your visual space, and so chairs are more salient than the other furniture.

Third, visual salience depends on the physical characteristics of the objects. Following the Gestalt theory, the most salient form is the 'good form,' i.e., the simplest one, the one requiring the minimum of sensorial information to be treated. This principle has been first illustrated by Wertheimer [25] for the determination of contours, but it is also suitable for the organization of forms into a hierarchy. Nevertheless, when the same form appears several times in the scene, one of the instances can be significantly more salient than the others. The salience of an object then depends on a possible peculiarity of this object, which the others do not have, such as a property or a particular disposition within the scene. Basically, those peculiarities can be summarized as follows:

1. classification of the properties that can make an object salient in a particular visual context:
   (a) category (in a scene with one square and four triangles, the square is salient),
   (b) functionality, luminosity (in a room with five computers, with one of them being switched on: this one is salient),
   (c) physical characteristics: size, geometry, material, color, texture, etc. (in a scene with one little triangle and four big triangles, the little one is salient, etc.),
   (d) orientation, incongruity, enigmatic aspect, dynamics (object moving on the screen)...
2. salience due to the spatial disposition of the objects: in a room containing several chairs, a chair which is very near the participant may be more salient than the distant ones, and an isolated chair may be more salient than the others if these ones are grouped.

When no salient object can be identified by means of the previous methods, visual salience also depends on the structure of the scene, i.e., the frame, the positions of the strong points in it, and the guiding lines that may restrain the gaze movements. The strong points are classically the intersections of the horizontal and vertical lines at the 1/3–2/3 of the rectangular frame. If the perspective is emphasized, vanishing points can also be considered as strong points. If the scene presents a symmetry or balance which hinges upon a particular place, this very place becomes a strong point. As a whole, the objects that are situated at strong points are usually good candidates for being salient. If they can be identified (from continuities in the disposition of the objects), the guiding lines go from salient objects to salient objects. Salience can thus be propagated.

The four stages that we have identified in this section correspond to the four stages of the algorithm we propose to automatically detect salient objects in a visual context. If a given stage cannot lead to significant results, the next stage is considered. Each result must be associated with a confidence rate (for example the number of characteristics that distinguish the salient object from the others). When no result is found, the whole visual context has to be taken into account.

3.2 Grouping

Following the Gestalt theory [25], the major principles to group objects are proximity, similarity and good continuation. From the list of visible objects and their coordinates, algorithms can build groups, which allows the system to have an idea of the user's global perception of the scene. An example of such algorithm is given by Thórisson [24]. The notion of salience can be extended from an object to a group. When the user sees a scene for the first time, one group may attract his attention more than the others and may be perceived first. According to our definition, this group will be salient. Based on proximity and similarity, the algorithm of Thórisson produces groups ordered according to goodness, and therefore according to salience.

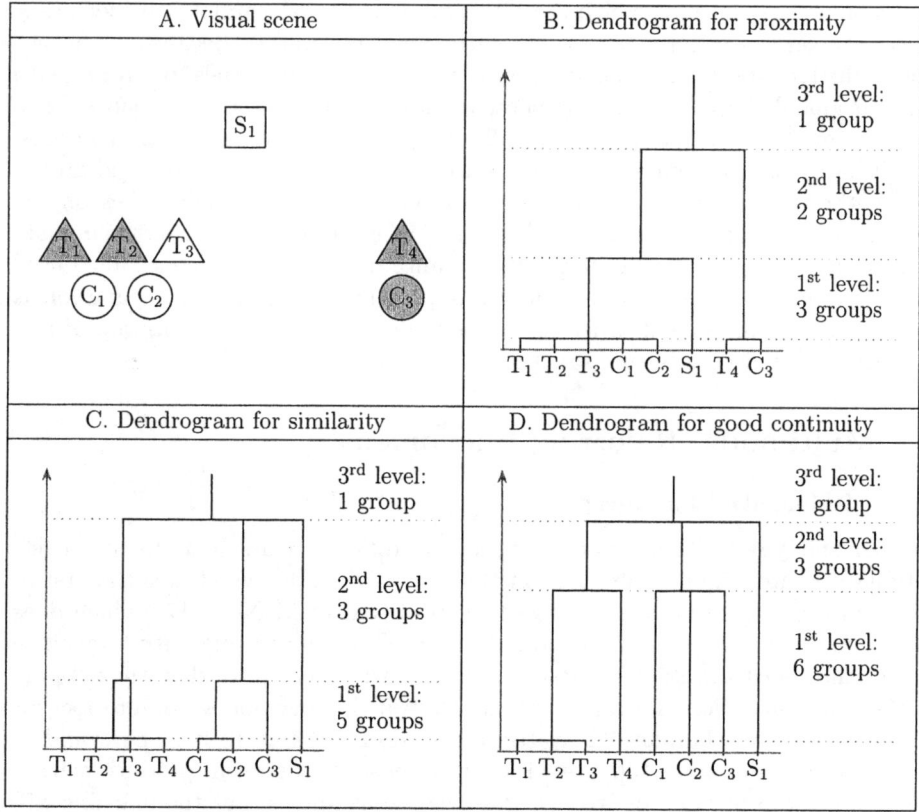

Fig. 3. Grouping objects using a dendrogram for each grouping factor

Grouping on the sole basis of the proximity principle amounts to the computation of distances between objects. Applying a classic algorithm of automatic classification, we obtain a hierarchy of partitions of the objects in groups, each group being characterized by a compactness score (see Figure 3-B). When a 2-D display of a 3-D scene is made, for example with a virtual environment displayed on a screen, grouping can be done in 3-D, or in 2-D with the coordinates of the projections of the objects. Strictly following the Gestalt theory, this second solution is in line with the application of proximity principle at the retina level. An experiment of Rock and Brosgole [21] shows however that users restore the third dimension, and that grouping is done at a later level than the early processing of retina information. Rock and Brosgole introduce the notion of phenomenal proximity, and the relevance of grouping objects in the underlying 3-D representation.

Grouping by taking into account the good continuation principle can be done by means of a recursive processing: groups are built from each single object and are extended to their nearest proximity, and so on until the whole space has been covered. Continuities are identified by doing linear regressions. Grouping with one Gestalt criterion or another leads us to different results (Figure 3). Moreover, only considering the proximity criterion produces various results depending on the compactness level at which the hierarchy is read. We cannot consider priorities between the criteria (as we did with salience criteria), because we do not know when it is better to consider groups with a high compactness or groups with a linear global shape. For the moment, we have to manage several results. Each of them must be associated with a confidence rate, for example the compactness.

Visual reference domains can be built on by using these focusing and grouping methods. The existence of a strong visual reference domain relies on the demarcation of a group in the dendrograms. The grouping factor of the domain will be the combination of criteria (for instance, proximity plus continuity) used when grouping. When a salient object is present in the group, a partition is created where this salient object is focused. The differentiation criterion of this partition is labeled as 'visual salience.'

4 Multimodal Referring Phenomena

4.1 Referential Gestures

Cosnier and Vaysse [3] propose a synthesis of different classifications of conversational gestures, taking into account the one of Efron [6], which was the first to focus on the referential aspect of gesture, and that of McNeill [17], which does so in a more thorough manner. How the fact of communicating with a machine incites the user to restrict his gestures on his own, especially when the support of the communication is a touch screen? Even if the machine as an interlocutor is symbolized by a human-like avatar, a user does not talk to it as he would to an actual human being [10]. Likewise, we suppose the user will produce neither synchronization nor expressive gestures because he knows that the machine will not perceive or be sensitive to them. As a general rule, we suppose that the user will produce only informative gestures, as opposed to gestures that facilitate the speech process, such as 'beats' and 'cohesives' [17]. For the moment, we focus our work on the design of systems with a touch screen. See the work of Bolt [2] for the origin, and for instance the work of Wolff et al. [26] for a more recent work. In such an interaction mode, the user may be conscious that touching the screen must be informative. Even when not explicitly prohibited from doing so, he will not produce gestures that do not convey meaning. He will also leave out gestures which require anything beyond 2-D, in particular 'emblems' [6] and a lot of 'iconic' and 'metaphoric' gestures [17]. Of the remaining gesture types, we are left with deictic, some iconic and some metaphoric gestures. We note here that these gestures are all referential, which emphasizes on the problem of reference.

The most frequent referential gesture in communication with a touch-screen is the deictic one [26]. What are its functions and the condition of its production, in term of effort (or cost)? As demonstratives or indexicals in language, deictic gesture is an index, i.e., an arbitrary sign that has to be learned and whose main function is to attract the interlocutor's attention to a particular object. A deictic gesture is produced to bring new information by making an object salient which is not already so [15]. Moreover, deictic gestures, as iconic and metaphoric ones, are often produced when a verbal distinguishing description is too long or too complicated, in comparison with an equivalent multimodal expression (a simple description associated with a simple gesture). A distinguishing description has a high cost when it is difficult to specify the object through its role or its properties in the context. It is the case for example when other objects have the same properties: the user has to identify another criteria to extract the referent from the context. He can use a description of its position in the scene, that leads to long expressions like "the object just under the big one at the right corner." Deictic gesture has a cost as well. It depends on the size of the target object and, in 3D-environments, its distance from the participant. Fitts' Law [7], a score that can be computed from these two parameters, is an indicator of the effort in pointing. Another indicator is given by the disposition of the objects in the scene. If the target object belongs to a perceptual group, it is more difficult to point out it than if it is isolated from the other objects. A score can also be computed to quantify the aggregation of the perceptual group. If several Gestalt criteria are simultaneously verified, this score will be high. Then, a gesture whose intention is to extract an object from this group will have a high cost, proportional to the difficulty of breaking the group. On the contrary, a gesture whose intention is to point the whole group will have a low cost.

As a pointing gesture on a single object can be extended to a group, it seems, from the system point of view, that several interpretations are often possible. What are the possible forms of a deictic gesture, and what are the possible interpretations that can be done considering the visual context? On a touch screen, deictic gestures can take several forms: dots ('pointing'), lines, opened or closed curves, 'scribbling.' Trajectories can pass between objects, in order to separate some of them (generally by surrounding them) from the other ones ('circling'), or pass on the target objects ('targeting'). Pointing, scribbling, circling and targeting were the four categories of trajectories extracted from

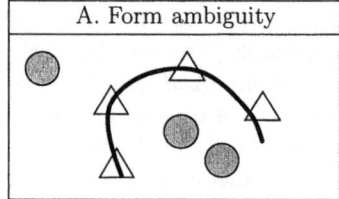 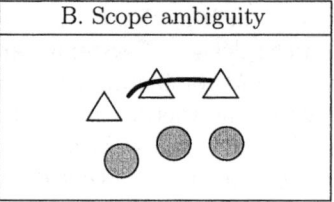

Fig. 4. Form and scope ambiguity

the corpus study by Wolff *et al.* [26]. This study leads to strategy ambiguity (individual reference opposed to group reference), as we already discuss, and to form ambiguity and also to scope ambiguity. There is a form ambiguity when the same trajectory, for example an unfinished circling curve, can be interpreted as a circling or as a targeting, as shown on the first scene of Figure 4 (the gesture can target the triangles, can surround two circles, or, following a mixed strategy, can point out all of them). There is a scope ambiguity when the number of referents can be larger than the number of target objects, as shown on the second scene of Figure 4 (the gesture can target two or three triangles).

These possible ambiguities emphasize an additional problem, that the target objects (the referents of the gesture) are not always the referents of the multimodal expression. In the next section we explore the links between speech and gesture and we characterize the links between the referents of the gesture and the referents of the multimodal expression. We then deduce a list of clues that the system may exploit to interpret the reference.

4.2 Gesture Referent and Multimodal Referent

We have seen that the verbal referring expression guides the interpretation of gesture. This can be illustrated by considering the possible expressions "these triangles" and "these circles" in the first scene of Figure 4, and by considering "these two objects" and "these three objects" in the second. In these expressions, only one word, the category in the first case and the numeral in the second, is sufficient to interpret the gesture and then to identify the referents. The demonstrative indicates the presence of a gesture in the referring action, that is if no set of triangles or circles is salient in the dialogue history (possibility of an anaphora). Nevertheless, if the gesture makes one object very salient, a definite article might be used instead of the demonstrative. This situation, more frequent in French than in English, happens in particular during the acquisition of the articles functions by children [12] and can be observed in some spontaneous dialogues (examples can be found in the studied corpus). Another example of the relaxation of linguistic constraints is the use of "him" ("lui" in French) or "he" ("il" in French) with a gesture. In some situations, "il" can be associated with a gesture instead of "lui," which is the usual word to focus on a person [15]. A third example in French is the use of deictic marks. When several objects are placed at different distances, "-ci" in "cet objet-ci" ("this object") and "-là" in "cet objet-là" ("that object") allow the interlocutor to identify an object closer to or further from him. When a gesture is used together with "-ci" or "-là," the distinction does not operate any more (a lot of examples can be found in the studied corpus).

The referents of some expressions are different from the referents of the associated gesture. It is the case of expressions like "the N_2 preposition this N_1" with a gesture associated with "this N_1." It can be expressions like "the color of this object" (an equivalent of "this color") or spatial expressions like "the form on the left of this object." Their common point is that their interpretation presents two stages, the first (the only one that has an interest here) being the multimodal reference of N_1, and the second being the use of this first identification to resolve the reference of the complete expression, by extracting a characteristic of the referent in the first case, by considering it as a site for the identification of N_2 in the second case.

Fig. 5. Generic interpretation

One of the classical aspects of reference is the possibility of a specific interpretation and of a generic one. It seems that every multimodal referring expression like "this N" with a gesture, can refer to the specific object that is pointed out, or to all objects of the N category. Sometimes there is a clue that gives greater weight to one interpretation. For example, an unambiguous gesture pointing out only one object will lead to the generic interpretation if it is produced with "these forms," where the plural is the only clue (Figure 5. This interpretation is confirmed by the presence of other objects with the same form, and by the fact that being in a perceptual group these objects need a high cost to be pointed out. On the contrary, the use of a numeral will reject the generic interpretation. When no clue can be found, the task may influence the interpretation (some actions must be executed to specific objects), and, for this reason, we do not settle here. To summarize, we propose the following list of clues:

- the components of the nominal phrase: the number (singular or plural, eventually determined by a numeral or a coordination like in "this object and this one" with one circling gesture); the category and the properties (to filter the visible objects and to count the supposed referents);
- the predicate: its aspect and its role considering the task (to reinforce the specific interpretation);
- the visual context: the presence and the relevance of perceptual groups (to interpret a scope ambiguity); the presence of similar objects (to make the generic interpretation possible).

These clues show that the multimodal fusion is a problem that occurs at a semantic level and not at a media level, as it is considered in many works ([2] is a famous example that is still followed).

4.3 Referent and Reference Domain Identification

We show in this section how the reference resolution goes through the identification of the referents and of the context from which these referents are extracted. We first demonstrate the importance of taking this context into account, and, second, we expose the possible links between a gesture trajectory and the context demarcation.

In the first scene in Figure 6, a triangle is pointed out by an unambiguous gesture associated with a simple demonstrative expression. Supposing that the next reference will be "the circle," it is clear that such a verbal expression will be interpreted without difficulty, designating the circle just under the triangle of the last utterance. Whereas two circles are visible on the scene, the one being in the same visual reference domain than the precedent referent will be clearly identified. This is one role of the proximity criterion of the Gestalt Theory, as we have precised it in section 3.2. If the reference domain is implicit in the first scene of Figure 6, it is explicit in the second scene. In this case, the expression "the triangle" has the role to extract the referent from the domain delimited by the gesture. Thus, Figure 6 shows the two main roles of gesture: delimitating referents or delimitating a domain.

As in Figure 6, we begin to study examples where the gesture is unambiguous, generally when it has a circling form that can not be interpreted as a targeting one. When the set of target objects is identified, it is compared to the linguistic constraints of the referring expression. These constraints are the category and properties filters, and the functionality of the determiner. Following Salmon-Alt [22], the use of a demonstrative implies the focus on some objects in a domain where other objects with the same category are present. This focus is done by salience, and particularly by the salience due to gesture. The use of a definite article implies an extraction of objects of a given category in a domain where some objects of another category may be present (but not necessarily).

These linguistic constraints give evidence for the role of the gesture. In the second scene of Figure 6, the target objects are not all "triangles." The use of the definite article "the" implies a domain containing triangles and other forms

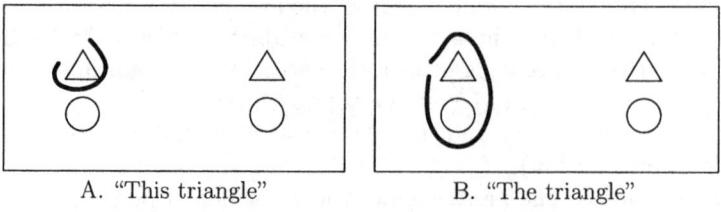

A. "This triangle"   B. "The triangle"

Fig. 6. Referent and domain delimitation

of objects. This domain is clearly the set of target objects. As the expression is singular and as there is one triangle in this domain, the extraction of the referent leads to the unambiguous identification of this triangle. In contrast, the target object in the first scene is a "triangle." As the expression is singular, the multimodal referent may be this target object, and the domain has to be identified. For that, we search a domain containing another triangle. The whole visual context is such a domain. It allows one to interpret the next reference "the other one" as "the other triangle in the domain." There is here a problem: at the beginning of this section, with Figure 6-A, we construct with the proximity criterion the perceptual group at the left of the scene, and we exploit this group, which can be seen as a reference domain, to interpret the next reference "the circle." But this reference domain hypothesis does not fit well with the demonstrative of "this triangle" because it does not contain any other triangle. Our model will handle both hypotheses, to make all interpretations possible. But the reference domain corresponding to the whole visual context will be labeled with a better relevance, and will be tested first in the interpretation process.

Another example where the gesture is not ambiguous but where the identification of the reference domain is complex is given in Figure 7. The hypothesis of a gesture delimitating the reference domain is impossible, and so the set of target objects may be the multimodal referents. For the identification of the possible reference domains, we must take "the most clear" into account. The hypothesis of the whole visual context is impossible because the three circles are lightly gray whereas the two squares are perfectly white. The proximity criterion gives a solution, by constructing a reference domain including the three circles and the three triangles. In this domain, the "forms which are the most clear" are the circles indeed.

When the gesture is ambiguous, a way to proceed is to test all the mechanisms seen above. With the example of a pointing gesture that can designate one object or a perceptual group, the use of a definite determiner will give greater weight to the hypothesis of the perceptual group as the reference domain. With the example of a gesture that can target two or three objects, the presence of other objects of the same category will influence the identification of reference domain. Considering the expression "the triangles" with the gesture of Figure 4-B, the hypothesis of the whole visual context will be relevant as reference domain and

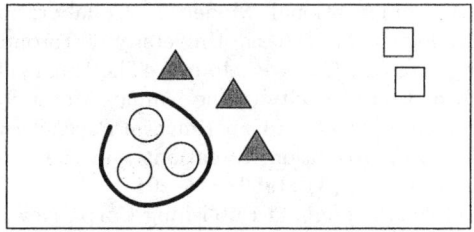

"These forms which are the most clear"

Fig. 7. Gesture initiating a domain

the referents will be the three triangles. On the other hand, using the demonstrative "these triangles," we restrict the referents to the two triangles under the trajectory, thus leaving the third triangle in the reference domain, and allowing for the demonstrative mechanism to be applied.

5 Conclusion

Reference to objects in multimodal dialogue systems can take several forms which are not linked to particular mechanisms of identification. As opposed to approaches like the one of Kehler [13], we consider that simple algorithms are not sufficient for a multimodal system to identify the referents. As we have seen with our examples, the gesture does not always give the referents, and the components of the verbal expression are not sufficient to distinguish them. But the combination of these ostensive clues with inferred contextual considerations does. Then, the question is: how can we combine all the clues to lead to the right interpretation? To answer this question, we investigate in this paper the 'multimodal reference domain' model, whose aim is to formalize the clues into homogeneous structures (reference domains) and then to combine these clues by comparing and merging reference domains. Our study is based on the concrete examples we found in the corpus of Wolff *et al.* [26] and in linguistic classical works like [18], [20] or [23]. As it is showed with the implementation of reference domains in two multimodal dialogue systems (in MIAMM and OZONE project, see section 2), our model appears to be relevant for different kinds of interaction modalities and for different kinds of applications.

References

1. Beun, R.-J., Cremers, A.H.M.: Object Reference in a Shared Domain of Conversation. Pragmatics and Cognition **6(1/2)** (1998) 121–152
2. Bolt, R.A.: Put-That-There: Voice and Gesture at the Graphics Interface. Computer Graphics **14(3)** (1980) 262–270
3. Cosnier, J., Vaysse, J.: Sémiotique des gestes communicatifs. Nouveaux actes sémiotiques (geste, cognition et communication) **52** (1997) 7–28
4. Dekker, P.: Speaker's Reference, Descriptions and Information Structure. Journal of Semantics **15(4)** (1998) 305–334
5. Edmonds, P.G.: A Computational Model of Collaboration on Reference in Direction-Giving Dialogues. Ms. Thesis, University of Toronto, Canada (1993)
6. Efron, D.: Gesture, Race and Culture. Mouton, The Hague (1972)
7. Fitts, M.: The Information Capacity of the Human Motor System in Controlling Amplitude of Movement. Journal of Experimental Psychology **47** (1954) 381–391
8. Grosz, B.J., Sidner, C.L.: Attention, Intentions and the Structure of Discourse. Computational Linguistics **12(3)** (1986) 175–204
9. Itten, J.: The Art of Color. Reinhold Publishing Corp., New York (1961)
10. Jöhsson, A., Dählback, N.: Talking to a Computer is not like Talking to your Best Friend. In: Proceedings of the Scandinavian Conference on Artificial Intelligence. Tromsø (1988)

11. Kamp, H., Reyle, U.: From Discourse to Logic. Kluwer, Dordrecht (1993)
12. Karmiloff-Smith, A.: A Functional Approach to Child Language. Cambridge University Press (1979)
13. Kehler, A.: Cognitive Status and Form of Reference in Multimodal Human-Computer Interaction. In: Proceedings of the 17th National Conference on Artificial Intelligence. Austin (2000)
14. Kievit, L., Piwek, P., Beun, R.-J., Bunt, H.: Multimodal Cooperative Resolution of Referential Expressions in the DenK System. In: Bunt, H., Beun, R.-J. (eds.): Cooperative Multimodal Communication. Springer, Berlin Heidelberg (2001) 197–214
15. Kleiber, G.: Anaphores et pronoms. Duculot, Louvain-la-Neuve (1994)
16. Landragin, F.: Dialogue homme-machine multimodal. Hermes Science Publishing, Paris (2004)
17. McNeill, D.: Psycholinguistics: A New Approach. Harper and Row, New York (1987)
18. Moeschler, J., Reboul, A.: Dictionnaire encyclopédique de pragmatique. Seuil, Paris (1994)
19. Olson, D.R.: Language and Thought: Aspects of a Cognitive Theory of Semantics. Psychological Review **77** (1970) 257–273
20. Reboul, A., Moeschler, J.: Pragmatique du discours. Armand Colin, Paris (1998)
21. Rock, I., Brosgole, L.: Grouping Based on Phenomenal Proximity. Journal of Experimental Psychology **67** (1964)
22. Salmon-Alt, S.: Reference Resolution within the Framework of Cognitive Grammar. In: Proceedings of the International Colloquium on Cognitive Science. San Sebastian, Spain (2001)
23. Sperber, D., Wilson, D.: Relevance. Communication and Cognition (2nd ed.). Blackwell, Oxford UK Cambridge USA (1995)
24. Thórisson, K.R.: Simulated Perceptual Grouping: An Application to Human-Computer Interaction. In: Proceedings of the 16th Annual Conference of the Cognitive Science Society. Atlanta, Georgia (1994)
25. Wertheimer, M.: Untersuchungen zur Lehre von der Gestalt II. Psychologische Forschung **4** (1923)
26. Wolff, F., De Angeli, A., Romary, L.: Acting on a Visual World: The Role of Perception in Multimodal HCI. In: Proceedings of AAAI'98 Workshop: Representations for Multi-modal Human-Computer Interaction. Madison, Wisconsin (1998)

Exploiting Rich Context: An Incremental Approach to Context-Based Web Search*

David Leake, Ana Maguitman, and Thomas Reichherzer

Computer Science Department, Indiana University, Lindley Hall 215,
150 S. Woodlawn Avenue, Bloomington, IN 47405, U.S.A
{leake, anmaguit, treichhe}@cs.indiana.edu

Abstract. Proactive retrieval systems monitor a user's task context and automatically provide the user with related resources. The effectiveness of such systems depends on their ability to perform context-based retrieval, generating queries which return context-relevant results. Two factors make this task especially challenging for Web-based retrieval. First, the quality of Web retrieval can be strongly affected by the vocabulary used to generate the queries. If the system's vocabulary for describing the context differs from the vocabulary used in the resources themselves, relevant resources may be missed. Second, search engine restrictions on query length may make it difficult to include sufficient contextual information in a single query. This paper presents an algorithm, IACS (Incremental Algorithm for Context-Based Search), which addresses these problems by building up, applying, and refining partial context descriptions incrementally. In IACS, an initial term-based context description is the starting point for a cycle of mining search engines, performing context-based filtering of results, and refining context descriptions to generate new rounds of queries in an expanded vocabulary. IACS has been applied in a system for proactively supporting concept-map-based knowledge modeling, by retrieving resources relevant to target concepts in the context of the rich information provided by "in progress" concept maps. An evaluation of the system shows that it provides significant improvements over a baseline for retrieving context-relevant resources. We expect the algorithm to have broad applicability to context-based Web retrieval for rich contexts.

1 Introduction

Many systems have been developed to aid users as they work, by performing automatic Web search for information to support tasks such as Web browsing, query generation, and document authoring, by mining the Web and other resources (e.g., [1, 2]). Reflecting context has long been recognized as important to realizing the potential of Web search in general [3], and context-sensitivity plays an especially crucial role in proactive retrieval systems: The extent to which the system can provide context-relevant information determines whether the system will be an aid or an annoyance. Unfortunately, fully

---

* This material is based upon work supported by NASA under award No NCC 2-1216. We would like to thank our collaborators Alberto Cañas and the IHMC CmapTools team for their many contributions to this project.

exploiting contextual information during Web search is challenging. In current search engines, there are strong limits on query length (e.g., Google's query length limit of ten terms), making it difficult to provide enough terms to describe rich contexts. Even if an adequate context description can be included within the limits, there is no guarantee that the vocabulary used to describe the context will match the vocabulary by which the resource is indexed.

This paper describes an approach which simultaneously addresses the problems of overcoming the variations in term-based context descriptions and reflecting rich context when mining search engines. It presents IACS (Incremental Algorithm for Context-Based Search), an algorithm which takes a novel incremental approach to mining search engines for context-relevant textual resources (such as html pages, pdf files, Word files, etc.), in light of continually refined context descriptions. IACS uses a cycle of characterizing context, generating search engine queries, performing context-based filtering of the results, and refining the context descriptions to emphasize terms discovered to be important, in order to describe the context for new rounds of queries and to accumulate resources relevant to the context as a whole.

We have tested our approach in the domain of proactive support for knowledge modeling. For some time, we have been investigating the development of intelligent support systems for aiding knowledge capture using concept maps, in collaboration with the CmapTools team at the Institute for Human and Machine Cognition [4]. Concept mapping has been extensively used for knowledge construction and sharing in education, and for the capture of expert knowledge by the experts themselves. Part of the CmapTools project focuses on facilitating this knowledge capture by generating context-relevant suggestions and aiding context-relevant search, to help the user decide which concepts to include in a concept map, to identify propositions to include about those concepts, and to find relevant resources to link to the current knowledge model [5, 6, 7]. When users request suggestions relevant to a selected concept in a concept map, the surrounding knowledge model—which may include hundreds of concepts—provides a rich source of contextual information to exploit during retrieval. IACS starts from this information and combines it with context-relevant information gathered incrementally to determine new query terms, extending the retrieval vocabulary beyond the terms in the concept map. Thus IACS mines search engines for resources at the same time it incrementally formulates and refines a context description to improve future search results.

The paper begins by examining the role of context in concept maps and presenting our goals for a proactive, context-relevant resource suggestion system to aid concept mapping. It next presents the IACS algorithm itself, followed by an evaluation comparing its performance to a baseline, non-incremental method. These results suggest that IACS provides significant improvements, both in terms of maintaining focus on context-relevant resources (measured by a generalization of precision) and in terms of retrieving resources providing good coverage of the context (measured by a generalization of recall). Because the algorithm itself relies only on the availability of a set of terms characterizing the context, and does not depend on any specific properties of concept maps, we consider the approach promising for exploiting rich contexts for other retrieval tasks as well.

2 Concept Maps and Concept Mapping

Concept maps [8, 9] are collections of propositions (simplified natural language sentences) displayed as a two-dimensional, visually-based representation of concepts and their relationships. Concept maps depict concepts as labeled nodes and inter-concept relations as labeled links, as illustrated in the sample concept map "Mars myth and science fiction" shown in Figure 1. Unlike semantic networks and other graph-based structures commonly used in artificial intelligence to perform automatic reasoning on the encoded knowledge, concept maps are "informal" knowledge representations that facilitate knowledge capture for human examination and sharing and enable students to learn "meaningfully" by connecting concepts held in long-term memory with new concepts and propositions.

Concept mapping is widely used in educational settings, in which teachers assign students to draw concept maps to encourage them to organize their knowledge and to make their understanding explicit for knowledge assessment and sharing. Studies show that students in a wide range of age groups, as early as in elementary school, can generate concept maps successfully. The naturalness of the concept mapping process makes it promising as a method for direct knowledge capture by experts themselves, and the conciseness and structure of concept maps assists understanding the captured information. To facilitate electronic concept map construction and sharing, the Institute for Human and Machine Cognition (IHMC) has developed CmapTools, publicly-available tools to

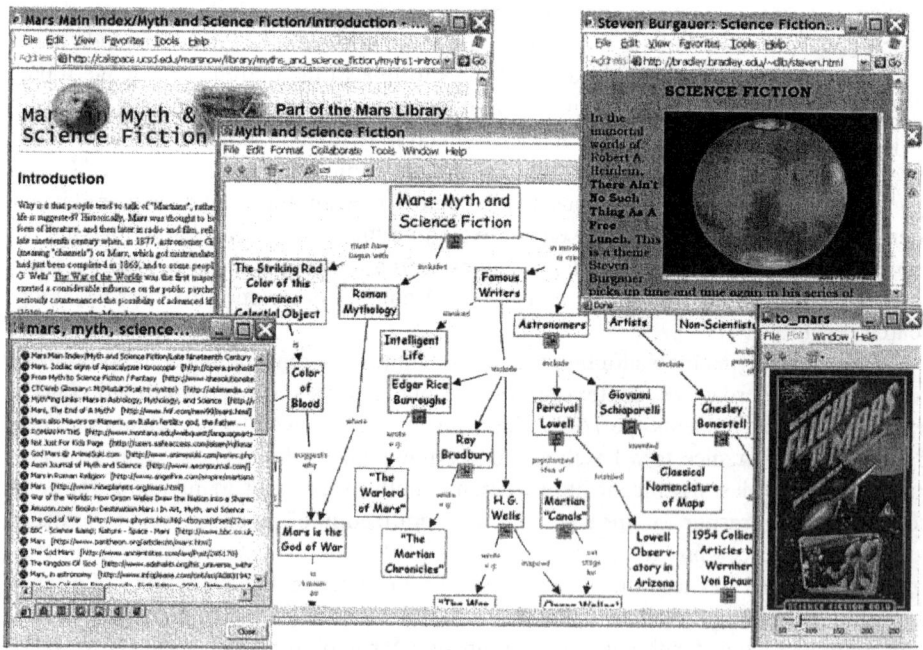

Fig. 1. The CmapTools Interface with the IACS resource suggestion window and related resources

support generation and modification of concept maps in an electronic form [4]. The CmapTools software enables interconnecting and annotating maps with material such as other concept maps, images, diagrams, and video clips, providing rich, browsable knowledge models available for navigation and collaboration across geographically-distant sites. CmapTools has been used for numerous projects including a large-scale initiative in modeling and sharing the knowledge of NASA experts on the planet Mars [10]. Figure 1 illustrates the interface's display of a sample concept map from that domain.

2.1 Adding Intelligent Suggesters

A goal of the CmapTools initiative is to empower experts to construct knowledge models of their domains without the need for a knowledge engineer's intervention, or to actively participate in knowledge modeling led by a knowledge engineer. While users find the interface itself natural and intuitive, part of the challenge of concept mapping is to determine the "right" concepts and relationships to include in the concept map. Informal studies show that users building concept maps often stop for significant amounts of time, wondering how to extend their models, and in some cases searching the Web to jog their memories or find new material to link to the current map. To support this process, a current effort augments the CmapTools interface with a family of "intelligent suggesters" to start from a concept map under construction, and propose context-relevant information to aid the user's knowledge capture and knowledge construction [11]. This paper focuses on one of those suggesters, a system that explores external resources on the Web to find related text documents that can be linked to the concept map or examined for additional information to be included into the concept map.

2.2 Contexts for Concept Mapping

In formal methods for knowledge capture, a goal is to associate each expression with a unique, context-independent meaning. Considerable effort and expertise may be required to train people to capture knowledge in such carefully-crafted forms. On the other hand, concept mapping tools are intended for "human-centered" knowledge capture, in which people express their knowledge informally, without a controlled vocabulary. Concept maps offer no assurance of unambiguous labels, but instead rely on the rich context of the rest of the map for disambiguation. For example, the concept label "Mars" might designate the planet Mars, the god Mars from mythology, or the Mars candy bar; the relevant meaning would be suggested by the context in which it was found. Consequently, to develop a suggester that retrieves resources relevant to a concept label, it is necessary for retrieval to reflect that concept's context in the knowledge model.

In concept map-based knowledge models, each concept can be seen as contained within several layers of context, as illustrated in Figure 2. We define the inner-most context layer C_1 of a target concept to be all concepts directly linked to the target concept in the concept map graph. This is the set of all concepts participating in propositions in which the target concept is directly involved. The second layer C_2 adds other

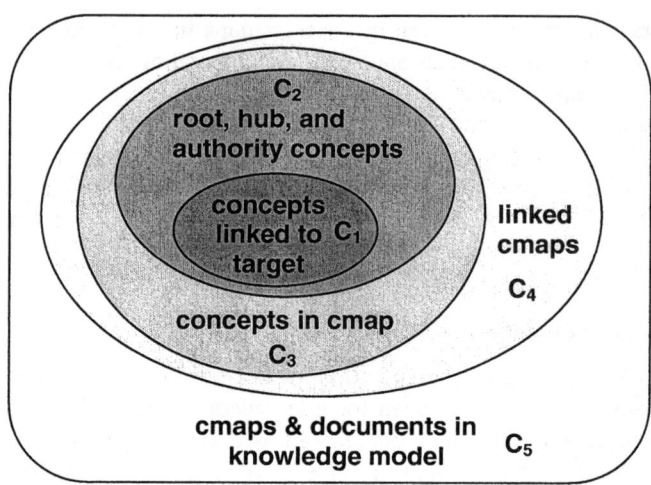

Fig. 2. Layers of context for a concept in a concept map

concepts that play a key role in describing the topic of the concept map as a whole. In previous research, we developed and tested a set of candidate models for predicting topic-important concepts, to select a model to use for weighting the importance of particular concept labels in generating a topic description [12, 13]. Our models assess each concept's role in describing a topic, based on the topological structure of the map. Human-subjects experiments showed a statistically-significant agreement between the predictions of our best model and the actual judgments of subjects who predicted concept map topics from the maps' structure.

For humans, the root concept, typically located at the top of a map, serves as a starting point to explore a map, thus providing a first hint as to what the map discusses. Important concepts for describing the context of a target concept in a concept map include the root concept and concepts with many incoming or outgoing links (*authority* and *hub* concepts, respectively).

Next, the layer C_3 of a concept's context is the set of all the concepts connected to the target concept within the boundary of the concept map. Fully developed concept maps contain a well connected set of concepts in which each concept is explained in terms of the relations and concepts directly connected to the concept. Because interpretations of each of these concepts are influenced by their own connections, any concept in the map may influence interpretation of the target concept.

The CmapTools software enables concepts to be linked to other concept maps, with a link analogous to a Web link, enabling users to jump from one map to another. However, unlike links in Web documents, links in concept maps may also allow users to navigate to the same concept discussed in different maps, with each map providing a different context for the concept. For example, the concept "rocket engine" may occur in a map on rocket architecture or in a map on rocket propulsion systems. Layer C_4 reflects this, extending the context of layer C_3 by also considering all the concepts in concept maps that are directly linked to the target concept. Finally, layer C_5, the most general context of a target concept, is the entire knowledge model consisting of a set of concept maps and annotations such as text documents, images, or other multi-media resources. All the concepts in the concept maps of a knowledge model share the same C_5 context.

Each of the layers of context could influence human judgments of resources' relevance to a target concept. Our current work focuses on exploiting information extracted automatically from C_2, in order to provide the user with suggestions of resources relevant to that context. IACS is applied in a system which describes contexts using a weighted set of terms, with term weights reflecting estimates of the terms' importances to characterizing the context. Initially, term weights for a given concept map are computed based on the structural analysis methods summarized previously. The next section describes how this initial context description can be incrementally refined and used to focus retrieval as new relevant material is retrieved from the Web.

3 An Incremental Strategy for Exploiting Rich Context

A limitation of current search engines is their restriction on query length, enabling only a small set of terms to be contained in any query. Consequently, to take advantage of the rich contextual information provided by a knowledge model, incremental approaches are needed to allow multiple queries to build up context-relevant information. In an incremental approach to Web search, contextual information can help to guide the exploration and discovery of relevant resources both at the moment a query is constructed (pre-query stage) and after an initial set of results have been obtained (post-results stage).

To retrieve resources relevant to a target concept in a concept map, IACS exploits the rich context of a surrounding concept map in three ways. First, it uses terms extracted from the concept map context C_2 to augment the initial search engine query. This is achieved by analyzing the concept map, identifying important terms and ranking them using the topological analysis methods sketched in section 2.2. The most highly-rated candidate terms are added to the terms of the concept label, reflecting the context in which the label occurs. This enables the use of limited context, but because of query length limits, few terms can be included, so it provides a coarse-grained starting point. Second, the context of the concept map is exploited after the initial set of results has been obtained, for filtering irrelevant material and ranking retrieved results based on their estimated context-relevance. This enables the rich context to help select relevant material.

Third, IACS exploits the context to generate new queries that go beyond the initial query, and that may even go beyond the vocabulary of the initial concept map. After the first set of results has been obtained, the search context is used to refine/extend the set of terms used for the context description. Terms that appear "often" in search results similar to the context tend to be good *descriptors* of the user's information needs. In addition, because these descriptors are expected to occur in a large fraction of the relevant material, they are useful as query terms when high recall is desirable. Likewise, terms that tend to occur "only" in results similar to the search context can serve as *discriminators*. When used as query terms, topic discriminators help restrict the set of search results to mostly similar material and therefore can help achieve high precision. A formal characterization of topic descriptors and discriminators as well as an evaluation of their usefulness as query terms can be found in [13].

IACS identifies topic descriptors and topic discriminators by analyzing the terms in retrieved documents. Consequently, descriptors and discriminators are not restricted to terms occurring in the originating search context, and if novel terms have high descriptive or discriminating power, they expand the initial vocabulary used to describe the context. Therefore, while the initial context only reflects the vocabulary of the originating concept map, new terms weighted as a function of their descriptive and discriminating power will be incrementally added to the search context. In IACS's incremental search process, the generation of second-round and subsequent queries can significantly benefit from a search context refined by the addition of good descriptors and discriminators.

Table 1. Pseudocode of the incremental algorithm for context-based search

PROCEDURE INCREMENTAL CONTEXT-BASED SEARCH
INPUT:
  M: source concept map
  s: number of iterations.
  n: number of search queries.
OUTPUT:
  A ranked list of resources related to **M**
BEGIN
  Use topological analysis to weight terms in M.
  Generate a set C of weighted terms (initial search context).
  $T[0] = \{C\}$ % *$T[i]$ is a set of sets of weighted terms.*
  $R = \emptyset$ % *Search results.*
  for (i=0; i < s; i++)
  do
    $T[i+1] = \emptyset$.
    for each set of terms $C \in T[i]$
    do
      Use the most important terms in C to form n search queries.
      Submit queries to a search engine.
      Use C to filter results and add them to R.
      Compare search results to C to identify best descriptors and discriminators.
      Weight terms as a function of their descriptive and discriminating power.
      Use best descriptors and discriminators to expand C.
      Use C to generate a set N of overlapping term clusters.
      T[i+1]= T[i+1] ∪ N.
    end do
  end do
  Clean and sort R.
  return R.
END

Table 1 presents an outline of the incremental algorithm for context-based search. The algorithm starts by applying topological analysis to a concept map to identify the most salient terms in the map. These terms define the initial search context, which is used to start the incremental Web search and context expansion/refinement process. Terms in the retrieved results are analyzed in light of the search context to refine the search context description, and the highest-ranked terms in the search context are used as query terms in subsequent Web queries.

For efficiency, IACS bases its processing on the short "snippets" of text returned for each page in the search engine results summary, rather than full pages. Results are filtered and weighted according to context. Filtering is done by comparing the set of keywords occurring in the snippets against the set of keywords associated with the current context. If the cosine similarity between the two sets is above a threshold (defined in terms of a "curiosity mechanism" described in detail in [14]) the results are added to the set of relevant material. Terms found in the search results are weighted according to their descriptive and discriminating power and used to refine the search context. The extended search context is clustered by a soft term clustering algorithm which we developed to facilitate the generation of cohesive queries in subsequent iterations [15]. Soft clustering algorithms generalize hard clustering algorithms by allowing cluster overlap (i.e. the same term may be part of more than one cluster). After all iterations have been completed, the collected search results are cleaned to eliminate redundancies and sorted and returned to the user.

4 Evaluation

4.1 Evaluation Criteria

To evaluate the performance of context-based retrieval for supporting concept mapping, we first had to develop evaluation criteria suitable for this task. We developed two criterion functions for evaluating retrieval performance: *global coherence* and *coverage* [14].

These two functions generalize the well known IR measures of precision and recall. However, in contrast to precision and recall, the measures of global coherence and coverage do not require that all relevant resources be precisely identified. Instead, these measures are applicable as long as an approximate description of the potentially relevant material is available. The relaxation of the requirement of a precise set of relevant resources makes these novel criterion functions suitable for the evaluation of context-based search on the Web, where a precise characterization of relevant resources is usually unavailable.

Assume $R = \{r_1, \ldots, r_m\}$ is a set containing approximate descriptions of potentially relevant material, where each r_i is a collection of keywords. Let $A = \{a_1, \ldots, a_n\}$ be the set of retrieved resources, with a_i also represented as a collection of keywords. A measure of *similarity* between a retrieved resource a_i and a relevant r_j can be computed using, for example, the *Jaccard coefficient*, defined as:

$$\textbf{Similarity}(a_i, r_j) = \frac{|a_i \cap r_j|}{|a_i \cup r_j|}.$$

Then, we can define the *accuracy* of resource a_i in R as follows:

$$\textbf{Accuracy}(a_i, R) = \max_{r_j \in R} \textbf{Similarity}(a_i, r_j).$$

When measuring the accuracy of a retrieved resource a_i, we obtain an estimate of the precision with which the terms in a_i replicate those of relevant resources.

We use the **Accuracy** function to define **Global_Coherence** as follows:

$$\textbf{Global\_Coherence}\,(A, R) = \frac{\sum_{a_i \in A} \textbf{Accuracy}(a_i, R)}{|A|}.$$

The **Global_Coherence** function measures the degree to which a retrieval mechanism succeeded in keeping its focus within the theme defined by a set of relevant resources. This is similar to the IR notion of precision, except that we use a less restrictive notion of relevance.

We note that a high global coherence value does not guarantee acceptable retrieval performance. For example, if the system retrieves only a single resource that is similar to some relevant resource, the global coherence value will be high. Because context-based suggesters should also maximize the number of relevant resources retrieved, we introduce a *coverage* factor to favor those strategies that retrieve many resources similar to a target set of relevant resources. We define a criterion function able to measure coverage as a generalization of the standard IR notion of recall:

$$\textbf{Coverage}\,(A, R) = \frac{\sum_{r_i \in R} \textbf{Accuracy}(r_i, A)}{|R|}.$$

4.2 The Performance Evaluation

A performance evaluation based on our criterion functions requires access to a set of terms taken to characterize the relevant resources (a target set R). For our task of suggesting information relevant to a concept-map-based knowledge model, we can define such a set based on an existing corpus of concept maps as follows.

Let $K = \{c_1, \ldots, c_m\}$ be a concept-map-based knowledge model, where each c_k is a set of keywords representing a concept map. Suppose c is a concept map in K and c is used for context-based retrieval. If the knowledge model K has been built by a reliable source and is sufficiently extensive, then, for evaluation purposes, the set K could act as a surrogate for R, the set of relevant resources. In our evaluations we use an expert-generated knowledge model on the Mars domain as our "gold standard" [10]. This knowledge model contains 118 concepts map, presenting an extensive description of the Mars domain.

In our tests the top-level concept map from the Mars knowledge model was used as the starting point (corresponding to the concept map under construction, for which related suggestions were sought) and IACS was used to search for resources on the Web, without access to any of the other maps in the knowledge model. As a baseline method for comparison, we implemented a simple non-incremental algorithm which constructs queries from the concept labels of the same concept map used as IACS's starting point, after stopword elimination. It submits these as individual queries to the Google Web API. For each query submitted by IACS, the baseline creates a query of equal size, using terms extracted from concept labels selected randomly from the source map. The baseline's queries include full concept labels when possible, but may use subsets to reduce query size or terms from additional concept's labels when needed, in order to as-

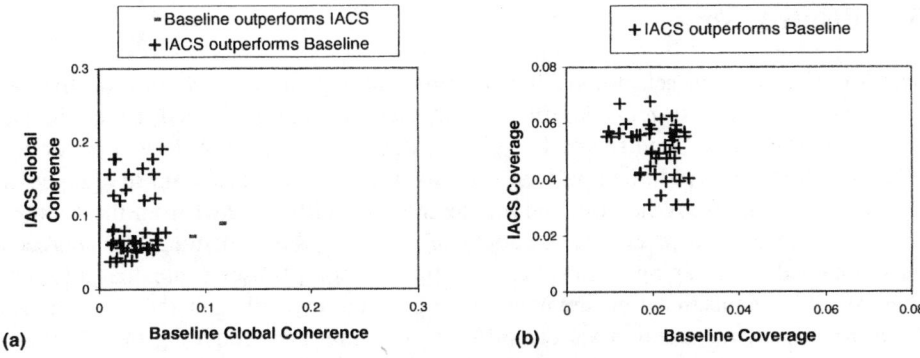

Fig. 3. IACS vs. Baseline: (a) Global Coherence and (b) Coverage

sure that neither method benefits from differences in query length. In contrast to IACS's incremental approach, the baseline constructs all its queries using terms that occur in the originating concept map. We expected IACS's incremental mechanism to provide results with superior global coherence and coverage for equal number of Web queries. When comparing the performance of our incremental search strategy against the baseline, we set the number of iterations to 3. Our evaluation involved 48 trials. Figures 3(a) and 3(b) compare the performance of the IACS algorithm to the baseline method in terms of global coherence and coverage. Each trial is represented by a point. The point's vertical coordinate corresponds to the performance of IACS for that trial, while the horizontal coordinate corresponds to the performance of the baseline method. The trials in which IACS outperforms the baseline can be identified as those points above the diagonal.

Table 2. Confidence intervals for the mean global coherence of the incremental algorithm for context-based search (IACS) and baseline

Method	N	MEAN	STDEV	95% C.I.
IACS	48	0.086	0.045	**(0.073, 0.099)**
Baseline	48	0.036	0.021	**(0.030, 0.042)**

Table 3. Confidence intervals for the mean coverage of the incremental algorithm for context-based search (IACS) and baseline

Method	N	MEAN	STDEV	95% C.I.
IACS	48	0.051	0.009	**(0.048, 0.054)**
Baseline	48	0.021	0.005	**(0.020, 0.022)**

In Tables 2 and 3 we present the number of trials (N), mean, standard deviation (STDEV), and mean confidence interval (CI) resulting from computing the performance criterion functions for IACS and the baseline. These comparison tables show that the proposed method results in statistically significant improvements over the baseline method.

5 Related Work

The use of context to select and filter information plays a vital role in proactive retrieval systems. Such systems observe user interactions, infer user needs for additional information resources, and search for relevant documents on the Web or other online electronic libraries. Traditionally, such systems find documents relevant to a target by augmenting terms from the target with indexing keywords selected from the context, to improve recall and precision. A variety of recent systems pursuing this approach have obtained encouraging results. For example, Watson [2] uses contextual information from documents that users are manipulating to automatically generate Web queries from the documents, using a variety of term-extraction and weighting techniques to select suitable query terms. Watson then filters the matching results, clusters similar HTML pages, and presents the pages to the user as suggestions. Another such system is the Remembrance Agent [1] which operates inside the Emacs text editor and continuously monitors the user's work to find relevant text documents, notes, and emails previously indexed. Other systems such as Letizia [16] and WebWatcher [17] use contextual information compiled from past browsing behavior—searches within the locus of a currently viewed Web page—to provide suggestions on related Web pages or links to explore next.

CALVIN [18, 19] is a context-aware system which monitors the user's Web browsing activity to generate a model of the user's task to use to retrieve relevant resources indexed in similar contexts. In addition, versions of the system provide capabilities for users to manually enter information about a variety of resources, such as descriptions of books or articles, and data on useful personal contacts. The gathered material is stored as contextualized cases recording information that users consult during their decision-making, and is suggested when the user context is similar to the one associated with the stored cases.

Except for Watson, these systems either suggest information previously indexed by the system or crawled from the currently viewed pages. In contrast, our system, like Watson, potentially considers the entire Web, using widely available search engines such as Google to search for related documents. The IACS approach differs from Watson in its incremental search, which refines the Web queries to find documents more closely related to the concept map in progress.

SenseMaker [20] is an interface that facilitates the navigation of information spaces by providing task specific support for consulting heterogeneous search services. The system helps users to examine their present context, move to new contexts or return to previous ones. SenseMaker presents the collection of suggested documents in bundles (their term for clusters), which can be progressively expanded, providing a user-guided form of incremental search. Our EXTENDER system [14], like IACS, also applies an incremental technique to build up context descriptions. Its task, however, is to generate brief descriptions of new topics relevant to the current concept map. Rather than providing documents, EXTENDER aims to jog the user's memory during the concept mapping process by presenting a set of keywords suggesting novel, diverse and relevant topics to start new concept maps that extend the knowledge model under construction.

While our work explores the use of the rich context provided by the structure and labels of a knowledge model under construction, other work has pursued retrieval based on other types of contextual information. For example, Suitor [21] is a collection of "attentive agents" that gather information from the users by monitoring users' behavior and context, including eye gaze, keyword input, mouse movements, visited URLs and software applications on focus. This information is used to retrieve context relevant material from the Web and databases. Outside of proactive retrieval systems, IACS' learning of new context-related terms may be seen as related to learning semantic correspondences, studied in Semantic Web research (e.g., [22]).

6 Conclusion and Future Directions

When rich contextual information is available, it provides a potential resource for improving the performance of proactive retrieval systems. However, it may be difficult to select terms to describe a context, and the descriptions may be difficult to apply in single search queries. This paper describes research on addressing these problems through an incremental algorithm, IACS, which successively retrieves relevant resources and refines the context description. IACS has been applied to the task of retrieving Web pages relevant to a concept in the context of a concept map, in order to aid the concept mapping process. In an evaluation using an expert-generated knowledge model as the basis for assessing relevance, the IACS approach outperformed a baseline in both coherence and coverage of the resources retrieved.

As discussed in section 2.2, concept-map-based knowledge models provide many different layers of context. The study reported in this paper examines the use of a single layer, the concepts judged important to the topic of the concept map. Consequently, an interesting followup study concerns developing strategies for including appropriate weightings of terms in other layers, and assessing the tradeoffs of expanded contexts in terms of global coherence and coverage.

The IACS algorithm is applicable to any domain for which it is possible to generate term-based characterizations of a context. Thus another interesting task is to study IACS for other task domains for which rich context is available. For example, IACS could be applied to retrieve resources relevant to an electronic document such as a report, an email message, a presentation, or a Web page as it is written or consulted. We expect incremental approaches to have broad potential applicability to exploiting rich contexts for context-relevant Web search.

References

1. Rhodes, B., Starner, T.: The remembrance agent: A continuously running automated information retrieval system. In: The Proceedings of The First International Conference on The Practical Application of Intelligent Agents and Multi Agent Technology (PAAM '96), London, UK (1996) 487–495
2. Budzik, J., Hammond, K.J., Birnbaum, L.: Information access in context. Knowledge based systems **14** (2001) 37–53

3. Lawrence, S.: Context in Web search. IEEE Data Engineering Bulletin **23** (2000) 25–32
4. Cañas, A.J., Hill, G., Carff, R., Suri, N., Lott, J., Eskridge, T., Gómez, G., Arroyo, M., Carvajal, R.: CmapTools: A knowledge modeling and sharing environment. In Cañas, A.J., Novak, J.D., González, F., eds.: Concept Maps: Theory, Methodology, Technology. Proceedings of the First International Conference on Concept Mapping. (2004)
5. Cañas, A., Carvalho, M., Arguedas, M., Eskridge, T., Leake, D., Maguitman, A., Reichherzer, T.: Mining the web to suggest concepts during concept map construction. In Cañas, A.J., Novak, J.D., González, F., eds.: Concept Maps: Theory, Methodology, Technology. Proceedings of the First International Conference on Concept Mapping. (2004)
6. Leake, D., Maguitman, A., Reichherzer, T., Cañas, A., Carvalho, M., Arguedas, M., Brenes, S., Eskridge, T.: Aiding knowledge capture by searching for extensions of knowledge models. In: Proceedings of the Second International Conference on Knowledge Capture (K-CAP), New York, ACM Press (2003) 44–53
7. Leake, D., Maguitman, A., Reichherzer, T., Cañas, A., Carvalho, M., Arguedas, M., Eskridge, T.: "Googling" from a concept map: Towards automatic concept-map-based query formation. In Cañas, A.J., Novak, J.D., González, F., eds.: Concept Maps: Theory, Methodology, Technology. Proceedings of the First International Conference on Concept Mapping. (2004)
8. Novak, J.: A Theory of Education. Ithaca, Illinois, Cornell University Press (1977)
9. Novak, J., Gowin, D.B.: Learning How to Learn. Cambridge University Press (1984)
10. Briggs, G., Shamma, D., Cañas, Carff, R., Scargle, J., Novak, J.D.: Concept maps applied to Mars exploration public outreach. In Cañas, A.J., Novak, J.D., González, F., eds.: Concept Maps: Theory, Methodology, Technology. Proceedings of the First International Conference on Concept Mapping. (2004) 125–133
11. Leake, D., Maguitman, A., Reichherzer, T., Cañas, A., Carvalho, M., Arguedas, M., Brenes, S., Eskridge, T.: Aiding knowledge capture by searching for extensions of knowledge models. In: Proceedings of KCAP-2003, ACM Press (2003)
12. Leake, D., Maguitman, A., Reichherzer, T.: Understanding knowledge models: Modeling assessment of concept importance in concept maps. In: Proceedings of CogSci-2004. (2004)
13. Maguitman, A., Leake, D., Reichherzer, T., Menczer, F.: Dynamic extraction of topic descriptors and discriminators: Towards automatic context-based topic search. In: Proceedings of the Thirteenth Conference on Information and Knowledge Management (CIKM), New York, ACM Press (2004) 463–472
14. Maguitman, A., Leake, D., Reichherzer, T.: Suggesting novel but related topics: Towards context-based support for knowledge model extension. In: Proceedings of the 2005 International Conference on Intelligent User Interfaces. (2005) 207–214
15. Maguitman, A.: Intelligent Support for Knowledge Capture and Construction. PhD thesis, Indiana University (2005)
16. Lieberman, H.: Letizia: An agent that assists Web browsing. In: Proceedings of the Fourteenth International Joint Conference on Artificial Intelligence (IJCAI-95), San Mateo, Morgan Kaufmann (1995) 924–929
17. Armstrong, R., Freitag, D., Joachims, T., Mitchell, T.: WebWatcher: A learning apprentice for the World Wide Web. In: AAAI Spring Symposium on Information Gathering. (1995) 6–12
18. Leake, D.B., Bauer, T., Maguitman, A., Wilson, D.C.: Capture, storage and reuse of lessons about information resources: Supporting task-based information search. In: Proceedings of the AAAI-00 Workshop on Intelligent Lessons Learned Systems. Austin, Texas, AAAI Press (2000) 33–37
19. Bauer, T., Leake, D.: WordSieve: A method for real-time context extraction. In: Modeling and Using Context: Proceedings of the Third International and Interdisciplinary Conference, Context 2001, Berlin, Springer-Verlag (2001)

20. Baldonado, M.Q.W., Winograd, T.: SenseMaker: an information-exploration interface supporting the contextual evolution of a user's interests. In: Proceedings of the SIGCHI conference on Human factors in computing systems, ACM Press (1997) 11–18
21. Maglio, P.P., Barrett, R., Campbell, C.S., Selker, T.: SUITOR: an attentive information system. In: Proceedings of the 5th international conference on Intelligent user interfaces, ACM Press (2000) 169–176
22. Doan, A., Madhavan, J., Domingos, P., Halevy, A.: Learning to map between ontologies on the semantic web. In: Proceedings of the Eleventh International WWW Conference, ACM Press (2002)

Context Adaptive Self-configuration System Based on Multi-agent[*]

Seunghwa Lee, Heeyong Youn, and Eunseok Lee

School of Information and Communication Engineering, Sungkyunkwan University,
300 Chunchun jangahn Suwon, 440-746, Korea
jbmania@selab.skku.ac.kr, {youn, eslee}@ece.skku.ac.kr

Abstract. This paper proposes an adaptive resource self-management system that collects *system resources*, *user information*, and *usage patterns* as context information for utilization in self-configuration. This system will ease the system maintenance burden on users by automating a large portion of the configuration tasks such as; *install, reconfiguration* and *update*, while also decreasing cost and errors. Working from the gathered context information, this system allows users to select and install appropriate components for their system context. This also offers a more personalized configuration setting by using user's existing application setting and usage pattern. To avoid a center server overload when transferring and managing related files, we employ Peer-to-Peer method. A prototype was developed to evaluate the system with a comparison study using the conventional methods of manual configuration and *MS-IBM* systems was conducted to validate the proposed system in terms of functional capacity, install time, etc...

1 Introduction

Recently, the advent of *Ubiquitous Computing* has been an influx of more diverse computing devices to the market and hence an increase in the subjects to be managed as well as complexity to be considered. Management of such systems has, until now, been handled manually and an increased number of human resources have had to be procured in step with these recent developments.

Take for example a system that manages and operates several hosts. Not only does each host need installation of distinct software, but also, the cost, time and labor needed to update these software applications are tremendously constraining for the maintenance personnel [1]. Novice end-users who are unfamiliar with their charge often fail to conduct critical maintenance procedures such as OS patches or vaccine program updates, a negligence that often leads to detrimental system defects. To cope with these problems, existing efforts are as follows: central integrated control of distributed resources, unattended remote installation, periodic automatic update,

---

[*] This work was supported by the Ubiquitous Autonomic Computing and Network Project, 21st Century Frontier R&D Program in Korea and the Brain Korea 21 Project in 2004. Dr. E. Lee is the corresponding author.

and etc... Like us, those are all intended to automate the configuration process, but they still require large amounts human participation while being uniformly performed without awareness of the context of each system and user. Thus, in this paper, we propose an adaptable self-management system that collects *system resources, user information*, and *usage patterns* as context information while reducing the system maintenance burden on users by automating a large portion of the configuration tasks such as; *install, reconfiguration*, and *update*. Concretely, for adaptive installation, this system generates personalized automatic answer files and environmental setting values by using user context data while performing personalized installation. For adaptive reconfiguration, this system controls the installed components options and their quality of service (QoS), including the killing and invoking of a specific processor according to the system context like memory capacity. For an adaptive update, this also controls the order and priorities of update candidates by using the context data such as a user's usage pattern and available memory capacity. We employ Peer-to-Peer method for efficient update file distribution, not conventional central distribution way.

A prototype was developed to evaluate the system while a comparison study with the conventional method of manual configuration and *MS-IBM* systems was conducted to validate the proposed system in terms of functional capacity, user's burden and install time. Section 2 classifies and defines the various functions of configuration (*install, reconfiguration, update*), introduces related studies on each of these functions, and addresses problems uncovered through analysis of previous studies. Section 3 explains the specifications of the proposed system, and Section 4 details its evaluation results. Finally, Section 5 discloses the study's conclusion and speculates on future action.

2 Related Work

This paper defines configuration as the installing of necessary components that need to be managed, subsequently reconfiguring these components for distinctive tasks. The configuration process can be sub-defined as follows:

- Install: newly installing necessary components (OS, software, etc.)
- Reconfiguration: reconfiguring installed components to fit unique situations
- Update: version management of applications or modification of components to correct defects. Also includes re-installation when parts of the configuration files have been corrupted due to virus attack or system error.

1) Install

Manual installation of system components is a consuming task in time and labor. In addition, serious problems may arise when there are a multiple of subjects.

The *Tivoli Configuration Manager* [2] by IBM was recently developed to resolve such issues. As a tool based on Microsoft's remote installation and other distribution technology [3], the administrator can remotely boot a connected system from a central server and send the previously saved image file of the OS or installation components for installation to the system. This tool allows integrated management of the mass system network and saves time and cost of installation.

When manually installing components, the administrator had to respond to numerous data queries such as user name and language. However, the *Tivoli* employed a script language to write an automated response file and thereby, realized self-installation. A certain amount of personalized installation was achieved by reflecting user information onto its script.

However, the *Tivoli* is not a fully customizable tool, as reconfiguration of the components must be manually attained to following the installation process.

The proposed system employs stored user data to automatically write customized response files and specific configuration values and sends them to the respective host, thus providing the user a more customized configuration.

The proposed system also collects system resources (available memory capacity) as a context, and is enabled with a script writing function that will automatically perform minimal installation when space is limited. This action is alerted and the final decision is left to the user. The agent then surveys and learns the user's response and reflects it on future activities. The proposed system will be discussed in further detail in Chapter 3.

2) Reconfiguration
Devices come in various system formats with countless configurations options with the installed components subsequently being composed of hundreds of configuration parameters. Personalizing and manually reconfiguring such components is a studious and complex task. Hence there is a need for a system that has optimal setup rules for each situation for use in automatic reconfiguration.

Such a response mechanism is a rule-based expert system that has long been studied in the field of artificial intelligence and is being applied to various middleware [4][5] with context awareness capabilities.

The proposed system collects the constantly varying memory as a context and adjusts application settings based on pre-defined rules to refine the QoS. This allows limited environments, such as of mobile devices, to retain a more stable performance. It also has the capability to automatically end processes that unnecessarily occupy resources and to restart them when resources are later made available.

3) Update
Novice users not fully educated in managing the devices in their care may cause serious malfunctions to arise within these systems by not conducting adequate or accurate patch or vaccine program updates in spite of the criticality of these maintenance tasks. Updating is an important responsibility in system management and may be tackled through a variety of methods. Microsoft's Windows Update is an integrated update method where the Windows Update's automatic updating agent regularly interacts with the server or informs the user to visit the website to perform an update. The user may personally specify each update but a full-file update is generally performed. Hence, unnecessary updates of rarely used applications are performed, consuming the use's valuable time.

The proposed system employs an agent that resides within the user's device and regularly (and continuously) communicates with the server to perform automatic updates. The frequency of a software use is collected and stored as a context. An update priority is determined based on this history. Thus, if storage space is limited or update time is lacking, only the most necessary updates are performed, thereby offering efficiency in the update process.

File distribution on existing configuration systems are performed via the central server. This function leads to server overload with a direct impact to the system possible arising from server malfunctions. The Astrolabe [6] of Cornell University was developed as a solution to the weaknesses of such centralized distribution. The various systems managed by the Astrolabe form hierarchical zones with distinct DNS codes and the file list of each host belonging to a specific zone are collected by the main host of their respective zone. Hosts that need specific files query their main host for this file and the main host connects with a host with this file and receives the file via peer-to-peer. This system has remedied many of the outstanding issues of the centralized distribution system. The proposed system has applied and integrated the many strengths of the Astrolabe system.

3 Proposed System

3.1 System Objectives

The system administrator has had to perform individual configuration tasks (*install*, *reconfiguration*, *update*) on numerous systems of various formats. But such tasks have been time and cost consuming, not to mention labor intensive. Several research projects have attempted to resolve these issues by developing an integrated and centralized management system. However, many tasks are still left to the system administrator for manual handling. A customized configuration system that reflects comprehensive context has not yet been fully realized.

This study proposes a *Context adaptive self-configuration system* that employs multi agents to collectively gather contexts based on the system resources and users' system usage patterns, analyzes the collected context, and performs automatic configuration as needed. This system will allow not only automation of the previously manual tasks but also a more customized configuration.

Moreover, the system also integrates the strengths of Cornell University's Astrolabe System, thus addressing the various weaknesses of conventional centralized distribution systems in decreasing the central server overload and allowing quick file distribution.

3.2 System Structure

1) Component Agent
Monitors the memory or performance status of the subject, checks user's application preference data and frequency, and transmits the gathered information to the server regularly (continuously). The Component Agent interacts with the configuration system for; install, reconfiguration, or update tasks and adjusts the application options as needed during reconfiguration. Upon installation, rules and knowledge of each

application for the option reconfiguration task are transmitted to the subject along with the application. The Component Agent manages these.

The Component Agent supports self-installation by providing packages or configuration files received from the Configuration System or a host in its zone. The Agent also analyses post-installation user history and relays it back to the Context Synthesizer.

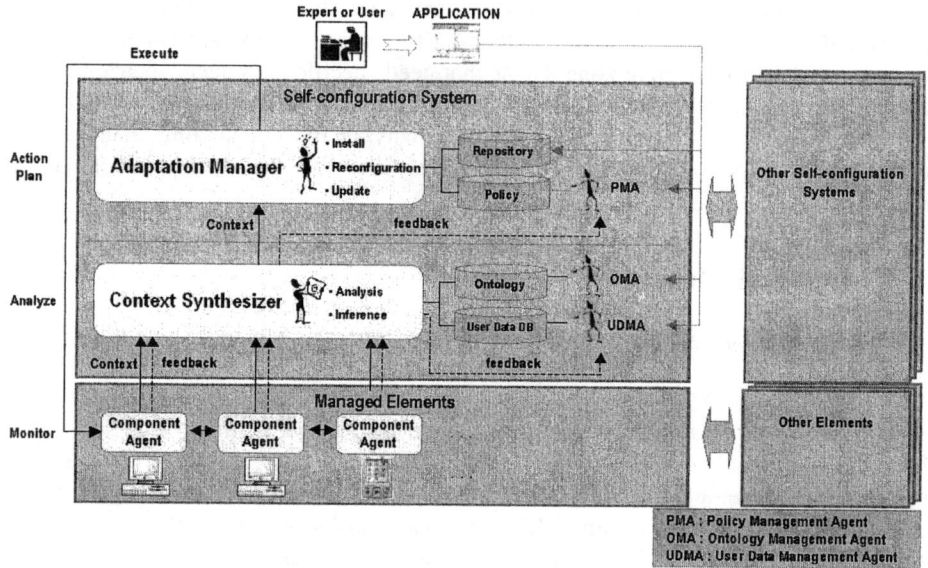

Fig. 1. Architecture of Proposed system

2) Context Synthesizer
Gathers the context provided by the Component Agent and stores the basic user data, application preferences and usage frequencies in the User Data DB. The Context Synthesizer analyzes and deducts the current status by using the stored user data and ontology (various expressions on preference data of the applications), relaying the results to the Adaptation Manager.

3) Adaptation Manager
Decides and performs appropriate tasks based on policies pre-defined through the data provided by the Context Synthesizer. The Adaptation Manager installs components stored in the Repository or performs configuration tasks such as reconfiguration and updates. An installation image on user data and customized 'auto response files' for installation is also automatically created.

The system administrator or the user can adjust and manage the components, policies, ontology and user data to be stored in the Repository. The proposed system can also communicate and share data with other external Configuration Servers.

3.3 System Behavior

1) Install
The proposed system will automatically write customized response files and specific preference values based on stored user data (user preferences deducted from personal information, preference data of existing programs, and usage patterns).

For instance, the proposed system sets as default fonts those frequently used by the user in similar applications when installing a new word processing program. (The Ontology server stores the different expressions with identical meanings for each application and uses the data to identify preferences for other applications.) The system identifies automatic backup frequencies of other applications to apply to the new word processing application preferences. User-specific preferences such as opting not to create application icons on Windows and other GUI OS desktops are also reflected in the automated scripts for new applications.

The proposed system identifies user preferences by analyzing preference values of similar applications and performs a customized installation process which is done by directly writing and distributing an automated script or preference value, as described above. The system also observes user activity on a continual basis to see whether such settings are altered and if so, performs *Reinforcement learning* [7] by adjusting its policies following such negative feedback.

It is also capable of writing a script to automatically select minimal installation when sufficient space is found to be lacking through gathered context data on system resources. The user is alerted and final decisions made, through which, the system learns and then reflects onto future tasks.

When available memory is secured at a later time, the system proposes full-file installation of the application to the user, subsequently learning and reflecting these user actions in future tasks as well.

2) Reconfiguration
The proposed system gathers continually varying memory capacity as a context to adjust the application settings based on pre-defined rules while refining service quality. This capability offers mobile devices and other devices with limited performance environments enhanced stability features. The rules and knowledge for adjusting the options of each application is sent to the subject along with the application upon installation and is managed by the Component Agent.

The Component Agent shuts down processes with high resource usage and restarts the process when sufficient resources become available. The processes are listed according to identified priority based on application usage frequency and their correlation with other applications. These are updated on a regular basis through interaction with the Configuration System to establish the most optimal rules.

3) Update
The Component Agent gathers application usage frequency as a context and stores the collected data in the User Data DB. Automatic updates, based on a priority determined through history, are performed through frequent dialogue with the server. The subject system may thus receive an efficient update that takes into account system storage space and time availability.

The update feature of the proposed system also reinstalls parts of component files that are corrupted due to virus or system error. The corrupted files will be identified through analysis of event messages, and copies of the files, in reflection of the Astrolabe [6] system infrastructure, will be transferred peer-to-peer from the local host.

The hierarchical structure of the subjects are as <Fig.2>, each with a unique ID value in URL format. The main host of the zone will gather each file list belonging to the hosts within that particular zone. When a host submits a request for a file to the main

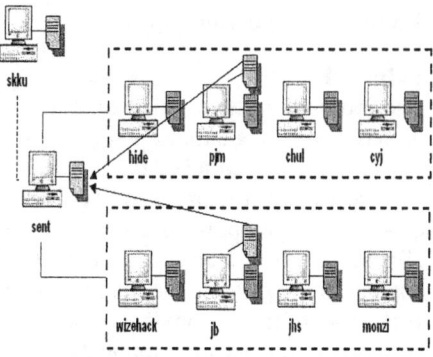

Fig. 2. Host Hierarchy for Peer-to-Peer Transfer

host, the main host identifies the location and transfers the file peer-to-peer to the requester. This format decreases the central server's overload, is error resistant and offers faster file copying capabilities.

4 System Implementation and Evaluation

4.1 System Implementation

Each module of the prototype system was developed with JADE (Java), while DB was with Oracle 9i. All subjects were PCs running Windows 2000 or Windows XP, main OS platforms of Microsoft. The context on Windows registry and application data was gathered in proprietary *.ini files using C.

During an application installation the above user preferences gathered ware referred to shown in <Fig.3>. User's preferences (font face or point

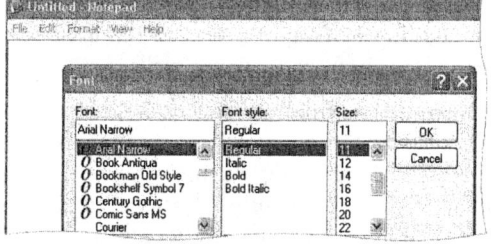

Fig. 3. Application with context based customized preference settings upon installation

size settings for similar applications, etc.) were used to configure the face and size of fonts for the newly installed Notepad. This was achieved by changing the registry values for IfFaceName and iPointSize.

Reconfiguration was performed by adjusting application options based on the continuously varying system context (i.e., available memory capacity) and according to pre-defined critical mass and rules. The rules and knowledge for option adjustments were transported to the subject upon installation. This process was subsequently managed and performed by the Component Agent. Based on pre-defined memory critical mass, application options are adjusted and memory is secured as can be seen in <Fig.4>.

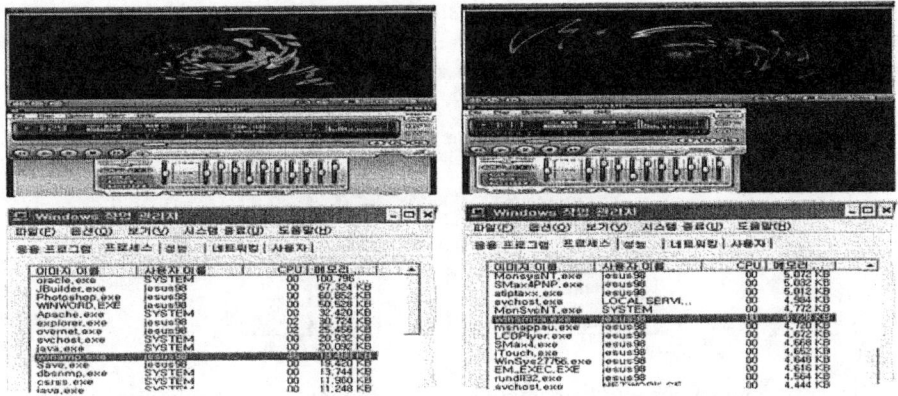

Fig. 4. System Context (available memory capacity) Adaptive Option Adjustment and Subsequent Memory Variations

To test the efficiency of the update, 10 update files were randomly selected and usage frequency recorded. The Component Agent regularly compared the version of the update file with the server, found a host in its zone that has the file, and copied the file peer-to-peer. The sequence order of the update was done based on usage frequency. The update files copied within the zone are checked for consistency through size comparison.

4.2 System Evaluation

The system evaluation was performed by comparing method A (manual installation of components on distributed systems – So far, most offices use this method.), method B (remote management system through networks such as MS-IBM), and method C (proposed system with context awareness functions) through their functions and processing time. The functional comparison of each method is described as below.

Table 1. Comparison in Characteristic features

	Function		A	B	C
Characteristics	Install	Customized Install	△	X	O
	reconfiguration	QoS, Resource management	△	X	O
	update	Update by priority	△	X	O
		P2P	X	X	O

As shown in Table 1, the proposed system has the following discriminated characteristics: Customized install, QoS, Resource management, Update by priority, efficient file copy by P2P.

Also, our system reduces the administrator's burden by employing customized 'auto response files' for installation and centralized management. These are weak points found in manual installation, as shown in Table 2.

Table 2. Comparison on Administrator's Burden

	Function	A	B	C
Admin's burden	Unattended install	X	O	O
	Central management	X	O	O

The next comparison study centered on installation time employing 3 different methods. The installed application was a 510Mbyte MS Office (script and preference files were under 1Mbyte) with an average installation time of 8 minutes varying by the test subject's performance capacity. Peer-to-peer transfer time via LAN recorded an average of 80 seconds. The average installation time is shown in formula (1). (The number of managed elements: N)

$$Average\_install\_time = \frac{Install\_time_1 + Install\_time_2 + \cdots Install\_time_n}{N} = \frac{\sum_{i=1}^{N} Install\_time_i}{N} \quad (1)$$

Method A (manual installation): The system administrator manually installs applications to each system under his care. The total install time is calculated by add the average customized setting time to formula (1) and multiply the number of managed element.

$$Total\_install\_time = (Average\_install\_time + Average\_customized\_setting\_time) \times N \quad (2)$$

Method B (remote installation): The application is remotely transferred through a network connected with the central server. The total install time is calculated by adding the average customized setting time, transport time and α (network overload) to formula (1).

$$Total\_install\_time = Average\_install\_time + Average\_customized\_setting\_time + \alpha \cdot transport\_time \quad (3)$$

Method C (proposed system): The total install time is calculated by adding transport time, α (network overload) to formula (1).

$$Total\_install\_time = Average\_install\_time + \alpha \cdot transport\_time \quad (4)$$

The install time comparison graph for each method is depicted in <Fig.5>. In method A's case the system administrator manually installs the application to each host when the number of systems to manage is small, preferably allowing fast installation by exception for transport time. As the number of systems increase, so to does the install time. However, installation by remote management provides adequate install time, even when the number of systems to be managed increases. Thus, we determine the profit to time and effort with the automation of existing manually installation.

The proposed system not only offers context adaptive customized configuration for distributed systems but is also fail free compared to the performance capacity of central systems managing distributed systems in terms of install time. Moreover, in the case of updates, the proposed system speeds up the file transfer process by reducing the server overload gained by copying files peer-to-peer instead of using the central distribution method, hence enabling a shorter update time than method A and B. The achieved time reduction was verified during method comparison.

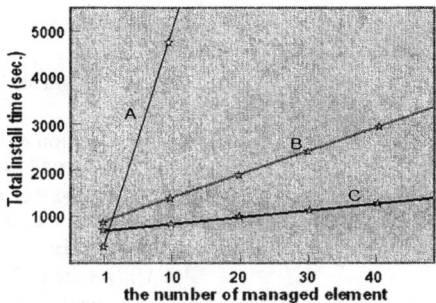

Fig. 5. Install_time Comparison Graph

5 Conclusion

The proposed system provided a function of context adaptive self-configuration that collects system resources, user information, and usage patterns as context data. It has been applied to system *installation*, *reconfiguration*, and *update*. From the experiments, we can verify its effectiveness in resource management by minimizing human interference to decrease work overload. Peer-to-Peer file transferring as a solution to the various weaknesses of centralized distribution systems could decrease central server overload while allowing quick file distribution for an efficient update.

This system, along with the above described self-configuration features, offers a differentiated configuration process that employs system and user contexts. These features are expected to enhance usability and user satisfaction, relieving system administrators of the load they bear in this ubiquitous environment, and offering enhanced computing convenience.

References

1. Paul Horn, "Autonomic Computing: IBM's Perspective on the State of Information Technology", IBM White paper, Oct.2001
2. http://www-306.ibm.com/software/tivoli
3. http://www.microsoft.com/technet/prodtechnol/winxppro/deploy/default.mspx
4. John Keeney, Vinny Cahill, "Chisel: A Policy-Driven, Context-Aware, Dynamic Adaptation Framework", In Proceedings of the Fourth IEEE International Workshop on Policies for Distributed Systems and Networks, Jun.2003
5. Anand Ranganathan, Roy H. Campbell, "A Middleware for Context-Aware Agents in Ubiquitous Computing Environments", In ACM/IFIP/USENIX International Middleware Conference 2004, Jun.2004
6. Robbert van Renesse, Kenneth Birman and Werner Vogels, "Astrolabe: A Robust and Scalable Technology for Distributed System Monitoring, Management, and Data Mining", ACM Transactions on Computer Systems, Vol.21, No.2, pp.164-206, May 2003

Effect of the Task, Visual and Semantic Context on Word Target Detection

Laure Léger[1], Charles Tijus[1], and Thierry Baccino[2]

[1] Laboratoire Cognition & Usages, Université de Paris VIII,
2 rue de la Liberté 93526, SAINT DENIS cedex, France
laure.leger@cognition-usages.org
charles.tijus@univ-paris8.fr
[2] Laboratoire Psychologie Expérimentale Quantitative, LPEQ, EA 1189,
Université de Nice Sophia Antipolis,
24, Avenue des Diables Bleus 06357 NICE, cedex 04 France
baccino@unice.fr

Abstract. Although being a daily task, the search for a word among others words is a new research domain we investigated in order to find the kinds contextual factors that can facilitate semantic oriented visual search. We report two experiments assessing task context, visual context and semantic context. Some of our results are found to be those of classical non-semantic visual search, while others show the impact of the semantic context. Basic recommendations can be find out for Human-Computer conception and cognitive chronometry methodology.

1 Introduction

This article is about the factors that could facilitate the detection of a word among others. There is a practical question if we want to facilitate the rapid and successful finding by a user of the information s/he is looking for when facing Web sites. For instance, if you search the schedule of a film on a cinema web site, how must be semantic and visual information arranged in order to facilitate the visual search activity. But there is also a theoretical question about how much semantic features facilitate visual search. Studies of visual search conducted in cognitive psychology about attention and visual processes consist to show the participant a visual scene composed of no semantic visual stimuli such as letters, digits, squares, circles, triangles, ... The task of the participant is then to detect a particular target among others stimuli. This target can be well defined (to find the square) or can be ill defined (to find the intruder) as in oddity search task.

Treisman and her colleagues ([1] and [2]) had tested the effect of different visual features (such as color, orientation, form,) on target detection efficiency. With a paradigm that consists to vary the number of the stimuli surrounding the target, they distinguished two visual search processes: a parallel search and a serial search. According to Treisman and Gelade [1] and Treisman and Gormican [2], the features (color, orientation, form…) of objects are first processed automatically and in parallel. Next stage following this parallel processing is to build the unicity of the object with focal attention on localization: the different features that are at the same localization are conjoined to form a unitary object, with sometimes illusory conjunction of features. If the object the participant is looking for (a red X) has a distinctive feature (red) in such a way this feature is in isolation (a red X among blue Xs), then this target will be found during the parallel processing stage: the feature characterizing the target stimulus is detect pre-attentively and then « calls » attention to the position of the target stimulus in the visual field at the next stage. On the opposite, if the target can be distinguished from the process of a conjunction of features (a red X among red Os and blue Xs) then a serial search is applied at the second stage and consists inspecting the background items that form the context, one by one, until the target is found or until the participant decides that the target is absent. The number of non-target stimuli on the visual array has no effect on time detection of a target being distinguished by a single feature from the non-target stimuli: the target captures directly attention and this phenomena produces what is called "pop out". When the target is differentiated by a conjunction of features, in the serial search stage, the response time is dependent of the number of non-target stimuli: when the number of items in the context increases the time spent to detect the target increases because it is then necessary to conjoin the features (color and form for the above example) and to scan the whole scene, item by item, to find the target.

Efficiency of the visual search appears to be strongly dependant on the similarity between the target and its context (its background composed of others stimuli). According to Duncan and Humphreys [3], the more the target is similar to its context, such as a conjunctive target, the target sharing one attribute with one kind of context stimuli and one other attribute with a second kind of stimuli, the more difficult is its detection. Targets "pop outs" in case of large dissimilarity. Neisser [4], for instance, observed that it is easier to detect a V among round letters such as O, P, D, G than among angular letters such as N, L, M, X.

Many factors could influence similarity between the target and the contextual non-targets and influence visual search. Sharing visual properties with the objects of the context is one of the factors that bring similarity. Another factor might be the nature of the shared or distinctive features. For instance, perceptual features might not be equivalent for the target detection. Treisman & Gormican [2] find that when the target is to be in a pop out situation (the only item with a particular attribute), a deviant attribute (such as magenta for the color dimension) allows detecting the target more rapidly than a standard attribute (such as red). For these authors, this difference is due to a difference of activation between these two types of attributes: a deviant attribute produces more activation than a standard attribute. However, when the target is a feature conjunction (identifying a red X among red Os and blue Xs), a target sharing standard attributes is detected more rapidly than a target sharing deviant attributes [5].

A third factor, studied in the literature, is the proportion of each kind of stimuli in the context when the target is defined by a conjunction of properties shared with the context objects. This ratio is generally fifty-fifty: as much red Os and blue Xs to detect a red X. Poisson and Wilkinson [6] and Shen, Reingold and Pomplum [7] had shown that response time to detect a conjunctive target depends on the ratio of the two kind of non-target stimuli: target detection is facilitated when the two categories of stimuli that form the context do not appear in equal number in the visual display; as the number of each category approach equivalence, response time increases.

A fourth factor is the semantic similarity between the target and the non-targets in its context. White [8] showed that semantic categories could play a facilitative role in target detection; thus detecting the letter "O" is easier when surrounded by numbers rather than surrounded letters. In the same way, it is easier to detect the number "O" (zero) from a display of letters rather than a display of numbers. White's study was of interest because the semantic category was under the experimenter's control; the form of the target (O) remained constant yet its meaning changed from the letter "O" to the number "O". In other words, detection is made easier when the target belongs to a different semantic category than the non-target stimuli.

In the visual search literature, the purpose of different studies is to determine factors that could facilitate or disturb visual search. In these studies, the material is simple: geometric shapes (such as squares, circles, triangles, bars, ...), digits or letters. Our study is for determining the factors that facilitate or disturb the detection of a word surrounded by others words. We reasoned that visual similarity effects might compete with more complex kind of features such as the semantic properties we get when using words as materials for visual search. This is also an ecological study, since semantic properties of perceived words might influence detection of a word-target in situations such as scanning an index, a newspaper or a web page. Whenever a group of words is perceived, these words could be semantically classified allowing a semantically contrasted target to be distinguished from other stimuli.

Two experiments were conducted to study the visual and semantic discrimination of a word-target from its context made of other words. In the first experiment, we examined the effects of the task context, of the semantic context surrounding the target and of the number of stimuli simultaneously displayed.

The task context is defined through the knowledge the participant has about the identity of the target: they know or they don't know its super-ordinate category.

The semantic context effect is explored by varying the semantic distance between the target and the non-target stimuli. For Rips, Shoben & Smith [9] the semantic distance means *"that when the subset was used as the predicate noun, the memorial representations of the subject and predicate nouns (ROBIN and BIRD) were closer together in some underlying semantic structure than when the superset was used as the predicate noun"*. For example, ROBIN is more semantically distant to ANIMALS than BIRDS. But this definition doesn't allow comparing two concepts that aren't on the same axis. For example, we can't evaluate the distance between TOYS and VEGETABLES. Then, we define semantic distance by the approximate number of superordinate categories between the target category and the superordinate category of both the target and the non-targets. The higher is the number of categories necessary to identify the common super-ordinate category of two concepts, the longer is the semantic distance between these two categories. For example, for the categories FISHES and BIRDS, it is easy to find a super-ordinate category ANIMALS that is their direct super-ordinate category. So FISHES and BIRDS are closely semantically related together. In opposite, for the categories FISHES and MANUFACTURED TOOLS, it is more difficult to find a category super-ordinate. So for these two categories (FISHES and MANUFACTURED TOOLS), we evaluate that they are semantically distant.

The effect of the number of stimuli in the context is studied by increasing the number of non-target words around the target-word.

In the second experiment, we examined the effect of the number of non-target words sharing a visual attribute with the target, the kind of visual property and the semantic typicality of the target. The effect of the number of non-target words is studied by increasing the number of non-target words that have the same color (red or black) or the same font (italic or not) than the target. We contrasted color and font in order to examine the effect of the kind of visual properties. The effect of semantic typicality was studied by using typical exemplar vs. non-typical exemplar of a category as being the target-word. It is strongly accepted in cognitive psychology that all exemplars of a category aren't equivalent: some are more representative of the category than others. For example, "robin" and "sparrow" are more typical of birds than "ostrich" or "penguin".

2 Experiment 1

This experiment investigated the effect of the semantic and visual background context, and of the task context on word visual detection.

The effect of semantic context is investigated by varying the semantic relatedness between the words surrounding the target. According to White [8], semantic differentiation between the target and the non-target facilitates detection. We reasoned that the detection of a target that is semantically distant to non-target words should be easier (higher success rate and shorter response time) than the detection of a target that is semantically close to the non-target words.

The visual context is investigated by varying the number of words surrounding the target. Because a word is a more complex item than a simple geometric form, we reasoned that the search for a word among others words would be serial. So we expect that increasing the number of non-target words would increase response time to detect the target.

The task context is investigated by providing or not providing the participant the semantic category of the target ("is there a animal" or "is there a word different to others"). Treisman and Sato [5] observed that when the target label wasn't given (the consign being to detect the intruder, an oddity search task), search is more difficult than when it was provided the participant: the search became a serial search as indicated by increase in response time as a function of increasing the number of stimuli simultaneously displayed. So, we reasoned that not providing information about the target would weaken performance (lower success rate and longer response time) than when providing information (the target' category label).

2.1 Method

Participants. The 54 participants were first-degree cycle students recruited in the psychology department. They did participated to another experiment on visual search. They were native French speakers or well mastered in French.

Stimuli. The experiment is computer-driven (FRIDA software). Stimuli are French words from 15 categories: flowers, vegetables, fruits, fishes, birds, insects, containers, tools, weapons, musical instruments, professions, toys, vehicles, sports, trees. The number of stimuli simultaneously displayed is of 9, 17 or 25, randomly posited in a matrix of five rows and five columns. For half of the trials, the target is present. A word of the same category than the non-target words is used when the target is absent.

When displayed among distractors (non-target words), the target is semantically distant to the non-target words for half of the trials and semantically close for other trials. Note that, although orthographic and phonologic similarities might play an important role, we didn't compute these factors, assusming that they should be counterbalanced across experimental variations. Something we will further control. The words-stimuli were written in Arial Police, in black and with a 16 points size on a white background. They appeared on a screen with 800x600 resolution.

Procedure. Participants are distributed either on the well-defined condition group or on the ill-defined condition group. The experiment starts with the instructions provided to the participant that present the type of task (detection of a word among others), the response modalities. In addition, participants are asked quick answers without mistakes about if "yes" or "no" the target is present in the display. Before experimental trials, each participant makes 10 training trials.

A trial is searching for a word, the participant being instructed as follows: *"Is there an exemplar of (semantic category (i.e. animal))?"* for the well-defined target condition and *"Is there a word different than others?"* for the ill-defined target condition. When the participant has read and understood the question, s/he has to press the space key that makes the words being displayed. When the participant finds the target, s/he has to press the "m" key, then to enter the name of the target with the keystroke. If the participant does not find any target, s/he has to press the "q" key. There were 72 experimental trials, for which 72 sets of 9, 17 or 25 words were randomly displayed. Recorded data for each of the 54 participants are, for each of the 72 trials, the yes/no response, the word typed on the keyboard in case of Yes response, and the response time.

Experimental Design. $S_{24}<C_2>*D_2*N_3$ where S_{24} corresponds to 24 participants per group: C_2 (well-defined target versus ill-defined target), D_2 corresponds to the two semantic distance between the target and the stimuli (close versus distant) and N_3 correspond to the 3 size of non-target contextual stimuli: 8, 16 or 24 stimuli.

Analysis. Positive trials (target is present) were retained for analysis. Success rate was computed by averaging the number of hits (to press "m" key and to give the right word) over the number of corresponding trials. Response time was computed only for hits. One participant of the well-defined target group and two participants of the ill-defined target group have their data suppressed for the response time analysis because they did not get at least one hit for each of the 6 experimental conditions. Thus, ANOVA analysis of success rate was made on 27 participants per group and analysis of response time was made on 26 participants for well-defined target and of 25 participants for ill-defined target.

2.2 Results and Interpretation

First, the type of task had no significant effect on success rate (well-defined: mean: .83, *SD:* .22, ill-defined: mean: .79, *SD:* .23; $F(1,52)=1,25$; $p=.27$, ns). However, as predicted an ill-defined target (mean: 7.92, *SD: 3.71*) is detected with longer response-time ($F(1,49)=47,94$; $p<.01$) than a well-defined target (mean: 4.36; *SD: 1.8*). Providing the participants the category of the target facilitates its detection. Second, a target that is semantically distant to the non-target words is detected with a higher success rate (mean .87, *SD: .18* versus .75 *SD: .23*; F(1,42)=30,26, p<.01) and with a shorter response time (mean: 5.54, *SD: 2.73* versus mean 6.73, *SD: 3.88*; F(1,49)=51,64; p<.01) than a target that is semantically close to the non-targets words. Semantic distinctiveness between the target and non-target words facilitates target detection. Third, the number non-target words had a significant effect on success rate (F(2,104)=19,99; p<.01) and response time (F(1,52)=38,9; p<.01). A target is detected with higher success rates when there are 8 (M : .87, *SD: .19*) and 16 non-target words (M: .84, *SD: .2*) than when there are 24 non-target words (M: .71, *SD: .26*) (F(1,52)=38,8; p<.01). Increasing the number of the non-target words in the context of the target increases significantly response time (8 stimuli : M: 4.70, *SD: 2.04*; 16 stimuli: M: 5.67, *SD: 2.48* ; 24 stimuli: M: 8.04, SD: *4.33*). As predicted, the less there is non-target words in the target context, the better its detection is.

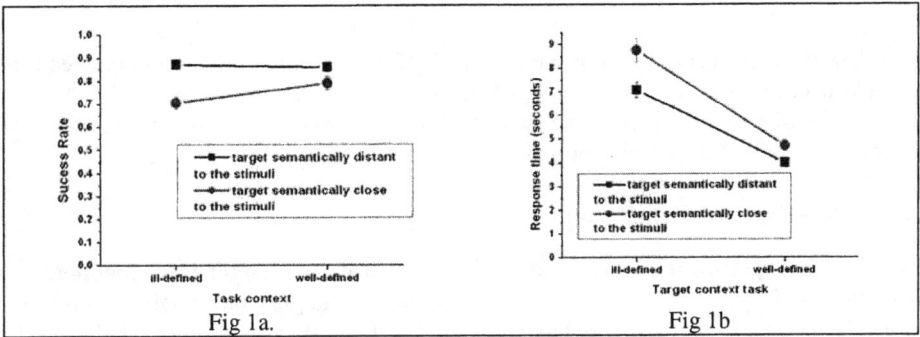

Fig. 1. Success rates (fig 1a) and response times (fig 1b) as a function of semantic context (close vs. distant) and of task context (ill defined versus well defined target)

The interaction between the task context (well vs. ill defined target) and the semantic context (close vs. distant) had a significant effect on success rate (fig. 1a, F(1,52)=5,18; p=.03) and on response time (fig 1b , F(1,49)=4,16; p<.01). As shown in the figure 1, the difference between the two semantic distances is weaker when the target is well defined than when it is ill defined. So, providing the target its category label facilitates the detection of a target that is semantically close to the non-target words.

The interaction between the task context, the semantic distance context and the number of non-target words had no significant effect on success rate (F(2,104)=0,67; p=.51, ns) but a significant effect on response time (F(2,98)=3,23; p=.04). The interaction between the semantic context and the task context had different effects as a function of the number of the non-target words. When there are 8 non-target words this interaction had no significant effect on response time (F(1,48)=0,05, p=.82, ns). The effect of semantic context between the target and non-target words is the same whatever the task context. When there are 16 non-target words (figure 2a), the effect of the semantic relatedness between the target and the non-target words depends on the task context (F(1,48)=7,44; p<.01): if the task is to detect the exemplar of a semantic category (well-defined target), response times are found shorter with targets that are semantically close than with targets that are semantically distant to the non-target words. Such a difference is not observed when target are ill-defined.

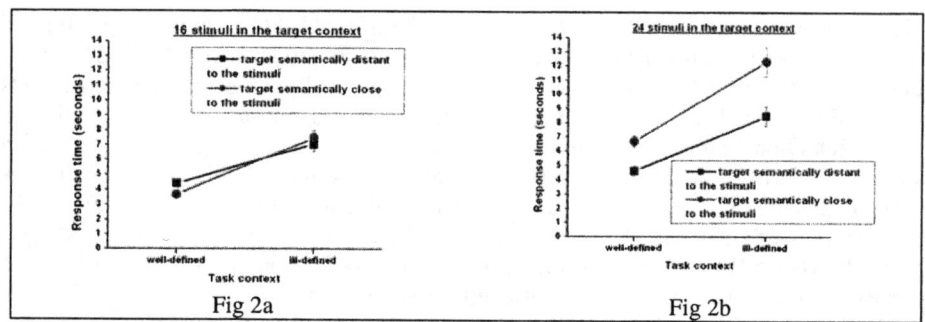

Fig. 2. Response time as a function of semantic context (close vs. distant) and of the task context (well vs. ill defined target) when the context is of 16 stimuli (fig 3a) and 24 stimuli (fig 3b)

When the context is of 24 non-target words (figure 2b), the interaction between the context task and the semantic context had an effect on response time ($F(1,48)=4,61$; $p=.04$): the difference between the two semantic contexts is more important when the target is ill-defined than well-defined.

2.3 Discussion and Conclusion

In this first experiment, we showed at first that providing target label facilitates its detection. This result isn't very surprising: when the category target label is provided (well-defined task), participants have to map top down data (the target label) and the bottom up data (the words in the visual scene), processing each displayed word and comparing its category with the category label of the target. When the target label isn't provided (ill-defined task), participants in addition of processing each displayed word, have to find what relates all the displayed words (a common category), except one that is the target.

As we predicted, semantic distinctiveness plays a role on target detection: it's easier to detect a target semantically distant to non-target words than a target semantically closed to them. There is also an effect of the number of non-target words: increasing the number of non-target words increases the participants' response time to detect the target.

Of interest is the result that semantic distinctiveness interacts with the task context (providing or not the category target label) and with the number of non-target words. It helps having higher success rate when the category target label is unknown (fig. 1.a) and reducing response times when there are many non-targets (fig. 2b). Thus, processing the semantic context appears to play a role in visual search for a word among other words when the target word is ill-defined and when the task is difficult due to a large number of candidate words.

In the following experiment, in addition of semantic context, we explored the effect of target typicality, another semantic factor, and the effect of the number of words sharing a visual attribute with the target, a no-semantic factor. To examine in details the effect of these factors, we used the eye movements' record technique with the assumption that semantic dissimilarity (having a target surrounded by semantically distant non-targets words) might decrease rejection times of non-target words.

3 Experiment 2

Longer response time to detect a semantically close target word to non-target words (as found in experiment 1) could come to the process of rejecting the fixated word as not being the target. Eye tracking enables the evaluation of semantic requirements, since items that are more difficult to process involve longer fixation times [10]. Thus, we predicted that trials in which non-target words were closely semantic related to the target would produce longer fixation times than those where non-target words were distantly semantically related to the target. Longer fixation times on trials where non-target words are closely semantically related to the target would demonstrate a greater level of difficulty for deciding to reject a fixated item as not being the target item.

All of the exemplars of a semantic category aren't equivalents: some objects are more typical of the semantic category than others: sparrow is a *better* bird than ostrich. In addition to semantic context, we reasoned that typicality factor might have an effect on detection performances. We predicted that a typical target is detected easier than a non-typical target (higher success rate and shorter response time). Difficulty to detect a non-typical target should come from a difficulty to access the target meaning, its super-ordinate category or, even if so, the participant might verify if another word in the background better corresponds to the target category. The first difficulty (difficulty to attribute the good category to the target) should generate weaker success rate. The second (background verification) should generate longer response times.

In experiment 1, we observed that the number of non-target words, in the target context, influenced visual search time. In experiment 1, all words were of same visual attributes: same font, same color, same size... Experiment 2 investigates the effect of the number of words sharing a visual attribute with the target when the number of non-target words is invariant. The rational of experiment 2 is as follows.

According to Poisson and Wilkinson [6] and Shen, Reingold and Pomplum [7], we reasoned that a target sharing a visual attribute with a small set of non-target words will be detected with shorter response times than a target sharing a visual attribute with a large set of non-target words. We predicted that increasing the number of words sharing a visual attribute with the target would increase response time to detect the target. In addition, for a conjunctive target (which have two attributes of the property: conju**n**ctive or conju*n*ctive), we predicted that the more the ratio tends to equivalence (fifty-fifty), the more response times would increase.

3.1 Method

Participants. 40 native French speakers capable of reading from a viewing distance of 0.50 m without needing spectacles or contact lenses were recruited.

Apparatus. The oculometer used to measure eye movements was a device using corneal reflection (ASL 5000 model). This technique involves illuminating the participant's eye using infrared light and collecting reflections from the cornea and pupil. The position of the eye in x and y coordinates is sampled every 20ms. Ocular fixation is defined using a minimum of five sampled points separated from each other by at least 0.5° of visual angle. This apparatus also enables measurements for each trial of response time, number of fixations, average fixation time and saccadic amplitudes.

Trial presentations were generated and response time was measured using a microcomputer. Experimental stimuli were presented on a flat 21-inch monitor screen (resolution: 1280 x 1024 pixels; color: 32 bytes).

Stimuli. According to the visual dimension under investigation color (black vs. red) or shape (normal vs. italic shape), the display was made of either has 31 words with black or red color, or with normal or italic font. An additional word is a conjunctive word: half-red half-black (detection) for the color dimension or half-italic font half-normal font (detection) for the shape dimension. Except the target, all words belong to the same semantic category (e.g. 31 words are kinds of birds). The target is a word is of another category (e.g., vegetables). To evaluate the target typicality, we collected from 70 voluntary participants exemplars of the 24 semantic categories we used: the same used in the experiment 1 and the followingd: animals, furniture, salt food, games, mammals, buildings, sweet foods, drinks, clothes. When less than 3 over 35 participants named an exemplar, it was coded as being "non-typical", when more than 10 over 35 participants named an exemplar, it was coded as being "typical".

For each visual dimension (shape font or color), five visual displays were built by varying the number of words with normal font (shape dimension) or of black word (color dimension): 1, 7, 16, 24, or 30. The others words were italic (shape dimension) or red (color dimension). For example if the display was of 7 black words, it also had 24 red words and one conjunction color word. The figure 4 shows the visual context when varying the shape dimension. Equivalent context for color dimension is build substituting italic words by red words.

For each dimensional visual display, the target was either in italic font, normal font, or half normal and half italic for the shape dimension. For the color dimension, the target was either red, black, or red and black color conjunction.

Words are in Arial font, in 14 points size. The position of the target was randomized, but counterbalanced across different areas designated by the columns and rows of the visual display.

Procedure. Each participant was seated comfortably at a viewing distance of 0.50m from the monitor screen and the ocular camera, with her/his chin stabilized in a chin rest. The experimenter read out the instructions that informed the participant that her/his ocular movements would be recorded and that s/he should avoid any further head movements. The instructions also described the experimental task to be carried out: the category of the item to be detected will appear in the center of the screen, when you left-click using the mouse, a group of words will appear on the screen; your task is to find the target, as quickly as possible and without making mistakes; as soon as you have found it, press the key and say aloud which word you have found; if you cannot find it, say "no". Next the experimenter began calibrating the oculometer; this process used a calibration card composed of nine black colored dots on a white background. After calibration was complete, six practice trials were presented to the participant before the experimental trials began.

On each trial, the participant clicked using the mouse and the category of the target to be detected was displayed at the center of the screen. When the participant clicked again, the 32 words appeared on the screen. As soon as s/he had detected the target, s/he clicked again and said aloud the detected word. The experimenter then recorded this word. If s/he could not detect the target, s/he clicked and then said "no".

The order of presentation of different trials for each of the different experimental conditions was randomized. Each participant made a total of 120 trials.

When all trials were over, the experimenter obtained the ocular data for each trial. Each block of data included "yes/no" response, and the word said in case of Yes response, response time (as determined by the participant's click), the number of fixations and mean fixation time.

Note that participants didn't know exactly what is the target: they didn't know its identity and its perceptual feature. They only knew its super-ordinate category.

Experimental Design. The experimental design was $S_{20} <G_2> * P_5 * A_3 * D_2 * T_2$, where S_{20} was the 20 participants per group (G_2: g1: color dimension, g2: shape dimension), P_5 was the number of words of the same visual attributes (1, 7, 16, 24, or 30), where A_3 was the target visual feature (black, red, half black - half red conjunction for the color dimension, normal font, italic font, half normal – half italic conjunction for the shape property), where D_2 was the semantic context (distant or close semantic distance between the target and the non-target words) and where T_2 was the typicality degree of the target (typical versus non-typical).

Analyses. Data were analyzed using an ANOVA statistical test. Only when the target was detected (correct responses) were response times and ocular measurements included in the analysis. The mean fixation duration corresponds to the average time per fixation before the target selection (total fixation duration over the number of fixations).

3.2 Results

We report separately results for each independent group (visual dimension) since effects of visual dimensions differ.

Italic Font Group. Analysis for success rates was made upon 20 participants. Analysis for response time was made upon participants who had a minimum of one success per experimental condition: 19 participants.

As shown in figure 3a, whatever the type of the target, the increase of the number of words sharing a same font with the target didn't generate a decrease of performances (less success and longer response times).

Contrary to our predictions, when the varied visual dimension is shape, target detection isn't facilitated by the decrease of the number of non-target words sharing the same shape than the target. Thus, for shape, a disproportionate ratio of distractors doesn't facilitate word target detection.

Color Group. Four participants were eliminated from the response time analysis because they didn't have at least one success by experimental condition. So, the response time analysis was made on 16 participants.

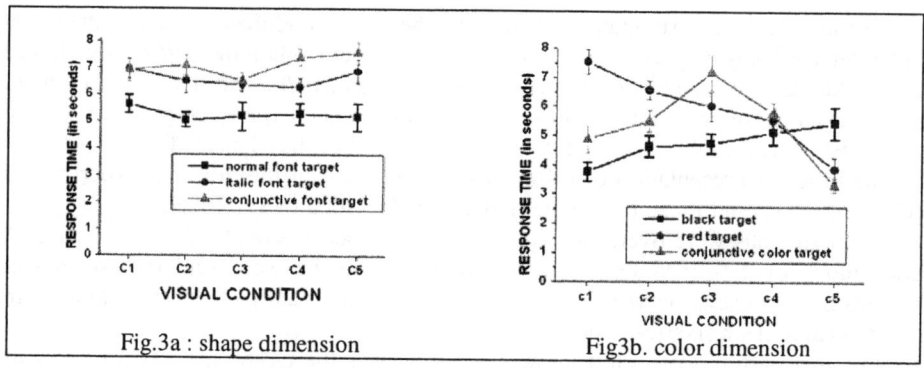

Fig.3a : shape dimension   Fig3b. color dimension

Fig. 3. Response times (in seconds) as a function of the shape of the target word (fig 3a) or the color of the target (fig. 3b) and of the number of non-targets words

As shown in figure 3b, an increase in the number of black words caused a success a response time increase (F(4,60)=3,15 ; p=.02) to detect a black target and a response time decrease (F(4,60)=16,57 ; p<.01) to detect a red target. For a conjunctive color target, a display with one black word provided lower success rates than others displays types (F(4,72)=7,60 ; p<.01). On the other hand, the more the distractor ratio tended to equivalence and the more the response time increased (F(4,60)=14,82 ; p<.01). As we predicted, an increase in the number of non-target words sharing color with the target causes a reponse times increase. The detection of a conjunctive target was also facilitated by a disproportionate ratio of the two kinds of non-target words.

Typical targets were detected with higher success rates (mean .85 *SD* .17; F(1,18)=78.96, p<.01) and shorter response times (mean 4.49, *SD* 1.46; F(1,16)=34.09, p<.01) than non-typical targets (success rate : mean .66, *SD* .19; reponse time: mean 5.31, *SD* 1.99). As we predicted, detection of typical targets was found to be easier than detection of non-typical targets.

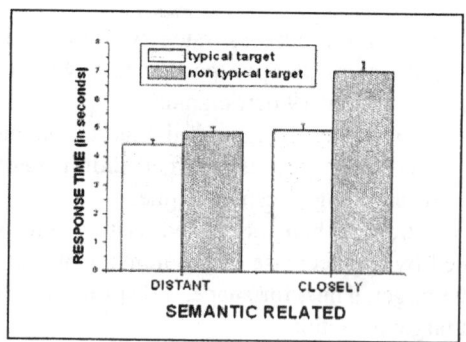

Fig. 4. Response times as a function of semantic context (distant vs. close) and of target typicality

A target word being semantically distant to the non-target words was detected with higher success rates (mean 0.82, *SD* *.17*; F(1,18)=66.03, p<.01) and with shorter response times (mean 4.33, *SD* *1.52*; F(1,16)=46.37, p<.01) than a target being semantically close to the non-target words (success rate: mean 0.70, *SD* *.21*; response time: mean 5.47, *SD* *1.86*). This shorter response time was associated with a shorter mean fixation time (semantically distant : mean 431ms, *SD* *64*; semantically close: mean 458ms *SD* *66*; F(1,16)=16.66, p<.01). According to our predictions, detection of a target was found to be easier when the target word is semantically distant to non-target words rather than semantically close to non-target words.

Significant interaction effects were found for response time (F(1,16)=43,25, p<.01). That is, participants took significantly more time and more fixations to detect a non-typical target surrounded by semantically close non-target words (see figure 4).

3.3 Discussion and Conclusion

Experiment 2 shows that to detect a word target, all visual features aren't equivalent. Varying the shape visual feature (normal font vs. italic font) produces no discrimination effect. This leads us to suppose that a shape feature doesn't allow discriminating the two groups of non-target words. Otherwise, for the color dimension, it seems that participants begin by glancing over the words group that have less representants before glancing over the words group which have more representants. Shen, Reingold and Pomplum [7] did find similar results. They observed that participants direct their first eye movements preferentially to the shorter distractors group before the larger distractors group. This search strategy is also used to detect a word cued by its superordinate category and for which the participant hadn't perceptual index. As we predicted, the semantic distance between the target and others words has an effect that is due to the facilitation or not to reject the fixated word as not being the target. As we predicted, target typicality plays also a role on target detection, but this is observed especially when the target word is semantically close to the non-target words.

4 General Discussion and Conclusion

Experiments 1 and 2 show that target's context influences detection. Here, the target context is made of 3 components: (i) the task – knowing or not knowing the category of the target –; (ii) the visual background of the display – the number of words surrounded the target, or the number of words sharing the same color than the target – and (iii) the semantic relatedness between the target word and the non-target words – the semantic distance between the target category and the category of the non-target words surrounded it –. Each of these components of the target context influences alone and in interaction the detection of a target word.

With the experiment 1, we observed that target detection is easier when the participant knows in advance the target category. As discussed above, this result results from a direct mapping between the top down and bottom up data. We also observed that when the target is hard to detect some contextual properties could facilitate the detection. When the target word category is unknown or/and when it's surrounded by many of non-target words, having a semantically close target to the non-target words facilitates visual search and detection. When the participant knows the target category, decreasing the number of non-target words sharing the color with the target word facilitates also its detection. When the target is non-typical of its semantic category, providing a distant semantic context did facilitate its detection.

Using an oculometer, experiment 2, helps finding that these factors influence differently eye movements. The number of non-targets words sharing the target color, as well as the typicality of the target, influence the number of fixations, indicating a search difficulty, whereas the semantic distance between the target and the non-target words influences the mean fixation duration, indicating visual search difficulties.

In summary:

- Visual dimensions and visual attributes aren't equivalent for target detection [2].
- The number of words surrounded the target have an important effect on detection when the target is similar to the non-target words ([1], [3]) even if this similarity is semantic.
- The ratio of the two groups of non-target words influences target detection [6], [7].
- Semantic differentiation influences target detection [8].
- This search is a first step of studies we intend to pursuit in order to explore the effect of word properties on the detection of a lexical target. We find that the detection of target words among other words is guided by a serial search because it's necessary to inspect one by one each item (increasing the number of non-targets increase response time is what is expected fromserial search) and we identified some factors that could facilitate this one by one inspection.

This research also provides some basis for web interface ergonomics. A daily task is searching for information on the web. This search can be oriented either by precise information (search for the program of a cinema), or by a more global demand (search for information about horses). And in order to get this information, a user is facing several web pages that present words lists. This study shows that a word list presentation could respect some recommendations to facilitate the user's search, such as using visual discriminating features when there are some perceptual groups of words, taking few words by perceptual group, using words which have large semantic distance (don't belong to the same super-ordinate category) and using typical words.

This study is also about a mental chronometry in word visual search. It might be that in the first step, visual attributes of the visual scene (such as the number of red and black words) guide participant eye movements on objects that are of a high degree of distinctiveness. In a second step, the participant processes the fixated word: *is it the target or not?* And in a third step, either this word is the target and the search stops, or this word isn't the target and the eyes are moved towards another word that is distinctive from others and so on until the target-word is found or evaluated as being absent. This mental chronometry is similar to the guided visual search developed by Wolfe [11]. For Wolfe, the level of activation of each item of the background guides visual search: one by one item inspection from the item with higher activation to item with weaker activation. The activation value depends of the bottom up processes (resulting of the display appearance) and of the top down processes (resulting from the target knowledge such as its color, its identity, ...). We can assume a similar process when the participant doesn't know in advance what the target is going to be: the eye movements of the participant are leaded by perceptual attributes of the visual scene and semantic features are used for making decision.

Studies centered on the effects of contextual semantic features on visual search, such as experiments 1 and 2, could be complete what we already know about visual search.

References

1. Treisman, A. & Gelade, G. (1980). A feature-integration theory of attention. *Cognitive Psychology, 12,* 97-136.
2. Treisman, A. & Gormican, S. (1988). Feature analysis in early vision: Evidence from search asymetries, *Psychological Review, 95,* 15-48.
3. Duncan, J. et Humphreys, G.W. (1989). Visual search and stimulus similarity. *Psychological Review, 96,* 433-458.
4. Neisser, U. (1963). Decision time without reaction time: Experiments in visual scanning. *American Journal of Psychology, 76,* 376-385.
5. Treisman, A. et Sato, S (1990). Conjunction search revisited. Journal of Experimental Psychology: Human Perception and Performance,16, 459-478.
6. Poisson, M.E. et Wilkinson, F. (1992). Distractor ratio and grouping processes in visual conjunction search. *Perception, 21,* 21-38.
7. Shen, J., Reingold, E.M., & Pomplum, M. (2000). Distractor ratio influences patterns of eye movements during visual search. *Perception, 29,* 241-250.
8. White, M.J. (1977). Identification and catégorization in visual search. *Memory and Cognition, 5,* 648-657.
9. Rips, L.J, Shoben, E.J. & Smith, E.E. (1973). Semantic distance and the verification of semantic relations. *Journal of verbal learning and verbal behavior, 12,* 1-20.
10. Just, M.A. & Carpenter, P.A. (1980). A theory of reading: From eye fixations to comprehension. *Psychological Review, 87,* 329-354.
11. Wolfe, J.M. (1994). Guided Search 2.0: A revisited model of visual search. *Psychonomic Bulletin and Review, 1,* 202-238.

Ontology Facilitated Community Navigation – Who Is Interesting for What I Am Interested in?

Nils Malzahn, Sam Zeini, and Andreas Harrer

Universität Duisburg-Essen, Collide Research Group,
Lotharstr. 63/65, 47057 Duisburg, Germany
{harrer, malzahn, zeini}@collide.info

Abstract. Networks have been a common way to show pathways for supporting communities. Usually the local focus of the actors in these networks does not allow them to perceive the broader context of their current interest. In this paper the authors propose an ontology-based approach to link people who have no explicit relation in the network despite potential common interests. With respect to the results we extend the concept of integrating ontologies into social networks to shared information spaces.

1 Introduction

Networks have been a common way to show pathways for supporting communities. Social Network Analysis (SNA), for example, has become popular for analysing communication structures within computer-mediated communities. Some approaches for using SNA in a Computer Supported Cooperative Work (CSCW) context have reached productive adoption. SPOKE[1] can be mentioned as an example for software that uses SNA to provide sales teams with a representation of their relationships inside and outside of their enterprise [1]. SNA based approaches can also be found within the Computer Supported Collaborative Learning (CSCL) community to support teachers and learners by supplying analysis methods for reflection (eg. [18]).

On the other hand theoretical approaches in human computer interaction (HCI) and especially for collaborative or cooperative systems stress the need of inclusion of objects within the theoretical frameworks. Therefore tools and media are commonly embedded in processes and structures of interaction. Engeström's activity system [11] based on the activity model [17] has been more or less intensively discussed in both CSCL and CSCW communities. Objects in the sense of non-persons are also well known to social network analyses: both persons and objects are considered actors. Especially the concepts of multi mode and affiliation networks distinguish between different sets of actors and provide consequently the aggregation of human and non-human actors [25]. Recent research addresses "Hybrid Networks" [15]: De Sousa et al. [9] demonstrate, for

---

[1] http://www.spoke.com

example, the relationship of technical and social dependencies as a social network graph by analysing a CVS directory.

Although embedding objects in the context of interactional processes can be seen as a value added to the SNA discussion within the CSCW community there are still questions to be asked. In our case the problem is that the representations of networks are difficult to interpret without domain specific knowledge. There are two promising solutions here to be mentioned. The first one is surely to embed the problem in a theoretical framework. This will open discussions with other researchers who are working on similar problems and to conduct a communicative validation for the formulated problem. The Actor Network Theory [15] is promising because mixed mode networks are well known to this theory. Furthermore there is a relationship to the Activity Theory as shown by Engeström and Escalante [10].

The second possible and our preferred solution is to include the domain-specific knowledge of the users. Therefore the user shall be enabled to express his knowledge in terms of an ontology by weighting the relations between concepts. According to Rosch et al. those concepts "[...] reflect both real world correlational structure and the state of knowledge of that structure of the people doing the categorizing"[20]. From a modelling perspective this means that the role of the interpreter can be congruent with the role of the user.

We propose to use combination of SNA and user weighted ontologies as a reflection method with the intention to take the network participants' "world view" into account.

In the following sections we will first discuss for which purpose such reflection can be used. After that we will present an approach using ontology as a basic method for interlinking personal networks. Finally we present the CONAVI (Community Navigation Visualizer) tool where we demonstrate some of the implications of our work.

2 Concept-Oriented Community Navigation

The "social navigation" approach [12] addresses the social dimension as a separate dimension [6], co-existing with spatial or semantic dimensions in the context of shared information rooms. Based upon the long tradition of awareness research some concepts of social navigation became popular using the power and potential of the fact that humans (in this case the users) are social beings. An example for the success of a system which uses techniques for social navigation is an online bookshop that uses the information about other users' behavior to recommend other books to the buyer. This example implies that social navigation can be employed by recommender systems.

Within social navigation we can distinguish between direct and indirect social navigation. While indirect social navigation aims to provide information about what other people do or have done, direct navigation involves direct communication between two ore more people [7]. Possible design issues for direct social navigation are turn taking, finding people who can help or give directions, etc.

2.1 Tracking Experts

In a general sense finding expertise is an important goal of recommender systems. This goal is also stated as an open issue in social navigation. Dieberger mentions that "identifying a user as an expert is especially important in direct social navigation situations" [7]. In this case we could run into social dilemmas as well as into the problem that finding expertise is more complex than just tracking down an expert using a given definition of expertise. The first is well known as "free rider" problem, i.e. there are persons who take more than their fair share from the network's benefit without contributing to the networks excess value. It is easy to comply that this problem is likely to occur within shared information spaces.

The tracking problem is caused by the fact that the amount of communication messages sent in the network need not necessarily correspond to the expertise. (ibid., [26]).We assume that this causes at least a bias in the weighting of expertise.

Based on the assumption that reliable indicators for expertise are hard to find, since different communities may have their exclusive definitions of expertise, we propose to use actor-specific ontologies to enable the communities to define their own indicators of interest between actors. Using interest will also allow exposing implicit expertise within a community.

2.2 Navigation Through Conceptual Objects

Filtering by interest can be characterised as an interactional approach that includes objects within the social dimension. The Actor Network Theory mentioned above conceptualises objects as actors which are able to translate the interests of other actors. Such objects can be observed and described as "epistemic objects" between different perspectives [16].

In the course of our empirical studies in the field of virtual communites we often observed situations in which interviewed persons told us about the importance of face-to-face meetings. Since this is surprising in the context of virtual communities, we decided to investigate these face-to-face situations further.

During the meetings objects like "mapping tables", "powerpoint presentations" or physically shared computer workspaces (two people looking at the same monitor while talking) between the "faces" were observed. So most of the time the so called face-to-face meetings were not characterised by direct communication but rather by object-mediated communication.

At this point we assume according to Star's [21] "boundary objects" that usually there is more than one object of interest for different human perspectives. Labels, for example, are a type of boundary object which express keywords or conceptual categories. Persons refer to these keywords or categories, but they perceive them with different significance.

This idea leads us to utilise ontologies to link epistemic objects of interest and outline knowledge maps based upon semantic networks. Looking onto bigger forums or the usenet different sub-categories in these forums can be considered defining a community. These communities live inside the boundaries of their sub-category. A common phenomenom in forums are crossposts. Crossposts are articles crossing the boundaries of the communities because the poster thinks that the article is interesting or fitting to more than one community. Elaborating on this, these are indicators that a member of one community can also be an expert for topics interesting to another community without directly participating in that community. They are linked by boundary objects (here: the crossposts). Ultimately they are linked by concepts.

Our approach is to use potential boundary objects represented by ontologies to stimulate links between members of different communities and to make them visible to each other.

3 How to Identify Persons and Topics of Interest

Several ways to suggest automatically objects of likely interest to users [2][3][4][14] have been proposed in recent years. Probably the most popular algorithm is the PAGERANK-algorithm [3] used in Google.

So far the presented approaches deal with the internet as a source for hyperlinked documents or with persons represented – if represented at all – by their homepages.

Looking at forums, especially on large forums, it is easy to identify persons who write articles in the same thread. Popular forums like phpBB[2] can be divided into categories to structure these threads. These categories are built of more or less broad topics. If the forum is large, i.e. concerning the amount of writers and categories, most persons know a subgroup of persons directly involved in topics where they write themselves. Other persons are potentially interesting to get to know because they share similar interests or they are known as experts in other parts of the forum.

In the case that a forum usually consists of subgroups which do not know or do not communicate with each other except through a small number of persons belonging to both groups, traditional social network analysis will fail to identify the important persons to get to know for other persons because the persons behind these connectivity cutpoints [25] will be invisible to the person. It is easily conceivable that persons who have similar interests participate in different parts of a large forum without having any direct or indirect connection in a communication-network. To be able to find a link between those persons an additional network-structure must be used. We propose that ontologies should be used to add missing links between persons who should be aware of each other.

Ontologies are a "systematic arrangement of all of the important categories of objects or concepts which exist in some field of discourse, showing the relations between them"[3]. The association between topics implied by the ontology can be used to connect topics in a topic-person network with additional links derived from the semantic network of the ontology (cf. Fig. 1).

---

[2] http://www.phpbb.org
[3] from the Collaborative International Dictionary of English v.0.48.

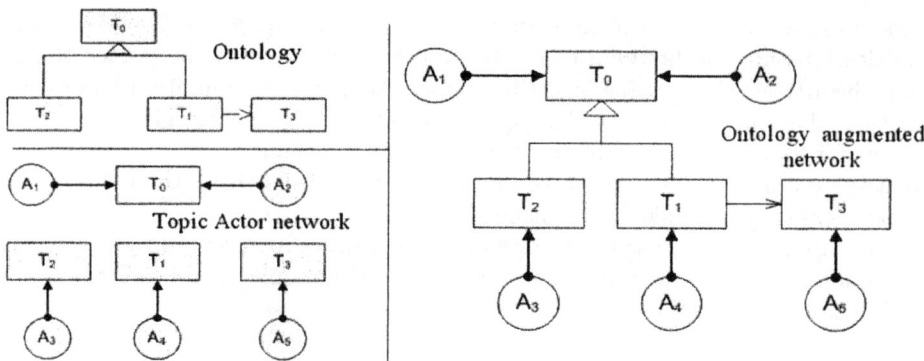

Fig. 1. Example for integrating an ontology into a topic (T) actor (A) network

Enriched with the information from the ontology a person-person network can be extracted from the person-topic network. Because of the new ontology-implied links in the communication network more persons will be regarded as potential candidates for suggestions who to contact because of congruence of interests (cf. [19]).

3.1 Influence of Ontologies on Importance of Persons for Other Persons

"An ontology necessarily entails or embodies some sort of world view with respect to a given domain. The world view is often conceived as a set of concepts (e.g. entities, attributes, processes), their definitions and their inter-relationships; this is referred to as a conceptualisation.

Such a conceptualisation may be implicit, e.g. existing only in someone's head, or embodied in a piece of Software." [23]

This constructivist definition of ontologies states that the conceptualisation of a world view depends on the specific ontology a person bears in mind. This implies that there are difficulties to create ontologies for other persons. But on an abstract, technical level there are ontologies which can be used to categorise threads in a forum. Nevertheless to achieve the goal to propose persons of similar interests to other persons based on topics, it is important to use an ontology on which the persons in question agree. That is especially important for associations in the ontology which are not only variants of "is-a"-relations but more complex relations like "implies" or "deals with". These are the relations even experts will argue about.

Keeping in mind that the links in the ontology are projected into the given person-topic network (see Fig. 1), the quality of the resulting person-person-network depends heavily on the quality of the ontology.

As described before the ontology determines which links between topics have to be inserted into the person-topic network. These new links are exploited to determine the congruence of interests of two persons or groups. If the ontology does not represent a conceptualisation of a common world view, the suggestion to the respective groups will be useless.

Even if the two groups in question agree on a common ontology the importance of specific topics/relations may differ between people's world view. Weighting these relations addresses their individual perspectives and needs. Intuitively it is clear that topics where both persons participate directly seem to be of common interest. As stated before: that is not the interesting configuration. They will know each other already.

If the ontology contains hierarchical structures like topic-subtopic relations or more generally speaking "is-a"-relations there are several possible derivations concerning how interest relates to these relations. To narrow we state two assumptions in this paper:

1. A person who is interested in a general topic, is also interested in the subtopics of this topic
2. The interest decreases with the path distance between two nodes, that means the interest in a direct sibling of a node is higher than in a later descendant.

Thinking about the different types of associations that are possible in an ontology, there has to be a way of handling the differences between them. The first approach is to weight these types differently. That means that certain types of associations lead to faster decrease in interest.

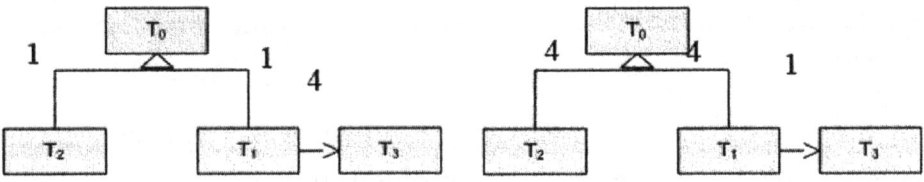

Fig. 2. Ontology with different users' weightings

While the subtopic to supertopic implications on interest may be common sense, the weighting of the interest decrease will definitely depend on the specific persons. Knowing that any tool, which should help actors to get to know interesting persons, has to be adaptable to the ontology ratings of the particular person, the system shall provide an ontology in which the users can adjust the weight of the connections between the concepts corresponding to their personal interests and their personal representation of the world.

To discuss the implications of the weighting further, we assume that every "is-a" relation has the weight 1 and every other relation has the weight 4 (cf. Fig. 2 l.). That means that a topic which is the fourth ancestor of another topic is as interesting as a topic that has another relation with a second topic (see assumption 2). Another actor may expect a different weighting (see fig. 2 r.) where direct associations are interpreted as stronger than generalization.

So the distribution of weights influences subsequent queries directed at the given person topic network. The network answers with a list of persons who should be contacted because of potential congruence of interests with the querying person inferred by the used ontology.

4 Implementation

Formally the presented approach analyses a given bi-partite network or graph. From this network we derive the affiliation matrix (actor topic matrix) for the relations between the two types of nodes. In the following paragraphs the first type of nodes will be called agents (e.g. persons), while the second will be called conceptual objects (e.g. topics in a forum).

After computing the affiliation matrix the ontology-matrix is created. This matrix is an adjacency matrix for the graph of the ontology with the user-given interest values as weights. If the user does not enter any value default values are used for each type of relation between two concepts (high values imply large distances).

In complex ontologies it is possible that a conceptual object can be reached from another conceptual object in multiple ways. To determine the decrease rate of interest only the shortest path is considered, because the "shortest" path represents the strongest relation.

Assume that (o_{ij}) represents the shortest path in the ontology matrix, (a_{ij}) is the actor topic matrix and (r_{ij}) the target actor topic matrix. The transformation function from the actor topic matrix to the ontology enhanced actor topic matrix is shown in Fig. 3.

$$r_{zs} = a_{zs} + \sum_{i=1, i \neq s}^{\|o\|} \frac{1}{o_{si}+1} \cdot a_{zi}$$

(o_{ij}) represents the shortest path in the ontology matrix
(a_{ij}) derived actor topic matrix
(r_{ij}) target actor topic matrix

Fig. 3. Flattening of the actor-topic matrix using an ontology

The general idea is to "fold" the weights corresponding to connections between actors and topics: topics that are related to the currently focussed one are folded onto that. The connections from the folded topics are transferred to the target topic. Since the "folded connections" are not direct links their link weight is reduced by a damping factor. At the moment we use the reciprocal value of the shortest path in the ontology matrix as the damping factor. This ensures that conceptionally farther topics do not have such a strong influence as tightly connected topics.

The resulting topic-person matrix is multiplied with its transposed matrix to compute the corresponding person-person relations (cf. [25]).

To enable the user to evaluate the results, the matrix is visualised as a graph in CONAVI. CONAVI is based on Touchgraph[4]. The original implementation was extended (cf. [22]) in order to distinguish two types of nodes to be able to represent hybrid networks. In addition to that the extended version now shows edge weights and lays out dense networks more clearly because of redesigned edges.

5 Example

The presented algorithm was evaluated on a forum intended for community support of freelancers. The forum is operated by an enterprise that wants to enable the members of the forum to exchange knowledge and form bonds for prospective virtual enterprises[5].

For members of this community it is very useful to know who has the same interest or a complementary interest.

For evaluation purposes three main topics of the forum are examined. The threads' topics were manually filed into the ontology's categories. The ontology was formed from topics and categories of the given forum.

Approximately 300 threads of 400 different persons were filed. Figure 4 shows two cut-outs from the network. The person P300 only contributed to T21 dealing with "dismissal". Directly known to these persons is P76 who also contributed an article to T21. The other cut-out shows the contributions to T18 which is about "human resource management" together with some other topics of the surrounding area. To this topic many articles were contributed by a number of different people. According to the used ontology shown in Fig. 4 the topic "dismissal" is a subtopic of "human resource management". If the additional links that can be derived from the ontology are ignored, only P76 will be presented to P300 as a person with potential congruence of interests. That is nothing new to P300 presuming that he has kept track of the forum topic he has posted himself. On the other hand lots of people who have been posting to T18 will be totally unaware of him as a potential candidate for interesting discussion and knowledge on the topic of "dismissals". Combining the ontology with the given network the relations from the ontology are transferred into the hybrid network and new prospects are revealed. As shown in Fig. 5 node P300 now has links to several other persons besides P76. So these links can be exploited to recommend P300 to other persons who have posted in T18 and vice versa. Of course not all of the proposed persons are equally important to P300 but the edges in this network are weighted so that a certain threshold can be applied to filter the resulting network further. This is especially useful for large networks or large ontologies with many relations between the correspondent nodes causing an increase of "weak" links between the nodes.

---

[4] http://touchgraph.sourceforge.com
[5] A virtual enterprise consists of several small enterprises or freelancers which cooperate loosely coupled to achieve a common goal. e.g. to get an order they would not get as a single enterprise.

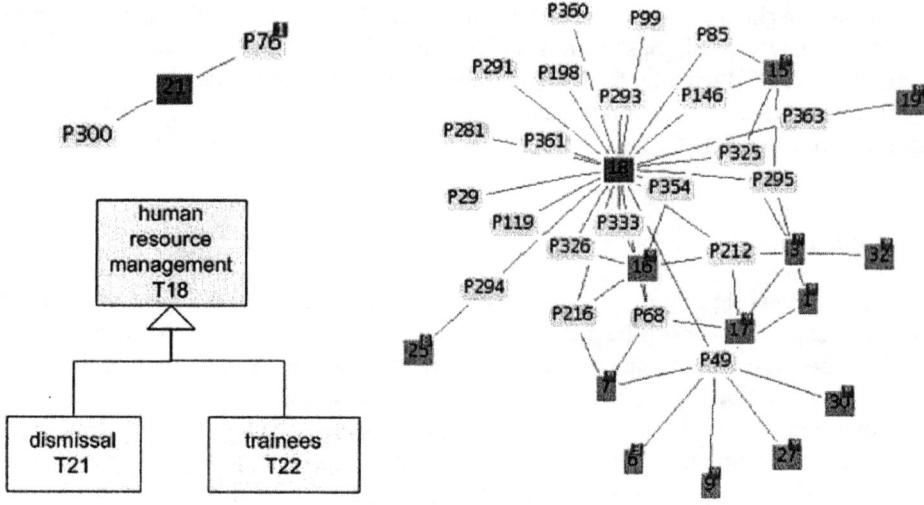

Fig. 4. Cut-outs from the given network and ontology

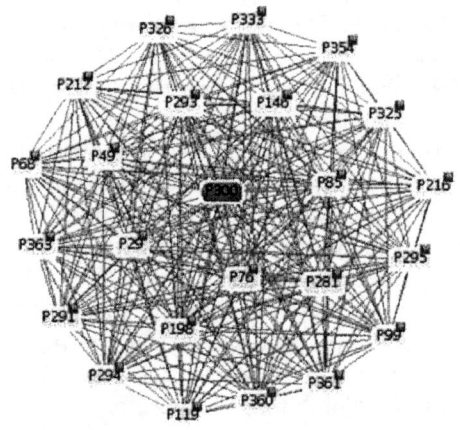

Fig. 5. Resulting person-person network

6 Conclusion

This article presented an ontology-based approach to community navigation. The proposed method was successfully applied to an existing forum finding out new potential contacts for the participants. These results can be exploited in several ways. First of all the system points out users to other users based on interest congruence. A second way of using the system is to enable third party persons, i.e. persons who are not participating in the particular forum, to form teams by virtually participating in the corresponding concept and edit the ontology for the project's needs. The proposed algorithm will identify a set of persons who are potentially relevant to the project's goals and needs.

Looking onto the promising results with forums we expect that the concept of integration of ontologies into social networks can be extended to shared information spaces on a general level. This can be done by generalising the concept of topics to conceptual objects. These conceptual objects are used in different domains like learning environments (as Learning Object) and knowledge construction (as Knowledge Object). Thus the same algorithm enables users of complex working or learning environments to reflect their situation within the particular network. On the implementation level the ontologies can be applied on the objects' metadata to assign the objects to specific concepts of the ontology.

The concept of conceptual objects allows transferring the results to theoretical frameworks like actor network theory or multiple activity systems (cf. [10]). These theoretical concepts help to stimulate the exchange between scientific communities and may also provide design principles for computer systems.

7 Future Work

While the data processing for the presented pilot study was done manually in the course of a bachelor thesis the next step is to analyse and process the conceptual objects automatically. The biggest difficulty will be the mapping between objects and ontology nodes. This will be done using well-known classification algorithms like SVMs [13][24] or k-means.

In the broader context of the VIP-NET project the CONAVI system shall be extended to allow real time group reflection of the users' interaction.

Acknowledgments

The work presented in this paper was funded by the German Federal Ministry of Education and Research. Project no. 01HU0128 (VIP-NET).

The authors are grateful for the work done by Svenja Schröder while implementing the prototype of the CONAVI (Community Navigation Visualizer) and collecting the data needed for the evaluation.

References

1. Prasad Akella, Mark Interrante, and Mark Granovetter. Selling the spoke way. In CSCW'04 Workshop on Social Networks, 2004.
2. C.C. Aggarwal, J.L. Wolf, and P.S. Yu. Horting hatches an egg: A graph-theoretic approach to collaborative filtering. In Proceedings of the Fifth ACM SIGKDD International Conference on Knowledge Discovery and Data Mining (KDD'99), pages 201–212, San Diego, 1999.
3. Sergey Brin and Lawrence Page. The anatomy of a large-scale hypertextual Web search engine. Computer Networks and ISDN Systems, 30(1–7):107–117, 1998.

4. Soumen Chakrabarti, Byron E. Dom, S. Ravi Kumar, Prabhakar Raghavan, Sridhar Rajagopalan, Andrew Tomkins, David Gibson, and Jon Kleinberg. Mining the Web's link structure. Computer, 32(8):60–67, 1999.
5. Chia-Hui Chang, Ching-Chi Hsu, and Cheng-Lin Hou. Exploiting hyperlinks for automatic information discovery on the www. In the Proceedings of the tenth IEEE International Conference on Tools with Artificial Intelligence, Chien Tan Youth Activity Center, Taipei, Taiwan, November 1998.
6. P. Dourish and M. Chalmers. Running out of space: Models of information navigation. In Proceedings of HCI '94, Glasgow, 1994.
7. Andreas Dieberger. Social connotations of space in the design for virtual communities and social navigation. In Kristina Hook, Benyon, and Alan J. Munro, editors, Designing Information Spaces: The social Navigation Approach, pages 293–314. Springer, 2003.
8. Paul Dourish. Where the footprints lead: Tracking down other roles for social navigation. In Kristina Hook, David Benyon, and Alan J. Munro, editors, Designing Information Spaces: The social Navigation Approach, pages 273–291. Springer, 2003.
9. Cleidson de Souza, Paul Dourish, David Redmiles, Stephen Quirk, and Erik Trainer. From technical dependencies to social dependencies. In CSCW'04 Workshop on Social Networks, 2004.
10. Yrjö Engeström and Virgina Escalante. Mundane tool or object affection? The rise and fall of the postal buddy. In Context and Conciousness - Activity Theory and Human-Computer Interaction, pages 325–373. MIT Press, 1996.
11. Yrjö Engeström. Learning by Expanding: an activity theoretical approach to developmental research. PhD thesis, University of Helsinki, 1987.
12. Kristina Höök, David Benyon, and Alan J. Munro, editors. Designing Information Spaces: The social Navigation Approach. Springer, 2003.
13. Thorsten Joachims. Text categorization with support vector machines: Learning with many relevant features. In Proceedings of the European Conference on Machine Learning. Springer, 1998.
14. Jon M. Kleinberg. Authoritative sources in a hyperlinked environment. Journal of the ACM, 46(5):604–632, 1999.
15. B. Latour. Science in action. How to follow scientists and engineers through society. Milton Keynes: Open University Press, 1987.
16. Bruno Latour. We Have Never Been Modern. Hemel Hempstead: Harvester Wheatsheaf, 1993
17. A. N. Leont'ev. Activity, consciousness, and personality. Prentice-Hall, 1978.
18. A. Martinez, Y. Dimitriadis, B. Rubia, E. Gomez, and P. de la Fuente. Combining qualitative evaluation and social network analysis for the study of classroom social interactions. In Actas del Workshop Documenting collaborative interactions in Computers and Education, special issue on Documenting Collaborative Interactions, August 2003.
19. Hiroaki Ogata, Kenji Matsuura, and Yoneo Yano. Visualizing knowledge awareness in a web-based CSCL environment. In Proceedings of WebNet - World Conference on the WWW and Internet, pages 927–932, 2001.
20. Eleanor Rosch, Carolyn B. Mervis, Wayne D. Gray, David M. Johnson and Penny Boyes-Braem, Basic objects in natural categories. In: Cognitive Psychology (8):3, pages 382–439, 1976

21. Susan Leigh Star. The structure of ill-structured solutions: boundary objects and heterogeneous distributed problem solving. In M. N. Huhns and L. Gasser: Distributed artificial intelligence: vol. 2, pages 37-54, Morgan Kaufmann Publishers Inc., 1990.
22. Svenja Schröder. Konstruktion und Visualisierung von Soziogrammen aus thematischen Bezügen am Beispiel eines Forums. Bachelor's thesis, University of Duisburg-Essen, November 2004.
23. Mike Uschold and Micheal Gruninger. Ontology: Principles, methods and applications. Knowledge Engineering review, 11(2), June 1996.
24. Vladimir N. Vapnik. The Nature of Statistical Learning Theory. Springer, 1995.
25. Stanley Wasserman and Katherine Faust. Social Network Analysis: Methods and Applications. Cambridge University Press, 1994.
26. Steve Whittaker, Loren Terveen, Will Hill, and Lyn Cherny. The dynamics of mass interaction. In Proceedings of the 1998 ACM conference on Computer supported cooperative work, pages 257–264, New York, USA, 1998. ACM Press.

Contextual Information Systems

Carlos Martín-Vide[1] and Victor Mitrana[1,2]

[1] Research Group in Mathematical Linguistics, Rovira i Virgili University,
Pça. Imperial Tarraco 1, 43005, Tarragona, Spain
carlos.martin@urv.net
[2] Faculty of Mathematics and Computer Science, University of Bucharest,
Str. Academiei 14, 010014, Bucharest, Romania
vmi@urv.net

Abstract. A rather common way of formalizing contexts as first class objects starts from the basic relation $ist(c,p)$ which asserts that the proposition p is true in the context c. However, the space in which terms take values may itself be context-sensitive. Our aim is to introduce contexts as abstract mathematical entities in a more general framework which includes context-sensitivity, namely knowledge represented by contextual information systems. Making use of some concepts from the Rough Set Theory we refine two relations: the indiscernibility relation between the objects and the similarity relation between the contexts within a contextual information system. Both relations are illustrated with examples showing how contextual information systems can express in a natural way a very few well known phenomena. Based on these relations we propose a simple strategy for decreasing the ambiguity of contextual information systems.

1 Introduction

This work tries to propose neither a definition nor another formalism for contexts, but it proposes an information-theoretic approach to contexts which is in the spirit of formal concept analysis. Our goal is to introduce contexts as abstract mathematical entities in the framework of knowledge represented by information systems [18]. By means of context, information systems are extended to account for different phenomena by object evaluation in different contexts. However, this work is tentative and incomplete; it is just a first step, coming from the information theory, towards a better understanding of this very rich concept that is the context.

Before doing this, let us take a snapshot of some pioneering formal approaches to contexts rooted in computational linguistics. Following *Cambridge International Dictionary of English*, in a text/speech approach context is "the text or speech that comes immediately before and after a particular phrase or piece of text and helps to explain its meaning". According to this possible definition, S. Marcus formally defined a context as a pair of words; however the term *context* is often replaced by *environment* or *neighborhood*. The idea of a context conceived as a pair of words belongs to the very beginning of descriptive distributional linguistics. Starting from this definition of the notion of context, S. Marcus introduced in [11] a new language generating mechanism that is not using auxiliary symbols called *contextual grammar*. This mechanism, based on one of the fundamental linguistic operations, namely the acceptance of a context by a word [12], differs from the other ones for which rewriting is the main operation. The interaction word-context appears as a basic ingredient of well-formedness in linguistic analysis. The mathematical theory of contextual grammars is now a mature topic in formal language theory, surveys can be found in [4, 16, 17].

Now we recall some quotations from the article about *context* in the Pergamon Press *Encyclopedia of Language and Linguistics* [19]: "'Context' is one of those linguistic terms which is constantly used in all kinds of contexts but never explained.(...) The central role that context plays in linguistic theorizing may be inferred from the fact that one could systematize a history of recent linguistics by describing the role of context in successive theoretical systems. (...) In the course of its movement from a peripheral to a more central position in linguistic theorizing, context itself developed from a somewhat static conception in terms of one or more related utterances towards a more processual notion. Context is viewed as being built up in production as well in comprehension of text." This article is mainly devoted to the semantic and pragmatic role of context in linguistics. In this note, we are especially concerned with syntax not necessarily related to linguistics.

Our approach fits with this definition of contexts taken from *Cambridge International Dictionary of English*: "the influences and events that helped cause a particular event or situation to happen". Our attitude is a theoretical computer science one. We do not claim that this work will result in a unique view of what context is. We believe that various definitions might be useful. Contexts are abstract objects; instead of offering a formal definition we prefer to propose a possible description. This description is done by enhancing the information systems with the capacity of manipulating contexts. Contexts can control actions, act themselves, specify, explain, etc. Some contexts may be very complex objects, therefore our description is necessarily incomplete. Consequently, a theory which proposes gradual relations, like the Rough Set theory, seems more appropriate for our investigation.

The paper is organized as follows. We first introduce the contextual information system, the main concept of the paper, as a framework which allows us to describe contexts and relationships among them by means developed in the Rough Mereology formalism [18]. We first discuss the indicernibility relation between the objects and the similarity relation between the contexts of a contextual information system. Making use of some concepts from the Rough Set Theory both relations will be refined. All concepts are accompanied with examples with some relevance in computational linguistics. Based on the similarity relation between contexts we propose a simple strategy for decreasing the ambiguity of contextual information systems. The papers ends by an example of a contextual information system modeling a biological phenomenon "in vitro", namely the DNA recombination under the enzymatic influence.

2 Contextual Information Systems

Following a series of papers, see, e.g., [6, 13, 14, 2], a possible way of formalizing contexts as first class objects starts from the basic relation $ist(c,p)$, introduced in [6], which asserts that the proposition p is true in the context c. Further, these statements can be nested, that is statements like $ist(c, ist(d,p))$ can be used to contextualize the interpretation of contexts themselves. The system is always in an outermost context which can be transcended by creating a new outermost context. The system can enter and exit contexts. Since contexts are defined as first class objects in the domain, one can quantify over them, have functions whose range is a context, etc. However, there are situations in which besides this relation one needs formulas like $value(c, term)$, where $term$ is a term. The interpretation of $value(c, term)$ involves a context c which might be used for making assertions about a particular situation of $term$. More generally, the space in which terms take values may itself be context-sensitive. We propose here a discussion in a more general framework which includes this aspect, namely *contextual information systems*. Actually, the formula $ist(c,p)$ is not very relevant for our formal construction. This is an easy way to define the indiscernibility relation between two objects in a considered context. However, ist modality might be replaced by other formalisms which consider the contextual dependency of terms as well as contextual dependency of formulae like [5].

A contextual information system is a quadruple

$$\Gamma = (U, A, V, C),$$

where U is a set of objects (the universe of the system), A is a set of attributes, V is a set of attribute values, and C is a set of contexts. Each attribute $a \in A$ is defined as a mapping from $U \times C$ to V. Informally speaking, $a(u, c) = v$ means that the attribute a associated with the object u has the value v in the context c. In a more general setting, $a : U \times C \longrightarrow 2^V_f$, where 2^V_f is the set of all finite parts of V. However, in any case, all attribute mappings have to be computable. If an attribute of a given object takes more values in a given context, then we say that the given object is ambiguous; moreover the cardinality of the set $a(u, c)$ expresses the degree of ambiguity of u in the context c. If the objects are utterances in a natural language discourse, this might be viewed as the degree of lexical ambiguity [9]. As it was expected, a context might or might not be able to disambiguate a utterance or to decrease the degree of its lexical ambiguity. The construction we propose works entirely well for contextual information systems having attribute mappings which can take more than one value. However, for sake of simplicity, we restrict all attributes in contextual information systems to attribute taking unique values. The *information set* of the object u in the context $c \in C$ denoted by $Inf(u, c)$ is defined by

$$Inf(u, c) = \{(a, a(u, c)) \mid a \in A\}.$$

In words, the information set of the object u in the context c is the set of all attributes of u together with the values they take in the context c.

The information set of the object u is naturally defined as the set of all attributes of u and the values they take regardless the current context. Formally, this set is defined by:
$$Inf(u) = \bigcup_{c \in C} Inf(u,c).$$
Two object $u, w \in U$ are said to be *indiscernible in the context* $c \in C$ if $Inf(u, c) = Inf(w, c)$. Note that two objects can be indiscernible in some context but one can distinguish them in another context. The objects u and w are *indiscernible* if they are indiscernible in every context, that is $Inf(u, c) = Inf(w, c)$ for all $c \in C$. In the formalism presented in the beginning of this section, u and w are indiscernible in the context c and indiscernible if and only if

$$ist(c, Inf(u,c) = Inf(w,c)) \quad \text{and} \quad \forall_{c \in C}\ ist(c, Inf(u,c) = Inf(w,c)),$$

respectively Note that $Inf(u) = Inf(w)$ holds for any pair of indiscernible objects u, w but the converse does not hold. This follows from the fact that in $Inf(u) = Inf(w)$ one can have a pair that comes from the evaluation of u and w in different contexts. The following problem naturally arises: Given two objects and a context can we algorithmically decide whether or not the two objects are indiscernible in the given context? Clearly, one can construct such an algorithm as soon as the sets of attribute and attribute values are finite. Indeed, under these circumstances the information sets of the two objects in the given context are finite so that one can check the equality of the two sets. The situation is definitely more intricate when at least one of the two sets is infinite. In this case, each information set is computed by a program and we have to check the equivalence of the two programs which might or might not be decidable. However, for all practical implementations, all the parameters of an information systems are finite sets.

Obviously, the indiscernibility relation in the context c, $c \in C$, is an equivalence relation which partitions the universe set U. We denote by $[u]_c$ the indiscernibility class in the context c defined by $[u]_c = \{w \in U \mid Inf(u,c) = Inf(w,c)\}$. Clearly, $[u]_c$ collects all objects with no discernible differences to u in the context c. Furthermore, the indiscernibility relation is also an equivalence relation which partitions the universe set U. We denote by $[u]$ the indiscernibility class defined by $[u] = \{w \in U \mid \forall_{c \in C} Inf(u,c) = Inf(w,c)\}$. Obviously,

$$[u] = \bigcap_{c \in C} [u]_c.$$

Let us consider now an example from linguistics. Let V be a finite alphabet, V^* be the set of all utterances over V, L be a language over V. As we mentioned in the first section, a context is meant here as a pair of words/utterances over V. We define the contextual information system

$$\Gamma_L = (V^*, \{decision\}, \{0, 1\}, V^* \times V^*),$$

where

$$decision(u, (x, y)) = \begin{cases} 1, & \text{if } xuy \in L \\ 0, & \text{otherwise} \end{cases}$$

One may say that a context (x,y) "accepts" the utterance w if xwy is a well-formed phrase in the language L. The equivalence class $[u]$ is exactly a *distributional class* as it was proposed by Dobrushin in [3]. A distributional class is the set of all utterances accepted by the same context. This concept of distributional class, considered in American descriptive linguistics in the fifties and sixties, was the starting point towards one of the first model of morphological categories. More generally, if two words are indiscernible in the contextual information system Γ_L, then they have the same type of morphological homonymy in the language L.

We would like to emphasize that the language L in the above example, viewed as a formal language, is regular in the Chomsky hierarchy (it can be computed by a finite automaton) if and only if the number of indiscernibility classes is finite (the Myhill-Nerode theorem). Details can be found in [20]. Before discussing a possible refinement of the indiscernibility relation, we consider one more example. To this aim, let V be a finite alphabet, V^* be the set of all words over V, $U(V^*)$ be the set of all phrases with words from V^*, $U(V^*) \times U(V^*)$ be the set of contexts, (S, Σ) be a partial additive domain/semantics [10]. We define the contextual information system

$$\Gamma = (U(V^*), \{eval\}, S, U(V^*) \times U(V^*)),$$

with

$$eval(uw) = \Sigma_{x \in \{u,w\}} eval(x).$$

For each context $c = (x,y)$, the indiscernibility class of the phrase u in the context c is called the weak synonymy class of u in [1], while the indiscernibility class of the phrase u is called the strong synonymy class in the same work.

For a more accurate refinement of the indiscernibility relation we make use of the standard *rough inclusion function* μ ([15]) defined for two finite sets X and Y by

$$\mu(X,Y) = \begin{cases} \frac{card(X \cap Y)}{card(X)}, & \text{if } X \neq \emptyset \\ 1, & \text{otherwise} \end{cases}$$

It is plain that the function μ is computable. The above definition can be extended to infinite sets X and Y by

$$\mu(X,Y) = \begin{cases} 1, & \text{if either } X \cap Y \text{ is an infinite set or } X = \emptyset, \\ 0, & \text{if either } X \cap Y \text{ is finite and } X \text{ is infinite or } X \cap Y = \emptyset, \\ \frac{card(X \cap Y)}{card(X)}, & \text{in all the other cases.} \end{cases}$$

Now the computability of μ is reduced to the computability of the intersection of two infinite sets and two decidability problems: emptiness (is the set computed by a program empty) and finiteness. Happily, this is a theoretical discussion only, since as we assumed above contextual information systems with finite sets are sufficient for practical matters. Given a contextual information system $\Gamma = (U, A, V, C)$, an object u *contextually dominates* the object w *in a degree at most* r, $r \in [0,1]$, iff for each context c

$$\mu(Inf(u,c), Inf(w,c)) \geq r.$$

We now say that two objects u, w are *indiscernible in a degree at most* r iff the two objects contextually dominates each other in a degree at most r. Clearly, as each attribute takes exactly one value, contextual domination in a degree at most 1 coincides with the indiscernibility. If one allows more values, then the two relations do not necessarily coincide anymore. Obviously, it is preferable that each contextual information system to have as less as possible indiscernible objects and the degree of indiscernibility between any pair of objects to be as low as (close to 0) possible.

3 Relationships Among Contexts

We start this section with a brief discussion about the interplay between objects and contexts. Let $\Gamma = (U, A, V, C)$ be a contextual information system. Given a set of objects $W \subseteq U$ we define the set

$$\mathcal{C}(W) = \{c \in C \mid \forall_{a \in A} \forall_{b \in A} \forall_{u \in W} (a(u, c) = b(u, c))\}.$$

One of the most desired properties of a context requires to distinguish between different objects or different uses of the same object. A context that has not this property is said to be *constant*. Formally, each context in the set $\mathcal{C}(U)$ is a constant context.

Analogously, given a set of contexts $D \subseteq C$ we define

$$\mathcal{U}(D) = \{u \in U \mid \forall_{a \in A} \forall_{b \in A} \forall_{c \in D} (a(u, c) = b(u, c))\}.$$

Obviously, the set $\mathcal{U}(D)$ is the set of all objects which cannot be distinguished by any context in D.

It is plain that

$$W_1 \subseteq W_2 \text{ implies } \mathcal{C}(W_2) \subseteq \mathcal{C}(W_1)$$
$$D_1 \subseteq D_2 \text{ implies } \mathcal{U}(D_2) \subseteq \mathcal{U}(D_1)$$

hence both operators $\mathcal{C}$ and $\mathcal{U}$ are anti-monotonous. Moreover, $W \subseteq \mathcal{U}(\mathcal{C}(W))$ and $D \subseteq \mathcal{C}(\mathcal{U}(D))$. Therefore, the relationships between objects and contexts are ruled by a Galois correspondence or a Galois connection. It is worth mentioning that the composition $\phi(W) = \mathcal{U}(\mathcal{C}(W))$ is a generalization of the *Sestier closure* [21].

Clearly, if $\mathcal{C}(U) = D$, then $\mathcal{U}(D) = U$ and if $\mathcal{U}(C) = W$, then $\mathcal{C}(W) = C$, but none of the converse relations does necessarily hold.

Two contexts $c, d \in C$ are *similar* with respect to (w.r.t.) an attribute a if $a(u, c) = a(u, d)$ for every $u \in U$, they are similar if they are similar w.r.t. each attribute, i.e.,

$$\forall_{u \in U} \forall_{a \in A} (a(u, c) = a(u, d)).$$

We denote by $\langle c \rangle_a$ and $\langle c \rangle$ the class of all contexts which are similar to c w.r.t. the attribute a and the class of all contexts which are similar, respectively.

An analogous refinement of the similarity relation between contexts to that of the indiscernibility relation can be defined. More precisely, given a contextual information system $\Gamma = (U, A, V, C)$, a context c *dominates* the context d *in a degree at most* r, $r \in [0,1]$, iff

$$\forall_{u \in U} \ \mu(Inf(u,c), Inf(u,d)) \geq r,$$

where μ is the function defined in the previous section. We now say that two contexts c, d are *similar in a degree at most* r iff the two contexts dominates each other in a degree at most r.

The ambiguity of a contextual information system resides in similar contexts and indiscernible objects. In order to decrease the degree of ambiguity of a contextual information system, one may proceed by one of the following ways:

- *Abstraction/Aggregation:* Aggregate similar (similar in a given degree) contexts into one context. The attribute mapping for any object in the aggregate context is lifted from the values of the attribute mapping for the same object in the similar contexts which were aggregated. Another strategy, is to redefine the contexts and the attributes accordingly. Generally, this requires a very deep understanding of the key features of the phenomenon under investigation. We shall present such an approach in the example of the next section.
- *Specialization/Restriction:* In order to make the contents/domain of a context more specific (decrease the degree of indiscernibility of objects) create a subcontext. This is a more difficult task than the previous one since the domain (attribute mapping definition) of the new contexts depends very much on the underlying model of the information system. On the other hand, the number of new contexts should stay as low as possible in order to have good computational properties of the system (see our discussion about the computability matter in the previous section).
- *Removal:* In general, every constant context may be removed.

We think that these general principles which are rather obvious might allow the construction of meaningful contexts. A starting point for this process might be a contextual information system whose contexts and assumptions based on these contexts are defined by similarity to other systems. Later these contexts as well as assumptions can be abstracted, refined and corrected using the above principles, process based on learning and experience. Easy to state informally, hard to implement algorithmically. Clearly, the constant contexts can be easily removed. The difficulty partly resides in finding algorithmically the similar contexts. If one needs similar contexts in a given degree, a rather difficult task is to choose an appropriate threshold. In our view, as stated above the most difficult part is to create a subcontext and define its attributes.

4 A Final Example

Let us consider here a more intricate phenomenon and the associated contextual information system. The fundamental mechanism by which genetic material is merged is *recombination*. DNA sequences are recombined under the effect of enzymatic activities. In 1987, T. Head [7] introduced the *splicing* operation as a language theoretical approach of the recombinant behavior of DNA under the influence of restriction enzymes and ligases.

Roughly speaking, the main idea of the splicing operation is that two sequences are cut at specified sites, and the first subsequence of one sequence is pasted to the second segment of the other and vice versa. Let us be more specific, the reader interested in more details is referred to [7]. For example, let us consider the following three DNA molecules [8]:

$$5' - CCCCCTCGACCCCC - 3'$$
$$3' - GGGGGAGCTGGGGG - 5'$$

$$5' - AAAAAGCGCAAAAA - 3'$$
$$3' - TTTTTCGCGTTTTT - 5'$$

$$5' - TTTTTGCGCTTTTT - 3'$$
$$3' - AAAAACGCGAAAAA - 5'$$

Observe that the three DNA molecules are complete double stranded molecules. Observe also the Watson-Crick complementarity (A and T are Watson-Crick complementary as well as C and G) between each pair of nucleotides, one in the upper strand and the corresponding one in the lower strand. These molecules are objects in our contextual information system. Restriction enzymes which are physically DNA molecules and theoretically contexts in our contextual information system can recognize specific DNA subsequences in the above DNA molecules and cut them such that each of the new DNA molecules has a hanging end. For instance, the DNA sequences recognized by the enzymes *Taq*I, *Sci*NI and *Hha* are respectively:

$$T|CGA \quad G|CGC \quad GCG|C$$
$$AGC|T \quad CGC|G \quad C|GCG$$

The way each restriction enzyme cuts a DNA molecule in which the enzyme structure is present is illustrated above. The action of these enzymes on the DNA molecules listed above results in the following fragments with hanging ends:

$$5' - CCCCCT \quad CGACCCCC - 3'$$
$$3' - GGGGGAGC \quad TGGGGG - 5'$$

$$5' - AAAAAG \quad CGCAAAAA - 3'$$
$$3' - TTTTTCGC \quad GTTTTT - 5'$$

$$5' - TTTTTGCG \quad CTTTTT - 3'$$
$$3' - AAAAAC \quad GCGAAAAA - 5'$$

One can easily see that the presence of enzymes in the test tube is not sufficient for producing these fragments. It is necessary that the DNA molecules contain the subsequences recognized by the enzymes. In other words, a context should appear both outside and inside the objects. Note that all obtained fragments have identical hanging ends, that is CG read from 5' to 3', but there is an important difference depending on how these ends are oriented: to 5' or 3'. One can easily see that *Taq*I and *Sci*NI produce ends oriented to 5' while *Hha*I produces ends oriented to 3'. This makes that the first four fragments are compatible for ligation, namely under the influence of a ligase (another context) they can join producing either the original molecules or new molecules.

$$5' - CCCCCTCGCAAAAA - 3'$$
$$3' - GGGGGAGCGTTTTT - 5'$$

$$5' - AAAAAGCGACCCCC - 3'$$
$$3' - TTTTTCGCTGGGGG - 5'$$

Now we try to define formally a contextual information system modeling this phenomenon. We take the set U of objects as the union of two sets:

- U_1 is the set of all complete (without hanging ends) DNA molecules. They are represented by usual words over the DNA alphabet of the four nucleotides $\{A, C, G, T\}$. By the Watson-Crick complementarity between nucleotides linear words suffice. Indeed, if we have one word which represents the upper strand, the lower strand can be immediately found by taking the complements of the nucleotides in the upper strand.
- U_2 is the set of all pairs of DNA molecules with hanging ends. These molecules are represented by words of the form xy and $\overline{y}x$, where x represents a complete DNA molecule and y is a standard word over the DNA alphabet.

The set of context C contains all enzymes existing in the test tube; it is formally defined by $C = \{enz_1, enz_2, \ldots, enz_k\} \cup \{ligase\}$. Since the DNA molecules were represented in U by words over the DNA alphabet, it seems quite natural to do the same with the enzymes, that is each enz_i of C to be formally represented by a word over the DNA alphabet. One immediately encounters the following problem: two enzymes having slightly different actions are represented by the same word. By this reason, looking closer to the enzymes structure, we redefine the contexts as follows: each context, still representing an enzyme, is formally defined by a pair of words over the DNA alphabet. For example, the three enzymes listed above are represented by the contexts:

$$(T, CGA) \quad (G, CGC) \quad (GCG, C).$$

The set of attribute mapping contains one element, that is $A = \{splice\}$. Finally, the set V of attribute values has three elements *cut*, *paste*, and $\perp$. As one can see, each value is actually an action ($\perp$ is the null action).

Now the contextual information system is completely defined as soon as we define the attribute mapping *splice*. We first try with

$$splice(x, enz_i) = \begin{cases} cut \text{ if } x \in U_1, x = x_1\alpha\beta x_2, enz_i = (\alpha, \beta), \text{ (the DNA molecule} \\ \quad x \text{ contains the structure of enzyme } enz_i) \\ \bot, \text{ otherwise} \end{cases}$$

One can easily see that we came up to the following problem: the contexts (G, CGC) and (GCG, C), representing the enzymes *SciNI* and *Hha*, respectively, are similar since

$$splice(x_1, (G, CGC)) = splice(x_1, (GCG, C)) = \bot,$$
$$splice(x_2, (G, CGC)) = splice(x_1, (GCG, C)) = cut,$$
$$splice(x_3, (G, CGC)) = splice(x_1, (GCG, C)) = cut,$$

where x_1, x_2, x_3 are the words representing the three DNA molecules. Moreover, the output of the action *cut* cannot be unambiguously defined. More precisely, despite we can say where (in the upper or lower strand) there will be the hanging ends (and this specification makes that the two enzymes be not similar anymore), we cannot say how long are they. Therefore, each enzyme will be represented as a triple of words over the DNA alphabet. Thus, the three enzymes listed above are now represented by the contexts:

$$(T, \underline{CG}, A) \qquad (G, \underline{CG}, C) \qquad (G, \overline{CG}, C).$$

We now consider two attribute values cut_- and cut^- instead of cut and define

$$splice(x, (\alpha, \underline{y}, \beta)) = \begin{cases} cut_- \text{ if } x \in U_1, x = x_1\alpha y\beta x_2 \\ \bot, \text{ otherwise} \end{cases}$$

and

$$splice(x, (\alpha, \overline{y}, \beta)) = \begin{cases} cut^- \text{ if } x \in U_1, x = x_1\alpha y\beta x_2 \\ \bot, \text{ otherwise.} \end{cases}$$

The result of cut_- in the case above is formed by two words, namely $x_1\alpha\underline{y}$ and $\overline{y}\beta x_2$, representing two incomplete DNA molecules. Analogously, the result of cut^- in the case above is formed by two words, namely $x_1\alpha\overline{y}$ and $\underline{y}\beta x_2$, Finally, we define

$$splice((x, y), ligase) = \begin{cases} paste \text{ if either } x = u\underline{z}, y = \overline{z}v \text{ or } y = u\underline{z}, x = \overline{z}v \\ \bot, \text{ otherwise} \end{cases}$$

The result of *paste* in this case is formed by the word uzv representing a complete DNA molecule. The value of *splice* in all the other cases is $\bot$.

Now, the reader can easily check that this contextual information system models the DNA recombination by splicing.

5 Conclusion

We have introduced contexts as abstract mathematical entities in a rather general framework, namely knowledge represented by contextual information systems. Our choice was a theoretical computer science approach in which contexts were considered abstract objects; instead of offering a formal definition we preferred to propose a possible description. This description was done by enhancing the information systems with the capacity of manipulating contexts. We have discussed the indicernibility relation between the objects and the similarity relation between the contexts of a contextual information system. Making use of some concepts from the Rough Set Theory both relations were refined. The new concepts were accompanied with examples with some relevance in computational linguistics. Based on the aforementioned relations we proposed a simple strategy for decreasing the ambiguity of contextual information systems. Some items of this strategy were illustrated in a example of a contextual information system modeling a biological phenomenon "in vitro", namely the DNA recombination under the enzymatic influence. In our opinion, this work is tentative and incomplete; it is just a first step, coming from the information theory, towards a better understanding of this very rich concept that is the context.

References

1. Atanasiu, A., Martín-Vide, C., Mitrana, V.: On the sentence valuation in a semiring. Information Sciences **151** (2003) 107–124
2. Buvač, S., Buvač, V., Mason, I. A.: Metamathematics of contexts. Fundamenta Informaticae, **23**(1995)
3. Dobrushin, R. L.: The elementary grammatical category. Byulleten Objedinenija po Problemam Mashinogo Perevoda **5** (1957) 19–21 (in Russian)
4. Ehrenfeucht, A., Păun, G., Rozenberg, G.: Contextual grammars and formal languages. In: Handbook of Formal Languages (G. Rozenberg, A. Salomaa, eds.), vol. 2, Springer Verlag, Berlin, 1997
5. Ghindini, C., Serafini, L.: Distributed first order logics. In: Frontiers of Combining Systems 2 (Papers presented at FroCoS'98), Research Studies Press, Wiley, 1998
6. Guha, R. V.: Contexts: A Formalization and Some Applications. PhD Thesis, Stanford University 1991
7. Head, T.: Formal language theory and DNA: an analysis of the generative capacity of specific recombinant behaviours. Bull. Math. Biology **49** (1987) 737 – 759
8. Head, T., Păun, G., Pixton, D.: Language theory and molecular genetics. In: Handbook of Formal Languages (G. Rozenberg, A. Salomaa, eds.), vol. 2, Springer-Verlag, Berlin, 1997
9. Hirst, G.: Semantic Interpretation and the Resolution of Ambiguity. Cambridge University Press, 1987
10. Manes, E. G.: Additive domains. Mathematical foundations of programming semantics. Lecture Notes in Computer Science **239** (1986) 184–195
11. Marcus, S.: Contextual grammars. Rev. Roum. Math. Pures Appl. **14** (1969) 1525–1534
12. Marcus, S.: Algebraic Linguistics. Analitical Models. Academic Press, 1967

13. McCarthy, J.: Notes on formalizing context. In: Proc. of the Thirteenth International Joint Conference on Artificial Intelligence IJCAI'93, 1993, 555–560
14. McCarthy, J., Buvač, S.: Formalizing context. In: Computing Natural Language (A. Aliseda, R. van Glabbeek, D. Westersthét al., eds.), Stanford University, 1997, 13–50
15. Pawlak, Z.: Rough Sets: Theoretical Aspects of Reasoning about Data, Kluwer, Dordrecht, 1992
16. Păun, G.: Marcus contextual grammars. After 25 years. Bulletin of the EATCS **52** (1994) 263–273
17. Păun, G.: Marcus Contextual Grammar. Kluwer, Dordrecht, 1997
18. Polkowski, L., Skowron, A.: Rough mereology. In: Proc. ISMIS'94, Lecture Notes in Artificial Intelligence 869, Springer Verlag, Berlin 1994, 85–94
19. Quasthoff, U. M.: Context. In: Encyclopedia of Language and Linguistics (R. E. Asher, ed.), vol. 2, Pergamon Press, Oxford, 1994, 730–737
20. Rozenberg, G., Salomaa, A. (eds.): Handbook of Formal Languages. Springer Verlag, Berlin, vol. 1, 1997
21. Sestier, A.: Contributions à une théorie ensembliste des classifications linguistique. Aces du Premier Congrès de l'AFCAL, Grenoble, 1960, 293–305.

Context Building Through Socially-Supported Belief

Naoko Matsumoto and Akifumi Tokosumi

Department of Value and Decision Science,
Tokyo Institute of Technology,
2-12-1 Ookayama, Meguro-ku, Tokyo, Japan 152-8552
{matsun,akt}@valdes.titech.ac.jp
http://www.valdes.titech.ac.jp/~matsun/

Abstract. Belief, defined as a knowledge structure with a degree of subjective confidence that is gained through interaction with other agents in a community, can play a crucial role in cognitive processes. As paradigmatic applications of our socially supported knowledge-belief system, (a) a utterance understanding parser was designed and implemented in Common Lisp CLOS with a belief management system, and (b) a cognitive model of attachment is proposed based on the analysis of analyzing fan letters sent to a toy company.

1 Introduction

The utterance *"Your paper is excellent"* could be interpreted as either the professor is praising my report or the professor is being cynical if the hearer knows the professor seldom praises students. The range of possible interpretations for an utterance depends on the individual's beliefs, memory, and the social and cultural contexts. In addition utterance-level ambiguities, there are also word-level ambiguities too. For instance, the word "paper" has multiple meanings, such as paper = thesis, paper = the substance and paper = newspaper. If the hearer works at a paper manufacturing plant and the utterance is from the boss, the utterance would be interpreted as [the boss is praising me for creating high-quality paper], but if the hearer is a journalist, then the hearer would interpret the utterance as praise for his newspaper company.

Changing the topic for a moment, Japan is witnessing an attachment phenomenon where middle-aged people are going crazy over a talking toy doll called Primopuel. Within their own communities, the fans of the toy are sharing their positive emotional state for the toy with others (e.g. *"I presented the toy to my aunt because he (= the toy) is very lovely and makes me relax"*), or, though seeing attachment for the toy in others, are experiencing even more attachment themselves for the toy (*"When I found out that she also loves him (= toy), I felt my affection for him grow even stronger"*).

While these two things, utterance understanding and attachment development, may appear to be quite different kinds of phenomena, both are about cognitive transitions within the agent. In regarding both of these phenomena as kinds of cognitive transitions within the agent, we suggest that they share a common mechanism: the cognitive state changes depend on the context of other agents within a community. In the case of utterance understanding, the hearer chooses the utterance meaning based on his beliefs, such as the professor is a good person/cynical person, or according to the hearer's work position, as well as their relationship with the speaker. In the case of attachment development, the agent strengthens her attachment by adopting the beliefs of other agents in the fan community.

In the present paper, belief, defined as a knowledge structure with a degree of subjective confidence that is gained through communication with others, constructs a kind of context which plays a crucial role in cognitive processes.

In the first half of the paper, we discuss the relationship between cognition and context, and then outline our cognitive model. In the second half, we focus on the phenomenon that have shaped our model; namely, utterance understanding and attachment development.

2 Context Paradigms

2.1 Meaning in Context

In traditional linguistics, it is the language itself that is key to interpreting meaning (e.g. [1]), and language meaning is relatively fixed and static. In this framework, while context is seen as the collection of information that helps to interpret language, it is regarded as being a given, peripheral, and static. \On the other hand, a number of different paradigms emerged in the later 1970s and early 1980s within sociolinguistics, cognitive linguistics, and pragmatics. Here the trend was to regard language phenomenon as being embedded with the context, where context is simultaneously social, cultural, and cognitive (e.g. [2][3][4][9]).

In our research, rather than having independent meaning, utterance/word meaning is determined based on the agent's subjective belief structure, which draws on their social and cultural experiences. We propose a model that systematizes the relationship between the language phenomenon and social/cultural factors.

2.2 Mind in Context

Cultural psychology claims that, not only language cognition, but human cognition in general is context dependent (e.g. [5]). This framework tries to push forward a new cognitive revolution, that mind must be understood in-situ. In contrast to the view in more traditional cognitive science that mind is a device for symbol manipulation independent from the environment, the new position, which we may refer to as the context paradigm, mind is always embedded within a series of contexts, ranging from the situational, environmental, social, and culture. Cultural psychology insists that the nature of mind can only be fully grasped in its historical/cultural/social contexts.

Echoing the motivations of cultural psychology, we propose that the agent's cognitive system functions in ways that are closely tied into the agent's social/cultural context.

The idea within cultural psychology that mind and culture are interdependent is also a key idea for our research. Cultural psychology tries to propose a model which satisfies two propositions; (a) mind is constructed by culture, and (b) culture is constructed by mind.

This paper proposes a model to account for dynamic shifts in an agent's belief states in response to the beliefs of others. To the extent that our research deals with the interdependency between the individual belief structure and the belief structures of others, it is also modeling the mind-culture interdependency within cultural psychology.

2.3 Message in Context

Sperber and Wilson have proposed Relevance Theory ([9]) which takes a cognitive science approach to human cognitive principles. In this framework, *assumption* in the hearer's mind is a cue to utterance interpretation. When the hearer has some verbal communication input, the assumptions are dynamically changed. If we take the assumptions in the hearer's mind as a kind of created context built, then clearly the cognition of one individual will differ from others because they will possess different knowledge structures.

We agree with this view, but would point out that this treatment of context is rather too focused on the individual. Although Sperber and Wilson do not discuss shared context in any detail, clearly, there must be a certain level of shared context with others for successful conversation.

In this respect, Tanaka and Fukaya ([10]) portrays context has a form of social dynamism, with social interaction realized through conversation. To have a successful conversation, there must be a certain level of similarity between the knowledge structures held by each agent.

Our socially-supported context model encompasses both characteristics, (a) context is constructed in the agent's mind, and at the same time, (b) the agent's context is shared with other agents in the community. In our position, context is a mechanism of support that exists between the cognitive systems of the members of a linguistic community. Within language communities, relatively-stable states of language meaning (with some degree of ambiguity) are achieved, and the mutual transfer of ideas (as well as emotions) is made possible.

2.4 Decomposition of Context into Social Constructs

By integrating cognition and context in this way, we define context is a belief system that is constructed within the agent based on a social foundation. As we, humans, engage in social activities, context is mediated through social restrictions ([10]). At the same time, as relevance theory points out ([9]), the internal limitation of each agent also contribute to the interpretative processing of a given situation, leading to different understanding in the various participants. In this way, the nature of context must be both (a) understood as a social network, through which we communicate with others, and (b) perceived as having an internal dimension.

In this paper, we will discuss social foundations and individual belief systems in terms of our model.

3 Context Built Through a Socially-Supported Belief System

The *Socially Supported Belief System* (SSBS) is a computational model of utterance comprehension which incorporates the agent's dynamically-revisable belief system. The belief system models the agent's linguistic and world knowledge with a component representing the degree of support from other agents in the language community. A key concept in this system is the *socially supported belief* (**ssb**) — so named to emphasize its social characteristics.

The **SSBS** has its origins in Web searching research where, in a typical search task, no information about the context of the search or about the searcher is available, yet a search engine is expected to function as if it knows all the contextual information surrounding target words [(6)]. The main similarity between the **SSBS** and search engines is in the weighed ordering of utterance meaning (likely meaning first) by the system and the presumed ordering of found URLs (useful site first) by search engines. Support from others (other sites in the case of Web search) is the key idea in both cases. It makes it possible to construct a belief structure by socially referring to other's beliefs and by revising the belief structure dynamically.

4 Utterance Understanding in the Socially-Supported Belief System

4.1 Disambiguate Utterance Meaning

As a hearer model, the task for **SSBS** is to identify the intention of the speaker using its **ssb** database. When the speaker's intention is identified, the **SSBS** incorporates the intention within the belief structure for the hearer. Thus, the hearer model builds its own belief structure in the form of **ssb**s. In the **SSBS**, each belief (**ssb**) has a value representing the strength of support from others. The belief that has the highest level of support is taken as the most likely interpretation of a message, with the ranking order of a belief changing dynamically through interaction with others.

In the **SSBS**, an agent's belief reflects the level of support from other agents' obtained through verbal interaction. Our first natural choice for an application of this system is a dialogue parser, called Socially Supported Belief Parser (**SSBP**), which has been implemented in Common Lisp Object System (CLOS)[1]. By applying the system, it is possible to interpret utterances only with the agent's internal beliefs.

4.2 The Internal Structure of a Socially Supported Belief (ssb)

We designed the **SSBP** as a word-based parser, because this has an architecture that can seamlessly accommodate various types of data structures. The **SSBP** processes a sequence of input words using its word knowledge which consists of three types of knowledge; (a) grammatical knowledge, (b) semantic knowledge, and (c) discourse knowledge (Table 1). Each knowledge type is represented as a daemon which is a unit of program code executable in a condition-satisfaction mechanism ([6]).

The grammatical knowledge controls the behavior of the current word according to its syntactic category. The semantic knowledge deals with the maintenance of the **ssb** database derived from the current word. The **ssb** has a data structure similar to the *deep cases* developed in generative semantics. Although the original notion of deep case was only applied to verbs, we have extended the notion to nouns, adjectives, and

---

[1] The system is currently being developed by increasing the size of the vocabulary and the number of sentence patterns that it can handle.

adverbs. For instance, in a ssb, the noun *paper* has an evaluation slot, with a value of either *good* or *bad*. The ssbs for nouns are included with evaluation slots in the SSBP, as we believe that evaluations, in addition to epistemic content, are essential for all nominal concepts in ordinary dialogue communication. The discourse knowledge can accommodate information about speakers and the functions of utterances.

Table 1. Knowledge categorization for the SSBP

categorization	action	examples	implementation
grammatical knowledge	Specifying the category	Depending its category (ex) "paper" : noun daemon	[category daemon] [predictable category daemon]
semantic knowledge	Extracting the meaning	Extracting the first ssb of the current word (ex) "paper" : thesis (ssb)	[knowledge belief daemon]
	Changing the belief structure	Controlling the knowledge belief structure with new input belief	[knowledge belief management daemon]
	Predicting the following word	Predicting the deep case of the current word (ex1) "buy" (verb) : "Tom" as agent (ssb) (ex2) "flower" (noun) : "beautiful" as evaluation (ssb)	[predictable word daemon]
discourse knowledge	Inferring the user of the word	(ex) "pavement" : British (ssb)	[speaker information daemon]
	Inferring the	Inferring the utterance	[utterance function daemon]

The SSBP parses each input word using the ssbs connected to the word. As the proposed parser is strictly a word-based parser, it can analyze incomplete utterances. An example of a belief implementation in the parser can be seen in the following fragment of code:

```
;;;class object for 'paper. The word 'paper" belongs
;;;to "commonnoun."
(defclass paper (commonnoun)

;;;the category is commonnoun
((Cate :reader Cate :initform 'commonnoun)

;;;the word's ssb. The first ssb is 'thesis,' the second
;;;sskb is 'newspaper.'
```

```
(sskb :accessor sskb
    :initform '((7 thesis)(5 newspaper))))

;;;the predictable verb.
(predAct :accessor predAct
        :initform '((4 read)(3 write))))

;;;the user information of the word.
(userInfo :accessor userInfo
        :initform '((4 professor)(2 father))))

;;;the evaluation. If the word has evaluated,
;;;the slot is filld.
(EvalInducer :accessor EvalInducer
        :initform '((3 beautiful)(1 useful))))

;;;the utterance function.
(uttFunc :accessor uttFunc
        :initform '((6 order)(5 encourage))))))
```

Weighing of a **ssb** in the **SSBP** is carried out through the constant revision of its rank order. The parser adjusts the rank order of the **ssb** for each word whenever interaction occurs with another speaker.

4.3 Inferring the Speaker's Intention

In the **SSBP**, the speaker's intension is inferred by identifying the pragmatic functions of an utterance, which are kept in the **ssbs**. As utterances usually have two or more words, the parser employs priority rules for the task. For instance, when an utterance has an adjective, its utterance function is normally selected as the final choice of the pragmatic functions for the utterance. In case of the utterance *"Your paper is excellent,"* pragmatic functions associated to the words *"your"* and *"paper"* are superseded by a pragmatic function *"praise"* retrieved from a **ssb** of the word *"excellent."* Because of these rules, it is always possible for the **SSBP** to infer the speaker's intention only with the agent's internal beliefs.

4.4 Word Meaning

We want to insist that there is no static or literal word meaning. In other words, a word has meanings depending on the agent's beliefs. We can converse with others and communicate with others based on a common beliefs. People construct word meaning depending on their belief structures built with other's socially-supported belief. Word meaning is decided by the strength of socially-supported confidences. This can account for the phenomenon that the same word means the different things according to the social groups. For instance, while the word "paper" means an academic report in academic fields, it means money in a bank, or while the word "subway" means the underground railroad in the U.S., it means the underground street in U.K. Socially-support system has possibility to explain the arbitrary of language.

4.5 Emotion Elicitation

Emotion elicitation is one of the major characteristics of the SSBP. Ordinary models of utterance interpretation do not include the emotional responses of the hearer, although many utterances elicit emotional reactions. We believe that emotional reactions, as well as meaning/intention identification, are an inherent function of utterance exchange and the process is best captured as a socially-supported belief processing. In SSBP, emotions are evoked by ssbs activated by input words. When the SSBP determines the final utterance functions as a speaker's intention, it extracts the associated emotions from the utterance functions. For instance, an utterance function *"praise"* from the utterance *"Your paper is excellent,"* then, the SSBP searches the emotion knowledge for the word *"praise,"* it extracts the emotion *"happy."*

We propose that emotion elicitation mechanism is also one of human's cognitive processes, which can be supported by SSBS. In the next section, we try to adopt our proposed system to emotional model.

5 Socially-Supported Emotion Model: Attachment

We adopt our proposed model SSBS to the phenomenon which the emotion attachment is transmitted in a certain fan community. At first, the phenomenon occurred in Japan and the analysis method is described. Next, we draw the characteristics of attachment based on our previous research. After that, we focus on the social/network characteristics of attachment transmission in a community, which can be called other agent's context.

5.1 Craze for a Toy

Primopuel, a talking toy doll, is very popular among certain middle-aged Japanese users of the toy, as evidenced in the fan letters sent to the toy company describing the fan's attachment for the character. The manufacturer of the doll have sold more than 700 thousand units over the last four years in Japan alone.

This toy doll has some characteristics: Primopuel has some touch sensors, a sound sensor, a temperature sensor, and a calendar system. The character can utter about 250-280 utterances stored in its memory (e.g., *"I love you." "You are doing your best."*), in response to user actions. Moreover, Primopuel modifies the probabilities of using the utterances according to an easy learning system. Thus, it appears as if different moods or personalities are being formed.

The toy may be regarded as a communication toy, a simplified version of a communication toy. Our research deals with the phenomenon of adults experiencing strong emotion ties with an artefact based on a constant and positive emotional state.

5.2 Analysis Method

In the fan letters, the fans described their mental states and/or their actions towards the character. Analysis of the fan letters makes it possible to perceive the details of the emotion attachment. In our previous research, we have adopted a procedure to extract the fan's cognitive state from the textual data in 51 fan letters and 271 Web postings. After identifying propositions in fan letters, these are categorized according to a classification system (Table 2). Because of letters include pragmatic messages, we contrived to see both cognitive viewpoint (e.g. about what fans write) and pragmatic one (e.g. report, evaluation, gratitude) ([7]).

To test the reliability of the classification system, a reliability index was calculated. The obtained value, $\kappa = 0.80$, is above the level recommended by von Somren ([8]).

5.3 The Characteristics of Attachment

In our previous research, we have examined attachment for the talking doll from the viewpoint of cognitive science, attaining the following findings: (a) Fans who have attachment to the talking doll regard it as a cohabitant artefact, it means fans regard the toy as a artifact and as a cohabitant at the same time, (b) the positive mental and physical states of the fans are often attributed to the doll, and (c) the fans believe that the toy enhances interaction with family members and/or with friends.

Table 2. Classification system

cognitive	pragmatic	example data
addressee	message	hello, best regards
toys	description toys as an artifact	it always answers with the correct time
	description toys by personifying	he says leave him alone, he wants a scarf
	information related to toys	I can't get find out where to buy it
fan (user)	fan's life	we have no children
	fan's action to toys(=artifact)	I bought the toy "Primopuel"
	fan's action to toys(=personifying)	I hold him tightly, I made clothes for him
	positive evaluation for toy (all)	it's very lovely
	positive evaluation for toy (each function)	it is a reasonable price
	positive evaluation for toy (toy's personality)	a year ago, he became member of our family
	negative evaluation for toy (all)	I bought the strange thing
	negative evaluation for toy (each function)	the battery box is broken.
	negative evaluation for toy (toy's personality)	no examples
	positive emotions	I enjoy my life because of this toy
	negative emotions	I hate this stain
	positive change of the fan's state	It helps my rehabilitation
	negative change of the fan's state	I always quarrel with my mother about the toy
others	description related to others	my boyfriend gave it to me as a present

5.4 Attachment Propagation Within a Fan Community

In particular, we have focused on finding (c), and have developed a model of attachment for artificial cohabitants, SEM (Socially-supported Emotion Model), which explains how fans can strengthen their attachment for the artificial cohabitant through communication with others and adopting the beliefs of other about the artificial cohabitant ([7]) (Figure 1).

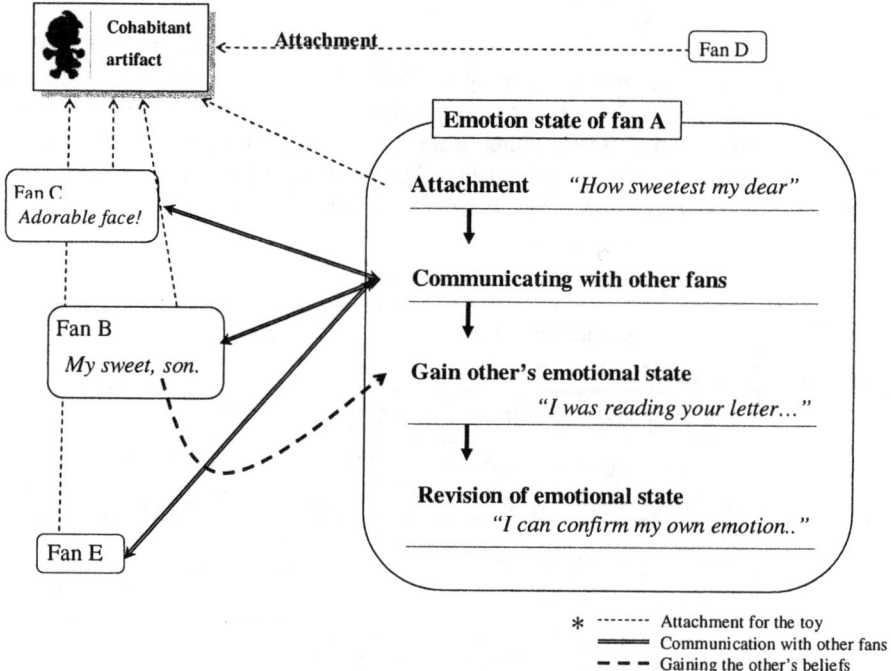

Fig. 1. SEM (Socially-supported Emotion Model) Fan A, who has already attachment for the cohabitant artifact, can communicate with other fans in a certain community. Thorough the communication, fan A knows other agent's emotional state. By gaining the other agent's beliefs, in other words, referring to other agent's context, fan A can revise her own emotional state

6 Summary

In this paper, we defined context as a belief system constructed within an agent based on a social foundation. In detail, (a) the belief system is constructed within the mind of the individual agent, while at the same time, (b) the agent's belief system is shared with other agents in the community.

Adopting a Web searching technique, we modeled SSBS (Socially-Supported Belief System) which can explain how the human belief system works in cognition through interaction with others. The system reflects the intentions of the other agent at each conversation turn, with each belief having a rank based on the support from others. We have designed and examined the SSBS model in (i) utterance understanding, and (ii) attachment development and its propagation within the community.

References

1. Chomsky, N. Language in a psychological setting. Sophia linguistica. XXII, Sophia Universigy, Tokyo (1987).
2. Grice, P. Studies in the Way of Words. Harvard University Press, Cambridge, Mass (1989)
3. Gumpers, J. J. Discourse Strategies. Cambridge University Press. (1982).
4. Halliday, M.A.K. and Christian, M.I.M. Matthiessen. Construing Experience Through Meaning: A Language-based Approach to Cognition. London: Cassell. (1999).
5. Kitayama, S., Markus, H. R., Kurokawa, M. Culture, emotion, and well-being: Good feelings in Japan and the United States, 14 (2000) 93-124.
6. Matsumoto, N., Tokosumi, A. A scially supported knowledge-belief system and its application to a dialogue parser. Vasile Palade, Robert J. Howlett, and Lakhmi Jain (Eds.) Knowledge-Based Intelligent Information and Engineering Systems,Lecture Notes in Artificial Intelligence, Subseries of Lecture Notes in Computer Science, LNCS/LNAI 2773, Springer-Verlag (2003) 778-784.
7. Matsumoto, N., Tokosumi A., and Hirai, Y. Affection for cohabitant toy dolls. *Computer Animation and Virtual World*, 15 (3-4) (2004) 339-346.
8. Someren M. van, W. F. Barnard, and J. A. C. Sandberg. The think aloud method: A practical guide to modeling cognitive processes. Academic Press, Amsterdam (1994).
9. Sperber, D., Wilson, D.: Relevance; Communication and Cognition. 2nd edn. Harvard Univ. Press, Cambridge, Mass. (1995).
10. Tanaka, S., and Fukaya, M. Development of SocioSemantics. (1998) (In Japanese).

A Quantitative Categorization of Phonemic Dialect Features in Context

Naomi Nagy[1], Xiaoli Zhang[2], George Nagy[2],
and Edgar W. Schneider[3]

[1] English Department, University of New Hampshire, Durham, NH 03824 USA
ngn@unh.edu
[2] DocLab, ECSE, Rensselaer Polytechnic Institute, Troy, NY 12180 USA
{zhangxl, nagy}@rpi.edu
[3] Department of English Linguistics, Regensburg University, Regensburg, Germany
edgar.schneider@sprachlit.uni-regensburg.de

Abstract. We test a method of clustering dialects of English according to patterns of shared phonological features. Previous linguistic research has generally considered phonological features as independent of each other, but context is important: rather than considering each phonological feature individually, we compare the patterns of shared features, or *Mutual Information (MI)*. The dependence of one phonological feature on the others is quantified and exploited. The results of this method of categorizing 59 dialect varieties by 168 binary internal (pronunciation) features are compared to traditional groupings based on external features (e.g., ethnic, geographic). The MI and size of the groups are calculated for taxonomies at various levels of granularity and these groups are compared to other analyses of geographic and ethnic distribution. Applications that could be improved by using MI methods are suggested.

1 Introduction

The way a given language is spoken by a particular group at a particular time is referred to as a dialect. Dialects can be grouped into categories in many different ways. Using *external features*, dialects may be grouped by geographic location (*e.g.*, Irish English), ethnic identity (*e.g.*, African-American Vernacular English), or social networks (*e.g.*, Liberian Settler English) of their speakers. Alternatively, using *internal features*, dialects may be grouped by shared features of pronunciation, vocabulary, or grammar. We explore quantitative approaches to see how similarly dialects cluster by these different methods.

How many dialects are there? For English, answers may range from 1 to ~341,000,000 (the number of mother-tongue speakers of English). Dialects can be as narrow as that of a single speaker (an *idiolect*), or as broad as a major language (*e.g.*, Chinese or Spanish). While no two speakers speak identically, speakers may be grouped into dialects by degree of similarity of vocabulary (lexical categories), grammar (syntactic categories), and/or pronunciation (phonological categories).

We exploit a data set showing variations in the pronunciation of a set of vowels and consonants to form dialect clusters by phonological categorization. From this perspective, a *dialect cluster* is the context that determines the variant (*allophone*) of each phoneme used by speakers of that dialect. We analyze pronunciation, rather than lexicon, in part because a more robust classification system should be possible due to the much smaller number (and concomitant higher frequency) of sounds than words in any given language.

Context requires both diversity and dependence. If all the varieties within a dialect cluster are phonologically similar, then there is no phonological context: how speakers pronounce one phoneme reveals nothing about how they pronounce another. Nor is there any context if the different speakers' phonological characteristics are statistically independent. While "context" has a broad range of definitions in logic, linguistics, philosophy, artificial intelligence, and sociology, inter alia [1, 2], we chose a narrow definition in order to operationalize it. We quantify context by *Mutual Information (MI)*, an information theoretic measure calculated from the joint and marginal probability distributions of the allophones of every pair of phonemes. MI is greatest when there is large and *consistent* variation among the phonological values of the varieties of the cluster. The strongest possible context among two features arises when their variants are all equally probable (and therefore most unpredictable in an information-theoretic sense) among the varieties, and statistically perfectly dependent. *Perfect dependence* means that knowing how a speaker pronounces one phoneme suffices to predict what variant of the other phoneme will be used by that speaker. This notion of context can be extended beyond pairs to any set of features, and to any number of varieties in a cluster.

The result of our analysis is a hierarchy of English dialect clusters with a measure of the MI at each level. Aside from its intrinsic interest in linguistics for comparison with alternative taxonomies, this approach may decrease error rates in automatic speech recognition (ASR) for dialectically homogeneous groups of speakers. Similar methods, based on *style context*, have met with some success in the recognition of hand-written digits and printed text [3, 4]. Whereas in speech the style context is provided by dialect, in hand-print it may be due to each form in a batch filled out by a single writer, and in printed text it may originate from a commonality of font, printer, scanner or copier.

In Sec. 2, we sketch the foundations of phonology, and the formation and characteristics of dialects and their taxonomies. Secs. 3 & 4 present the rationale and collection protocol for our phonological data. Although the clustering algorithm and probabilistic distance measure that we use are not new to computer and information scientists, we illustrate them with brief examples using phonological features. (Our contribution is the adaptation of *hierarchical clustering* and of a *measure of statistical dependence* to new linguistic data.) Secs. 5 & 6 present the groupings we obtained with their information-theoretic measure of context and a comparison of our dialect clusters with groupings obtained by alternative methods. Sec. 7 outlines applications in the domains of digital speech and automatic speech recognition that could be improved by using these methods.

2 Phonology, Language and Dialects

At the phonological level, the units of spoken language are phonemes, the smallest units of sound recognized as distinct by speakers of the language. For example, [t] and [d] are distinct phonemes in English—speakers recognize that they distinguish the words "try" and "dry" (as well as many other word pairs, like "lit" and "lid"). Both are articulated as alveolar stops, but the first is voiceless and the second voiced. These three articulatory features (place and manner of articulation and voicing) are used to uniquely identify the consonant phonemes of a language. Finer phonetic distinctions exist: phonemes may be pronounced differently depending on the context in which they are found. These different pronunciations are referred to as allophones of a particular phoneme, e.g., the difference between the aspirated [t^h] in "take" vs. the unaspirated [t] in "stake.". Native speakers of a language are rarely conscious of these subphonemic or allophonic differences, but they are important dialect markers.

Vowels are described in terms of tongue position, lip rounding and duration. Tongue height and backness, combined with duration, uniquely distinguish the vowels of English. Again, finer-grained distinctions exist below the level of consciousness of speakers. It is these phonetic distinctions that are described by the 168 binary features in our descriptive system.

A new language variety develops when a group of people maintain contact with each other over an extended time period and are isolated from other speakers. A variety which starts out as a dialect of a language may, given enough time, develop into a new language (no longer mutually intelligible). For example, English got its beginnings as Anglo Saxon, a Germanic dialect, when people from (present-day) Germany (Angles, Saxons, and Jutes) settled in (present-day) England. Due to a period of extended isolation from other German speakers, a non-mutually intelligible dialect eventually developed.

What is of importance here is that there is a (fuzzily) nested set of ways of speaking which, at one extreme of granularity, includes language families such as Germanic or Romance and, at the other end, includes idiolect. In between, we find languages (*e.g.*, English, German) and dialects (*e.g.*, Midwestern American English), with no clear-cut distinction between these two. In this paper we look at different size groupings of linguistic varieties within the English language.[1]

There have been several previous attempts at categorization of dialects. [5] describes varieties of American English in terms of lexicon and [6] does so in terms of phonology. [7] and [8] describe the dialects of British English. The aforementioned do not attempt quantified categorization. Recently, there have been sophisticated quantitative analyses of English dialect data [9], and other languages (Dutch, Norwegian, Chinese) [10-14], including some cluster analyses. None of these, however, consider the interrelationship of the phoneme variants across dialects.

3 Methods: Data Collection and Organization

Our database is a side product of a major publication project: *A Handbook of Varieties of English* [15] which describes the pronunciation variants of English in a great

---

[1] "Linguistic variety" is a cover term for idiolects, dialects, and languages.

many varieties (national, regional and ethnic dialects) from around the globe (see list in [16]). The database consists of a spreadsheet with possible pronunciation variants as rows, language varieties as columns, and information on whether or not the respective variant occurs in a given variety as cell entries (see partial example in Table 2).

For publication, the pronunciation variation needed to be systematized in order to produce categorical displays of certain phenomena as realized in specific varieties. To allow tabular and cartographic representations of the essentially infinite pronunciation variability world-wide, E.W. Schneider devised a scheme of distinct descriptive categories. One difficulty in this process was that the range of possible variants was not fully known in advance; another was to decide how finely to sub-categorize in order to remain both informative and descriptively adequate as well as manageable. Schneider set up a listing of 179 features of pronunciation (vowel, consonant and prosodic features) intended to represent the entire range of possible variants, each of which may or may not be used in each of the varieties under consideration (see Table 1).

The list of *vowel features* builds upon the "lexical sets" devised by [17], a system of distinct vowel types identified by certain key words (*e.g.* TRAP for the vowel in *cat* and *bad*; FACE for the vowel in *rain* or *gate*). 28 different lexical sets are considered, and for each of these 2-7 different realization possibilities (variants) are suggested by specifying articulatory features and International Phonetic Alphabet characters. For example, in features 1-4, possible variants of the vowel of KIT are identified as (1) "canonical" high front [ɪ]; (2) raised and fronted variant phonetically identified by the symbol [i], (3) centralized [ə], and (4) with an offglide, *e.g.* [ɪə/iə]. Thus, the 121 vowel features can be grouped together in 28 coherent sets of alternative realizations. At least one of these variants should apply to each of the language varieties under consideration. However, the variants need not be mutually exclusive: in many communities the degree of variability is high and more than one variant may occur. The *vowel distribution features* relate to so-called mergers, *i.e.*, the fact that certain vowel types sound alike (so, for instance, feature 131 applies if there is homophony between the vowels of LOT and STRUT). The *consonant features* include a tendency to delete word-initial *h-* (*'eart* 'heart'), or the rhotic realization of postvocalic /r/ ([bɑɹn] vs. [bɑːn] 'barn'). The last group includes *prosodic features*, like the deletion of word-initial unstressed syllables (*e.g.* *'bout, 'cept*) or the "high-rising terminal" contour, a tendency to raise one's pitch at the end of declarative statements.

Table 1. Summary of phonological data

Feature type	# features	# variants	Geographic distribution	
vowel	28	121	British Isles	9
vowel distributions	4	4	Pacific & Australia	10
consonants	32	38	Africa & Asia	19
prosody	5	5	Americas & Caribbean	21
(omitted--redundant	11)		TOTAL	59
TOTAL	69	168		

The authors of the Handbook chapters were asked to fill out the list of pronunciation variants for their respective regions, *i.e.*, to specify for each feature whether or not it occurs. To achieve a roughly even coverage of varieties, the regional editors filled in the feature lists as necessary. Altogether, the columns of the database used here represent 59 distinct varieties of English, divided into four major world regions. (See Table 1.)

In each of the 10,561 feature-by-variety cells, one of three codes originally appeared indicating that in the respective form of English, the respective feature is used (A) regularly, (B) in specific circumstances, or (C) not at all. For the present statistical analysis, binary features are used. "1" indicates that the variant is used regularly (originally A) while "0" indicates that it is used either sometimes (B) or never (C).

4 Methods: Clustering and Mutual Information

The completed data sheets described above were transformed into a binary observation array W, where each element w_{ij} corresponds to a variant of a phonological feature for variety V_i. There are 69 phonological features F_i (See Table 1), with 2-7 variants or possible values per feature. Thus, each binary feature vector $\mathbf{w}_i$ has 168 elements. Varieties with 1's in the same column of the array pronounce a given word in the same way, therefore an appropriate measure of the similarity of two varieties V_i and V_j is the number of 1's in the logical AND of their feature vectors, normalized by the product of their lengths. (See Table 3 below.) Then, the *dissimilarity* ρ_{ij} between two varieties is

$$\rho_{ij} = 1 - |\mathbf{w}_i \wedge \mathbf{w}_j|/|\mathbf{w}_i| |\mathbf{w}_j| = 1 - \cos(\mathbf{w}_i \mathbf{w}_j) . \qquad (1)$$

Our starting point for grouping varieties to form *dialect clusters* is a 59×59 element *dissimilarity matrix M*. We note that there is no general way of determining, from a similarity or dissimilarity matrix or from an array of feature vectors, how many clusters there are in a data set. Clustering, or "unsupervised learning," requires some external information, such as the maximum acceptable distance between patterns in the same cluster, or the minimum distance between patterns from different clusters, or the minimum or maximum number of patterns in a cluster. Clustering may be *hierarchical* or *flat*, *agglomerative* or *divisive*, and *crisp* (mutually exclusive clusters) or *fuzzy* (with a continuous cluster membership function). Dozens of clustering algorithms have been developed and applied [18-21]. Objects characterized by a similarity matrix can be transformed into a vector space by multi-dimensional scaling; conversely, the similarity of feature vectors can be obtained from their pairwise distance [22]. Current research focuses on *clustering ensembles, i.e.,* on combining the results of diverse clustering algorithms [23, 24], and on related algorithms for probabilistic Expectation Maximization [25, 26].

We performed clustering with the Complete Link Algorithm (hierarchical, agglomerative, and crisp), which can be found in many statistical data analysis packages [20]. At any given threshold, the Complete Link Algorithm forms clusters such that the maximum dissimilarity between any two varieties in the cluster is less than θ. Clusters are merged when the maximum dissimilarity between a variety in one cluster and a variety in the other cluster is less than θ. The resulting clusters are mutually exclusive, and completely exhaustive: at any given threshold, every variety belongs to exactly one cluster.

Initially, each variety is a distinct dialect cluster (an idiolect). The threshold is increased from 0 to 1 to decrease the number of clusters. At 1 (or at any value of the threshold greater than the dissimilarity of the least similar pair of varieties), all the varieties are merged into a single dialect (the English language). A simple example is given in [16].

The amount of context at level θ of the hierarchy is given by the average MI between pairs of features. This measure is based on the marginal and joint probabilities of the features within a cluster. It is equal to the relative entropy between the two distributions: it indicates how much each distribution reveals about the other. MI can represent non-linear statistical dependence, unlike the correlation coefficient. Its formula is:

$$I_{x,y} = H(x) - H(x|y) = H(y) - H(y|x) = \sum_{x,y} p(x,y) \log_2 \frac{p(x,y)}{p(x)p(y)} \quad (2)$$

where $p(x,y)$ is the joint probability distribution of features x and y, and $p(x)$, $p(y)$ are their marginal distributions. $H(x)$ and $H(y)$ are marginal entropies, and $H(x|y)$ is the conditional entropy. To illustrate, Table 2 shows the feature frequencies in a dialect cluster of 13 varieties for two phonemes. The first phoneme has three allophones, the second has two.

The Mutual Information $I_k(j,l)$ for a pair of phonological features F_j and F_l over all varieties in dialect cluster k at level K is

$$I_k(j,l) = \sum_m \sum_n p(F_{j,m} F_{l,n} | V_i \in C_k) \log_2 \frac{p(F_{j,m} F_{l,n} | V_i \in C_k)}{p(F_{j,m} | V_i \in C_k) p(F_{l,n} | V_i \in C_k)} \quad (3)$$

where $F_{j,m}$ is the mth variant of the jth feature of variety V_i in dialect cluster C_k.

Table 2. Feature frequencies for two words in 13 dialects of English

VARIETY	KIT			DRESS	
	raised	central	back	raised	central
Orkney & Shetland		1			1
North of England	1				1
East Anglia	1			1	
Philadelphia	1			1	
Newfoundland			1		1
Cajun English	1			1	
Jamaican Creole		1			1
Tobago Basilect	1				1
Australian Eng.	1			1	
Tok Pisin		1			1
Fiji English			1		1
Nigerian Pidgin	1				1
Indian S. African Eng.		1			1
Total	7	4	2	4	9

Table 3. Calculations of joint and marginal frequencies for two words in 13 dialects of English

			KIT		
			back	central	raised
			0.15	0.31	0.54
DRESS	central	0.69	0.15	0.31	0.23
	raised	0.31	0.00	0.00	0.31
		$I(x_i, y_j)=$	0.08	0.16	-0.16
			0.00	0.00	0.27

Table 3 shows the joint frequency ($p(F_{j,m} F_{l,n}|V_i \in C_k)$) and marginal frequencies ($p(F_{j,m}|V_i \in C_k)$ and $p(F_{l,n}|V_i \in C_k)$) of the two features. The six individual components of MI are shown below: they sum to 0.35.[2]

Since many previous dialectology studies focused on vowel features [27], we also calculated MI separately for the 28 vowel features and the 32 consonant features.

5 Results: Clustering

The dendrogram and tables in this section show the results of clustering at various levels. We show that this method, using only internal features, constructs clusters that are very similar to those that have been constructed by more traditional dialectology approaches, using both internal and external features. Fig. 1 shows the clusters achieved with varying thresholds, using all features. A line marks K=10. Table 4 lists the dialects in each cluster, and provides the thresholds (θ) at which the dialects fall into 10 clusters. (See [16] for dialect names.) Similar clusters were achieved using the subsets of only vowels and then only consonants, and at K=20 and K=30.

The resulting clusters are remarkably meaningful and homogeneous in a linguistic perspective. Based upon the analysis of all features, at K=10, clusters 1 through 5 are extremely tight-knit, and the following two clusters, while less obvious, also allow meaningful interpretation. Cluster 1, the biggest, comprises the Pacific contact varieties on the one hand (Bislama, Tok Pisin, Solomons Pijin, Hawaiian Creole, Fijian English) and a strong cohort of African pidgins and contact Englishes (of Nigeria, Ghana, Cameroon, East Africa, and black South Africa) on the other. Singaporean and Malaysian English, also strongly contact-shaped, also occur in this group. Cluster 2 unites the oldest colonial offspring of British English: Irish English and most varieties of American English (New England, New York, Philadelphia, Inland Northern, Western), including Canadian English and two American contact dialects, African American Vernacular English (AAVE) and Chicano English. Cluster 3 combines the Caribbean Creoles of Jamaica, Barbados, Trinidad and Tobago, as well as their closest kin in North America, Gullah; and in Britain, British Creole. In addition, a few non-contiguous contact dialects can be found in this cluster: Australian Aboriginal English, northern Nigerian English, and Indian English. Cluster 4 groups the so-called "southern hemisphere" dialects, namely Australian and New Zealand English, Maori English, White So. African English, and Cape Flats English.

---

[2] $I_{DRESS,KIT}=0.35 < H(x) = 0.89 < log_2 2 = 1.00; H(y) = 1.41 < log_2 3 = 1.58$.

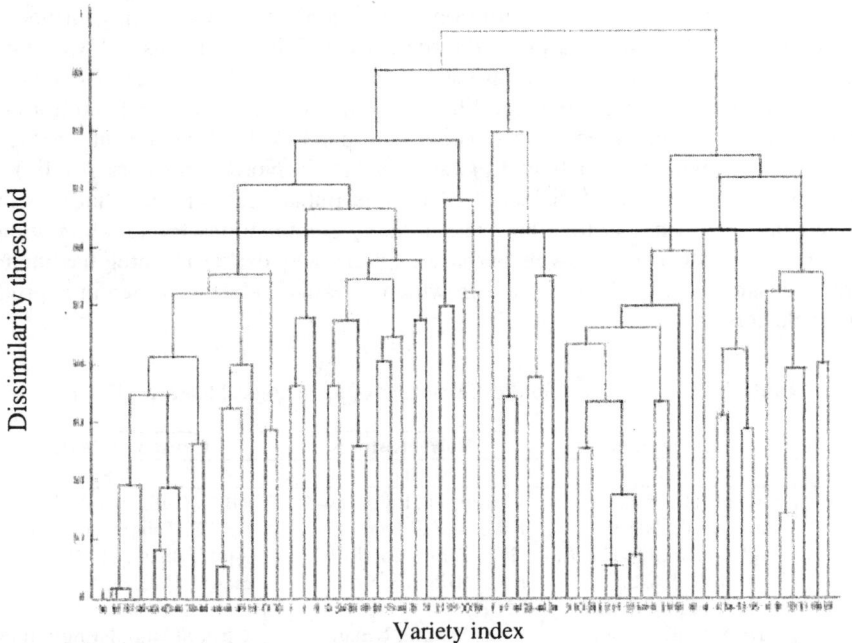

Fig. 1. Dendrogram of Complete Link Analysis for all features (horizontal line at K = 10)

Interestingly, it combines these with only one other dialect which has frequently been suspected to be a major donor of these colonial varieties, the dialect of East Anglia. Cluster 5 unites dialects with a historical or physical connection to the American South: urban and rural Southern, Cajun, Bahamian, and Liberian Settler English (the variety spoken by descendants of repatriated American slaves). Cluster 7, with Wales, Scotland, and Orkney and Shetland, unites some non-central dialects of Britain. Interestingly enough, RP (standard non-regional British English) stands on its own.

The clustering based on vowel features only is very similar in many respects — not surprisingly, given that this subset covers the majority of the features. In comparison with the "all features" categorization, Cluster 1 is identical. Cluster 2 brings out the unity of American English even more strongly, combining almost all American dialects including Southern. It is noteworthy that Irish English groups with the American dialects in all analyses. The only exception (also meaningful, as this is the only English-based creole on the North American mainland) is Gullah, which still groups with the Caribbean Creoles and other contact varieties, a group which now includes the North of England. One unexpected outcome is that Trinidad and Tobago join the Scottish-Welsh cluster.

Looking at consonants only, the resulting patterns are somewhat different. Cluster 1 unites the Caribbean creoles on the one hand with the Pacific pidgins on the other. Cluster 2 combines long-standing first language colonial varieties: Ireland, most dialects of American English (from Canada to the South, from New England to the West), and the antipodean varieties. Cluster 3 consists mostly of varieties of English as a Second Language with a strong focus on all parts of Africa, but including also

Singapore and, not fitting this description, Newfoundland. Cluster 4 has further African varieties, in addition to, surprisingly, the Channel Islands. In cluster 5 we find two pairs which are geographically relatively coherent internally but surprising in their mutual combination: the Orkney and Shetlands and Wales on the one hand, India and Pakistan on the other. Three more clusters with pairs of varieties are interesting linguistically: one with two distinct English dialects (7: North, and East Anglia), one with two historically related dialects (AAVE and Bahamian), and one that puts Scottish and Jamaican English together. The unlikely combinations found when comparing consonants rather than vowels perhaps explains why earlier research has included more discussion of vowels, a context in which external and external features produce similar clusters.

Table 4. Dialect clusters for 3 different sets of phonological features (K = 10)

K=10	All features	Vowel features	Consonant features
θ	0.63	0.6	0.8
1	{Bislm, TP, Pijin, HawC, FijE, NigES, NigP, GhE, GhP, CamE, CamPE/K, EAfE, BlSAfE, SgE, MalE}	{Bislm, TP, Pijin, HawC, FijE, NigES, NigP, GhE, GhP, CamE, CamPE/K, EAfE, BlSAfE, SgE, MalE}	{BrC, GulhE, CajE, JamC, T&TC, TobgB, SurC, AbE, AusC, Bislm, TP, Pijin, HawC, FijE, SgE}
2	{IrE, StAmE, NEngE, InlNE, NYCE, PhilE, WMwE, CanE, NfldE, AAVE, ChcE}	{IrE, StAmE, NEngE, InlNE, NYCE, PhilE, SE, UrbS, WMwE, NfldE, AAVE, CajE, ChcE, BahE, LibSE}	{IrE, StAmE, NEngE, InlNE, NYCE, PhilE, SE, UrbS, WMwE, CanE, ChcE, NZE, MaoE, AusE}
3	{BrC, GulhE, JamE, BarbE, T&TC, TobgB, AbE, NigEN, IndE}	{NE, BrC, GulhE, JamE, BarbE, AbE, NigEN, IndE}	{NfldE, NigES, NigP, LibSE, CamE, CamPE/K, EAfE, BlSAfE, IndSAfE, StHE, MalE}
4	{EA, NZE, MaoE, AusE, WhSAfE, CFE}	{Chanl, WhSAfE, IndSAfE, CFE, StHE, PakE}	{Chanl, NigEN, GhE, GhP, WhSAfE, CFE}
5	{SE, UrbS, CajE, BahE, LibSE}	{OrkS, ScE, WelE, T&TC, TobgB}	{OrkS, WelE, IndE, PakE}
6	{NE, Chanl, IndSAfE, StHE, PakE}	{EA, NZE, MaoE, AusE}	{ScE, JamE}
7	{OrkS, ScE, WelE}	{JamC, AusC}	{NE, EA}
8	{JamC, AusC}	{SurC, PhlE}	{RP, PhlE}
9	{SurC, PhlE}	{CanE}	{AAVE, BahE}
10	{RP}	{RP}	{BarbE}

6 Results: Mutual Information

While the clustering results illustrate the degree of consistency among dialects, MI shows, whenever there is variation across two dialects, how dependent the dialects are on each other. MI can be seen as an additional type of measure, besides similarity, that is valuable in distinguishing dialects.

Table 5. MI for 4 tense and 4 lax vowels, all dialects

F1 \ F2	lax vowels				tense vowels			
	KIT	DRESS	FOOT	THOUGHT	FLEECE	FACE	GOAT	GOOSE
KIT	2	0.41	0.58	0.33	0.52	0.61	0.69	0.51
DRESS		1.48	0.13	0.30	0.24	0.3	0.40	0.32
FOOT			1.4	0.28	0.48	0.58	0.53	0.29
THOUGHT				1.41	0.24	0.44	0.41	0.56
FLEECE					1.53	0.57	0.68	0.42
FACE						2.24	1.30	0.58
GOAT							2.33	0.57
GOOSE								1.56

Table 5 lists the amount of MI between each pair of phonemes in a subset of 8 features (4 tense and 4 lax vowels), with all dialects together in one cluster. Auto-comparisons are shaded. The 3 highest values are outlined—interestingly, all involve the GOAT vowel. These dependencies are not, to our knowledge, discussed in the dialectology literature. More generally, there is a degree of MI across every pair—any word recognition/ production application would be improved by including MI in its calculations.

Table 6 shows both clustering and MI results. This table considers the same 8 words as Table 5, but was calculated for K=10. Only the 6 clusters containing 5 or more dialects are shown. Again auto-comparisons are shaded. The 4 outlined cells illustrate the value of combining clustering and MI: these values are all greater within their clusters than for the 59 dialects as a whole (where MI=0.41). Thus, applications such as voice recognition systems would be improved by individually trained classifiers for each dialect cluster. The value of MI is affected by the number of values that occur per feature. This depends both on the selected pair of features, and on the varieties included in a cluster. Note that a really tight cluster would necessarily have low MI values—whenever the dialects share the same features, the variation in features cannot be used to predict patterns.

Table 6 shows that MI provides information useful in predicting pronunciation patterns—there are *no* cases of completely independent variation. Again, we see the value of including MI in speech recognition applications. The 0 values indicate a complete lack of variation among the dialects in that cluster for that vowel pair. That is, if there is complete predictability for one of the words, then knowing about the other cannot improve our predictions of the first. Aside from these cases of 0's, including MI would always improve performance. This finding is in keeping with what has been shown for MI as applied to handprinting recognition [4].

Finally, in an effort to determine the extent to which a subset of the phonemic features determines the dialect, and the remaining features, we automatically classified the varieties into clusters using a Nearest Neighbor classifier algorithm and a leave-one-out design for partitioning the samples [22]. We computed the error rate of misclassifying the dialect cluster of the variety, given the remaining varieties of that dialect cluster, at various threshold levels. With K=3, the dialects were classified with 6 errors, an error rate of 0.10.

Table 6. MI for 4 tense and 4 lax vowels, for 10 dialect clusters

K = 10, θ = 0.63	{Bislm TP Pijin HawC FijE NigES NigP GhE GhP CamE CamPE/K EAfE BlSAfE SgE MaleE}	{IrE StAmE NEngE InlNE NYCE PhilE WMwE CanE NfldE AAVE ChcE}	{BrC GulhE JamE BarbE T&TC TobgB AbE NigEN IndE}	{EA NZE MaoE AusE WhSAfE CFE}	{SE UrbS CajE BahE LibSE}	{NE ChanI IndSAfE StHE PakE}
KIT, KIT	1.16	0	0.92	1.92	1.37	1.37
KIT, DRESS	0.57	0	0.07	0.92	0.72	0.97
KIT, FOOT	0	0	0.25	0	0.17	0
KIT, THOUGHT	0	0	0.31	0	0.82	0
KIT, FLEECE	0.47	0	0.46	0.58	0.82	0
KIT, FACE	0.09	0	0.46	0.79	0.97	0.42
KIT, GOAT	0.04	0	0.46	1.58	1.37	0.97
KIT, GOOSE	0.13	0	0.20	0.32	1.37	0
DRESS, DRESS	1.55	0.44	0.50	1.25	0.72	1.52
DRESS, FOOT	0	0.01	0.04	0	0.07	0
DRESS, THOUGHT	0	0.11	0.5	0	0.72	0
DRESS, FLEECE	0.24	0.44	0.04	0.71	0.72	0
DRESS, FACE	0.11	0.01	0.04	0.46	0.32	0.17
DRESS, GOAT	0.05	0.03	0.04	0.92	0.72	1.12
DRESS, GOOSE	0.13	0.26	0.02	0.11	0.72	0
FOOT, FOOT	0	0.44	0.99	0	0.72	0
FOOT, THOUGHT	0	0.11	0.55	0	0.17	0
FOOT, FLEECE	0	0.01	0.53	0	0.72	0
FOOT, FACE	0	0.01	0.53	0	0.32	0
FOOT, GOAT	0	0.03	0.53	0	0.32	0
FOOT, GOOSE	0	0.06	0.50	0	0.17	0
THOUGHT, THOUGHT	0	1.68	1.45	0	1.37	0
THOUGHT, FLEECE	0	0.11	0.55	0	0.82	0
THOUGHT, FACE	0	0.11	0.55	0	0.57	0
THOUGHT, GOAT	0	0.24	0.55	0	0.97	0
THOUGHT, GOOSE	0	0.80	0.50	0	0.82	0
FLEECE, FLEECE	1.05	0.44	0.99	1.25	1.37	0
FLEECE, FACE	0.38	0.01	0.99	0.46	0.57	0
FLEECE, GOAT	0.03	0.03	0.99	0.92	0.97	0
FLEECE, GOOSE	0.10	0.26	0.50	0.11	0.82	0
FACE, FACE	0.70	0.44	0.99	1.46	1.52	0.97
FACE, GOAT	0.01	0.03	0.99	0.79	1.52	0.97
FACE, GOOSE	0.05	0.06	0.50	0.65	0.97	0
GOAT, GOAT	0.35	0.87	0.99	1.92	1.92	1.92
GOAT, GOOSE	0.02	0.49	0.50	0.32	1.37	0
GOOSE, GOOSE	0.72	1.49	0.50	0.65	1.37	0

7 Applications and Future Work

We have examined the phonological correlates of English dialects from the orthogonal perspectives of consistency (clustering) and context (MI). Hierarchical clustering organizes dialects with similar pronunciations. MI, on the other hand, reveals statistical dependence between alternative pronunciations of pairs of vowels within the same dialect cluster. This second aspect is novel. Its value must be assessed by further investigation: dialects are not traditionally characterized by their phonological context. Given access to appropriate data, perhaps from [10-12], we could test the method with other languages.

Ideally we would test these methods at all levels of the continuum from idiolect to language. The necessary data would include descriptions of many idiolects for each dialect, just as we have many dialects for the one language considered here. Once such a classification is obtained, we would be able to predict, for a partially unanalyzed dialect, what features it will exhibit based on knowledge of some subset of features that it does exhibit. This could be applied to speaker identification by permitting a stochastic description of a speaker's full dialect based on a sample which contains only a subset of the phonemes.

Phonological context may also find practical application in automated speech recognition (ASR). This technology has made good progress since the first attempts in the 1960s to recognize "yes" vs. "no" for accepting or declining a collect call. ASR has been deployed for telephone trees, directory assistance, and queries for stockmarket prices. Other restricted-vocabulary dialogs, for airline reservations and for hands-free operations like stock inventory and non-critical vehicular applications (radio, seat adjustment, cell-phone dialing), have also been developed. Large-vocabulary trainable dictation systems have been available for several years. In most of these applications, recognition accuracy could be raised by exploiting both the consistency and the statistical dependences in the pronunciation of speakers of a given dialect cluster.

One caveat is that this will be useful only if it can be verified from acoustic waveforms that most of the speakers of a variety actually pronounce the words in the ways that have been described, and if that can be reliably detected *automatically*. Multimodal Hidden Markov Models, widely used in speech recognition [28], would provide the appropriate framework for continuing this work with automated phonological characterization. Further interdisciplinary studies could render differences between dialects an advantage, rather than a detriment, to ASR.

References

1. Fetzer, A. *Recontextualizing context: Grammaticality meets appropriateness*. 2004. Benjamins: Philadelphia.
2. Giunchiglia, F. and P. Bouquet, Introduction to contextual reasoning. An Artificial Intelligence Perspective, in *Perspectives on Cognitive Science 3*, B. Kokinov, Ed. 1997, NBU Press: Sofia, Bulgaria.
3. Sarkar, P. and G. Nagy, Style consistent classification of isogenous patterns. *IEEE Trans. Pattern Analysis and Machine Intelligence*, 2005. 27(1):88-98.

4. Veeramachaneni, S. and G. Nagy, Style context with second order statistics. *IEEE Trans. Pattern Analysis and Machine Intelligence*, 2005. 27(1):14-22.
5. Carver, C.M. *American Regional Dialects: A Word Geography*. 1987. University of Michigan Press: Ann Arbor.
6. Labov, W., S. Ash, and C. Boberg. *Atlas of North American English*. 2005. Mouton de Gruyter: Paris.
7. Hughes, A. and P. Trudgill. *English Accents and Dialects: An Introduction to Social and Regional Varieties of British English*. 1987. Edward Arnold: London.
8. Trudgill, P. *The Dialects of England*. 1999. Blackwell: London.
9. Nerbonne, J. and P. Kleiweg, Lexical distance in LAMSAS. *Computers and the Humanities*, 2003. 37(3):339-57.
10. Gooskens, C. and W. Heeringa, Perceptive evaluation of Levenshtein dialect distance measurements using Norwegian dialect data. *Language Variation and Change*, 2004. 16(3):189-207.
11. Cheng, C.-C., Measuring Relationship among Dialects: DOC [Dictionary on computer] and Related Resources. *Computational Linguistics and Chinese Language Processing*, 1997. 2(1):41-72.
12. Heeringa, W. and A. Braun, The Use of the Almeida-Braun System in the Measurement of Dutch Dialect Distances. *Computers and the Humanities*, 2003. 37(3):257–71.
13. Heeringa, W., *Measuring dialect pronunciation differences using Levenshtein distance*. 2004, University of Groningen: Groningen.
14. Heggarty, P.A. *Measured Language: From First Principles to New Techniques for Putting Numbers on Language Similarity*. in prep. Blackwell: Oxford.
15. Schneider, E.W., et al., eds. *A Handbook of Varieties of English: A Multimedia Reference Tool*. 2005, Mouton de Gruyter: Berlin, New York.
16. Nagy, N., *Addenda to "Categorization of phonemic dialect features in context"*. http://pubpages.unh.edu/~ngn/papers/Context05/CONTEXT05_addenda. 2005.
17. Wells, J.C., ed. *Accents of English*. 1982, Cambridge University Press: Cambridge.
18. Kaufman, L. and P.J. Rousseeuw. *Finding Groups in Data: An Introduction to Cluster Analysis*. 1980. Wiley: Hoboken, NJ.
19. Day, W.H.E. and H. Edelsbrunner, Efficient algorithms for agglomerative hierarchical clustering methods. *Journal of Classification*, 1984. 1(1):7-24.
20. Jain, A.K. and R.C. Dubes. *Algorithms for Clustering Data*. 1988. Prentice Hall.
21. Theodoridis, S. and K. Koutroumbas. *Pattern Recognition*. 1999. Academic: NY.
22. Duda, R.O., P.E. Hart, and D.G. Stork. *Pattern Classification*. 2001. Wiley-Interscience: Hoboken, NJ.
23. Topchy, A., et al. Adaptive Clustering Ensembles. *Proc. ICPR*. 2004. Cambridge.
24. Jain, A.K., et al. Landscape of Clustering Algorithms. *Proc. ICPR*. 2004. Cambridge.
25. Redner, R.A. and H.F. Walker, Mixture densities, maximum likelihood, and the EM algorithm. *SIAM Review*, 1984. 26(2):195-235.
26. Topchy, A., A.K. Jain, and W. Punch. A Mixture Model for Clustering Ensembles. in *Proc. SIAM International Conference on Data Mining (SDM04)*. 2004. Florida.
27. Foulkes, P., Current trends in British sociophonetics. *Univ. of PA Working Papers in Linguistics: A Selection of Papers from NWAV 30*, 2002. 8(3):75-86.
28. Rabiner, L.R. and B.H. Juang. *Fundamentals of Speech Recognition*. 1993. Prenctice Hall: Englewood Cliffs, NJ.

Context-Oriented Image Retrieval

Dympna O'Sullivan[1], Eoin McLoughlin[1], Michela Bertolotto[1],
and David Wilson[2,*]

[1] Department of Computer Science, University College Dublin,
Belfield, Dublin 4, Ireland
[2] Department of Software and Information Systems,
University of North Carolina at Charlotte,
NC 28223-0001, USA
{dymphna.osullivan, eoin.a.mcloughlin, michela.bertolotto}@ucd.ie,
davils@uncc.edu

Abstract. In order to help address problems of information overload in digital imagery task domains, we have developed an interactive approach to the capture and reuse of image context information. Our framework models different aspects of the relationship between images and the domain tasks that they support by monitoring the interactive manipulation and annotation of task-relevant imagery. In particular, a strong focus on task context serves to ground image annotations in domain specific goals. This contrasts with prevalent annotation schemes that focus on what individual images contain but that provide no context for which, if any, of those aspects are important to users. Our work attempts to leverage a measure of the user's intentions with regard to tasks that they address. We analyze human-computer interaction information that enables us to infer why image contents are important in a particular context and how specific images have been used to address particular domain goals.

1 Introduction

Continuing advances in techniques for digital image capture and storage have given rise to a significant problem of information overload in imagery task domains (e.g., intelligence analysis). As a result it has become necessary to develop intelligent application support to help manage imagery tasks. The majority of current retrieval techniques retrieve images by similarity of appearance, using low-level features such as color and shape or by natural language textual querying where similarity is determined by comparing words in a query against words in semantic image metadata tags. The biggest problem arising from these techniques is the so-called semantic gap [1] — the mismatch between the capabilities of the retrieval system and user needs.

---

* The support of the Proof of Concept Fund of Enterprise Ireland is gratefully acknowledged.

In order to bridge this gap we have developed an application that can unite information about underlying visual data with more high-level concepts provided by users as they complete specified tasks. Capturing human expertise and proficiency allows us to understand why relevant information was selected and also how it was employed in the context of a specific user task. We use this context-specific information to facilitate communication between a user and the system. The approach allows us to capture and reuse best practice techniques by automatically constructing a knowledge base of previous user experiences (referred to as user sessions) using case-based reasoning techniques. This knowledge base can be exploited to improve future context-based query processing by retrieving and reusing similar experiences as similar tasks are no longer simply interpreted as a collection of natural language terms. Rather they are grounded in the context of the specific user domain goals. This approach has two main benefits. The first is that it allows for a dramatic reduction in both the time and effort required to carry out new tasks as amassed contextual knowledge is reused in support of the similar tasks. Secondly, there are benefits from a knowledge management standpoint in that contextual knowledge pertaining to particular tasks may now be stored and reused as an additional resource for support, training and preserving organizational knowledge assets.

Our approach incorporates task-based querying for image and session retrieval and image manipulation, as well as annotation tools that the user can employ to emphasize and organize relevant aspects of task-relevant imagery. Based on the architecture we have developed for users to provide information as they complete tasks, we have identified three essential notions of context that we use to perform similarity matching between the current context and previous user experiences.

The first of these is an annotational context where we capture each individual annotation applied by a user as they work with a set of imagery. This allows us to determine which individual image components are most comparable to the current task context, which facilitates individual image relevance (those with the greatest number of relevant annotations). It also enables selection of the most relevant pieces of information to form a visual summary for the user when interacting with similar user sessions.

The second idea of context the system captures is an image context, where for each image annotated as part of a user's task, we capture all entered metadata and task descriptions for the image as well as all applied annotations. By capturing the overall relationship between individual images and the current task context we can present contextually relevant individual images to users as a ranked list of task-specific information.

The final notion of context extracted by the application is a session context. For each user session in the knowledge base the system captures all metadata and task description queries, annotated and browsed session images and added contextual user annotations for the set of imagery employed for the duration of the user task. The acquisition of all of the contextual knowledge for each session allows us to make inferences about the high-level concepts that the user has in mind when carrying out their task. We can then present this in an intelligent manner to the current user so that they may exploit both the image interactions and the reasoning process of a previous user.

In addition to the ideas of image context, we maintain the notion of user context for interaction. This includes the current user goal, which arises from user inputs (e.g., captured queries) and captured user interaction information (e.g., queries, selected imagery and new annotations).

We have achieved the goal of capturing important contextual information by situating intelligent support for gathering it inside a flexible task environment. It is considered imperative that the information is collected implicitly to shield the users from the burden of explicit knowledge engineering. Collecting contextual information by observing and interpreting user actions (as they proceed with domain specific tasks) allows correlations and associations between complex imagery to be made. From the users' perspective, the tools developed for the image interaction environment support them in carrying out their task (e.g., producing a report) by facilitating operations such as selecting and highlighting relevant features, storing insights and summarizing aspects of their work progress. However, from a system perspective the tools are employed to monitor and record user actions and ultimately to capture fine-grained task knowledge that improves the ability of the application to make pro-active context-based recommendations to similar users.

This paper describes how our environment for image interaction incorporates various notions of image and task context and how that information is captured and leveraged to support just-in-time access to task-relevant imagery. It provides an evaluation that demonstrates the usefulness of a task context focus. The paper begins with a brief discussion of related research in Section 2, and it continues with a description of our contextual framework and task-based image retrieval facility in Section 3. Section 4 explains how we combine image interaction information with more high-level user concepts to retrieve complete task-based user sessions. In section 5 we describe the annotation tools available for image manipulation and capturing contextual task information. Section 6 outlines an evaluation of the system. We conclude with a discussion of future work.

2 Related Research

In this research we are working with large collections of experience and user context. As in [2], we believe that user interactions with everyday productivity applications provide rich contextual information that can be leveraged to support access to task-relevant information. As part of our work we aim to shield our users from the burden of making explicit queries. All contextual knowledge is gathered by the system using implicit analysis so that users are shielded from the burden of relevance feedback or other such explicit interest indicators [3, 4]. We observe how users proceed with their task, recording this as user context. By situating intelligent tools and support within task environments (e.g., [5, 6]), we can unobtrusively monitor, interpret and respond to user actions concerned with rich task domains based on a relatively constrained task environment model. We predict user goals by comparing their current context to that of a similar previous user.

We have drawn from work that extracts context relevant information during document browsing to support users in fulfilling tasks [7]. Examples of such systems include ARIA [8], an email agent that suggests images relevant to the user's text as they type a message and Mobile Media Metadata (MMM) [9] which is an application that infers social, temporal and spatial context for mobile phone users. It does not however consider task context. Furthermore, we have considered research into techniques for developing intelligent interfaces that provide users with proactive support and assistance [10].

Our methods for annotating multimedia are related to annotation for the semantic web [11] and multimedia indexing [12] where the focus is on developing annotated descriptions of media content. Multimedia database approaches such as QBIC [13] provide for image annotation but use the annotations to contextualize individual images. In this work we are concerned with a task-centric view of the annotations, where we employ annotations to tell us how an image relates to a current domain task by using image annotations to contextualize task experiences.

3 Task-Based Image Retrieval

Image storage and retrieval systems are typically characterized by one of two main approaches; they either support keyword-based annotations and indexing or a content-based approach where the retrieval of images is on the basis of features automatically extracted from the images themselves. It is, however, recognised that neither of these approaches are fully adequate for answering the complete range of user search questions. The keyword-based approach is a human-centric approach that depends on images being accompanied by textual descriptions. The indexes for such large image collections are time consuming to create and maintain, particularly since entries are not grounded in how the collections are being used (which exacerbates indexing subjectivity). Also, keyword indexing for images only provides hit-or-miss type searching as the range of successful queries is limited to the interpretation of the indexer.

Content-based approaches focus on the automatic extraction of low-level visual features. This computer-centric approach suffers from several disadvantages. Results can frequently be poor due to the semantic gap and the subjectivity of human perception. The first is the difference between the high-level concepts that users search for and the low-level features actually employed to retrieve the imagery. The latter addresses the fact that different people or the same person in different situations may judge visual content differently. Other disadvantages include complex computations that can lead to unacceptable response times and the need for complicated interfaces where average users can be bewildered by requirements to select features and weights for the query process.

In order to overcome some of these shortcomings we have developed a task-centric approach to image retrieval. We believe that in order to resolve the mismatch between user queries and image retrieval techniques we must have an understanding of how and why the user intends to employ the retrieved imagery. If image retrieval is employed as part of a greater task or goal we can capture other contextual information around the retrieval request that allows us to place the imagery in the context of the specific user need.

For example when a user employs the system we have developed to retrieve imagery corresponding to their current task needs they can structure their query so that it may consist of any combination of specified metadata, semantic task descriptions and a sketched configuration of image-objects. The system captures the user query and processes it, placing it in the context of previously entered user tasks. Our retrieval metrics are currently focused on textual metadata, queries and annotations. In performing similarity matching between the current context and previous user contexts the system employs three separate indices:

- Annotation Index, where the current context is compared to all applied contextual task-based annotations in the dataset
- Image Index, where the current context is compared to all the task-related information for each individual image in the dataset (metadata and task descriptions queries and added contextual annotations for each image)
- Session Index, where the current context is compared to all of the contextual information for each user session in the knowledge base (metadata and task descriptions queries, annotated and browsed session images and added contextual user annotations).

For a full description of how these indices are calculated and combined please refer to [14]. Returning to the system, say for example a user planning a holiday may be interested in retrieving imagery of their planned destination to help them organise their vacation. The user could outline a metadata query specifying the location that they are interested in visiting and a task description describing the kind of holiday activities they are interested in, the type of accommodation they are searching for and the time of year they would like to visit. For example a user interested in a holiday in France may have a task context such as *"Interested in a spring break in Paris, France. I would like to spend a few days exploring the city's famous sights and the museums as well as sampling the food and wine. I would like to stay in a hotel in central location"*. This current context (in this case the textual metadata and user query) is captured and compared against a combination of the annotation and image indexes for each image in the dataset. The results are returned to the user as a ranked list of matching imagery.

Figure 1 shows an example of the interface for retrieved imagery. In order to allow multiple candidates to be displayed we present the matching images as a ranked list of thumbnails with their associated matching percentage score. A subset of the most relevant metadata for each image is available as a tool tip text when mousing over the image. The user can browse the images in the results screen and select any images that are relevant to the task at hand for further manipulation. As the user selects images from the results screen the matching image scores are recalculated dynamically using a combination of the current task context and annotations previously uploaded to the particular image as new

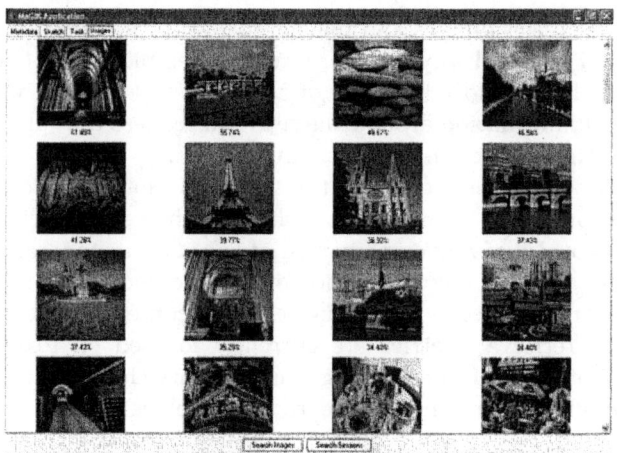

Fig. 1. Image Retrieval

parameters for matching. The interface is then redrawn to reflect the updated task context. All the selected images are then collected in the current user context and made available for task specific annotation.

4 Context-Based Session Retrieval

As the system builds up encapsulated contextualized user interactions, a knowledge base of user sessions consisting of both high-level user concepts and actual image interactions is continuously updated. This knowledge base can improve our context-based query processing by enabling the retrieval of entire previous task-based sessions. This allows a current user to look for the previous image analysis tasks that are most similar to the current task context, both to find relevant imagery and to examine the decisions and rationale that went into addressing the earlier task. We are particularly interested in three main benefits of this approach. Firstly by reusing the collective contextual knowledge in support of similar tasks the time required to carry out a new task can be significantly reduced. Secondly, this approach facilitates knowledge sharing by retrieving potentially relevant knowledge from other experiences. Finally from a knowledge management point of view, contextual knowledge relating to particular tasks may now be stored to create accessible organizational memory.

If when entering a query to the system, the user planning their holiday in Paris chooses to view information provided by similar users from a knowledge base of previous user experiences, they are presented with the interface shown in Figure 2. In designing this interface we endeavored to present an abstraction of all the contextual information we had recorded during the course of the similar user's task. A significant challenge here lies in how to present a condensed summary of a similar user's entire session in a limited space while retaining enough information so that our user can quickly discriminate potential relevancy.

We have developed an approach to mapping the most relevant aspects of user session context into summary visual context. Each row in the interface corresponds to one session result, ranked according to matching session score. Below is a detailed description of the information displayed in the interface for each session.

Percent Similarity Score: This score represents how comparable the user's query about a trip to Paris is to other user sessions in the knowledge base. This is an overall score calculated using similarity between user queries, task descriptions, images annotated and browsed and annotations applied.

The Most Discriminating Query Information: This includes metadata and task descriptions entered by the similar user. According to our similarity metrics these queries are similar to the queries entered by the current user. As there may be large volumes of text to be displayed in this column, a tool tip text is displayed when a user mouses over the space informing them that they may scroll down for all task descriptions. Our sample user can see from these previous sessions that other users have employed the system to plan a trip to Paris. Users from the most relevant previous sessions, shown in Figure 2, stated that they were also interested sightseeing and French cuisine.

The Most Important Annotations: Annotations may be textual in nature or they can be multimedia based. These annotations have been judged by the system to be most likely to help our user in fulfilling their task. Multimedia

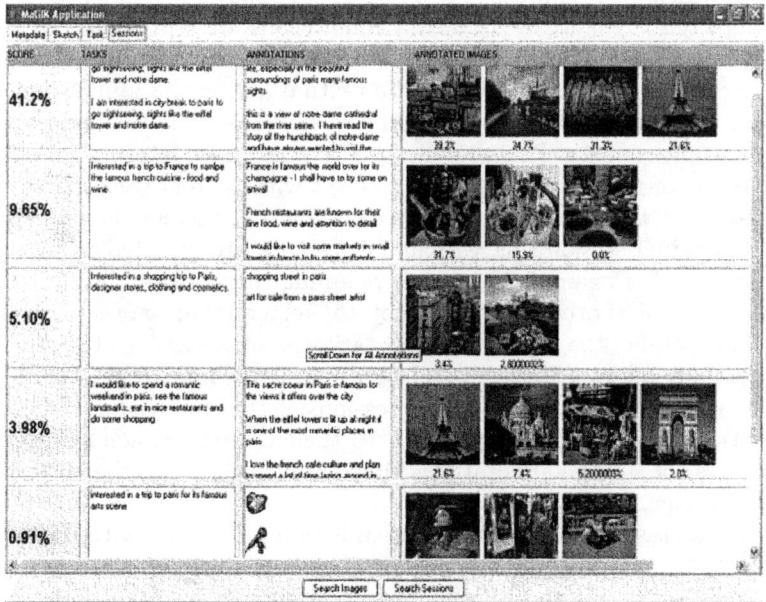

Fig. 2. Session Retrieval

annotations are represented in a succinct manner by media buttons. These media buttons, if clicked on, play any audio or video files uploaded during the session. From Figure 2 we can see that other system users have annotated images of Paris in the course of planning their trip. They have made multimedia and textual annotations concerning famous landmarks such as the cathedral of Notre Dame and the Eiffel Tower and also about French gastronomy. All textual annotations may be displayed in a web browser thereby giving the user the opportunity to read annotations uploaded from the World Wide Web and to link to any other relevant material from those documents. This type of navigation allows us to build context from web pages around the actual relevant images. By virtue of system interaction a link structure can be created between different types of knowledge. If a user clicks on a media button then the system deems that the user is interested in this work and that this annotation has a high similarity to the current context so the session scores are recalculated using the current context and the annotation just accessed as new parameters for similarity. Our adaptive similar sessions interface is redrawn to reflect the new task context.

Thumbnail Versions of the Most Important Images: These are images that have previously been annotated as part of the similar task. Beneath each image is an associated matching image score. These scores are calculated by comparing any textual annotations associated with these individual images to the current context. These scores therefore provide a measure of how relevant each individual image is to the current context. If the user wishes to view the annotations made to any of these images, they may do so by clicking on the thumbnail, which brings up the image and all its annotations in a screen such as the one depicted in Figure 3. All media annotations are represented by relevant icons when displayed on the images in order to prevent the images from becoming cluttered. The icons representing the textual and media annotations adapt their levels of transparency depending on how similar each individual annotation is to the current context. This provides a direct interface interpretation of relevance in terms of visual cues, which can adapt with current user context. If the current user selects an image for closer examination, the similar session scores and the individual session images matching scores are updated in response to this interaction. For example if our user planning their holiday in Paris browses the annotations of another user or annotates images that another user has already found useful in the context of their task, the scores assigned to any similar sessions and any relevant images within that session will increase. The user may further annotate images from similar sessions if they wish and/or retain the previous users annotations by adding it to their current session context. Once the user saves the desired annotations, they are transferred to the current user's view of the image.

Because the task-based retrieval system is tightly coupled with the activities that the user is performing, the system also has the capacity to make proactive recommendations in a natural and unobtrusive manner by monitoring the users current task context. Information requests, increments in the image information accessed and annotations provided are no longer taking place in an isolated en-

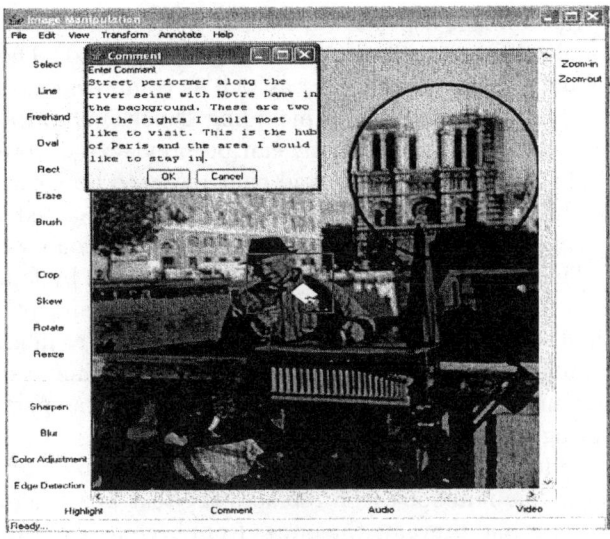

Fig. 3. Image Annotation

vironment. Rather they are grounded in the context of the activities the user is performing so the system can correspondingly anticipate and update what previous experiential knowledge would be relevant at that stage, making it available to the user. The knowledge is provided unobtrusively, so that it need only be accessed when required. Thus the process does not distract from the task at hand, yet makes relevant knowledge available just-in-time.

5 Annotation Tools for Capturing User Context

Our research is focused on capturing contextual task knowledge to perform more effective image retrieval by employing annotations in similarity retrieval metrics. To this end we have developed tools for direct image manipulation to assist the user in organizing information about relevant imagery and their task. These insights are then distilled into a form that captures high-level user concepts which allows associations or relations between images to be made without the need for content-based or keyword-based analysis. The tools for direct image manipulation include filters, transformation, highlighting, sketching, post-it type and multimedia annotations. They allow the user to identify regions of interest that can be linked to clarifications, rationale and other types of annotations. The manipulations and annotations do not alter the underlying raster data: rather they are layered to provide a task-specific view. This enables the capture of refinement of more general task-based ideas and rationale.

We have selected the kinds of manipulations that would be most useful in helping to analyze and focus on image features. All of the sketching manipulations can be performed in a variety of colors and brush styles. The user can then go on to add personal media annotations to the image as a whole or to particular highlighted image aspects. Currently, the system supports annotation by text, audio and video, though retrieval is focused on text. The system integrates real-time audio and video capture as well as compression. A wide variety of compression formats are supported, including QuickTime, Mpeg and H.263. All textual, audio and video annotations can be previewed before being incorporated as part of the knowledge base, and once recorded they can be saved and uploaded to the image context as a knowledge parcel associated with the task in question. A facility is in place that allows users to upload web documents as annotations. This allows further context to be extracted by following HTML links. The system also supports annotation by cut, copy and paste between a given image and other images in the dataset, as well as images in any application that supports clipboard functionality for the given operating system. Figure 3 shows how a user made use of the image annotation tools as part of their task of planning a spring break in Paris.

The user has added a textual annotation which is represented by the notebook icon on the image. The notebook icon replaces the text box where the user has typed their comments and prevents the image from becoming cluttered. The user can go back and access their annotation at anytime by double-clicking on the icon which will display the comment in the text box. The dark rectangle around the icon is the area associated with the annotation and is emphasized when the icon is moused over. The user comment mentions that this is an area they would like to visit as there are famous landmarks like the river Seine and the cathedral of Notre Dame. They also mention that they would like to find accommodation in this area as it is centrally located. The user has also highlighted Notre Dame cathedral by sketching an oval shape, indicating that it is a place in which they have an interest in visiting.

Once the user has finished interacting with the imagery their entire task process is stored as an encapsulated session case in the knowledge base. This includes the query entered, all images and sessions returned as well their associated matching score to the current task context. It also includes all annotations and actions performed by the user on returned imagery and sessions in the course of carrying out their task. This captured context provides the basis for computing similarity between new tasks and the goals of previous users.

6 Evaluation

The system was originally designed for the retrieval of geo-spatial imagery and an earlier evaluation of session retrieval indicated that sessions were useful as a basis for task-based retrieval [15]. In this evaluation we show that (1) the approach can generalize and scale to other types of image sets, and (2) that task-based annotation context has advantages for individual image retrieval. We evaluate the system using images from the Corel Image Dataset [16]. One of the major difficulties faced in this research has been the acquisition of a suitable dataset upon which to evaluate our system. Datasets for multimedia annotation typically focus on keyword type tagging of individual media elements, with no notion of usage context. An investigation into available image libraries found no suitable candidates for the type of retrievals being tested, and thus we have developed a task-aware annotated image dataset that approximates the task/annotation-based environment we have described throughout the paper. We hope our methodology in building this corpus will be useful to forward multimedia annotation evaluation in general, as well as helping to demonstrate the advantages of our approach.

The starting point in the construction of our annotated image dataset was the acquisition of the Corel Image Dataset which consists of 60,000 colour photographs. We took a subset of 500 of these images to begin building the image library. The 500 images corresponded to 5 countries (Australia, Ireland, England, France and Thailand), with 100 images per country. The aim was to annotate these images with information that would be useful for planning a holiday to the particular countries.

In the first order each of the 500 individual images was analyzed by hand and manually annotated with four or five keywords describing image objects, scenes depicted or actions taking place in the image. In order to extract contextual information for each image, the set of keywords for each image was entered as a query to web sites offering information on holidays in the outlined countries, for example http://www.australia.com/ and http://www.visiteurope.com/france.html.

The resulting web pages were analyzed and the most relevant paragraphs from these pages were applied as annotations to the images. These annotations were between 2 and 5 sentences in length. An example of an annotation for a photograph featuring the Sydney Opera House was "*The Sydney Opera House, designed by the Danish Architect Joern Utzon, is meant to resemble a giant sailing ship*". The purpose of these annotations was to act a baseline for comparison with task-based queries relating to holidays in the outlined countries. By employing these retrieved texts, we simulate annotations from a variety of users.

In order to demonstrate the task-based retrieval capabilities of the system it was necessary to outline several tasks relating to specific types of holidays in the countries in our dataset and to simulate user behavior by annotating relevant returned imagery with task-specific information. Some examples of the outlined tasks for holidays in Australia and England were "*Interested in a holiday in Australia. Fly into Sydney and travel up the east coast, interested in wildlife and adventure sports*", and "*Planning a trip to England, spend a few days in London, visit the markets, followed by a few days in the countryside*". An example of of an annotation applied to a photograph of a skydiver for the first task described above was "*An item high on my agenda is to do a skydive along one of the east coast beaches*".

Once the images had been annotated according to this first set of tasks another set of tasks corresponding to holidays in the countries were outlined. The purpose of this was to see if the contextual annotations added to the images for the first set of tasks could be re-used to find imagery relevant to the second set of tasks. Some examples of these tasks were "*Interested in a holiday in Sydney, Australia, to see the famous harbour, the opera house, the harbour bridge and to do some shopping"*, and *"Interested in a trip to the English lake district, hiking, walking and water sports."*.

We then created three searchable indices; the first containing only the keywords per image, the second containing only the baseline annotations per image and the third containing a concatenation of the baseline annotations and the added annotations from the first set of tasks. (In the graphs below these are referred to as Keyword Only, Baseline Annotations Only and Task-Based Annotation). In the first instance we were interested to see if the addition of contextual annotations can help users locate more images that are relevant to their task than retrieval based on keywords alone. We were also interested to see if the addition of further annotations (i.e. more annotation per image) can help users find more relevant image information.

The 500 images were then categorized by hand as either relevant or not relevant to each outlined task in the second category of tasks. Each task was entered as a query to the system and the returned images were recorded. A precision and recall analysis was then carried out on these images and the results are displayed in Figure 4.

From Figure 4(a) we can see that for finding a small number of contextually relevant images both the baseline annotation and task-based annotation approaches outperform the keyword-based approach. As the number of results increases, the different approaches begin to level out though the task-based retrieval remains the strongest across the dataset. From Figure 4(b) we can see that task-based retrieval finds many more relevant images across the dataset than either of the other approaches. Task-based retrieval provides much better

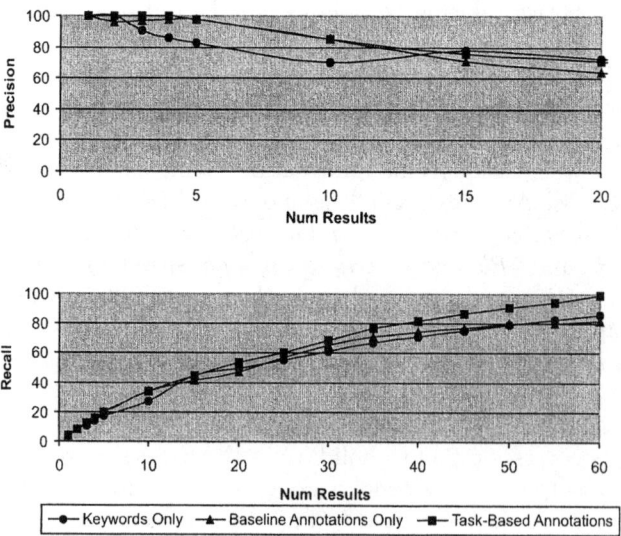

Fig. 4. (a) Precision and (b) Recall

coverage over the entire dataset than either keyword-based retrieval or retrieval based on baseline annotations, allowing the user to retrieve more contextually relevant images. This shows that the task-based annotations can indeed provide useful additional retrieval context.

The results on our small image library are very promising. We plan to expand our library both in terms of the number of images and also in terms of the number of applied task-based queries and annotations. We see this experiment as a starting point for automating the construction of datasets from where users can retrieve information based on their current context rather than just on keyword or content-based search. Based on our initial results we expect our context-based retrieval to improve as we continue to move in the direction of a more representative task-based dataset.

7 Conclusions and Future Work

We have introduced our approach to developing a context aware system with a task-based focus for retrieving imagery and knowledge. Our evaluation shows that task-based annotations can provide useful additional context in retrieval and that our system is performing as expected thus far. We plan to use these results as a baseline for future evaluation frameworks where more task-based annotations have been added. Our intention is to use these results as a basis for examining the relative contributions of the different notions of annotation context from individual annotation to complete session. We also plan to perform comprehensive user trials in the near future. We intend to extend our implicit knowledge acquisition techniques as well as to supplement our primarily text-based retrieval system by including multimedia annotational information for retrieval.

References

[1] Hollink, L., Schreiber, A., Wielinga, B. & Worring, M. "Classification of User Image Descriptions," International Journal of Human Computer Studies, *Elsevier Science* 66(5) (2004) 601-626
[2] Budzik, J. & Hammond, K.J. "User Interactions with Everyday Applications as Context for Just-in-Time Information Access," Proceedings of the 2000 International Conference on Intelligent User Interfaces, IUI-00, *Springer-Verlag* (2000)
[3] Claypool, M., Le, P., Waseda, M. & Brown, D. "Implicit Interest Indicators," Proceedings of ACM Intelligent User Interfaces Conference, IUI-01, *ACM Press* (2001) 33-40
[4] Goecks, J. & Shavlik J. "Learning users interests by unobtrusively observing their normal behavior," Proceedings of ACM Intelligent User Interfaces Conference, IUI-00, *ACM Press* (2000) 129-132
[5] Budzik, J., McLoughlin, L. & Hammond, K. "Information access in context," Knowledge Based Systems, 14(1-2), (2001) 37-53
[6] Lieberman. H "Letizia: An agent that assists web browsing," Proceedings of the Thirteenth International Joint Conference on Artificial Intelligence, (1995)

[7] Bauer, T. & Leake D. " WordSieve: A Method for Real-Time Context Extraction," Modeling and Using Context: Proceedings of the Third International and Interdisciplinary Conference, Context 2001, *Springer-Verlag* (2001) 30-44
[8] Lieberman, H. & Liu, H. "Adaptive Linking between Text and Photos Using Common Sense Reasoning," Proceedings of the 2nd International Conference on Adaptive Hypermedia and Adaptive Web Based Systems, (2002)
[9] Davis, M., King, S., Good, N. & Sarvas, R. "From Context to Content: Leveraging Context to Infer Media Metadata," Proceedings of 12th Annual ACM International Conference on Multimedia (MM 2004), *ACM Press* 188-195, (2004)
[10] Bauer, T. & Leake D. " Exploiting Information Access Patterns for Context-Based Retrieval," Proceedings of the 2002 International Conference on Intelligent User Interfaces, IUI-01, *ACM Press* (2001)
[11] Hollink, L. Schreiber, A., Wielemaker, J. & Wielinga, B. "Semantic Annotation of Image Collections," Proceedings of K-CAP 2003 Workshop on Knowledge Capture and Semantic Annotation, (2003)
[12] Worring, M., Bagdanov, A., Gemerr, J., Geusebroek, J., Hoang, M., Schrieber, A., Snoek, C., Vendrig, J., Wielemaker, J. & Smuelders, A. "Interactive indexing and retrieval of multimedia content," Proceedings of the 29th Annual Conference on Current Trends in Theory and Practice of Informatics, (SOFSEM) (2002) 135-148
[13] Flickner, M., Sawhney, H., Ashley, J., Huang, Q., Dom, B., Gorkani, M., Hafner, J., Lee, D., Petkovic, D., Steele, D. & Yanker, P. "Query by image and video content: The QBIC system," IEEE Computer, 28(9) (1995) 23-32
[14] O'Sullivan, D., McLoughlin, E., Wilson, D. C. & Bertolotto, M. "Capturing Task Knowledge for GeoSpatial Imagery," Proceedings of K-CAP 2003 (Second International Conference on KnowledgeCapture, *ACM press* (2003) 78-87
[15] O'Sullivan, D., McLoughlin, E., Bertolotto, M. & Wilson, D.C. "A Case-Based Approach to Managing Geo-Spatial Imagery Tasks," Proceedings of ECCBR 2004, Seventh European Conference on Case-Based Retrieval, *Springer-Verlag* (2004) 702-716
[16] Li, J. & Wang, J. Z. "Automatic linguistic indexing of pictures by a statistical modeling approach," IEEE Transactions on Pattern Analysis and Machine Intelligence, 25(9) (2003) 1075-1088

An Approach to Data Fusion for Context Awareness

Amir Padovitz[1], Seng W. Loke[1], Arkady Zaslavsky[1],
Bernard Burg[2], and Claudio Bartolini[2]

[1] Centre for Distributed Systems and Software Engineering, 900 Dandenong Rd.,
Caulfield-East, Victoria, Australia
{amirp,swloke,arkady.zaslavsky}@csse.monash.edu.au
[2] HP Labs, Palo-Alto
{claudio.bartolini}@hp.com

Abstract. We propose and develop an approach modeled with multi-attribute utility theory for sensor fusion in context-aware environments. Our approach is distinguished from existing general purpose fusion techniques by a number of factors including a general underlying context model it is built upon and a set of heuristics it covers. The technique is developed for context-aware applications and we argue that it provides various advantages for data fusion in context-aware scenarios. We experimentally evaluate our approach with actual use cases using real sensors.

1 Introduction

Various approaches attempting to partially reduce the incurred uncertainty associated with context and attempts to build effective models that describe context are still at an early stage. Context itself can reflect circumstances borrowed from a variety of domains, including social interactions and human perception, and reasoning about context often involves the task of identifying human users' situations. Subsequently, the diversity and complexity of the conditions involved in reasoning about context contribute to the inference challenge. One key ingredient in inferring and interpreting context is fusing together sensor data. The integration of sensed information, preferably coupled with other (perhaps higher-level) reasoning methods, provides a promising approach (although not complete) for reasoning about context.

In this paper we develop and propose a novel approach modeled with multi-attribute utility theory for fusing sensor data. Our approach is distinguished by its association with a general context model as the underlying description of context related knowledge. It attempts to incorporate various heuristics that should impact context inference, to produce a better fusion result. Our approach is specifically intended for use in context-aware applications, and exhibits many characteristics desirable in reasoning about context. The underlying general context model offers a new theoretical perspective on context that enables development of new reasoning techniques such as the fusion technique discussed in this paper.

We organize this paper as follows. In Section 2, we introduce our theoretical modeling approach referred to as the Context Spaces model, and discuss different concepts, which are used during the development of the fusion process. In Section 3, we present our fusion algorithm, as a practical extension of the Context Spaces model. We first discuss the heuristics we seek to cover in the fusion and then develop the technique itself. We experimentally evaluate our proposed approach in Section 4 in two stages. We use simulation experiments with which we analyze the performance of the fusion and the heuristics we seek to cover; and we provide results of our fusion technique in actual office-related use cases using combinations of real-life sensors. We conclude our paper in Section 5.

2 The Context Spaces Theoretical Model

Our objective is to develop an approach for inferring the occurrence of situations by a context-aware system for any type of situation it is familiar with. It should be able to provide a solution to a query such as: "Is situation X occurring?" (Or "Is a user U in situation X?"). To enable this capability for general use (i.e. by different applications for different types of situations and available information) we develop a formal model that represents perceived context and situations of interests of a system. We then develop a technique for fusing the modeled information to achieve situation awareness.

The paradigm of context-aware computing can be regarded as an attempt to obtain information with limited sensing capabilities, but which nevertheless reflects circumstances useful to the application at hand. Designers of a context-aware application would seek to define perceivable context, which reflects interesting circumstances or real-life situations, as accurately as possible, governed by limitations such as technology and cost. The fundamental nature of context (in context-aware computing) can therefore be regarded as a constrained view of a system over the world, which can either be immediately used (for triggering actions) or can represent real-life situations (in which case further computation may be required to relate the context to the situation, e.g. data fusion techniques).

The Context Spaces theory aims to model this fundamental nature of context and enable context and situation awareness for systems that are able to sense information with various degrees of imperfections at runtime. It is an attempt towards a general context model to aid thinking and describing context. The concepts use insights from geometrical spaces and state-space models - we hypothesize that geometrical metaphors such as states within spaces are useful to guide thinking about context, though the user will need to take the concepts and elaborate on them within the context of the application to be built.

In the process of developing our fusion technique we will make use of the following concepts.

2.1 Context Attribute

Definition 1. *A context attribute (denoted by a_i) is defined as any type of data that is used in the process of inferring situations. A context attribute is often associated with* sensors, virtual or physical, having the value of the sensor readings denote the context attribute value at a given time t (denoted by a_i^t).

2.2 Context State

Definition 2. *A context state describes the application's current state in relation to chosen context, and is denoted by a vector Si. It is a collection of context attributes' values that are used to represent a specific state of the system at time t. A context state is $S_i^t = (a_1^t, a_2^t, ..., a_N^t)$ where S_i^t denotes a vector defined over a collection of N attribute-values, where each value a_i^t corresponds to an attribute a_i's value at time t.*

2.3 Situation Space

Definition 3. *A situation space represents a real-life situation. It is a collection of regions of attribute values corresponding to some predefined situation and denoted by a vector space $R_i = (a_1^R, a_2^R, ..., a_N^R)$ (consisting of N acceptable regions for these attributes). An acceptable region a_i^R is defined as a set of elements V that satisfies a predicate P, i.e. $a_i^R = \{V \mid P(V)\}$.*

For example, in numerical form the acceptable region would often describe a domain of permitted real values for an attribute a_i. A region of acceptable values is defined as a set which satisfies some predicate; hence, it can contain any type of information, numerical or non-numerical.

We complete the basic model with a set of functions, which add more realism to the representation by relating interesting facts to the fundamental concepts. In this paper we will consider two of these functions, as follows.

2.4 Relevance Function

Definition 4. *Given a situation space $R = (a_1^R, a_2^R, ..., a_N^R)$, a relevance function ε_R associates weights $w_1, ..., w_N$ with regions of values $a_1^R, ..., a_N^R$ of Ri, respectively, where $\sum_{j=1}^{N} w_j = 1$. A weight $w_j \in [0,1]$ represents the relative importance of an attribute region a_j^R to other regions in the situation space's definition.*

In many cases some types of information are more important than others for inferring a situation, e.g., high body temperature may be a strong indication of a general sickness of a person while other attributes may not be so important to infer that specific situation. For example, high respiratory rate may be caused by and therefore also indicative of other situations, such as a person doing physical exercise. To model this difference in the importance of context attributes for inferring a situation, we define the relevance function, which assigns weights to context attributes. The weights reflect how important each attribute is (relative to other attributes) for describing a situation.

2.5 Contribution Function

Definition 5. *A function η_j assigns a contribution level $c \in [0,1]$ for each element (a value of a context-attribute) in a region (corresponding to that context attribute) of values j. The contribution level of an element in a region reflects how well that element associated with the modeled situation is.*

In the relevance function, we model the *relative* importance between the attributes of a situation space, whereas in the contribution function we model the individual contribution of elements within a specific region for inferring a situation, i.e. more than merely knowing that the sensed value is within or not within a region, if the value is within, we also consider the particular value itself: the fact that the value is within the region is indicative of the situation, and even more so indicative if the value is of some range (within the region) - how much so is what the contribution function represents. For example, it might be that some values are better reflecting the purpose of that specific region than others. E.g., for a context attribute of 'body temperature' in the definition of 'subject is healthy' the values between 36.5 and 36.7 would reflect a contribution of 1 (on a scale between 0 and 1) and values between 36.3 and 36.5 and between 36.7 and 36.8 would reflect a lesser degree of supportive contribution.

3 Data Fusion Algorithm for Context-Awareness

The most basic insight covered in the Context Space paradigm is that of representing context in terms of state and space. The fact of a context state being within some situation space indicates (to a certain degree) the occurrence of the situation represented by that space. So, indicators (or evidential support) for the occurrence of a situation are represented by the values (of context attributes) in the context state being within the accepted regions of the situation space (*and* quite strongly indicative if they also fall within particular ranges). We develop a data fusion technique, based on multi-attribute utility theory (MAUT) [6] that takes the information represented by the model (i.e. the condition of the context state and the definition of the situation space to be inferred) and computes a degree of support for the occurrence of that situation. The degree of support is then compared with support for other alternative situations or with a support threshold predefined by the system designers, to facilitate a decision regarding the occurrence of the situation.

The algorithm works by accumulating positive indicators for the occurrence of a situation. These indicators cover several heuristics that we believe are important to be considered in reasoning about context. They include the following:

(1) *Individual significance and contribution of context attributes -*
This information is modeled by the relevance and contribution functions in the model.

(2) *Completeness of containment -*
In context-aware computing application designers specifically choose a relatively small number of context attributes to represent a particular situation. Therefore it is

significant whether all or only some of the values of the chosen attributes correspond to the situation definition.

(3) *Different characteristics of context attributes and their effect over the fusion* - We distinguish between two types of context attributes in regard to the definition of a situation space, which have different effects over the fusion outcome, as follows.

Symmetrically Contributing:
Definition 4. A symmetrically contributing context attribute increases the support in a situation taking place if it is within the corresponding region, and decreases the confidence if it is outside that region.

Asymmetrically Contributing:
Definition 5. An asymmetrically contributing context attribute increases the confidence in a situation taking place if it is within the corresponding region but sensing values outside the accepted region would not decrease the computed confidence.

This distinction between different types of context attributes, therefore, exercises dissimilar impact over the final inference result. To clarify this distinction, examine the following intuitive examples. First consider a situation space defined for a situation of 'subject is healthy' with a symmetric attribute of 'Body Temperature' and acceptable region of values between 36.6 C and 37.2 C. Any value outside this region, say 38 C, would mean (with some degree of confidence) that the subject is unhealthy. Therefore the values of this type of attribute must always remain within the region of values specified for it in the situation space.

In contrast, when we use a context attribute 'PDA location' of a user to help us infer 'user in a meeting' situation, the accepted region for this attribute would be location information corresponding to the meeting room. Having indication of a PDA located in the meeting room will assist to infer the situation. However, if the PDA location is sensed to be elsewhere, the 'user in a meeting' situation may still be valid (e.g., the PDA was left at home or in the office). Therefore, this type of attribute would be referred to as asymmetric, since it only assists in inferring a situation but does not refute a situation if its values are outside the situation space.

(4) *Completeness of containment vs. individual contribution* -
There is a trade-off between ensuring complete containment of all symmetric attributes and their individual contributions. An important context attribute should produce greater impact on the final result of the fusion than one which is considered less important. In contrast, when several attributes are equally very significant for the evaluation of the real-life situation we may want to ensure that all of them are contained, in particular when they are explicitly chosen at design time. We therefore need to account for this type of trade-off in the fusion process.

(5) *Inaccuracies of sensors, yielding uncertain context attributes values* -
Inherent inaccuracies associated with sensor readings should affect our confidence in the outcome of matching state (defined over inaccurate context attributes values) and space. The true state of an attribute which value is outside some region of values

3.1 A Utility Based Fusion Approach

We make use of MAUT as the basis for deriving a measure that reflects the accumulated support for the occurrence of a situation represented by the situation space. The process considers the heuristics previously described and results in a single numerical measure ranging between 0 and 1.

MAUT provides a convenient way for combining together seemingly different contributions into a single measure, expressing the result in terms of utility [6]. In our case, we see utility (or contribution towards our goal of determining the occurrence of a situation) as the degree of evidential support given to the hypothesis of a situation occurring when a context attribute value is within the corresponding region. The more indicators we have that the context state matches the definition of a situation space, the greater utility is gained.

MAUT is considered an evaluation scheme, which provides a general evaluation function $v(x)$ over an object x to denote the overall object's utility. The evaluation function is traditionally defined as a weighted accumulation of evaluating the objects' value dimensions [6, 3], which represents a combination of different contributions relevant to the object. Each dimension can also be individually evaluated using the general approach of weighted contribution by evaluating its own value dimensions.

Below, we provide our way of representing combined evidential support from multiple context attributes. (If some heuristic is irrelevant then it should not be included; the principle of a utility combination approach remains, however, the same.)

We start by defining a dimension d_1, representing how much the individual context attribute values match a situation space definition.

$$d_1 = \sum_{i=1}^{n} \hat{w}_i \cdot p(a_i^t \in a_i^R), \qquad (1)$$

where $p(a_i^t \in a_i^R)$ is the estimated probability of having a context attribute a_i's value contained in its corresponding region of values a_i^R in the situation space, and $\hat{w}_i$ (where $\sum_{i=1}^{n} \hat{w}_i = 1$) is a weight measure expressing the relative importance of the specific context attribute evaluation to the overall utility. The calculations of $\hat{w}_i$ is done using a process of weights redistribution.

In general, in terms of MAUT, $p(a_i^t \in a_i^R)$ reflects the evaluation of a particular attribute containment; the higher the probability of an attribute value being within the region, the greater contribution is evaluated for the attribute. The more confident we are in the containment of an attribute, the more this attribute can contribute to support the situation. $\hat{w}_i$ reflects the individual weight or significance of the attribute in the overall evaluation process of d_1. The need to determine weights with additional computation ($\hat{w}_i$ rather than w_i) is the result of considering the characteristics of the context attributes, being either symmetric or asymmetric.

For example, an asymmetric attribute may be an important ingredient in inferring a situation and therefore should be associated with a high weight value. However, if this attribute does not correspond well to the situation space, its absence should not affect the inference result (since it is defined as asymmetric). This fact would affect the weight values of the remaining context attributes, which should still reflect the same relative importance between themselves.

The general steps for determining weights for equation (1) are the following.

Let $w_1, w_2, ..., w_k, ..., w_{k+m}, ..., w_N$ *denote initial weights associated with a given context state* S_i^t *for the specific fusion procedure, where weights are determined according to their relative importance in regard to a specific situation.*

1. Repeat for each asymmetric context attribute k:

  1.1 If $p(a_k^t \in a_k^R) < C$ *then* $w_k = 0$, *where C defines a threshold value*

2. Repeat for each context attribute i (both asymmetric and symmetric):

$$\hat{w}_i = w_i / \sum_{j=1}^{N} w_j$$

During redistribution of weights, the system examines how well an asymmetric attribute fits the definition of a situation space. If a sufficiently low confidence in the containment of that context attribute is gained (thereby reflecting the disassociation of the attribute with the space) it is ignored by assigning its weight to zero. Finally, after the adjustment of all asymmetric attributes, the overall remaining attributes' relative weights are recalculated.

Next, we define a dimension d_2, representing the contribution of the fact that the *entire* set of symmetric attributes matches the definition of the situation space. We compute d_2 as the conjuncture the containment of all symmetric attributes in the situation space, as follows.

$$d_2 = \bigwedge_{i=1}^{m} p(a_i^t \in a_i^R), \qquad (2)$$

where a_i is an symmetric context attribute.

We have so far defined two dimensions that express different aspects that are important in obtaining an accumulated support measure and contribute to an overall utility estimate (d_1 looks at individual attribute contribution and d_2 at having, for all attributes, the values of the context state within their respective regions). We now proceed and combine these two dimensions in a general utility function, as follows.

$$U = q_1 d_1 + q_2 d_2, \qquad (3)$$

where $q_1 + q_2 = 1$.

The overall utility U considers each dimension's relative weight in determining a combined result. The individual weight (i.e. q_i) of each of the two dimensions in the overall utility measure greatly depends on the situation at hand. More specifically, it depends on the distribution of the weights in equation (1).

An interpretation of d_1 and d_2 reveals a trade-off between the need to have all the attributes contained within a space (reflected by d_2), and the need to account for their individual contributions (d_1). If for example, all context attributes are equally significant (e.g., as an application developer determines), then it is sensible to put more emphasis on d_2, i.e. having, for all the attributes, values contained within the space. So that if an attribute is outside the situation space, an overall low confidence would be achieved. If, on the other hand, some context attributes are much less significant, then it would be more appropriate to give greater importance to d_1, i.e. give more weight to the significant attributes containment, rather than insisting on having, for all attributes, values contained within the respective regions.

Consistent with this line of reasoning, we set the overall utility weights by analyzing the differences between the weights defined in the situation space definition. Greater differences increase q_1 and smaller differences increase q_2. We suggest the following simple computation to determine the weights:

$$q_2 = \frac{\hat{w}_{min}}{\hat{w}_{max}}, \text{ where } \hat{w}_{min} \neq 0,$$

and $\hat{w}_{min}$ and $\hat{w}_{max}$ denote minimal and maximal weights after weights redistribution process, respectively. Alternatively, other approaches such as standard deviation analysis can be used.

3.2 Estimating Containment

Finally, let us refine our estimate of the sensor readings. As sensors can in general be inaccurate, the correct value of the phenomenon might be significantly different than what is reflected by the sensor readings. We would therefore need to estimate the probability of the true state being within the region of values, i.e. $p(s_i^t \in a_i^R)$, where s_i^t denotes the true value. (If we are unable to perform such estimation, then $p(a_i^t \in a_i^R)$ would represent complete confidence in the actual observation represented by the sensor readings).

Let us develop a general approach for deriving confidence in containment of the true state of a context attribute sensed value in a region of values, given a known distribution of the reading errors or some estimation for it. For simplicity we assume that changes in the context attribute values do not significantly change the errors' distribution behavior (i.e. their mean and standard deviation). This estimation is intended for sensor readings that correspond to continuous region of values.

Let $e_j = a_i^t - s_i^t$ denote an error of a sensor reading from the true state of the context attribute value it senses, where s_i^t represents the true state measured by a sensor for context attribute a_i and a_i^t denotes the sensor reading value. Note that the error is intentionally characterized as having signed value rather than absolute; (when a sensor is biased, reading errors may have significantly different densities for values higher or lower than the true state).

Let $a_{i_{min}}^R$ and $a_{i_{max}}^R$ denote minimum and maximum values of an accepted region of values a_i^R, which corresponds to a context attribute a_i.

Proposition. The probability of the true state s_i^t being contained within the region a_i^R can be computed by:

$$p(s_i^t \in a_i^R) = p(e_j \text{ between } (a_i^t - a_{i_{min}}^R) \text{ and } (a_i^t - a_{i_{max}}^R)).$$

Proof: For containment in a continuous region of values the context attribute true value must be greater or equal to $a_{i_{min}}^R$ and smaller or equal to $a_{i_{max}}^R$, i.e.:

(1) $a_{i_{max}}^R \geq s_i^t \geq a_{i_{min}}^R$. Given the definition $e_j = a_i^t - s_i^t$, than $s_i^t = a_i^t - e_j$.

Substitute s_i^t with $a_i^t - e_j$ to obtain: (2) $a_{i_{max}}^R \geq a_i^t - e_j \geq a_{i_{min}}^R$. Rewrite as:

(3) $a_i^t - a_{i_{min}}^R \geq e_j \geq a_i^t - a_{i_{max}}^R$

The practical implication of this is that given an estimation of the reading errors' distribution we can estimate the true state containment in the region by the observation (i.e. the inaccurate sensed value) alone. For example, if we know that the reading errors follows, say some *Normal* distribution estimation, than the containment of the state could be computed by: $p(e_j \leq a_i^t - a_{i_{min}}^R) - p(e_j < a_i^t - a_{i_{max}}^R)$.

4 Experimental Evaluation

We have described a utility based approach, which considers various heuristics that we believe are important when fusing together sensors data for reasoning about context. We proceed with a set of experiments that examine the behavior of our measure and its sensitivity to various changes in the environment. We separate the experimental evaluation into two parts. In the first, we use a simulation of a variety of sensor types, inaccuracies and scenarios. We seek to show how our heuristics are captured in the outcome of the support measure. In the second part, we provide results of an experiment, which uses real-life sensors, where we try to distinguish between three types of situations in real-life office settings, using the fusion process.

4.1 Performance Analysis

We compute the support measure in a setting of 'user in a meeting' situation, using a variety of simulated sensor types, accuracy levels and varying knowledge over the sensors accuracies. A description of the sensors used here is provided in Table 1. Each sensor activity is simulated using random value generation according to natural changes in the true event it senses and its defined inaccuracy. We allow different knowledge and type of inaccuracy for different sensor types. For example, we produce *Normal* approximated discrete samples to characterize location errors, assign Boolean result with fixed known accuracy for light detection and assume no knowledge whatsoever about the inaccuracy of noise level detection.

We assign importance on a scale between 1 and 5 and associate a relative weight for each sensor, corresponding to its significance for inferring the situation. Here, we seek to obtain a general sense of relative importance, using expert knowledge, rather than an exact resolution of the weights, which may be harder to achieve. For example, in estimating if a specific user is in a meeting, the fact that the user is located in the meeting room is more significant than the fact that noise is detected in the room.

Finally, we characterize each context attribute in the situation space definition as being either asymmetric or symmetric. For example, we assume that a user always carries his RFID tag but may leave his PDA behind, therefore refer to user location inferred by the PDA location as asymmetric. Similarly, a projector may or may not be active during a meeting. Its activity would contribute to the inference of a meeting but not the opposite.

Table 1. Sensors definitions for simulation of 'user in meeting'

attribute name	importance (1-5)	asymmetric	weight
User RFID Y Location	4	No	0.114286
User RFID X Location	4	No	0.114286
User PDA Y Location	3	Yes	0.085714
User PDA X Location	3	Yes	0.085714
MR Light Level 1	4	No	0.114286
MR Light Level 2	4	No	0.114286
MR Noise Level	2.5	No	0.071429
MR Motion Detected	2.5	No	0.071429
MR Projector Active	4	Yes	0.114286
MR Microphone Active	4	Yes	0.114286

First, we examine the effect of asymmetric attributes over the support measure. Figure 1 depicts the results of the measure for the 'user in meeting' situation for three scenarios. In the first scenario all sensors yield values that correspond to the definition of the situation space (i.e. all attributes values are contained in corresponding regions). This results in relatively high support for the specific situation (i.e. A) and varies in time according to inherent inaccuracies of sensors and changes in the experiment settings (such as user changes locations inside the meeting room). In the second scenario, we position the user's PDA in his office. As this attribute is defined as asymmetric the overall support remains nearly identical, yielding high support for the situation (i.e. B). Minor differences compared with the previous settings result from fewer number of attributes participating in the fusion.

In contrast, the effects of positioning the user's RFID tag outside the meeting room are significant. The support measure immediately drops to reject the 'user in meeting' situation. This corresponds to a scenario when the user is actually in his office but forgot his PDA for example, in the meeting room.

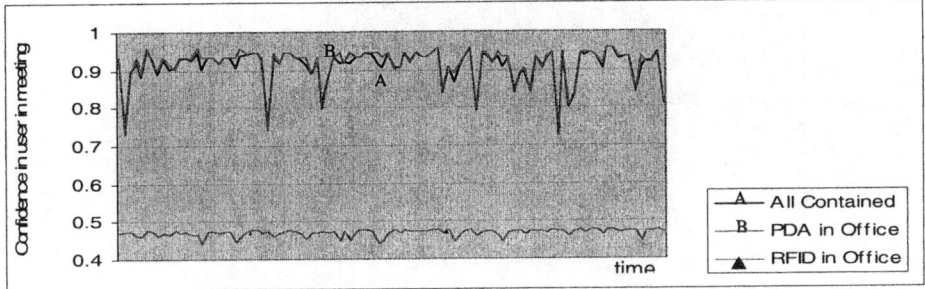

Fig. 1. Effects of asymmetric attributes

Next we examine the general behavior of the measure when we significantly change the readings of various sensors. Figure 2 depicts different support levels for the situation in three different scenarios. In the first scenario all sensors yield values, which are contained in their corresponding regions of values. In the second scenario, values are only partially contained in the regions, due to their inherent inaccuracy and experiment settings.

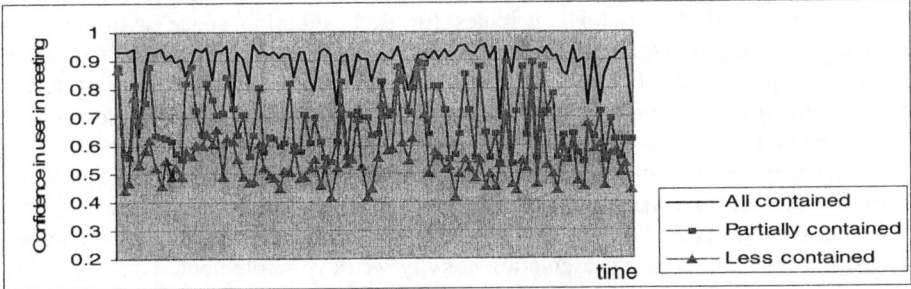

Fig. 2. Effects of sensors values

The last scenario, depicts situations where some of the events the sensors sense are truly outside the region of values. For example, when the user is in the corridor close to the meeting room and do not participate in the meeting. As expected, the results show degradation in support as sensed values reflect notably different situations.

Finally, we observe in figure 3 the effect of associated inaccuracies of sensors. The higher the inaccuracy of a sensor, a lower degree of support is gained for the situation and vice versa.

The experiment's results reafirm the expected behaviour of the fusion and the heuristics it attempts to capture.

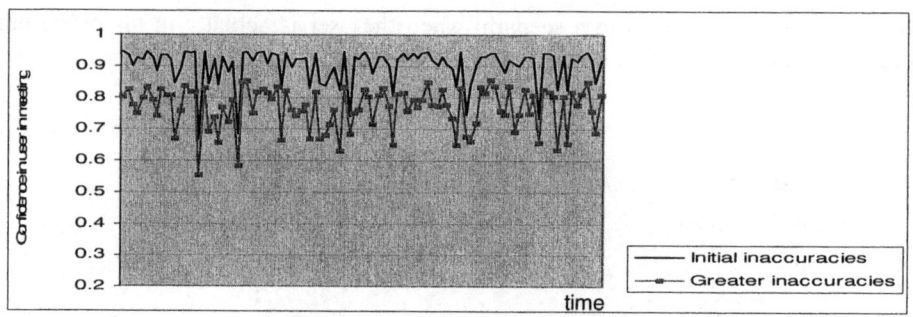

Fig. 3. Effects of sensors inaccuracies

4.2 Performance in a Real-life Use Case

We follow with a performance analysis of a real-life use case fusing together a blend of different real sensors and data sources. In the following experiment we were interested in identifying and distinguishing between three types of user activities, taking place in our smart meeting room, namely, (1) a user giving a presentation, (2) a user attending another's presentation and (3) a user attending a meeting.

To reason about these activities we have selected four basic context attributes physically positioned in different locations, specifying the user location, the meeting room light level, the user's notebook keyboard and mouse activity and presence of active presentation processes on the user notebook. We defined the corresponding region of values of the context attributes for each situation space with different appropriate values, including asymmetric and symmetric attributes and associated inaccuracies of readings. For user location we used Ekahau Positioning Engine [9] that tracks the user's personal devices such as her PDA and notebook. The positioning service computes spatial positions by analyzing wireless signal strengths and comparing them to previous calibration. It provides an associated confidence of its inferred location with a measure between 0 and 1. We used Berkeley Motes [10] for sensing and communicating light levels in the meeting room. For retrieving information about the user's presentation activity we have implemented a service that hooks to the notebook operating system and provides information on latest keyboard and mouse activity. We also provide a service that identifies active presentation processes in the notebook.

In the meeting room we use a portable light-weight presentation projector that is connected to the presenting user's notebook. It is a common practice and the assumption in this experiment that each presenter uses his personal notebook.

During experimentation we have revealed significant inaccuracy in location estimation due to less than optimal wireless network settings that affect signal strength and number of optimal access points. The Positioning Engine Server itself is located in a remote location and communicates with an agent that performs the fusion via wireless 802.11b infrastructure.

The functional architecture of the reasoning process is as follows. An agent working on behalf of the user and running on the user notebook is equipped with communication protocols for exchanging information with remote sensing services as well as with local independent processes that provide information on notebook activity. The agent performs fusion using the inputs of the incoming information and assigns a support measure to each of the candidate situation spaces, which represent the real-life activities. Before applying the fusion, information is pre-processed either by the agent or remotely, depending on the type of data. For example, light levels are sampled by the motes sensors a number of times and then averaged. The Motes Interface Service, which handles this information, matches the averaged result against predefined light levels and communicates back a predefined value. In contrast, interpretation of keyboard and mouse activity is preformed directly by the agent. Here, the amount of time lapsed since the last captured activity influences the confidence that the user is currently using his notebook. Various similar approaches of preprocessing raw sensor readings (also known as cooking the data or using cues) are common in fusion for context recognition and have been shown to significantly improve inference results (e.g. [2, 1]).

During actual experimentation we have switched between the activities and observed the results of the measures for the different situations. Figure 4 provides support measures for the three activities. In this experimental run we have started with the user presenting first for 15 minutes then attending a colleague's presentation for 15 minutes and finally participating in a discussion or general meeting on the topics presented (again for 15 minutes). Interpretation of the results reveals matching support levels with the actual activity taking place. At the time the user is presenting, the support for this particular activity averages around 0.9 whereas support for other situations is significantly lower. A change in the situation towards the user only attending another's presentation results in a drop of the 'User Presenting' situation to support levels below 0.4, and a rise in the support for the 'User attending a presentation' situation to levels around 0.9.

Similarly, when a discussion (equivalent to a meeting) over the presentations involving our user is starting right after the second presentation, the support for 'User in a meeting' situation rises to 0.9 and support for the former situation drops significantly.

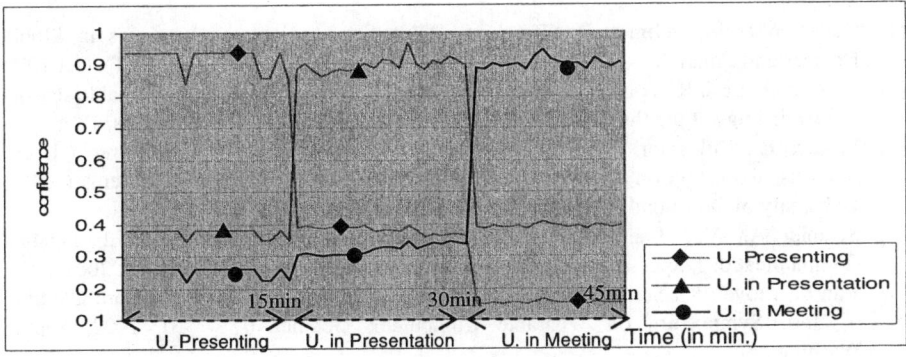

Fig. 4. Experimental run

The successful results of this experiment can be partly contributed to observing assumed behaviors associated with the defined activities, such as having the lights switched off during presentations and on during the following general discussion, or controlling the presentation shown from the user's notebook.

In order to relax these assumptions more context attributes need to be introduced, such as those discussed and demonstrated in section 4.1. Here, the choice and design of sensor networks clearly affects the quality of inference [8, 4, 5]. In this experiment we have shown that our approach is applicable for fusion in real-life scenarios using real sensors and data.

5 Conclusion

We have presented an approach for fusing sensor readings that considers heuristics that we believe are pertinent in reasoning about context. Our approach can be considered within the general field of sensor fusion modeled with MAUT for situation classification. We have also offered a unique theoretical perspective about what it means for a situation to occur: the occurrence of a situation is represented by a tuple of values (i.e., the context state comprising values obtained via sensors (and perhaps some reasoning)) being within a tuple of accepted regions (i.e. the situation space representing the situation, which is pre-defined). Building on this idea, we have designed a new measure to determine if a situation is occurring, which takes into account the relative importance of context attributes, the significance in context attributes being within particular ranges, the significance when the sensed values for *all* context attributes are within the scope of the situation, and uncertainties in sensed values. Our techniques can be applied to different context-aware scenarios.

Acknowledgments

We thank Hewlett-Packard (HP) for sponsorship of this work.

References

[1] Gellersen H.W., Schmidt A., Beigl M., Multi-Sensor Context-Awareness in Mobile Devices and Smart Artifacts, in Mobile Networks and Applications (MONET), Oct 2002.
[2] Golding A. Lesh R., N., Indoor navigation using a diverse set of cheap, wearable sensors, In Proceedings of the third International Symposium on Wearable Computers, 1999.
[3] Schafer R., Rules for Using Multi-Attribute Utility Theory for Estimating a User's Interests, workshop on Adaptivity and User Modelling in Interactive Systems (ABIS), University of Dortmund, Germany, October 2001.
[4] Schmidt A., Van Laerhoven K., How to build smart appliances. IEEE Personal Communications, Special Issue on Pervasive Computing, 8(4):66-71, August 2001.
[5] Van Laerhoven K., Schmidt A., Gellersen H.W. Multi-Sensor Context Aware Clothing, 6th Intl. Symposium on Wearable Computers, October 07 - 10, 2002, Seattle, Washington.

[6] Winterfeld D., Edwards W. Decision Analysis and Behavioural Research, Cambridge, England, Cambridge University Press 1986.
[7] Wu H., Siegel M., Steifelhagen R., Yang J., Sensor Fusion Using Dempster-Shafer Theory, IEEE Instrumentation and Measurement Technology Conference, Anchorage, AK, USA, May 21-23, 2002, May, 2002.
[8] Yoshimi B., On Sensor Frameworks for Pervasive Systems, Workshop on Software Engineering for Wearable and Pervasive Computing (SEWPC), ICSE 2000, Limerick, Ireland.
[9] Ekahau Positioning Engine, http://www.ekahau.com
[10] Information on Motes available at: www.tinyos.net/, www.xbow.com/

Dynamic Computation and Context Effects in the Hybrid Architecture AKIRA*

Giovanni Pezzulo[1,2] and Gianguglielmo Calvi[1]

[1] Istituto di Scienze e Tecnologie della Cognizione - CNR,
Via S. Martino della Battaglia, 44 - 00185 Roma, Italy
[2] Universitá degli Studi di Roma "La Sapienza", Piazzale Aldo Moro,
9 - 00185 Roma, Italy
giovanni.pezzulo@istc.cnr.it, gianguglielmo.calvi@noze.it

Abstract. We present AKIRA, an agent-based hybrid architecture designed for cognitive modeling. We describe some of the underlying ideas motivating its development, such as the possibility of exploiting distributed representations and performing parallel, dynamic and context aware computation. We illustrate its main components and capabilities and compare it with some related cognitive architectures, such as DUAL and Copycat. We present also a sample simulation in the visual search domain, exploiting AKIRA's peculiarities for cognitive modeling.

1 Introduction

Cognitive modeling is a vast area of research today; however, there are still few computational models that are both credible from the theoretical point of view and powerful from the technological one. The most notable exceptions are the generic, unified architectures SOAR [24] and ACT-R [1]; some other architectures, such as the DUAL/AMBR [14, 15] and Copycat [11] models of analogy, do not address all the cognitive processes, but have a general scope, because they capture some of the underlying principles of high-level cognition.

AKIRA [21, 28] is not a cognitive architecture in itself, but a framework for implementing and testing cognitive models at different levels of complexity and integration. We have tried to make the platform as versatile as possible: AKIRA includes a rich toolkit of mechanisms and algorithms from different paradigms, allowing developers to test, compare and possibly integrate many symbolic and connectionist models. However, even if AKIRA does not commit to a single cognitive model, some theoretical choices are in a sense embedded: our main source of inspiration are the Society of Mind [19] and the Pandemonium [12] models. From the architectural point of view, AKIRA is inspired by DUAL [14] and Copycat [11]. AKIRA integrates Multi-Agent and Pandemonium features. A server process (the *Pandemonium*) executes and monitors many agent instances (the *Daemons*); differently from standard MAS architectures, agents have links,

---

* This work is supported by the EU project **MindRACES**, FP6-511931.

forming an *Energetic Network*, that affords energetic exchanges. The objective is to create architectures performing parallel, dynamic and emergent computation by distributing the operations between many simple, interacting processes, each one carried on by an agent. The advantages of distributed representations are explored e.g. by [5]; they have been also shown [10, 15] to be very important for modeling contextual effects.

In AKIRA, as well as in DUAL and Copycat, context is accounted as a set of *pressures* introduced by (partially) active elements over the current computation. AKIRA and DUAL use a set of micro-agents, each having an activation level for each instant and each representing a single object, concept or process; the agents are hybrid, carrying on both *symbolic content* (e.g. a frame) and *connectionist elements* (an activation level, links to the other agents). For example, in one of the main applications of the architecture DUAL, the AMBR [15] model of analogical reasoning, each micro-agent includes a frame representing a single object, e.g. a cup, or a plate. While *objects* are thus localist representations, *situations* are fully distributed ones: e.g. *the cup is on the plate* is represented by a number of partially activated micro-agents; moreover, the patterns of activation change dynamically because of system evolution (e.g. decay) and thanks to new elements or events (e.g. the cup is broken). By exploiting distributed representations AMBR is able to perform analogical reasoning in a dynamic and context dependent way. Many analogy phases, i.e. analog access, mapping, and transfer are performed as as *parallel subprocesses* rather than serial stages. The result of the computation dynamically emerges from the parallel and concurrent activity of many partially active agents, carrying on portions of semantic information both in their symbolic and connectionist parts. AMBR models many other cognitive processes, including blending and intrusions, priming, re-representation of episodes, problem solving [10, 15], etc. Its results have been successfully compared to human data [16]. In a similar way, Copycat [11] performs analogical reasoning on the basis of a representation of the problem space that is not pre-fixed, but is computed and may change dynamically during the analogical processing. Representation-building and mapping (e.g. of two situations) run in parallel and influence each other.

AKIRA exploits a similar apparatus for context-aware goal processing. We introduce its main underlying theoretical assumptions in Section 2. In Section 3 we present its main components and features. In Section 4 we compare AKIRA with some related cognitive architectures: DUAL [14, 15], Copycat [11], IDA [9] and the Behavior Networks [17]. In Section 5 we present a sample simulation in the visual search domain.

2   Principles of AKIRA

Here we present the main inspiring principles of AKIRA, focusing on how distributed representation and control can be used for cognitive modeling.

Hybridism and Locality Principle. AKIRA is a hybrid system; differently from many others, AKIRA exploits agents that are *hybrid at the micro-level* [14]:

each agent has both a connectionist and a symbolic component. The connectionist component involves the activation level of the agent as well as the energy exchanges between the agents; moreover, agents can group and organize into higher level assemblies called *Coalitions*, that are able to solve together composite tasks that are impossible to solve alone. The symbolic component involves the set of operations an agent can perform; each agent is a specialized computational unit[1]. The operations can range from simple to very complex ones (e.g. AKIRA includes many mechanisms and algorithms). However, in accordance with the underlying distributed approach, each agent should be specialized, thus complex tasks are more likely performed by Coalitions of cooperating agents.

AKIRA also follows a *Locality Principle*: each interaction between the agents, both connectionist and symbolic, is implemented as a peer-to-peer operation, without centralized control. Hybridism and locality permit to design a continuum of computation styles, ranging from centralized, hierarchical control to distributed, emergent computation (e.g. in modular, layered or distributed architectures). The connectionist side of AKIRA endorses the emergent and distributed phenomena, exploiting the patterns of activation of the agents. The symbolic side permits both to introduce top-down drives and structure and to manage operations requiring semantic compositionality or explicit communication.

The Energetic Metaphor. Agents' symbolic activity is influenced by their connectionist side. The agents dynamics follow an Energetic Metaphor [14, 20]: *greater activation corresponds to a greater computational power*, i.e. speed. Each agent has an amount of computational resources that is proportional to its activation level (and is a measure of its relevance, both absolute and contextual, in the current computation). More active agents have a priority in their symbolic operations and their energetic exchanges. This mechanism permits to model a range of cognitive phenomena such as context and priming effects, because active agents are able to influence the others. As a consequence of the dynamics of system (priority is related to the activation) and energetic exchanges between related agents, in facts, each agent introduces a *contextual pressure* over the computation even without explicit operations, but only being active; for example, if an agent that recognizes a given visual feature is active, it will activate or inhibit other visual agents, or it will influence the way other agents operate on the same stimulus. Moreover, the results of its computation can be available to other agents (as in the Pandemonium model), e.g. to higher-level feature-integration agents; this capability will be used in the example in Section 5. The converse is also true: agents that are salient in the same context evolve stronger links and are able to recruit more energy. Many kinds of contextual pressures can be naturally modeled using this schema, e.g. perceptual, goal-driven, cultural, conceptual, memory-based, etc.

---

[1] It is important to notice that even subsymbolic mechanisms can be used, e.g. neural networks. We call them "symbolic" since they are discrete processing unities, playing identifiable roles: they are *used as symbols* by the rest of the system.

3 Structure and Components of AKIRA

AKIRA integrates features from MAS and Pandemonium. The agents are not isolated but related each other and to a central resource; they can share energetic resources; they can form assemblies called Coalitions; they can exchange explicit messages via a Blackboard (as in MAS) and even exploit an *implicit* form of communication [4], that consists in the *observation of the activity* of another agent, that is routinely notified to the Blackboard; this feature is mainly exploited for Pandemonium-like models. Here we briefly introduce the main components of AKIRA: the kernel, called Pandemonium; the Agents, called Daemons; the Blackboard. For full reference, see [21, 28].

The kernel is called Pandemonium; it acts like a management structure, performing a number of routine actions such as managing the Blackboard and monitoring the state of the agents. The agents are called Daemons; each one has a thread of execution, a concurrent access to the Energy Pool and a functional body where its behavior is specified. Daemons are not isolated: they can pass messages, spread activation via the energetic links (that can be both predefined and evolved) and on-the-fly join Coalitions. Each Daemon has a predefined sequence of execution, repeated for each cycle; this includes the connectionist operations (*Tap, Spread, Join, Pay*, that will be introduced later) as well as the symbolic one (*Execute*, that encapsulates the Daemon's specific behavior). Each Daemon has a private memory space for data and processes, and a queue for incoming messages. In accordance with the Energetic Metaphor, its resources and the priority of its thread are related to its current activation.

Even if each single Daemon can be programmed to realize arbitrarily complex behavior, a central requirement for successfully modeling dynamic and contextual effects is distributing the information and the control structure throughout many Daemons and exploiting the built-in features of AKIRA such as the energetic dynamics; thus complex ones should be realized either by high-level Daemons exploiting the results of low-level ones, or by a whole Pandemonium. AKIRA does not define any specific agent architecture but provides a set of prototypes that can be extended for realizing agents having different design (e.g. reactive, deliberative, layered) and capabilities. Their behavior is fully customizable; moreover the AKIRA Macro Language provides a toolkit of widely used algorithms and mechanisms to be exploited for agent programming, including BDI [23], Behavior Networks [23], etc.

AKIRA has a Blackboard, a shared data structure, that is used both as a message dispatcher between the Daemons and as a common workspace where the Daemons notify their current activity and activation, even without an explicit receiver; this feature can be exploited for implicit communication [4], as we will show later. It has also an Energetic Network, involving all the links between the Daemons. The Network is labeled, too, and it is possible to specify the links behavior depending on the label.

3.1 AKIRA Energetic Model (AEM)

A major difference exist with many neural-like architectures: AKIRA has a custom energetic model, the AKIRA Energetic Model (**AEM**, [21]), exploiting some ideas from Boltzmann Machines [18]. There is a centralized pool of resources, the *Energy Pool*, that gives an upper bound to the resources that the Daemons can tap. If a Daemon taps some resources, these are not available to the others until they are released; the Daemons compete for energy (access to the Energy Pool) and resources (e.g. access to the effectors, if included). Spreading activation has a special meaning here: the receiver takes it and the giver loses it; this mechanism is similar to the Behavior Networks one [17] and it is mainly used as a form of "weak delegation": an agent spreads activation to another one that is able to fulfill one of its needs in order to successively exploit its results.

Performing a symbolic operation has a cost in energy for the Daemon, that is released to the Energy Pool before executing (this operation is called *Pay*): thus the Daemons have to accumulate a certain amount of energy before really operating[2]. The cost of an operation should be set in accordance with its complexity and urgency: less cost means more easily activated. Fast and urgent behaviors such as like stimulus-response ones can be represented by low-cost operations. More complex cognitive operations are slower, since they need to recruit a lot of energy; moreover, often they have to exploit operations by other agents, or to join Coalitions.

For each cycle, if a Daemon successfully executes its symbolic operation, it Pays some energy, that comes back to the Energy Pool and becomes ready to be tapped by other Daemons; it also notifies its success to the Blackboard (this operation is called *Shout*). Shouting is not only used by other Daemons (as in a Pandemonium model) as an information, but also for creating new links; it is a request to other Daemons to be linked by them (or to reinforce existing links). If a Daemon does not successfully execute its symbolic operation, it *Spreads* its activation to its linked Daemons, that are more pertinent (or are able to help it, realizing the previously described "weak delegation"), and weightens its incoming links. Both successful and unsuccessful Daemons can reply to a Shout and link with successful Daemons; this operation is called *Join* and it is the basis for the formation of assemblies of Daemons called Coalitions. As a result of Shouting and Joining, without any centralized control, the energy is conveyed from unsuccessful to successful Daemons. Moreover, Daemons that are active and salient in they same situation evolve stronger links, since they write and read more often, and at the same time, from the Blackboard.

The activation of each Daemon is dynamically calculated for each cycle (this operation is called Tap) and results from the sum of three elements: *Base Pri-*

---

[2] Separating the activation of a Daemon from its possibility to really act allows to retain the contextual relevance of partially activated agents while preventing too many Daemons to fire actions in parallel (thus simplifying the control structure); moreover, the cost mechanism prevents active Daemons to operate the same operation twice.

ority, *Energy Tapped* and *Energy Linked*. The Base Priority should be set in accordance with the importance of a given Daemon (or class of Daemons); it is private and not shared neither with the Energy Pool nor with the other Daemons. For example, a Daemon representing a concept can have as a default more activation than a Daemon representing a feature. The Energy Tapped depends on the Tap Power attribute of each Daemon. For each cycle the Daemon tries to tap a correspondent amount of energy from the Energy Pool (say 50); however, the Pool could have less energy available (say 30), so the Energy Tapped indicates only the energy really tapped. The Energy Linked is tapped from the incoming links, providing that some other Daemons spread it. The network is not a simple medium, but it actually contains some energy that is accessed concurrently by the Daemons through the Tap and Spread operations. All the links are weighted and this influences how much energy can be tapped. As an example, the Daemons A, B and C have 50 energy units; they are all linked with one another, and each link has weight 0.5. If A and B both spread before C taps, at the end A and B will have 25 energy units and C 100 as Energy Linked. Base Priority and Tap Power are conceived for modeling *absolute* relevance of a process. For example, when a Daemon with a strong Tap Power is activated, he grows (energetically) faster than others. This feature is useful for implementing high priority processes (such as *alarms* in [26], that quickly have to obtain resources for e.g. fleeing). Energy Linked indicates instead the *contextual* relevance of a Daemon, since the network is dynamically rearranged in accordance with the contingent situation.

As a consequence of the AEM, the whole system is *homeostatic*: the resources are bound to a limit and influence the computation. As we will discuss later, this permits not only to endorse concurrence, but to model the concept of *Temperature* (used in Copycat) and some dynamics of Baars' *Global Workspace Theory* [2] (used in IDA). The AEM constrains the models that can be implemented using AKIRA; in this sense, AKIRA is midway between a general framework and a cognitive architecture. Of course the AEM, as well as many other components, is only one of the available options; it can be replaced by other models (for example, spreading activation in [7]); however, here we assume it as the default.

4 Comparison Between Architectures

Here we compare AKIRA with the cognitive architectures DUAL [14], Copycat [11], IDA [9] and with the Behavior Networks [17] action selection model.

4.1 AKIRA and DUAL

AKIRA and DUAL are both directly inspired by the Society of Mind. AKIRA borrows from DUAL a number of ideas and technical solutions: e.g. micro-level hybridization, Coalitions organization, variable speed of the parallel processors.

DUAL is thought as a cognitive architecture, so all models built on it should use the same mechanisms (or a subset of them) as well as the same parameter tuning. One of the most important claim underlying dynamic and emergent cognitive modeling is that the whole set of micro-agents represent Long Term Memory, while currently active ones represent Working Memory. Active agents dynamically modify and operate on the working memory in highly context-dependent way. DUAL agents, from the symbolic point of view, have all the same structure: they are micro-frames, having general slots (e.g. *type* and *ISA* relations) and frame-specific slots (that depend on the concept represented, e.g. indicating one of its features or attributes). DUAL agents have weighted and labeled links that are the medium for spreading activation; they mainly connect related slots in the micro-frames. There is also the possibility to build temporary links, to exchange symbolic messages and to use a marker passing mechanism.

AKIRA is not a cognitive architecture in itself and it is not committed to a single agent model; it includes a rich macro language with many resources for agent modeling (including BDI, fuzzy systems, neural networks, etc.). It is being used for different tasks, such as social and socio-cognitive simulations [8], decision making under uncertainty [22] and as a development tool for BDI agents [21]. AKIRA is also suited for Pandemonium-like models; Section 5 provides an example of this mechanism.

The Energetic Model. AKIRA and DUAL follow the same underlying principle, the Energetic Metaphor [20]; however, there are many differences between the two energetic models. DUAL implements the energetic metaphor by delayed computations between the agents (in precedent versions [14] time sharing was used). It exploits the spreading activation mechanism of [7] for passing energy between the agents; there is also a decay mechanism that periodically lowers the energy of the agents. DUAL agents also consume energy for their tasks and each agent has a specific *consumption* depending on its conceptual complexity [20]. Each DUAL agent has an individual *start-executing threshold* which is a subject of learning. Thus micro-agents corresponding to important and often used processes will have a low threshold and will start running even at low energy levels, while others may require high energy levels; some may be able to run only when they are in the focus of attention. Moreover, if the energy of an agent falls below a certain threshold, its work is lost.

AKIRA exploits the AEM (introduced earlier). Differently from DUAL, in AKIRA there is not a mechanism controlling the energetic level of the agents (e.g. thresholds for starting or stopping them), but each agent has its own thread of execution and its activation is directly mapped on the thread priority; thus, a Daemon never really stops running. There is a cost mechanism that is similar to consumption in DUAL; Pay is a specific phase in the Agent life cycle; there is not a decay phase (but the same result can be obtained by specifying a little default cost for each cycle in the Pay phase).

The two systems can address the same class of phenomena. AKIRA, being not a cognitive architecture, is more open and allows programmers to plug-in different options. The more important difference is the energetic model: AEM is conceived for modeling control structures and it is similar to some models in behavior-based robotics, where decentralization of control and data structures is very important.

Coalitions. Coalitions are interconnected sets of agents (via permanent or temporary links) that can exchange energy and symbolic content. DUAL assembles Coalitions by recognizing "time bindings" between Daemons: a specialized Agent called the Coalition manager is responsible for forming them by monitoring the activation level of all the micro-agents. In AKIRA Coalition formation is instead a distributed activity exploiting the (*Shout* and *Join*) operations of the Daemons. Coalitions can not only be "flat", i.e. composed of Daemons at the same level of complexity, but can have hierarchies and layers; we call the former *Bands* and the latter *Hordes*. While a Band mainly emerges as an auto-organization of many agents at the same level of complexity, an Horde mainly arises thanks to the top-down pressures of a "leader" Daemon called *Archon*[3]. Hordes can be exploited as organizing elements; for example, they can simulate a K-line in the Society of Mind model (in this case the Archon is a kind of "handler" that only activates other Daemons). Hordes can also be used for introducing top-down pressures; this is the case of meta-processes in Pandemonium style, where an Archon monitors and drives the activity of many Daemons (see the example in Section 5). Differently from DUAL, where learning is mainly realized at the network level, introducing hierarchies of Coalitions permits a form of episodic learning that is close to case-based reasoning: a new Archon is built when a Band meets some requisites (e.g. it is stable, new, etc.)[4].

4.2 AKIRA and Copycat

Differently from DUAL and AKIRA, that are hybrid at the micro level, Copycat (and similar models in [11]) keep the knowledge and the procedure levels separated, having both a semantic network and a procedural memory. It introduces the Slipnet, a kind of semantic network where the links between the concept are labeled by other concepts inside the same network; thus, when a concept becomes relevant, it strengthens the associated links. Procedures are called Codelets and they are kept separated from the semantic network; their priority depend on their activating concepts (into the Slipnet). They run concurrently, but in a stochastic and not parallel way: each one has a probability

---

[3] The distinction between Bands and Hordes resides mostly at the design level. We use the term top-down for stressing the fact that the control flow is more driven from the higher level. Of course top-down and bottom-up processes coexist and smoothly interact into the framework. As a design rule, powerful Archons should be avoided, because their top-down pressures can have strong centralization effects.

[4] Some implementation details: DUAL is implemented by using an extension of LISP called S-LISP (suspendable LISP) [20]; it is not really parallel, but permits to simulate a parallel process by using delayed evaluation of each symbolic operation carried on by the agents. The same mathematical model can be in principle extended to a parallel architecture and to a parallel hardware. AKIRA is fully written in C++; it integrates many open source libraries and runs under Linux platform. The system is parallel and works both with serial and parallel hardware. It is also possible to have many kernels implementing multiple societies that interact, using a client-server scheme and to interface external components (e.g. agents or objects).

to be chosen for activation depending on their current priority (otherwise they reside in a *stochastic coderack*, the procedural memory); this feature introduces non determinism and variability in the system behavior, performing a special context-aware search that is called *parallel terraced scan*. In AKIRA the declarative and procedural parts can both be included into the Daemons; however, the labels of the Energetic Network can be exploited for building semantic networks such as the Slipnet by using as nodes Daemons containing only semantic content and no operations. Copycat introduces also the concept of *Temperature* of the system: it is represented by the currently used energy; it can increase and decrease over time and it is proportional to how far the system is from a solution (e.g. a well formed analogy). In AKIRA Temperature is en emergent property of the system and depends on how much energy is tapped from the energy Pool; as in Copycat, the assumption is that many Daemons and Coalitions can compete as concurrent hypothesis (e.g. different analogies for a situation); a hot system is far from stabilization and performs quick-and-dirty computation, with rapid hypothesis shift; a cold system indicates stability and successful solutions, computing in a more conservative way.

4.3 AKIRA and IDA

IDA [9] (Intelligent Distribution Agent) is a complex cognitive architecture developed for the US Navy. It includes a number of systems: the Sparse Distributed Memory [13], used as an associative work memory; a perceptive module based on the Copycat architecture; an action selection module using Behavior Networks (with the significant addition of variables); an emotive control module using neural networks [18]; a meta-cognitive module using fuzzy classifiers (an extension of [3]); a Pandemonium model [12] for deliberation and passage to voluntary action. IDA includes in a single framework many theoretical approaches; however, it does fully integrates them at the low level, e.g. by using the same underlying mechanisms, but the systems are kept quite independent and juxtaposed as modules. This feature introduces a certain amount of rigidity, because they can influence each other only in a limited way. IDA is based on the *Global Workspace Theory* (GWT) of Baars [2], where many subconscious processes compete for the access to "consciousness": processes having more priority are explicitly selected for "entering the consciousness spot", where they can broadcast messages to all the other processes (they can ask help to all the other processes for resolving their tasks). Daemons and Coalitions in AKIRA compete for resources and even for an attentional spot (we prefer not to call it consciousness); however, there is no extra mechanism for this, because the energetic dynamics make some processes more influent than others. Moreover, in AKIRA more active Daemons have more frequent access to the Blackboard, so their requests are sent more frequently and can reach more other Daemons; this default functionality can be used in substitution of broadcasting. Some AKIRA local dynamics resemble those proposed in the GWT (and are similar to Temperature): e.g. interesting, unsolved problems lead to "hot" Coalitions, with many Daemons joining them. Solved problems require no more attention (except perhaps for learning).

4.4 AKIRA and the Behavior Networks

Behavior Networks [17] are an action selection mechanism inspired by the *distributed control* and *situated action* paradigms of behavior-based robotics. BNs include *goals, facts* and *competence modules*. Modules and goals are in concurrence as in a constraint satisfaction network and there is activation flow between them; depending on the goals as well as on the bottom-up pressures of active facts, the more contextually appropriate module is selected for activation. In this way planning is never predetermined, but it is on-line and context sensitive. Subgoaling is implicit: if a module has a false fact as a precondition, it spreads activation to another module that is able to make it true (subgoals are not explicitly represented). However, all these dynamics are realized with a centralized mechanism: for each cycle, the BN is run and the more active module is selected. If its activation is over a certain threshold, it is executed; if not, the threshold is lowered and the cycle restarts.

AKIRA uses a similar energetic model (e.g. spreading activation means giving energy), but with an extra feature, the Energy Pool, that gives an upper bound to the resources that can be used. Moreover, AKIRA is realized in a fully parallel way and there is not a cycle of control similar to the BNs. In order to be more adaptive in dynamic environments, the BN do not store representations (e.g. variables in modules), but each action is totally specialized. This feature seriously weakens their scalability; some models have been proposed using instead deictic representations (a single variable, the "attentional focus" is passed between the modules) or variables [9]. AKIRA actions are carried on by Daemons, that are like nodes in the BN sense; however, they are much more complex and can store (and pass) any kind of variable or object.

5 A Sample Simulation in the Visual Search Domain

Here we describe a sample simulation in the *Visual Search* domain [27]; this example is provided not for its cognitive plausibility (and it is not compared with human data), but in order to show how it is possible to exploit the features of AKIRA for cognitive modeling. In fact, even if by using AKIRA it is possible to realize the same model in many different computational ways, we want to stress the use of some of its peculiar features, such as the competition between the agents produced by the Energy Pool. In this simulation an agent is realized involving the whole Pandemonium: the task is performed by a number of Daemons operating concurrently and not by a centralized process. Daemons with different specialization reside in different layers; some of them are sensitive to the environment; some others to the activity of the Daemons in the lower layers. The task involves two visual features: two colors and two letters; the agent has to find the red "T" in a picture (the environment) containing many *distractors* [27] (green "Ts" and red "Ls"). The agent can not see the picture all together, but only the content of its movable spotlight, consisting in three concentric spaces having good, mild and bad resolution (a very simplified model of human fovea). The involved Daemons are divided into five layers:

1. **Full Points Detectors.** Each Daemon monitors a point of the spotlight, e.g. the left corner. There are Daemons for the inner, central and outer spotlight; inner ones are more numerous and have more Tap Power; central ones have less, and outer ones the worse one. They directly ask to the environment: *is this point full or empty?* and notify the result to the Blackboard.
2. **Color Recognizers.** In visual search tasks there can be processes specialized for different features (size, orientation, location, etc.), but for this sample simulation we only use Color Recognizers. They monitor the activity of the Full Points Detectors; if they find a "full" point, they ask e.g. *is this point red?* or *is this point green?* and notify the result to the Blackboard.
3. **Line Recognizers.** They recognize particular shapes (sequences of points having the same color) as lines. They have not a permanent memory and they are not able to store the position of the points and to build a map of the environment; they only can concatenate on-line two or more consecutive points of the same color and communicate their position to the Blackboard.
4. **Letter Recognizers.** They are more complex shape recognizers, using the information provided by the Line Recognizers for assembling "Ls" or "Ts" (even having different orientations).
5. **Spotlight Mover.** A single Daemon receives commands from all the other ones (e.g. move to the left) and consequently moves the center of the spotlight (the area of influence of the Full Points Detectors is of course affected, too).

The left part of Figure 1 shows the Daemons involved into the simulation; the layers are numbered. The simulation starts by setting a *Goal Daemon*, representing e.g. "Find the Red T". It is actually an Archon and it spreads activation to the "Red Recognizer" and the "T Recognizer" (the arrows represent the links), introducing a strong goal directed pressure: at the beginning of the task some Daemons are more active than others (dark and white circles). The dotted lines represent instead the monitoring activities performed by the Daemons: if a Daemon successfully matches, some Daemons in the higher layers can exploit its activity. Thus, during the simulation there will be more or less active Daemons influencing the overall process. Successful Daemons send commands to the Spotlight Mover, too; it dynamically sums them up (there is also a certain inertia), and the spotlight traces a trajectory (starting from the cen-

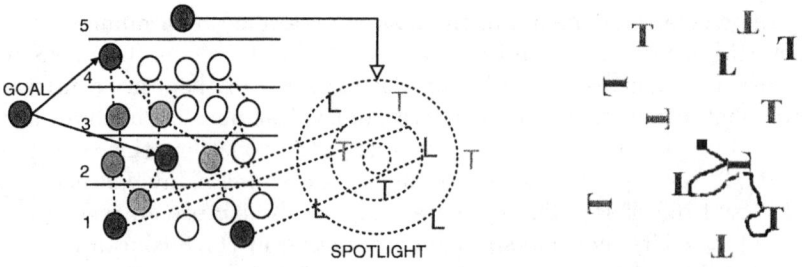

Fig. 1. Left: the components of the simulation. Right: a sample trajectory

ter), as illustrated in the right part of Figure 1. Each Daemon tries to move the spotlight where it anticipates there is something relevant for its matching operation. For example, if the Red Recognizer matches something in a certain point, it tries to move there the center of the spotlight; the Green Recognizer does the contrary, but with much less energy, because it does not receive any activation from the Goal Daemon. The Line and Letter Recognizers try to move the spotlight in the surroundings of already matched points in order to verify if there is a complete line or letter and its position. The simulation ends when the Goal Daemon receives simultaneous success information from the two Daemons it controls. The search process results thus from an interplay of bottom-up pressures (e.g. Full Points Detectors that succeed attract the attention of Color Recognizers) and top-down search strategies (e.g. Letter Recognizers trying to complete their complex pattern matching). The whole process is context aware: there is of course a goal context, introduced by the Goal Daemon; and in fact different goals lead to different trajectories. However, all the features in the picture can in principle introduce a little pressure over the whole computation. For example, a Red Recognizer Daemon is attracted by all the red elements and tries to move the spotlight consequently; a green element creates an "avoidance zone". The main point of the model is that the context of the task is not explicitly represented as a single entity, but it influences the activity of all the Daemons: the colors influence the Color Recognizers; the shape, the orientation and the size of the letters influence the complexity of the pattern matching of the Line and Letter Recognizers; the position, both absolute and relative, of the letters influences the Full Points Detectors and in consequence all the Daemons; etc.[5]

The simulation uses the AKIRA Energetic Model; Daemons exchange activation with the Energy Pool and the priority of their symbolic operations depend on their activation; they also evolve temporary links. Daemons in the upper layers have more Base Activation, reflecting their power of introducing top-down perssures; since they perform more complex symbolic operations, they have even higher costs. The Spotlight Mover receives commands from all the other Daemons and sums them up, thus the movement of the spotlight depends on all their pressures; but Daemons that succeed in their operations and are thus more relevant can write more times on the Blackboard, so they have more influence on the Spotlight movement. It is not an aim of the present paper to discuss the cognitive plausibility and the limitations of the model; we only used it for showing how to exploit the peculiarities of AKIRA for cognitive modeling. However, it is worth noticing that some visual search phenomena such as the *pop out effect* [27] in the single feature case (if there were only green Ls) and the *interference* [27] between two features (in the example, colors and letters) are easily modeled by using distributed representations and processes in Pandemonium style.

---

[5] In a sense there are not explicit representations at all, but only distributed processes that are interpreted as representations. For example, an active Red Recognizer can be interpreted as "there is a red point here" by another Daemon monitoring it.

6 Conclusions and Future Work

We have introduced AKIRA, a platform for cognitive modeling; we have explained some of its main principles, including how it exploits distributed representations and processes and how it models the Energetic Metaphor. We have shown its components and described its features for agent modeling. We have discussed the main differences with related models such as DUAL, Copycat, IDA and the Behavior Networks. Up to the moment, AKIRA has been successfully exploited for a number of projects, including socio-cognitive simulations [8], goal oriented processing [21], decision making under uncertainty [22].

In the sample visual search simulation at the moment there is no memory of past searches: the Pandemonium starts anew for each simulation. We are now improving the model by storing the learned links; in this way it should be able to account for some implicit learning phenomena. For example, in the *Contextual Cueing* paradigm [6] the subject (in repeated experiments) learns to use some contextual information, such as the relative position, orientation and distance of the letters, without being able to explicitly report them. Permanent links, that can of course cross the layers, can be created or strengthened when a successful solution is achieved, thus reinforcing a certain search path, because good trajectories are rewarded. Links can also be created or reinforced when a Daemon exploits an information produced by another one, thus reinforcing a certain pattern of activation of the Daemons; e.g. if a certain Full Points Detector is useful for a Color Recognizer, the latter will spread it some energy in order to exploit successively its results. By now the Energetic Network is only exploited by the Goal Daemon that links some Recognizers; after the learning phase it should work also as an associative memory, implicitly representing some information (e.g. *left corners are uninteresting*) and rules (e.g. *when there is a green zone, move to the left* or *when you find a red spot awaken the letter recognizers*). This capability should make the model able to be primed and to recall past pictures, extracting from them some cues for the new search.

References

1. Anderson, J. R., Lebiere, C. 1998. The atomic components of thought. Mahwah, NJ: Erlbaum.
2. Baars, B. J. (1988). A Cognitive Theory of Consciousness. New York: Cambridge University Press
3. Booker L. B., Goldberg D. E., and Holland J. H., Classifier Systems and Genetic Algorithms, Artificial Intelligence, vol. 40, pp. 235-282, 1989
4. Castelfranchi C., Silent Agents: From Observation to Tacit Communication, In: Proceedings of MOO 2004
5. Chalmers, D., 1992, Subsymbolic Computation and the Chinese Room, in J. Dinsmore (ed.), The Symbolic and Connectionist Paradigms: Closing the Gap, Hillsdale, NJ: Lawrence Erlbaum
6. Chun, M. M. (2000). Contextual cueing of visual attention. Trends in Cognitive Science, 4 (5).

7. Collins, AM, and Loftus E.F. (1975) A spreading-activation theory of semantic processing, Psychological Review 82, 407-428.
8. Falcone R., Pezzulo G., Castelfranchi C., Calvi G. (2004). Why a cognitive trustier performs better: Simulating trust-based Contract Nets. Proceedings of AAMAS 2004.
9. Franklin, Stan, Kelemen, Arpad, and McCauley, Lee. (1998). IDA: a cognitive agent architecture. Proceedings of the IEEE Conference on Systems, Man and Cybernetics, 2646-2651.
10. Grinberg, M., Kokinov, B. (2003). Analogy-Based Episode Blending in AMBR. In: Kokinov, B., Hirst, W. (ed.) Constructive Memory. Sofia: NBU Press.
11. Hofstadter, D. R., Fluid Concepts and Creative Analogies: Computer Models of the Fundamental Mechanisms of Thought. NY: Basic Books, 1995.
12. Jackson J. V., Idea for a Mind. Siggart Newsettler, 181:23-26, 1987
13. Kanerva, P. 1988. Sparse Distributed Memory. Cambridge MA: The MIT Press.
14. Kokinov B. N., The context-sensitive cognitive architecture DUAL, in Proceedings of the Sixteenth Annual Conference of the Cognitive Science Society, Lawrence Erlbaum Associates, (1994).
15. Kokinov, B., Petrov, A. (2001). Integration of Memory and Reasoning in Analogy-Making: The AMBR Model. In: Gentner, D., Holyoak, K., Kokinov, B. (eds.) The Analogical Mind: Perspectives from Cognitive Science, Cambridge, MA: MIT Press
16. Kokinov, B., Zareva-Toncheva, N. (2001). Episode Blending as Result of Analogical Problem Solving. In: Proceedings of the 23rd Annual Conference of the Cognitive Science Society. Erlbaum, Hillsdale, NJ.
17. Maes P., Situated Agents Can Have Goals. Robotics and Autonomous Systems, 6 (1990).
18. McClelland, J. L., Rumelhart, D. E. (1988). Explorations in Paralell Distributed Processing: A Handbook of Modles, Programs and Exercises. MIT Press, Cambridge, MA.
19. Minsky M. The Society of Mind. Simon and Schuster, N. Y. 1986
20. Petrov, A., Kokinov, B. (1999) Processing symbols at variable speed in DUAL: Connectionist activation as power supply. Proceedings of the Sixteenth International Joint Conference on Artificial Intelligence (vol. 2, pp. 846-851).
21. Pezzulo G., Calvi G. Designing and Implementing MABS in AKIRA, in P. Davidsson et al. (Eds.) MABS 2004, LNAI 3415, pp. 49-64, 2005
22. Pezzulo G., Lorini E., Calvi G. (2004). How do I Know how much I don't Know? A Cognitive Approach about Uncertainty and Ignorance. Proceedings of COGSCI 2004.
23. Rao A., Georgeff M., BDI Agents from Theory to Practice, Tech. Note 56, AAII,1995.
24. Rosenbloom, P. S., Laird, J. E., Newell, A. (1992) The Soar Papers: Research on Integrated Intelligence. Volumes 1 and 2. Cambridge, MA: MIT Press.
25. Rumelhart D. E., Mc Clelland J. L. and the PDP Research Group, Parallel distributed processing: explorations in the microstructure of cognition. Vol. I, 1986.
26. Sloman, A. 1999. What Sort of Architecture is Required for a Human-like Agent? In Foundations of Rational Agency, ed. M. Wooldridge, and A. Rao. Dordrecht, Netherlands: Kluwer Academic Publishers.
27. Wolfe, J. M. (1996). Visual search. In H. Pashler (Ed.), Attention. London, UK: University College London Press
28. www.akira-project.org

Goal-Directed Automated Negotiation for Supporting Mobile User Coordination

Iyad Rahwan[1,*], Fernando Koch[2], Connor Graham[3], Anton Kattan[3], and Liz Sonenberg[3]

[1] Institute of Informatics, The British University in Dubai,P.O.Box 502216, Dubai, UAE
  iyad.rahwan@buid.ac.ae
[2] Institute of Information and Computing Sciences,
  Utrecht University, Utrecht, The Netherlands
  fkoch@acm.org
[3] Department of Information Systems, University of Melbourne,Parkville, VIC 3010, Australia
  {cgraham, kattan, l.sonenberg}@unimelb.edu.au

Abstract. While interacting with other users in dynamic use contexts, one often aims at coordinating activities as events unfold. Such coordination can often be *unplanned* or *impromptu*. There are opportunities for supporting impromptu coordination among mobile individuals by representing and processing contextual information. In this paper we present a novel technique, based on goal-oriented automated negotiation, to enable computational agents acting on behalf of users to automatically negotiate opportunities for coordination. Our focus is on the technology 'under the skin' that can represent, analyse, and integrate information to support the user's tasks in a timely and appropriate way. An implemented prototype is demonstrated via a scenario, which is based on a workday narrative.

1 Introduction

The use of mobile and handheld computing devices in everyday life is increasing largely due to the advancement of enabling technologies [20] and increasing efforts to improve usability [21]. In order to cope with the dynamism resulting from user mobility during task execution, providing *context-sensitive* support becomes crucial. Dey [6] describes *context* as "any information that can be used to characterize the situation of an entity. An entity is a person, place or object that is considered relevant to the interaction between a user and an application, including the user and the application themselves." For the purpose at hand, context-sensitive support includes performing background computation to identify the situation, in order to take action on behalf of the user, and to represent appropriately to the user the state of any such external interaction.

In this paper, we are concerned with forms of interaction in which multiple users, with differing agendas and interests, may realise opportunities for useful coordination of

---

* This work was initiated while Iyad Rahwan was at the University of Melbourne. Iyad was supported by a University of Melbourne Research Scholarship. We also thank Hewlett Packard's Philanthropy Division for donating equipment.

their activities. This interaction occurs in a dynamic environment and hence automated support for this interaction has to be context-sensitive, taking into account contextual variables such as location, time, and the task(s) at hand. As a result, mobile coordination is often *unplanned*, and therefore takes place *impromptu* as users happen to be in a context where coordination opportunities arise. We are interested in supporting this type of interaction and, therefore, not concerned with cooperation in a classic "groupware" sense. The proposed technology discussed here is not targeted at supporting "groups of people engaged in a common task (or goal)" [7–page 40] or providing "an interface in a shared environment" (ibid). We are concerned with communication among multiple entities and coordination, in a non-team-driven sense. We use the term *coordination* to capture the kinds of activity we have in our focus, adopting Malone and Crowston's definition of coordination as "managing dependencies between activities" [13].

We explore scenarios where the user is engaged in goal-oriented task execution where they could exploit impromptu coordination with others with overlapping task structures. Hence, we consider the user's *task structure* as part of the user's context. We present a novel framework (which we sketched earlier in [17]) for reconciling task structures of different users using *interest-based negotiation* (IBN) [16], a form of automated negotiation that exploits an explicit representation of task structures. We also present a prototype system based on the framework.

The paper advances the state of the art in two ways. To our knowledge, it is the first attempt at using automated negotiation to support non-routine coordination of mobile users. Moreover, the paper introduces a novel coordination architecture, that integrates context-aware networked devices, agent-based reasoning, and automated negotiation. This approach may be used for building a variety of mobile coordination-support systems that suit domains beyond that of the simple narrative used here for illustration.

We begin by discussing the characteristics of mobile coordination in the next section. This helps clarify the features required by potential supporting technology. In Section 3 we present a framework for automated coordination support that provides some of these features, followed by a description of the implemented prototype in Section 3.4. We discuss related work in Section 4 and conclude in Section 5.

2 Problem: Impromptu Coordination While Mobile

We begin with some observations regarding the role of technology in facilitating interactions through which multiple users, with differing agendas and interests, may realise their opportunities for useful coordination of their activities. To better understand the opportunities for technology intervention, we "examine activities in which people engage with others when they are 'mobile' and how various tools and artifacts feature in those activities" [12]. To this end, we analyse a 'Day-in-the-life' scenario [19] to distill essential characteristics of mobile use.

Scenario: Supporting Impromptu Coordination

In the settings of interest, the user is mobile, connected, and engaged in complex interactions. This creates an opportunity for technology to support the user. A basic approach would be to provide connectivity, e.g. using mobile telephones. In this case, users would need to keep track of all changes to their context "in their heads," manage the complexity of identifying opportunities as events unfold, deal with multiple interaction partners, and so on. This places great cognitive load on mobile users, and it is precisely for this reason that support software such as calendar applications are appropriate tools.

When a mobile phone is endowed with a calendar functionality, the user can *outsource* the storage of large amounts of information about activities (meetings, special occasions etc.) to his/her device. This representation of *individual* activities can then be used to help a user coordinate with others. Applications allowing for group task representation go a step further by providing users with global activity representations.

One could envisage device support not only through *representation* of individual and group activities, but also *automation* to support the cognitive processes that exploit and manipulate those representations. Such automatic processes would use the available information about the user's context as well as information available about other users in order to automatically negotiate agreements over collaboration and coordination of activities. We now turn to look in more detail at types of situations where such support could be advantageous - with a view to illustrating the settings in which negotiation can effectively facilitate impromptu coordination. "Mobility" [15] naturally creates such opportunities as we shall see. The analysis below emerged from discussions in a multi-disciplinary focus group and from a narrative based on a diary of an actual PhD student renamed Fred, generated over a period of three days. The narrative approach has been used in order to understand individual mobile activities in other projects, such as ActiveCampus [8]. An approach grounded in broader and more systematic data collection would be desirable in the future, c.f. [10].

> I realized I had not set up a lift home so I called my wife. I couldn't get through, so I left her a message and asked her to call me when she was close. While waiting for her to reply, I continued work. Then Jack gave me a call to discuss our Wednesday meeting. Jack asked if I could get him a book from the university library, which he needs for an assignment. I declined because I needed more time to finish my work. But Jack happened to be planning to head home to study at the same time I wanted to leave the University. I managed to get myself a lift home by offering to help him out with his assignment, in which case he no longer needed the book.

From the analysis of this scenario, four distinctive characteristics of mobile work emerged. We argue that "negotiation" has the potential to address these characteristics.

Fluidity. Kakihara and Sorenson [11] describe how interaction experienced by mobile individuals is "fluid". They describe how "human interaction is becoming ambiguous and transitory. The patterns of social interaction are dynamically reshaped and renegotiated through our everyday activities significantly freed from spatial, temporal and contextual constraints" [ibid]. Fluidity in mobility suggests that interaction can be occasional in mobile use situations, since the context in which these portable devices operate changes more frequently than with stationary computers. Thus, well-established, long-term relationships, in which task structures are well-defined and agreed upon, are less likely (e.g. due to the dynamism of resource availability). Negotiation is one way to reach temporary agreement in such dynamic settings, as Fred did in the above scenario.

Impromptu Coordination. For mobile users, opportunities for collaboration arise more frequently than with static users due to the more diverse forms of context change, such as change in the user's location or the proximity of multiple users. Such opportunities usually cannot be anticipated a priori. In the narrative above, Fred being connected to Jack was critical to him being able to capitalise on the opportunity presented by Jack's proximity. The phone did not allow him to predict the possible chances of the success of this opportunistic interaction through a representation of Jack's goals or tasks. Negotiation is a way of dynamically realising and taking advantage of such opportunities.

Heterogeneity. When the modelling of context is to take into account varying location, time, user profiles, tasks, interaction history etc., we are confronted with a much greater variety of agent (and user) types. Each individual agent may achieve tasks in a different way. It is unlikely that information about this heterogeneity will be available a priori. Negotiation is a natural way to exchange information and reach useful agreement or compromise with collaborators (or in collaboration settings) not known before. In the above narrative, Fred's coordination could have been made easier by accessing a representation of Jack's activities or, at the very least, his availability.

Privacy and Connectivity. Mobile users are constantly confronted with different interaction partners that want to obtain information about them. Users may be unwilling to disclose all the information required to run a centralised algorithm for coordinating joint activity. They may be willing to do so only when interacting with particular partners, or when the they realise the potential benefit of exchanging such information. Negotiation is a natural way to reconcile one's own wish to protect private information with the potential benefit of interacting with others.

3 Solution: Negotiating Impromptu Coordination

In the previous section, we argued that impromptu mobile coordination requires the ability to represent information about the tasks of different users, and the ability to interactively process this information. We now present a framework, based on goal-directed automated negotiation, that addresses some of these requirements.

3.1 Conceptual Model

The conceptual framework for mobile user coordination through automated negotiation is illustrated in Figure 1. An agent running on a user's mobile device acts as an intermediary between the user and other potential collaborators. The agent gathers contextual information from the environment (e.g., lecture times, location of user and colleagues) and from the user (e.g., availability, goals). The agent then uses this information, as well as domain-specific knowledge (e.g., procedures for borrowing books from the library) in order to negotiate with agents representing other users. Negotiations are motivated by the user's goals and aim at achieving "deals" with other users. If negotiation results in potentially useful deals (e.g., appointment, lunch, lift home), these are proposed to the respective users, who might accept, reject, or modify these deals as they see suitable.

Note that agents may have incomplete information about each others' plans and desires, and about the environment and the appropriate planning procedures within it. In

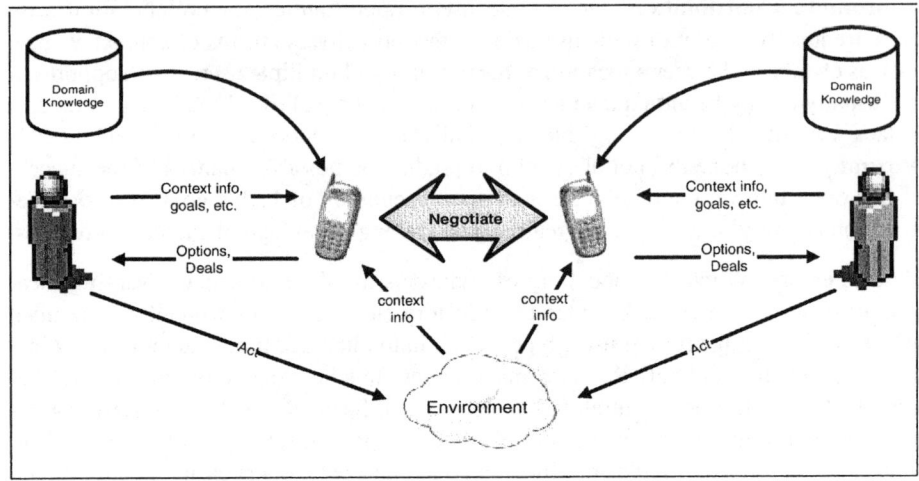

Fig. 1. Conceptual framework for automatically negotiated mobile coordination

order to address this issue, the automated negotiation framework must enable agents to exchange information about these underlying notions. Since the negotiation mechanism required must exploit representations of users' tasks and goals, interest-based negotiation seems an appropriate choice.

3.2 The Architecture

At it's core, our solution is based on an extended implementation of *3APL* [9]. 3APL is a logic-based agent programming language which provides constructs for implementing agents' beliefs, goals and capabilities as explicit run-time entities. It uses practical reasoning rules in order to generate plans for achieving agents' goals and for updating or revising these plans. Each 3APL program is executed by means of an interpreter that deliberates on the cognitive attitudes of that agent. Due to space limitation, we shall not discuss 3APL in detail, and suffice by introducing the general idea behind it.

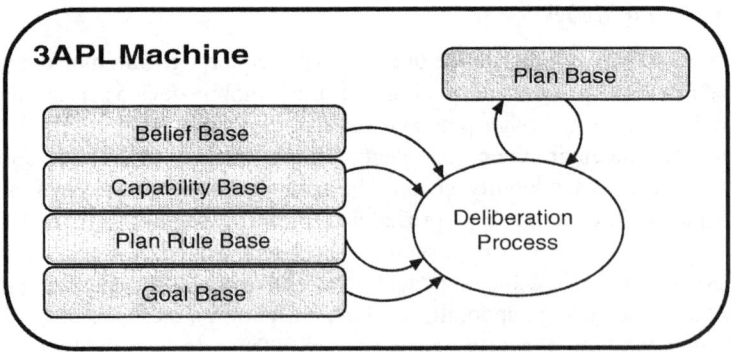

Fig. 2. 3APL Architecture

Figure 2 describes the abstract architecture of 3APL. Each computational agent has explicit representations of its goals in the *goal base*, which contains goals in the form of ground predicates. For example, the goal to finish an assignment may be represented with the predicate *finish(assignment)*. In order to achieve its goals, the agent decomposes these into less abstract *sub-goals* using planning rules from the *plan rule base*, which may themselves be decomposable into other sub-sub-goals, until concrete *actions* are reached (i.e., physical actions agents may execute directly in the world). This results in a hierarchical structure in which the top-level root nodes represent desires, intermediate nodes represent abstract goals, and leaf nodes represent concrete actions to be executed. During plan generation, the agent takes into account its *belief base*. Beliefs take the form of predicates and represent contextual information the agent may use in its reasoning. For example, the predicate *location(jack, bldgA)* could denote that the agent believes Jack is currently located in building A. In a mobile coordination scenario, beliefs may be acquired by sensing the environment (e.g. using a location system) or directly through user input. The *capability base* describes possible actions by the agent and user. A planning rule takes the form *head ← guard|body*, and means that if the agent has goal g that unifies to the head of the plan *head* and the condition declared in *guard* is satisfied (i.e., it unifies to the contents of the belief base), then goal g can be achieved by executing the sequence of actions (or set of sub-goals) listed in *body*. Following is an example planning rule which states that if one wants to finish an assignment, and is currently outside, then one needs to collect the relevant book and go home to study:

$$finish(assignment) \leftarrow location(me, out) \mid collect(book), go(home)$$

where *go(home)* may have to be decomposed to further sub-goals, such as catching a taxi or taking the train.

In the initial 3APL version presented by Hindriks et al [9], the deliberation cycle is fixed and agents generate their plans by choosing the first applicable rule that matches a particular goal/desire. This means that an agent generates only one plan for each goal, and only generates other plans if the initial plan fails. This deliberation cycle is depicted in Figure 3.

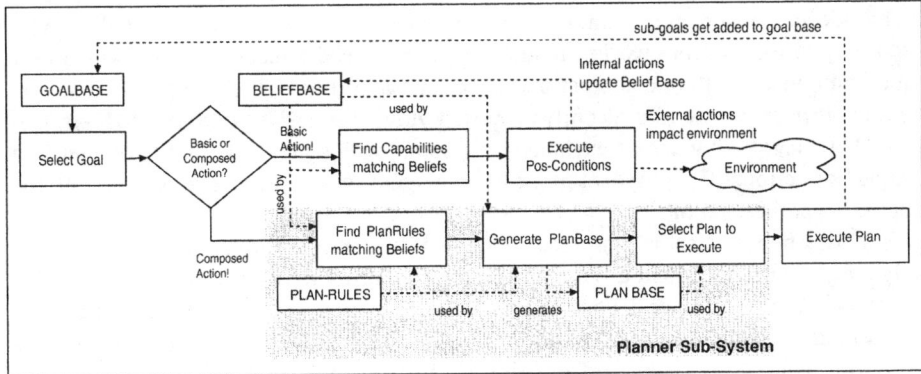

Fig. 3. Basic Deliberation Cycle

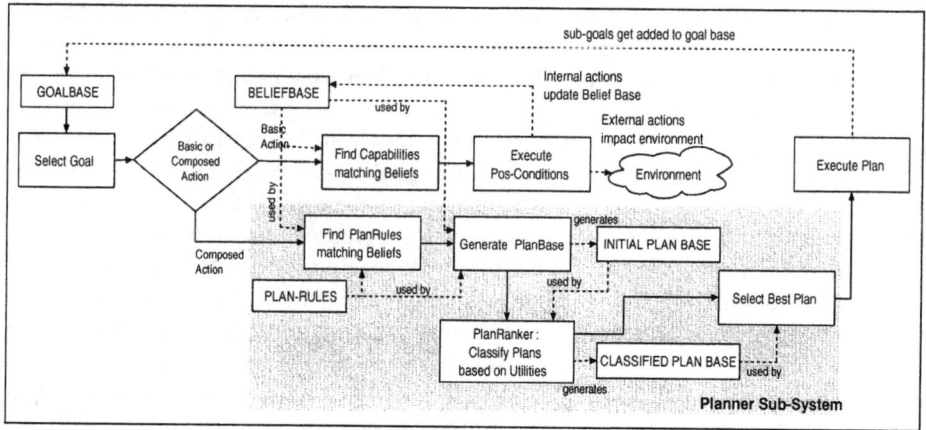

Fig. 4. Extended Deliberation Cycle

We want our agents to generate all possible plans and select the best plan using some appropriate criteria (perhaps based on user preferences), which is performed by the *plan ranker*. Currently, the plan ranker works by calculating the *utility* of different plans based on the *costs* of executing the plan and the *worth* of the goals achieved. This process results the *plan base*, a list of plans generated by the deliberation process. This list is available during the processing phase and can be manipulated by an external logic (supplied by the developer). This allows the creation of external Plan Analysers, that can be use to improve the deliberation cycle (for example, classify plans per utility). This *Deliberative 3APL* implementation is based on an abstract modified reasoning cycle presented by Dastani et al [5]. The deliberation cycle is described in Figure 4.

Each user in our framework has a 3APL agent running on his/her device. The agent has a representation of the user's goals, a set of predefined capabilities, and a set of beliefs that represent the current context. We assume this knowledge is captured through a separate mechanism (see Section 5 for more discussion). The agent must be capable of receiving information from the user and environment, and taking action externally (e.g. activating a reminder alarm or requesting information from the user). This is catered for by adding sensor and actuator interfaces that allow the integration of the 3APL machinery to the external world. In addition, agents need a mechanism by which they can communicate with one another and negotiate opportunities for coordinated action. For that purpose, we use our recently proposed *interest-based negotiation* (IBN) framework [16]. Negotiating agents exchange *proposals*, which are suggested "exchanges of favours" or actions. If a proposal made by one agent is accepted by another agent, the proposal becomes a *deal*.

The idea behind IBN is that agents can also exchange information about their underlying goal structures and use this information in order to discover better deals. This communication takes place according to a protocol, which is discussed in detail in [16]. This extended architecture is described in figure 5. Basically, the IBN protocol provides a way for agents to exchange information about their own goal structures, and to exchange new planning rules and beliefs. This enables agents to discover opportunities

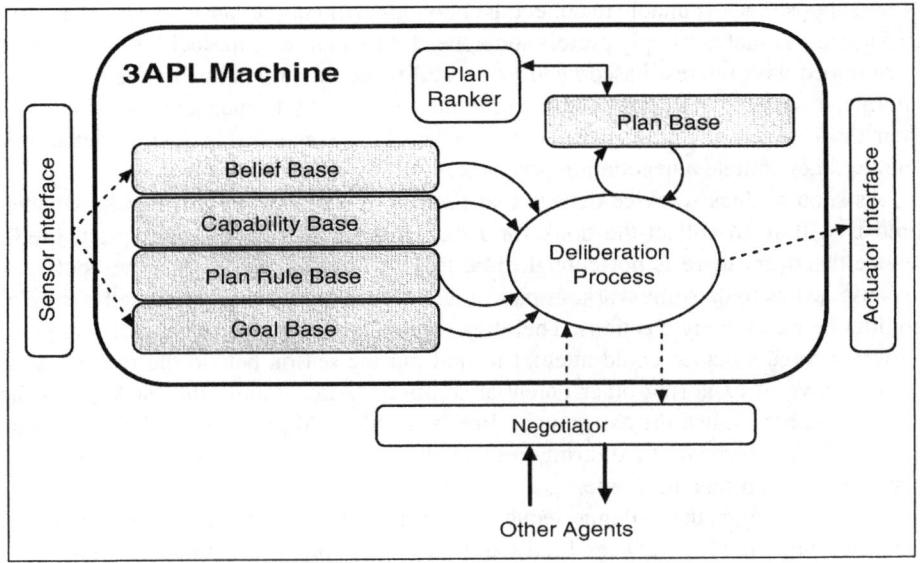

Fig. 5. 3APL-M and IBN

for reconciling their individual plans for mutual benefit. The way this works will be clarified in the following subsection.

3.3 Illustrative Example

Let us revisit the narrative from section 2. Recall the situation where Fred fails to get in contact with his wife to secure a lift home. Suppose Fred has a permanent desire to save money whenever possible, and this is stored in his device in the form of a desire generation rule for which the rule body is always true. One way of saving money is by getting a lift home rather than taking the train or a taxi. This is represented on the device as a planning rule. The two rules are encoded as follows, where $\mathcal{G}^i$ denotes the goals of user i and $\mathcal{P}^i$ denotes the planning rules i's agent is aware of.

$$\mathcal{G}^{Fred} = \{save(fred))\}$$
$$\mathcal{P}^{Fred} = \{save(Y) \leftarrow \text{true} \mid GetLift(X, Y)\}$$

Recall also that Jack would like to finish his assignment, and in order to achieve that, he believe he needs to collect a book and go home.[1]

$$\mathcal{G}^{Jack} = \{finish(assignment)\}$$
$$\mathcal{P}^{Jack} = \{finish(assignment) \leftarrow \text{true} \mid Collect(book), Go(home)\}$$

---

[1] Note that we do not have an explicit representation of time and temporal constraints. This is mainly to simplify implementation. A realistic implementation would of course need to account for temporal aspects, integrate tasks in a proper calendar application etc.

Now, suppose Jack is unable to collect the book himself (say because he has other tasks to perform) or that he simply prefers someone else to do it for him. Jack's device, which is equipped with interest-based negotiation ability, could automatically attempt to find alternative ways to get a lift home by searching for nearby friends and checking (with their devices) for potential collaboration. When Fred's and Jack's devices detect one another, they initiate a negotiation process.

As soon as Jack's device detects that Fred is in a nearby area, it checks whether Fred is willing to collect the book for Jack. Upon inspection of the request, Fred's device discovers there is not enough time to go to the library to pick the book, say because he has to do some work, as part of another task; i.e., that there is some form of conflict between the two actions. Therefore, Fred's device would reject Jack's request. However, Fred's device could attempt to find out the reason behind the request, with the objective of exploring other potential solutions. After finding that Jack wants the book in order to finish the assignment, Fred's agent could propose an alternative way to achieve the objective, by offering Fred's help. This is not for free, though, as it is in exchange for getting a lift home.

The following is the dialogue sequence just described, encoded using the IBN protocol presented in [16], between Fred's and Jack's negotiation-enabled mobile devices. Locutions PROPOSE(.), ACCEPT(.) and REJECT(.) allows agents to propose, accept and reject deals, respectively. The locution ASSERT(.) allows an agent to make assertions about its own beliefs, goals, sub-goals or planning rules. Finally, locution REQ-PURPOSE(.) allows an agent to ask another for the higher-level purpose of a particular request (e.g. "why do you need the book?"). The special predicate prule(.) is used to exchange planning rules among agents. Another modality $instr(X, Y)$ is used by one agent to indicate that goals X are adopted because they are *instrumental* towards achieving higher-level goals Y. Part (a) in Figure 6 shows a sketch of (parts of) the plan structures for Fred and Jack from the narrative. Part (b) shows Jack's modified plan, which can achieve his goals while also helping Fred.

JACK: PROPOSE(*jack,fred*, **do**(*fred, Collect(book)*))
FRED: REJECT(*fred,jack*, **do**(*fred, Collect(book)*))

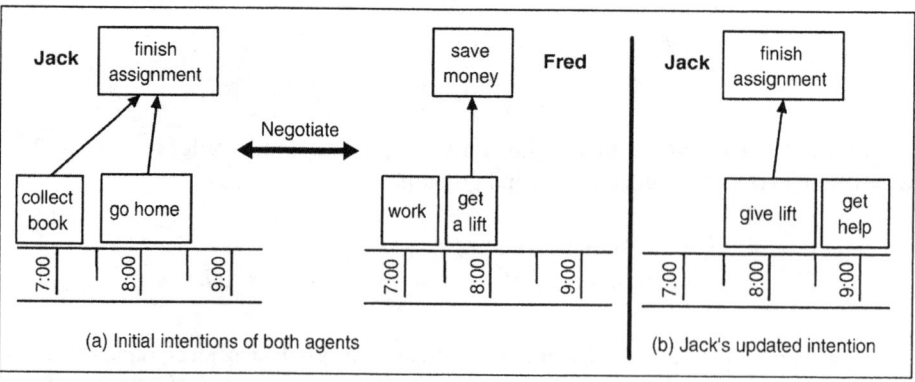

Fig. 6. An abstract view of negotiation

JACK: PASS(*jack*)
FRED: REQ-PURPOSE(*fred, jack, Collect*(*book*))
JACK: ASSERT(*jack, fred,* instr(*Collect*(*book*), *finish*(*assignment*)))
FRED: ASSERT(*fred, jack,* prule(*finish*(*assignment*) ←
       true | *Help*(*assignment*), *GiveLift*(*jack, fred*))
JACK: PASS(*jack*)
FRED: PROPOSE(*fred, jack,* **do**(*jack, GiveLift*(*jack, fred*)),
       do(*fred, Help*(*assignment*)))
JACK: ACCEPT(*jack, fred,* **do**(*jack, GiveLift*(*jack, fred*)),
       do(*fred, Help*(*assignment*)))
FRED: OK(*fred, jack,* **do**(*jack, GiveLift*(*jack, fred*)),
       do(*fred, Help*(*assignment*)))

There are other types of arguments that Fred and Jack could exchange. For example, Jack could ask Fred why he needs the lift, and after finding out that is is purely for saving, he could offer to take him out for dinner instead, in exchange for collecting the book. Not all such arguments may be realistic. A sensible approach might be to encode a set of typical "attitudes" towards cooperation (represented as negotiation strategies), from which the user could choose for his/her agent, or which could be automatically learned by observing the user's response over a period of time.

3.4 Results

We have implemented a prototype system as a proof-of-concept for our framework. The prototype uses our implementation of *3APL-M* [1] using Java Micro Edition. The deliberation part of 3APL-M makes use of *mProlog*, a scaled-down Prolog engine. It is a sub-product of this project and packaged with the 3APL-M application library. Sensors and Actuators are developed in Java and attached to the 3APL-M machinery through special methods in the API. Figure 7 shows sample screen shots of the prototype. On the left is the interface of Jack's mobile device, showing the list of active tasks. The user can add new tasks or edit existing tasks. The right hand side of the figure shows Fred's mobile device, with an instant messaging dialogue with another user named "Bill."

The "Deliberate" button on Jack's device triggers the deliberation process, which is described in figure 8. In this figure, the normal arrows denote messages that immediately follow an action by the user. For example, after Fred clicks "Yes" on the top-right screen, a REJECT(.) message is sent. Dotted lines, on the other hand, denote messages exchanged by the agents without an explicit user action. For example, the REQ-PURPOSE(.) message from Fred's agent to Jack's agent did not require a confirmation or explicit request by the user, but is rather triggered automatically as part of Fred's agent's built-in coordination strategy.

In this particular implementation, Jack's agent selects Jack's task to finish the assignment, and by applying the planning rule mentioned above, creates a plan that involves collecting a book and then going home. After searching for nearby friends, Jack's agent discovers Fred's presence and suggests to Jack to ask Fred to collect the book on his behalf. If Jack clicks on "Yes" (part (a) in figure 8), his agent initiates a negotiation dialogue with Fred's agent by proposing that Fred pick the book. Fred's device

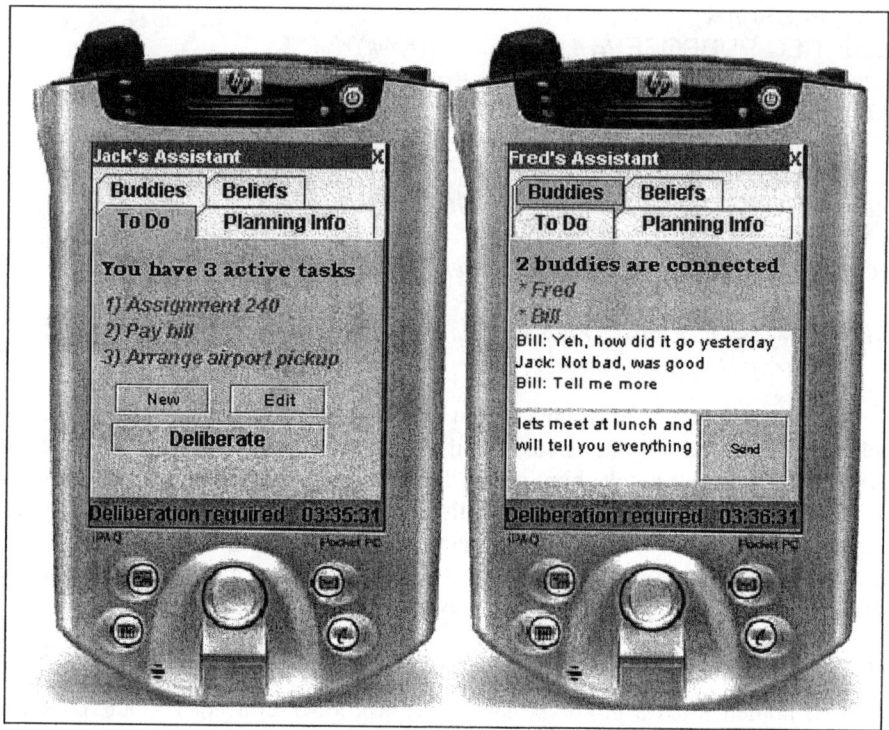

Fig. 7. Screen shot of the prototype implementation

presents this proposal, which Fred rejects by clicking "No" (part (b)). However, Fred's agent attempts to discover the reason behind Jack's request, and based on that information, presents an alternative. Then Fred's agent asks Fred whether he would be interested in getting a lift home (part (c)). Fred clicks "Yes," after which Fred's agent sends a proposal to Jack's agent. Upon receipt, Jack's agent presents the proposal to Jack, who accepts the proposal by clicking "Yes" (part (d)). This approval is sent to Fred's agent, which informs Fred of the deal and sends a confirmation back to Jack (part (e)).

4 Related Work

A growing body of research looks at the importance of unintended interactions in task-oriented (typically workplace) settings, and on systems that partially support discovery of opportunity for such interactions [8, 10]. For many activities involving coordination between mobile users, geographic co-location is a key indicator of opportunity, and this frequently applies also to the problems we are interested in. Increasingly, location aware devices can provide robust access to such data. The ActiveCampus project [8] is one example of such an approach – deployed in a university setting and mediating serendipitous learning opportunities.

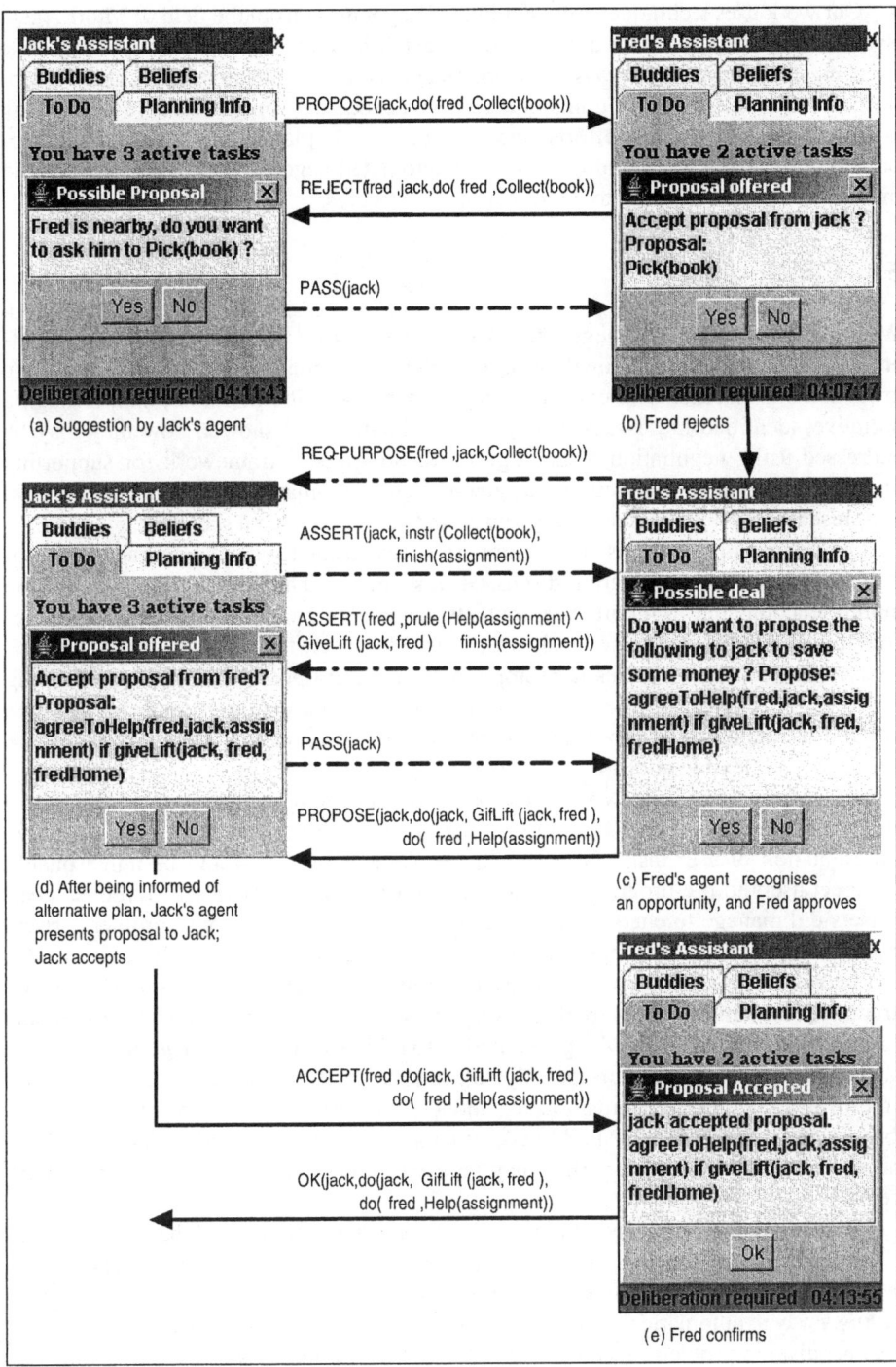

Fig. 8. Screen shots of the scenario run

Our work uses techniques from automated negotiation from the field of Multi-Agent Systems (MAS) [22]. Other attempts to use agent-based techniques to support mobile users exist. However, most existing work on agents for mobile devices focuses on supporting single users [18] or collaborative teams executing routine tasks [3]. Our work is complementary to these efforts since we focus on unplanned, opportunistic collaboration. Paurobally and Jennings [14] uses automated negotiation for negotiating paid mobile services as opposed to coordination of activities.

5 Conclusions

We argued that automated negotiation technologies, from the multi-agent systems literature, could support coordination of activities in dynamic, mobile use contexts. We grounded our discussion in current views of mobility in the literature, and, through a narrative, identified key issues of mobile coordination and showed how they may be addressed using negotiation technologies. We presented a framework for supporting unplanned coordination based on our goal-directed automated negotiation framework. We presented a pilot application and illustrated its use through a scenario.

Our framework supports the features required to deal with the characteristics of impromptu mobile coordination discussed in section 2. The framework caters for the fluidity encountered in mobile use contexts, since coordination does not assume pre-determined and pre-negotiated task structures. Moreover, the focus on tasks and their underlying goals also enables impromptu realisation of opportunities for coordinating activities. By expressing the resources and objectives explicitly, it becomes possible to build technology that processes this information in order to "allow more effective planning land flexible allocation of resources" [15].

A very important issue worth mentioning is "knowledge elicitation." The scenario implemented here was based on a set of hard-coded rules, based on a relatively naive representation of user tasks. In a realistic implementation, it would be unreasonable to expect users to pre-program their task structures into their devices. Moreover, even if users did manage to encode their task representation, an 'ontology' problem arises, since different users may represent the same tasks differently, and adequate translation between these description is likely to be nontrivial. However, in relatively well-structured domains, such as truck delivery, pre-programming typical task representations may be practical. Knowledge elicitation to initialise automated negotiating agents has begun to receive attention recently [2]. In less structured domains, it may be possible to use knowledge acquisition techniques based on machine learning [4] to enable agents to automatically build task structures by observing user behaviour. Once these structures are available, IBN may be used to automate interaction that exploits them.

Future work can be carried out in two directions: the technical and the user-oriented. Technical challenges include integrating our system with existing calendar and organiser applications that run on common mobile devices. Another technical challenge is finding a way to automatically capture the task knowledge in order to address the knowledge acquisition problem. From a user-oriented perspective, it is essential to study the design issues surrounding the interaction between the device and the user, as well as those relating to social interaction.

References

1. 3APL-M, 2004. http://www.cs.uu.nl/3APL-M.
2. J. J. Castro-Schez, N. R. Jennings, X. Luo, and N. Shadbolt. Acquiring domain knowledge for negotiating agents: a case study. *Human-Computer Studies*, 61(1):3–31, 2004.
3. H. Chalupsky, Y. Gil, C. A. Knoblock, K. Lerman, J. Oh, D. V. Pynadath, T. A. Russ, and M. Tambe. Electric elves: Applying agent technology to support human organizations. In H. Hirsh and S. Chien, editors, *Proc. IAAI-2001*. AAAI Press, 2001.
4. S. F. Chipman and A. L. Meyrowitz. *Foundations of Knowledge Acquisition: Machine Learning*. Kluwer Academic Publishers, 1993.
5. M. Dastani, F. de Boer, F. Dignum, and J.-J. Meyer. Programming agent deliberation: an approach illustrated using the 3APL language. In *Proc. AAMAS*. ACM Press, 2003.
6. A. Dey. Understanding and using context. *Personal and Ubiquitous Computing*, 5(1), 2001.
7. C. A. Ellis, S. J. G. G. L, and Rein. Groupware: Some issues and experiences. *Communications of ACM*, 34(1):38–58, 1991.
8. W. G. Griswold, R. Boyer, S. W. Brown, T. M. Truong, E. Bhasker, G. R. Jay, and R. B. Shapiro. Using mobile technology to create opportunitistic interactions on a university campus. In *W.shop on Supporting Spontaneous Interaction in Ubiquitous Comp. Settings*, 2002.
9. K. V. Hindriks, F. S. de Boer, W. van der Hoek, and J.-J. Meyer. Agent programming in 3apl. *Autonomous Agents and Multi-Agent Systems*, 2(4):357–401, 1999.
10. E. A. Isaacs, J. C. Tang, and T. Morris. Piazza: a desktop environment supporting impromptu and planned interactions. In *Proc. ACM conf. on CSCW*, pages 315–324. ACM Press, 1996.
11. M. Kakihara and C. Sørensen. Mobility: An extended perspective. In R. S. Jr, editor, *Proc. 35th Hawaii Int. Conf. on Systems Sciences*. IEEE Press, 2002.
12. P. Luff and C. Heath. Mobility in collaboration. In *Proceedings of the 1998 ACM conference on Computer Supported Cooperative Work*, pages 305–314. ACM Press, 1998.
13. T. W. Malone and K. Crowston. The interdisciplinary study of coordination. *ACM Computing Surveys*, 26(1):87–119, 1994.
14. S. Paurobally, P. J. Turner, and N. R. Jennings. Automating negotiation for m-services. *IEEE Trans. on Systems, Man and Cybernetics (Part A: Systems and Humans)*, 33(6), 2003.
15. M. Perry, K. O'Hara, A. Sellen, B. Brown, and R. Harper. Dealing with mobility: understanding access anytime, anywhere. *ACM Trans. on Comp. Human Interaction*, 8(4), 2001.
16. I. Rahwan. *Interest-based Negotiation in Multi-Agent Systems*. PhD thesis, Department of Information Systems, University of Melbourne, Melbourne, Australia, 2004.
17. I. Rahwan, C. Graham, and L. Sonenberg. Supporting impromptu coordination using automated negotiation. In *Proc. Pacific Rim Int. Workshop on MultiAgent Systems*, 2004.
18. T. Rahwan, T. Rahwan, I. Rahwan, and R. Ashri. Agent-based support for mobile users using AgentSpeak(L). In P. Giorgini, B. Hederson-Sellers, and M. Winikoff, editors, *Agent Oriented Information Systems*, volume 3030 of *LNAI*, pages 47–62. Springer Verlag, 2004.
19. R. Reimann and E. Bacon. A scenario-based approach to creating interaction frameworks. In *Proc. Workshop on Tools, Conceptual Frameworks, and Empirical Studies for Early Stages of Design*, 2001.
20. N. M. Sadeh. *M–Commerce Technologies, Service, and Business Models*. Wiley, 2003.
21. S. Weiss. *Handheld Usability*. Wiley, Hoboken NJ, USA, 2002.
22. M. J. Wooldridge. *An Introduction to MultiAgent Systems*. John Wiley & Sons, 2002.

In Defense of Contextual Vocabulary Acquisition
How to Do Things with Words in Context

William J. Rapaport

Departments of Computer Science and Engineering and Philosophy;
Center for Cognitive Science,
State University of New York at Buffalo, Buffalo, NY 14260-2000, USA
+1-716-645-3180x112; fax +1-716-645-3464
rapaport@cse.buffalo.edu
http://www.cse.buffalo.edu/~rapaport/CVA/

Abstract. Researchers in "contextual vocabulary acquisition" differ over the kinds of context involved in vocabulary learning, and the methods and benefits thereof. This paper presents a computational theory of contextual vocabulary acquisition, identifies the relevant notion of context, exhibits the assumptions behind some classic objections, and defends our theory against these objections.

1 A Computational Theory of Contextual Vocabulary Acquisition

Contextual vocabulary acquisition (CVA) is the deliberate acquisition of a meaning for a word in a text by reasoning from context, where "context" *includes*: (1) the reader's "internalization" of the surrounding text, i.e., the reader's "mental model" of the word's "*textual* context" (hereafter, "co-text" [3]) integrated with (2) the reader's prior knowledge (PK), but it *excludes* (3) external sources such as dictionaries or people. CVA is what you do when you come across an unfamiliar word in your reading, realize that you don't know what it means, decide that you need to know what it means in order to understand the passage, but there is no one around to ask, and it is not in the dictionary (or you are too lazy to look it up). In such a case, you can try to figure out its meaning "from context", i.e., from clues in the co-text together with your prior knowledge.

Our computational theory of CVA—implemented in a the SNePS knowledge representation and reasoning system [28]—begins with a stored knowledge base containing SNePS representations of relevant PK, inputs SNePS representations of a passage containing an unfamiliar word, and draws inferences from these two (integrated) information sources. When asked to define the word, definition algorithms deductively search the resulting network for information of the sort that might be found in a dictionary definition, outputting a definition frame whose slots are the kinds of features that a definition might contain (e.g., class membership, properties, actions, spatio-temporal information, etc.) and whose slot-fillers contain information gleaned from the network [6–8,20,23,24].

We are investigating ways to make our system more robust, to embed it in a natural-language-processing system, and to incorporate morphological information. Our research group, including reading educators, is also applying our methods to the development of what we hope will be a better pedagogical curriculum than the current state of the art for teaching CVA.

To this end, we have been studying the CVA literature from a variety of disciplines that, generally speaking, seem to ignore each other's literature (including computational linguistics, reading education, second-language education, and psychology [22]). Two often-cited papers by reading scientists [2,26] have claimed that not only are certain contexts less than useful for doing CVA, but that most "natural" (as opposed to artificial) contexts are not helpful at all. Their arguments make several assumptions inconsistent with our computational theory. Thus, their objections do not apply to us.

2 Are All Contexts Created Equal?

2.1 The Role of Prior Knowledge

Beck et al.'s [2], subtitled "All Contexts Are Not Created Equal", claims that "it is not true that every context is an appropriate or effective instructional means for vocabulary development". They begin by pointing out that "the context that surrounds a word in text can give *clues* to *the word's meaning*" (my italics). But a passage is not a clue without some other information to interpret it as a clue. Therefore **(A1) Textual clues must be supplemented with other information in order for a meaning to be computed.** *This supplemental information must come from the reader's PK.* Such PK (which need not be true) might include general "world" or cultural knowledge, "commonsense" knowledge, specialized "domain" knowledge, and perhaps the "background" knowledge the author assumes the reader will have. However, not all of the reader's PK may be consciously available at the time of reading, and each reader will bring to bear upon his or her interpretation of the text *idiosyncratic* PK [10,12].

I will use 'co-text' to refer to the text surrounding an unfamiliar word, reserving 'context' or 'wide context' to refer to the reader's available PK "integrated" with the reader's "internalization" (or "mental model") of the co-text. Its integration with the reader's PK involves belief revision: New beliefs would be inferred as conclusions from arguments in which at least one premise comes from the internalized co-text and at least one premise comes from PK. Typically, withdrawn beliefs are PK beliefs inconsistent with co-text propositions [20]. The "context" that the reader uses to compute a word's meaning is not just the *co-text* but this *wider* context.

The reader's internalization of the text involves some interpretation (e.g., resolving pronoun anaphora) or the immediate and unconscious drawing of an inference (e.g., that 'he' refers back to a male or that 'John' is a proper name typically referring to a male human) [10]. Consider the following natural passage (my italics): "The archives of the medical department of Lourdes are filled with *dossiers* that detail well-authenticated cases of what are termed miraculous healings" [17]. Is this to be understood as saying (a) that the archives are filled with dossiers, and that *these dossiers* detail cases of miraculous healings? Or is it to be understood as saying (b) that the archives are filled with dossiers, and *dossiers in general* are things that detail cases of miraculous healings? The difference in interpretation has to do with whether "detail ... miraculous healings" is a restrictive relative clause (case (a)) or a non-restrictive relative clause (case (b)).

Arguably, it should be understood as in (a); otherwise, the author should have written, 'The archives are filled with dossiers, *which* detail miraculous healings". But a reader might not be sensitive to this distinction. Misinterpretation cuts both ways: The author might not be sensitive to it, either, and might have written it one way though intending the other. It makes a difference for CVA. A reader unfamiliar with 'dossier' might conclude from the restrictive interpretation that a dossier is something found in an archive and that these particular dossiers detail miraculous healings, whereas a reader who internalized the non-restrictive interpretation might conclude that a dossier is something found in an archive that (necessarily) details miraculous healings. (We have anecdotal evidence that at least some readers of this passage interpret it in the latter way.)

Even a common word can mean different things to different people: Something that looks like a sofa but seats only one is a 'sofa' in Indian English but a 'chair' in American English. Thus, two fluent English speakers might interpret a passage containing the word 'sofa' differently: The text is the same in both cases, but the readers' *internalized* texts will differ. It can also involve simple misreading: I read the sign on a truck parked outside one of our university cafeterias, where food-delivery trucks usually park, as "Mills Wedding and Specialty Cakes". Why had I never heard of this local bakery? Why might they be delivering a cake? So I re-read the truck's sign more carefully. It actually said, "Mills Welding and Specialty Gases"! A related modifying influence stems from reading difficulties that might circumscribe the amount of co-text that the reader can understand and therefore integrate into his or her mental model [30].

2.2 Do Words Have Unique Meanings?

The assumption—inconsistent with our theory—underlying [2]'s use of the phrase 'the word's meaning' is **(A2) A word has a *unique* meaning.** The definite description '*the* word's meaning' or '*the* meaning of a word' is ubiquitous but worth avoiding, for it incorrectly suggests that a word has a unique meaning. Perhaps what's normally intended by this phrase is "the meaning of a word *in the present context*": "[C]ontext always *determines* the meaning of a word, it does not always *reveal* it" (Deighton, cited in [26]). But it follows from our observations about **(A1)** that the reader will supplement the co-text with idiosyncratic PK; consequently, each reader will interpret the word slightly differently. Deighton is still essentially correct: Wide context determines *a* meaning for the word, though it requires further processing to reveal that meaning.

The need for further processing underlies [14]'s observation that we don't store definitions, even of words we understand. It also undercuts pedagogical strategies for CVA that instruct the reader merely to "guess" the meaning [4]. Nation [19] even boasts that his guessing strategy "does not draw on background content knowledge" since "linguistic clues will be present in every context, background clues will not". But background knowledge (PK) is essential and always used; it is unavoidable, even in Nation's own strategy: Where he says "Guess" (the entire step 4 in his 5-part strategy!), he must in fact mean "make an educated guess", i.e., an inference, but that inference must rely on more premises than merely what is explicit in the text; such premises come from PK [20,29].

2.3 Do Words Have Correct Meanings?

A closely related assumption that many authors make is **(A3) There is such a thing as "the *correct* meaning of a word"**. "[E]ven the appearance of each target word in a strong, directive context [i.e., a context conducive to figuring out "a correct meaning"] is far from sufficient to develop *full knowledge* of word meaning" [2; my italics].

Perhaps what is meant by the "correct" meaning is that there is a certain meaning that *the author* intended. But if we are concerned with a word's meaning *as determined by* the reader's internalized co-text integrated with the reader's PK, then it might very well be the case that the *author's* intended meaning is *not* thus determined. Our investigations suggest that this is almost always the case. The best that can be hoped for is that a reader will be able to hypothesize or construct *a* meaning *for* the word, rather than *the* meaning *of* the word (i.e., the reader *gives* or *assigns* a meaning *to* the word).

If the meaning that the reader computes *is* the intended one, so much the better. If not, has the reader then *mis*understood the text? Misunderstanding is not necessarily a bad thing: If no one ever understood texts differently from other readers or from the author's intended meaning, there would be little need for reading instruction, literary criticism, legal scholarship, etc. Because of individual differences in our idiosyncratic conceptual meanings, we *always* misunderstand each other [21]. This is the mechanism that makes conversation and the exchange of information possible [25]. The important question is not whether a reader can compute *the correct* meaning of a word, but whether the reader can compute *a* meaning for the word *that is sufficient to enable him or her to understand the text*. The reader need not understand the text "perfectly", but merely *well enough to continue reading*. We don't normally have, *nor do we need*, full, correct definitions of the words we understand [14].

Consider the following passage:[1] "All chances for agreement were now gone, and compromise would now be impossible; in short, an *impasse* had been reached" [5]. A reader might compute a meaning for 'impasse' from this text thusly: A compromise is an agreement. If all chances for agreement are gone, then agreement is impossible. So both conjuncts of the first clause say almost the same thing. 'In short' is a clue that what follows means almost the same as what precedes it. So, to say that an impasse has been reached is to say that agreement is impossible. And (perhaps with a bit more plausible PK) that means that an impasse is a *dis*agreement. *Is* it? At least one dictionary defines it simply as a "deadlock". Suppose that "deadlock" is "the correct meaning" of 'impasse'. If the reader decides that 'impasse' means "disagreement", not "deadlock", has the reader misunderstood the passage? Consider the following scenarios: (1) The reader never sees the word 'impasse' again. It then hardly matters whether she has not "correctly" understood the word (though, in the case of this particular bit of CVA, she has surely computed *a* very plausible meaning). (2) The reader sees the word again in a context in which "disagreement" is a plausible meaning. Since her PK now includes a belief that 'impasse' means "disagreement", this surely helps in understanding the new passage. (3) The reader sees the word again in a context in which "deadlock", not

---

[1] From an article detailing teachable contextual clues for CVA; so this might be a "pedagogical", not a "natural", passage. Our project reports are at [http://www.cse.buffalo.edu/~rapaport/CVA/].

"disagreement", is the "best" meaning. E.g., she might read a text discussing operating-system deadlocks, in which a particular deadlock is referred to as an "impasse". Here, it *might* make little sense to consider the situation as a "disagreement", so: (3a) The reader might decide that this occurrence of 'impasse' could not possibly mean "disagreement". Again, there are two possibilities: (3a-i) She decides that she must have been wrong about 'impasse' meaning "disagreement", and she now comes to believe (say) that it means "deadlock". (3a-ii) She decides that 'impasse' is polysemous, and that "deadlock" is a second meaning. (Cf. the polysemous verb 'to dress' [23]; a reader might firmly believe that to dress is to put clothes on but, from co-texts such as "King Claudas dressed his spear before battle", infers that to dress is *also* to prepare for battle.) (3b) Or the reader might try to reconcile the two possible meanings, perhaps by viewing deadlocks as disagreements, if only metaphorically [1].

2.4 Two Kinds of Textual Context

Beck et al. are interested in using *co-text* to help *teach* "the meaning" of the word. We, however, are interested in using *wide* context to help compute a meaning for an unfamiliar word, for the purpose of *understanding the passage* containing it. These two interests don't always coincide, especially if the former includes as one of its goals the reader's ability to *use* the word. From the fact that a given co-text might not clearly convey a word's "correct" meaning, it does not follow that a useful meaning cannot be computed from it (especially since the wider context from which a meaning is computed includes the reader's PK and is not therefore restricted to the co-text). Some co-texts certainly provide more clues than others. But should all CVA be spurned because of less-helpful co-texts?

Their classification divides all co-texts into *pedagogical* and *natural*. The former are "specifically designed for teaching designated unknown words". It will be of interest later that the only example they give of a pedagogical co-text is for a *verb*: "All the students made very good grades on the tests, so their teacher *commended* them for doing so well."

By contrast, "the author of a natural context does not intend to convey *the meaning of a word*" (my italics). Note the assumptions about unique, correct meanings at work. In contrast, and following Deighton (§2.2, above), the author of a natural co-text *does*—no doubt, unintentionally—convey *a* meaning for the word in question. And that meaning is the only one that a reader might be expected to compute. [2] goes on to observe that natural "contexts will not necessarily provide *appropriate* cues to the meaning of a particular word" (my italics). This does not mean that no cues (or clues) are provided. It may well be that clues *are* provided for *a* meaning that helps the reader understand the passage. Note that the pedagogical-natural distinction may ultimately be hard to maintain: A passage produced for pedagogical purposes by one researcher might be taken as "natural" by another (see §2.6, below).

2.5 Four Kinds of (Natural) Co-texts

Misdirective Co-texts. Natural co-texts are divided into four categories. "At one end of our continuum are misdirective contexts, those that seem to direct the student to an *incorrect* meaning for a target word" (my italics). Some co-texts may indeed be misdirective. But [2]'s sole example does not inspire confidence: "Sandra had won the dance contest and the audience's cheers brought her to the stage for an encore. 'Every step she takes is so perfect and graceful,' Ginny said *grudgingly*, as she watched Sandra dance." Granted, a reader might incorrectly decide from this that 'grudgingly' meant something like "admiringly". But there are three problems with this example: (1) There is no evidence that this co-text is natural. This is minor; many such allegedly misdirective co-texts could be found "in nature". (2) If it is natural, it would be nice to see more of it. Many CVA researchers assume that **(A4) co-texts have a fixed, usually small size.** But there might be other clues, preceding or following this short co-text, that would rule out "admiringly". Perhaps we know or could infer from other passages that Ginny is jealous of Sandra, or that she is inclined to ironic comments. Strictly speaking, one could logically infer from this passage a disjunction of possible meanings of 'grudgingly' and later rule some of them out as more occurrences of the word are found. (3) Most significantly, 'grudgingly' is an adverb. Another assumption is **(A5) All words are equally easy (or difficult) to learn.** But adverbs, adjectives, and other modifiers are notoriously hard cases for CVA and for first-language learning [11].

Thus, the evidence provided for the existence of misdirective co-texts is weak, primarily since there should be *no* limit on the size of a co-text (see §3.2, below) and since the only example concerns an adverb, which can be difficult to interpret in any context. There is no "limit" on the size of the *wide* context. Certainly a reader's PK (which is part of that wide context) might include lots of beliefs that might assist in coming up with a plausible meaning for 'grudgingly' in this passage.

Beck et al. conclude, "[I]ncorrect conclusions about word meaning are likely to be drawn" from misdirective co-texts. This assumes **(A6) Only one co-text can be used to compute a meaning for a word.** Granted, if a word only occurs once, in the most grievous of misdirective co-texts, then it is quite likely that a reader would "draw an incorrect conclusion". But, in such a case, it does not matter if the reader even concludes anything at all, for it is highly unlikely that anything crucial will turn on such a word. More likely, the reader will encounter the word again, and will have a chance to revise any initial hypothesis about what it might mean.

The task of CVA is hypothesis generation and testing, a fundamentally scientific task of developing a theory about a word's (possible) meaning. It is not mere guessing. It is like detective work: finding clues to determine, not "who done it", but "what it means". And, like all hypotheses, theories, and conclusions drawn from circumstantial evidence, it is susceptible to revision when more evidence is found.

Admittedly, all of this assumes that the reader is consciously aware of the unfamiliar word and notes its unfamiliarity. It also assumes that the reader remembers the word and its hypothesized meaning (if any) between encounters. None of these further assumptions are, unfortunately, necessarily the case.

Nondirective Co-texts. "[N]ondirective contexts, ... seem to be of *no* assistance in directing the reader toward any particular meaning for a word" (my italics). Here is [2]'s example: "Dan heard the door open and wondered who had arrived. He couldn't make out the voices. Then he recognized the *lumbering* footsteps on the stairs and knew it was Aunt Grace." Again, the evidence is underwhelming, and for the same reasons: no evidence of the sole example being natural, no mention of any larger co-text that might provide more clues, and the word is a modifier (this time, an adjective). I suggested that the reader could ignore a single unfamiliar word in a misdirective text. The same is true of a non-directive text. But could an author use a word uniquely in such a way that it *is* crucial to understanding the text? Yes—authors can do pretty much anything they want. But, in such a case, the author would be assuming that the reader's PK includes the author's intended meaning for that word. As a literary conceit, it might be excusable; in expository writing, it would not be.

Syntactic Manipulation. *All* co-texts (even misdirective and non-directive) are capable of yielding a clue. The technique for squeezing a clue out of any co-text is to syntactically manipulate it to make the unfamiliar word its focus, much as one syntactically manipulates an equation in one unknown to turn it into an equation with the unknown on one side of the equals sign and its "co-text" on the other. For example, from the above "misdirective" text, we could infer that, whatever else 'grudgingly' might mean, it could be defined (if only vaguely) as "a way of saying something" (and we could list all sorts of such ways, and hypothesize that 'grudgingly' is one of them). Moreover, it could be defined (still vaguely) as "a way of (apparently) praising someone's performance" (and we could list all sorts of such ways, and hypothesize that 'grudgingly' is one of them). I put 'apparently' in parentheses, because some readers, depending on their PK, might realize that sometimes praise can be given reluctantly or ironically, and such readers might hypothesize that 'grudgingly' is that kind of way of praising. Similarly, from the "lumbering" passage, a reader might infer that lumbering is a property of footsteps, or footsteps on stairs, or even a *woman's* footsteps on stairs ([19] makes similar remarks).

General Co-texts. Not all co-texts containing modifiers are mis- or nondirective: "general contexts ... provide enough information for the reader to place the word in a general category". E.g., "Joe and Stan arrived at the party at 7 o'clock. By 9:30 the evening seemed to drag for Stan. But Joe really seemed to be having a good time at the party. 'I wish I could be as *gregarious* as he is,' thought Stan." Note that this adjective is contrasted with Stan's attitude. From a contrast, much can be inferred. In our research, several adjectives that we have computed meanings for occur in such contrastive co-texts: "Unlike his brothers, who were noisy, outgoing, and very talkative, Fred was quite *taciturn*" [5] (though this is probably not a natural co-text, or else it is a "directive" co-text.)

Directive Co-texts. Their fourth category is "directive contexts, which seem likely to lead the student to a specific, correct meaning for a word". But, here, their example is that of a noun: "When the cat pounced on the dog, he leapt up, yelping, and knocked over a shelf of books. The animals ran past Wendy, tripping her. She cried out and fell to the floor. As the noise and confusion mounted, Mother hollered upstairs, 'What's all the *commotion*?' " Again, it's not clear whether this is a natural co-text. More importantly, the fact that it is a noun suggests that it is not so much the co-text that is helpful as it is the fact that it is a noun, which is generally easier to learn than adjectives and adverbs. (Note, too, that this text is longer than the others!)

2.6 CVA, Neologisms, and Cloze-Like Tasks

Beck et al. conducted an experiment involving subjects who were given passages from basal readers. The researchers "categorized the contexts surrounding target words according to" their four-part "scheme", and they "then blacked out all parts of the target words, except morphemes that were common prefixes or suffixes ... . Subjects were instructed to read each story and to try to fill in the blanks with the missing words or reasonable synonyms". Independent of the results, there are several problems with this set-up:

(1) The passages may indeed have been found in the "natural" co-text of a basal reader, but were the stories in these anthologies written especially for use in schools, or were they truly natural? (Remember: One reader's natural co-text might be another researcher's pedagogical one.) (2) How large were the surrounding co-texts? Recall that a small co-text might be nondirective or even misdirective, yet a slightly larger one might very well be directive. (3) It is unclear whether the subjects were given any instruction on how to do CVA before the test. Here we find another assumption: **(A7) CVA "comes naturally", hence needs no training.** Our project, by contrast, is not focused on incidental CVA, but on deliberate CVA, carefully taught and practiced.

(4) Another problem arises from the next assumption: **(A8) Cloze-like tasks are a form of CVA.** A "cloze-like" task involves replacing certain words in a passage with blanks to be filled in. This is not CVA. A serious methodological difficulty faces all CVA researchers: If you want to find out if a subject can compute a meaning for an unknown word from context, you don't want to use a word that the subject knows. You could filter out words (or subjects) by giving a pretest to determine who knows which words. But then those who don't know the test words will have seen them at least once before (during the pretest), contaminating the data. Finding obscure words (in natural co-texts, no less) that are highly unlikely to be known by any subjects is difficult; in any case, one might want to test familiar words. Two remaining alternatives—replace the word with a neologism or a blank—introduce complications: We have found that, when students confront what they believe to be a real (but unknown) word, they focus their attention, thoughts, and efforts on meaning (i.e., what could this word mean?), but when obvious neologisms or blank spaces are used, readers focus on "getting" the word, not on expressing its possible meaning. These tasks are related, yet distinctly different.[2] Schatz & Baldwin [26] also claim that "Using context to guess the meaning of a semantically unfamiliar word is essentially the same as supplying the correct meaning in a cloze task." But this is not the case: In cloze-like tasks, the reader is invited to guess (rather than compute), and there is a unique, correct answer, whereas, in CVA as we see it, the goal is to compute a meaning sufficient for understanding the passage.

I have no clever solution to this methodological problem. My preferred technique for now is to use a plausible-sounding neologism (with appropriate affixes) and then to inform the subject that it is a word from another language that might not have a single-word counterpart in English, but that in any case the subject's job is to compute what it might mean, not necessarily find an English synonym, exact or inexact.

---

[2] I am grateful to my co-researcher, Michael Kibby, for this insight.

2.7 Beck et al.'s Conclusions

Beck et al. claim that their experiment "clearly support[s] the categorization system" and "suggest[s] that it is precarious to believe that naturally occurring contexts are sufficient, or even generally helpful, in providing clues to promote initial acquisition of a word's meaning". However, "Only one subject could identify any word in the misdirective category". This is significant because it suggests that CVA *can* be done even with misdirective co-texts, supporting *our* theory, not theirs.

They conclude that "Children most in need of vocabulary development—that is, less skilled readers who are unlikely to add to their vocabularies from outside sources—will receive little benefit from such indirect opportunities to gain information". An assumption underlying this is: **(A9) CVA can be of help only in vocabulary acquisition.** But another potential benefit far outweighs this: CVA strategies, if properly taught and practiced, can improve general reading comprehension. This is because the techniques that our computational theory employs and that, we believe, can be taught to readers, are almost exactly the techniques needed for improving reading comprehension: careful, slow reading; careful analysis of the text; a directed search for information useful to computing a meaning; application of relevant PK; application of reasoning for the purpose of extracting information from the text.

3 Are Context Clues Unreliable Predictors of Word Meanings?

3.1 Schatz & Baldwin's Argument

Schatz & Baldwin [26] takes the case against context a giant step further, arguing "that context does not usually provide clues to the meanings of low-frequency words, and that context clues actually inhibit the correct prediction of word meanings just as often as they facilitate them".

In summarizing the then-current state of the art, they ironically note that "almost eight decades after the publication of ... [a] classic text [on teaching reading] ..., publishers, teachers, and the authors of reading methods textbooks have essentially the same perception of context as an *efficient* mechanism for inferring word meanings" (my italics). Given their rhetoric, the underlying assumption here appears to be: **(A10) CVA is *not* an efficient mechanism for inferring word meanings.** Their argument seems to be, roughly, that *co-text* can't help you figure out "the" correct meaning of an unfamiliar word; therefore, CVA is not "an effective strategy for inferring word meanings". In contrast, I am arguing that *wide* context *can* help you figure out *a* meaning for an unfamiliar word; therefore, CVA *is* an effective strategy for inferring (better: computing) word meanings. Insofar as the *purpose* of CVA is thought of as getting "the correct meaning", it is ineffective. But insofar as its purpose is to get a meaning sufficient for understanding the passage in which the unfamiliar word occurs, it can be quite effective, even with an allegedly "misdirective" co-text.

Perhaps CVA is thought to be too magical, or perhaps too much is expected of it. Schatz & Baldwin claim that "context clues should help readers to infer the meanings of ... [unfamiliar] words ... *without the need for readers to interrupt the reading act* with diversions to ... dictionaries, or other external sources of information" (my italics). This could only be the case if CVA were completely unconscious and immediate, as if one could read a passage with an unfamiliar word and instantaneously come to know what it means. (This *may* hold for "incidental" CVA [18], but not for "deliberate" CVA.) In contrast, our theory requires interruption—not to access external sources—but for conscious, deliberate analysis of the passage. Computer models that appear to work instantaneously are actually doing quite a lot of active processing, which a human reader would need much more time for.

In any case, stopping to consult a dictionary does not suffice. With the exception of learner's dictionaries designed primarily for ESL audiences, most dictionaries are notoriously difficult to use and their definitions notoriously difficult to interpret [16]. More importantly, *CVA needs to be applied to the task of understanding a dictionary definition itself*, which is, after all, merely one more co-text containing the unfamiliar word [27]. Indeed, CVA is the base case of a recursion one of whose recursive clauses is "look it up in a dictionary".

3.2 Schatz & Baldwin's Methodology

Nouns and Verbs vs. Modifiers. Schatz & Baldwin offer several experiments to support their claims. As with [2]'s experiments, there are a number of apparent problems with their methodology. Their first experiment took 25 "natural" passages from novels, selected according to an algorithm that randomly produced passages containing low-frequency words. But consider some of their words: 'cogently', 'cozened', 'ignominiously', 'imperious', 'inexorable', 'perambulating', 'recondite', 'salient'. Note that 4 are adjectives, 2 are adverbs, 1 is a verb ('cozened'), and 1 ('perambulating') might be either a verb or an adjective, depending on the co-text. These are only "examples"; we are not given a full list of words, nor told whether these statistics are representative of the full sample. If they are, then fully 75% of the unfamiliar words are modifiers, known to be among the most difficult of words to learn. Their example passages consist of an adverb ('ruefully'), three adjectives ('glib', 'pragmatic', 'waning'), and four nouns ('yoke', 'coelum', 'dearth', 'ameliorating'). This brings the statistics to around 67% modifiers, 27% nouns, and 6% verbs (not counting 'perambulating'). Of these, two of the nouns ('dearth', 'ameliorating') are examples of words occurring in "facilitative" co-texts. Their example of a "confounding" co-text is for an adjective ('waning').

These examples raise more questions than they answer: What were the actual percentages of modifiers vs. nouns and verbs? Which lexical categories were hardest to determine meanings for? How do facilitative and confounding contexts correlate with lexical category? They admit that "a larger sample of words would certainly be desirable" but that their selection of "70 items ... offer[s] a larger and more representative sample than most studies of context clues". A representative sample of co-texts? Of words? The sort of representativeness that is needed should (also) be a function of the variety of lexical category. What would happen with natural co-texts of, say, all four of [2]'s categories with nouns, verbs, adjectives, and adverbs in each such co-text (i.e., 16 possible types of co-text)? Schatz & Baldwin's (and [2]'s) results may say more about the difficulty of learning meanings for modifiers than they do about weaknesses of contexts.

CVA vs. WSD. Moreover, in two of their experiments, subjects were *not* involved in the task of CVA. Rather, they were doing a related—but distinct—task known as "word-sense disambiguation" (WSD [13]). The CVA task is to figure out a word's meaning "from scratch". The WSD task is to choose a meaning for a (typically polysemous) word from a list of possible meanings for the word in different contexts. In [26]'s experiment, the subjects only had to replace the unfamiliar word with each multiple-choice meaning-candidate (each of which was a proposed one-word synonym) and see which of those possible meanings fit better; no real CVA was needed.

In the third experiment, real CVA was being tested. However, **(A3)** raises its head: "we were interested only in full denotative meanings or accurate synonyms". There is no reason to believe or to expect that CVA will typically be able to deliver on such a challenge. But neither is there any reason to demand such high standards; once this constraint is relaxed, CVA is a useful tool for vocabulary acquisition and general reading comprehension.

Space and Time Limits. The smaller the co-text, the less chance there is of figuring out a meaning, because there will be a minimum of textual clues. The larger the co-text, the greater the chance, because a large enough co-text might actually include a definition of the word! (Recall that CVA needs to be applied even in the case of an explicit definition!) What is a reasonable size for a co-text? Our methodology has been to start small and work "outwards" to preceding and succeeding passages, until enough co-text is provided to enable successful CVA. 'Successful' only means being able to compute *a* meaning enabling the reader to understand enough of the passage to continue reading; it does *not* mean figuring out "the correct meaning of" the word. This models what readers can do when faced with an unfamiliar word in normal reading: They are free to examine the rest of the text for possible clues. In contrast, [26] arbitrarily limited their co-text size to only 3 surrounding sentences. An inability to do CVA from such a limited co-text shows at most that such co-texts are too small, not that CVA is unhelpful.

Also, [26] observes that "All students finished in the allotted time". But real-life CVA has no time limits (other than self-imposed ones). CVA might extend over a long period of time, as different texts are read.

Teaching CVA. Finally, there was no prior training in how to use CVA: "we did not control for the subjects' formal knowledge of how to use context clues". Their finding "that students either could not or chose not to use context to infer the meanings of unknown words" ignores the possibilities that the students did not know that they *could* use context or that they did not know *how* to. Granted, "incidental" (unconscious) CVA is something that we all do; there appears to be no other explanation for how we learn most of our vocabulary [18]. But "deliberate" (or conscious) CVA is a skill that, while it may come naturally to some, can—and needs—to be taught, modeled, and practiced.

Thus, their conclusion that "context is an ineffective or little-used strategy for helping students infer the meanings of low-frequency words" might only be true for untrained readers. It remains an open question whether proper training in CVA can make it effective and can add it to the reader's arsenal of techniques for improving reading comprehension (though there is some positive evidence [9,15]). [26] disagrees: "[I]f the subjects had been given adequate training in using context clues, the context groups in these experiments might have performed better. We think such a result would be unlikely because the subjects were normal, fairly sophisticated senior high school students. If students don't have contextual skills by this point in time, they probably are not going to get them at all." (**A7**) is at work again. But students are not going to get "contextual skills" if they are not shown the possibility of getting them. Moreover, the widespread need for, and success of, critical thinking courses strongly suggests that students need to, and can, be educated on these matters.

3.3 Three Questions About CVA

In their general-discussion section, [26] raises three questions: (1) "Do traditional context clues occur with sufficient frequency to justify them as a major element of reading instruction?" This is irrelevant *if* CVA can be shown to foster good reading comprehension and critical-thinking skills. For clues need not occur frequently in order for the techniques for using them to be useful general skills. CVA can foster improved reading comprehension, but more research is needed. Traditional context clues do occur and—augmented by the reader's PK and training in CVA techniques for developing revisable hypotheses about an unfamiliar word's meaning—are justified as a major element of reading instruction.

(2) "Does context *usually* provide accurate clues to the denotations and connotations of low-frequency words?" This is also irrelevant under our conception of CVA: We are not interested in "accuracy". Moreover, a "denotation" (in the sense of an external referent of a word) is best provided by demonstration or by a graphic illustration, and a "connotation" (in the sense of an association of the unfamiliar word with other (familiar) words) is not conducive to the sort of "accuracy" that [26] (or [2]) seem to have in mind. Context *can* provide clues to revisable hypotheses about an unfamiliar word's meaning.

(3) Are "difficult words in naturally occurring prose ... usually amenable to such analysis"? Such words are always amenable to yielding at least some information about their meaning, as discussed in §2.5, above.

4 Conclusions: A Positive Theory of Computational CVA

Progress is often made by questioning assumptions. This essay has questioned the assumptions underlying [2]'s and [26]'s arguments and experiments challenging CVA. Their papers are best read as asserting that, *given those assumptions*, CVA is not as beneficial as some researchers claim it is. We conclude by presenting our theory's contrasting beliefs. (Details are in [23,24].)

(**B1**) Every co-text C can give some clue (even minimally) to a word w's meaning (at the very least, its "algebraic" meaning obtained by rephrasing C to make w the subject). But w will also have a meaning that is partly determined by reader R's accessible PK, which may be time-dependent. None of the meanings R computes for w is necessarily "the" meaning (in either a dictionary sense or that of a reading teacher).

(**B2**) w's co-text gives clues to w's meaning that must be supplemented by the reader's PK in order for a meaning to be computed. There is no such things as "misdirective", "non-directive", "general", or "directive" co-texts. A co-text's value depends on the reader's PK and ability to use clues and PK together.

(**B3**) CVA is distinct from cloze-like tasks.

(**B4**) Co-texts can be as small as a phrase or as large as an entire book, with no arbitrary space or time limits.

(B5) Many co-texts may be required before CVA can "asymptotically" approach a "stable" meaning for a word.

(B6) A word does not have a unique meaning, even in directive and pedagogical co-texts.

(B7) A word does not have a (single) correct meaning, not even in directive and pedagogical co-texts. Nor does it *need* a correct meaning in order for a reader to be able to understand it (in context). Even a familiar and well-known word can acquire a new meaning in a new co-text. In fact, each new C and each new R can yield a new meaning, so meanings are continually being extended (as when words are used metaphorically [4]).

(B8) Some words are harder to compute meanings for than others (e.g., nouns are easiest).

(B9) CVA is an efficient method for inferring word meanings.

(B10) CVA can improve general reading comprehension.

(B11) CVA can (and should) be taught.[3]

References

1. Budiu, R. & Anderson, J.R. (2001), "Word Learning in Context: Metaphors and Neologisms", *Tech. Rep. CMU-CS-01-147* (Carnegie Mellon Univ., School of Comp. Sci.).
2. Beck, I.L.; McKeown, M.G.; & McCaslin, E.S. (1983), "Vocabulary Development: All Contexts Are Not Created Equal", *Elementary School Journal* 83: 177–181.
3. Brown, G., & Yule, G. (1983), *Discourse Analysis* (Cambridge Univ. Press).
4. Clarke, D.F., & Nation, I.S.P. (1980), "Guessing the Meanings of Words from Context", *System* 8: 211–220.
5. Dulin, K.L. (1970), "Using Context Clues in Word Recognition and Comprehension", *Reading Teacher* 23: 440–445, 469.
6. Ehrlich, K. (1995), "Automatic Vocabulary Expansion through Narrative Context", *Tech. Rep. 95-09* (SUNY Buffalo Comp. Sci.).
7. Ehrlich, K., & Rapaport, W.J. (1997), "A Computational Theory of Vocabulary Expansion", *Proc. 19th Annual Conf. Cog. Sci. Soc.* (Erlbaum): 205–210.
8. Ehrlich, K., & Rapaport, W.J. (2004), "A Cycle of Learning: Human and Artificial Contextual Vocabulary Acquisition", *Proc. 26th Annual Conf. Cog. Sci. Soc.* (Erlbaum, 2005): 1555.
9. Fukkink, R.G., & De Glopper, K. (1998), "Effects of Instruction in Deriving Word Meaning from Context", *Rev. Ed. Res.* 68: 450–458.
10. Garnham, A., & Oakhill, J. (1990), "Mental Models as Contexts for Interpreting Texts", *J. Semantics* 7: 379–393.
11. Gentner, D. (1982), "Why Nouns Are Learned before Verbs", in S.A. Kuczaj (ed.), *Language Development: Vol. 2* (Erlbaum): 301–334.
12. Hobbs, J.R. (1990), *Literature and Cognition* (Stanford, CA: CSLI).
13. Ide, N.M., & Veronis, J. (eds.) (1998), Word Sense Disambiguation, *Comp. Ling.* 24(1).
14. Johnson-Laird, P.N. (1987), "The Mental Representation of the Meanings of Words", *Cogn.* 25: 189–211.

---

[3] I am grateful to Albert Goldfain, Michael W. Kibby, Jean-Pierre Koenig, Shakthi Poornima, Stuart C. Shapiro, Karen M. Wieland, and the SNePS Research Group. A longer version of this paper is at [http://www.cse.buffalo.edu/~rapaport/Papers/paris.pdf].

15. Kuhn, M.R., & Stahl, S.A. (1998), "Teaching Children to Learn Word Meanings from Context", *J. Lit. Res.* 30: 119–138.
16. Miller, G.A. (1986), "Dictionaries in the Mind", *Lang. & Cog. Procs.* 1: 171–185.
17. Murphy, J. (2000), *The Power of Your Subconscious Mind, Rev. Ed.* (Bantam).
18. Nagy, W.E.; Herman, P.A.; & Anderson, R.C. (1985), "Learning Words from Context", *Reading Res. Qtly.* 20: 233–253.
19. Nation, I.S.P. (2001), *Learning Vocabulary in Another Language* (Cambridge Univ. Press).
20. Rapaport, W.J. (2003a), "What Is the 'Context' for Contextual Vocabulary Acquisition?", in P.P. Slezak (ed.), *Proc. ICCS/ASCS-2003*, Vol. 2: 547–552.
21. Rapaport, W.J. (2003b), "What Did You Mean by That? Misunderstanding, Negotiation, and Syntactic Semantics", *Minds & Machines* 13: 397–427.
22. Rapaport, W.J. (2004), "Bibliography of Theories of Contextual Vocabulary Acquisition", [http://www.cse.buffalo.edu/~rapaport/refs-vocab.html].
23. Rapaport, W.J., & Ehrlich, K. (2000), "A Computational Theory of Vocabulary Acquisition", in L.M. Iwańska & S.C. Shapiro (eds.), *Natural Language Processing and Knowledge Representation* (AAAI/MIT Press): 347–375.
24. Rapaport, W.J., & Kibby, M.W. (2002), "Contextual Vocabulary Acquisition: A Computational Theory and Educational Curriculum", in N. Callaos et al. (eds.), *Proc. SCI-2002*, Vol. II: 261–266.
25. Russell, B. (1918), "The Philosophy of Logical Atomism", in *Logic and Knowledge* (Capricorn, 1956): 177–281.
26. Schatz, E.K., & Baldwin, R.S. (1986), "Context Clues Are Unreliable Predictors of Word Meanings", *Reading Res. Qtly.* 21: 439–453.
27. Schwartz, R.M. (1988), "Learning to Learn Vocabulary in Content Area Textbooks", *J. Reading*: 108-118.
28. Shapiro, S.C., & Rapaport, W.J. (1995), "An Introduction to a Computational Reader of Narrative," in J.F. Duchan et al. (eds.), *Deixis in Narrative* (Erlbaum): 79–105.
29. Singer, M., et al. (1990), "Bridging-Inferences and Enthymemes", in A.C. Graesser & G.H. Bower (eds.), *Inferences and Text Comprehension* (Academic).
30. Stanovich, K.E. (1986), "Matthew Effects in Reading", *Reading Res. Qtly.* 21: 360–407.

Functional Model of Criminality: Simulation Study

Sarunas Raudys[1,2], Aini Hussain[3], Viktoras Justickis[1,*],
Alvydas Pumputis[1], and Arunas Augustinaitis[1]

[1] Mykolas Romeris University
[2] Institute of Mathematics and Informatics,
Ateities 20, Vilnius-2057, Lithuania
[3] Department of Engineering, National University of Malaysia
justickv@takas.lt

Abstract. We describe a new point of view that uses contextual information of the interrelated conditions in which criminality or deviant occurs by means of computer simulation to examine its role in a progressively changing society. Based on the functionalist point of view, we model society as a multi agent system and criminality as a noise. A noise is injected as training signals to the single layer perceptrons that functions as agents (i.e. society) and all agents must comply the fitness function. Failure to comply will result in the agent being removed from the "society" and be replaced by a "newborn" that inherits some "upbringing" information from its "mother's agent". Our simulation studies point toward the constructive effects of criminality. The major contribution in our study is that new paradigm can show and measure criminality effects as in contrast to mere verbal descriptions of the social sciences. The new paradigm could be used to provide the functional explanation of atypical trends in criminality in East- and Middle European and some other countries and provide "artificial experimental data" for future studies.

1 Introduction

Criminality is a complex multi-faceted social phenomenon and it is generally admitted that it brings immense economical, moral and psychological damages. Lack of reliable theory mainly leads to failure to fight it [1- 3]. Several different explanations were proposed and it was found that criminality is highly affected by unemployment, urbanization, migration of the population, etc. Thus far, all investigations saw the increase in criminality rate only from its unconstructive side - merely as a manifestation of profound and highly dangerous disorganization of society caused by its transition from socialism to capitalism. We, on the contrary, try to investigate the *functional* side of criminality and provide realistic, *double – sided* scientific view of this highly important social phenomenon.

In an effort to understand and comprehend some of the social processes that are affecting criminality trends in the modern world, we imitate the explanation of criminality proposed by very eminent criminological and sociological "functionalist

---

[*] Corresponding author.

theorists" [3-5]. They were an English philosopher Herbert Spencer (1820-1903) and a French sociologist Émile Durkheim (1858-1917), Americans Talcott Parsons, Robert Merton, Jeffrey Alexander and German Miklas Luhman. The basis of functional theory is the idea of *"structural context"* in which the society is seen as a large self-regulating system that consists of many subsystems. Such subsystems are defined as individuals, social institutions, enterprises, corporations, political or social groups etc. Each has a variety of its possible reactions to changes that occurs in society and surrounding world. All these reactions and a given social situation dictate which reaction is preferable for adaptation to create the desired structural context. The task of science is to analyze this context and to find out how much operation of society in this context promotes discovery of most suitable reactions.

Functional approach turned out to be especially fruitful in the analysis of positive roles of "negative" and "destructive" social phenomena like revolutions, wars, social conflicts and social crises. They were viewed in terms of their functions in adaptation of the whole society. Analogically to explanation of such "negative biological events" like illness or death in terms of their meaning for survival of species, this approach in sociology and criminology was able to see these supposedly totally "negative social events" in terms of their meaning for survival and stability of a society[1]. To mimic such situations, we perform computer simulation of populations of intelligent agents that ought to readapt to permanently and unexpectedly changing environments [6].

Simulations studies help to create a variety of imitation situations for studying the context [7]. In our approach, we model society as a population of agents. Each agent is trained with partially incorrect training directives corrupted by the criminal activity. The agent instead of being encouraged is sometimes punished by criminal activity. We investigate the role of dysfunctions, amongst others defined as spontaneous criminality and illegal entrepreneurship, and the adaptation of social agents towards social and economic changes. The dysfunctions are imitated by injecting noise into the perceptron models to yield the desired outputs. By varying the magnitude and frequency of environmental changes via simulation, we generate a context. The context is used to describe the dynamic processes involving set of changing situations/relationships of criminality and society. By this, we propose that the context is a set of changing relationships/situations that may be shaped by the history of those relationships. As a result, we observe interesting and informative knowledge in terms of patterns in the agents' behavior that are useful for our future practical predictions in criminality control and investigations. Thus making it easier to explain the national trends in criminality of several selected nations recorded in the world wide United Nation Organization and on European Sourcebook of Crime and Criminal Justice statistical surveys [8 - 11].

2 Changes in Recorded Crime Rate in Different Countries

As aforementioned, this work utilizes the United Nation Organization (UNO) research statistics on crime rate for the period of 1980-1997. UNO periodically conducts the

---

[1] The survival means some level of functioning of an agent, its ability to fulfill its functions. We have to stress that we do not model the welfare. The welfare and survival in changing environments are two contradicting concepts.

most representative and comprehensive world survey and the national trends of recorded crime are one of the outcomes. In this work, we have used the data of the sixth world survey, which is made of 108 nations [8] and the European data [10, 11]. We concentrated mainly on those nations that have (i) recently endured revolutionary changes and (ii) the rather stable, highly developed and prosperous nations. As such, ex-socialist countries in Eastern and Middle Europe e.g. Hungary, Poland, Ukraine, Lithuania and the rapidly developing Asian countries like Malaysia and Singapore were selected to represent the first category of nation whilst countries such as USA, Denmark and Germany to represent the second.

We noted two distinct characteristics amongst the recorded crime patterns.

(i) A sudden increase of recorded crime (or the "jump" region) mainly in the first group of countries.
(ii) A fluctuation period of criminality level followed by notable downward trends in some of these countries (Fig.1).

Several researches [9 - 11] also noted this jump in criminality in East and Middle Europe. However, the increase of criminality is explained only as destruction - a manifestation of profound and highly dangerous disorganization of society caused by its transition from socialism to capitalism. Thus far, there was no attempt to provide a functional explanation of the jump in the criminality in these countries. As a result existing general criminological ideas on transitional society remain one-sided and distorted. It is especially bad that our knowledge on ways in which a society can get out of this extreme situation stays incomplete hence reducing our ability to help such society.

3 The Model of Criminality Factors

Society – In our simulation experiments, a society (or a country) is imitated by a population of intelligent agents. The agents ought to solve a sequence of diverse problems interpreted as the pattern recognition tasks. The adaptive nonlinear SLP is selected to model the intelligent agents for at least two particular reasons. Firstly, agent adaptation in a real society and the neural network learning process for pattern recognition task proceed in a very similar manner in which the agents improve their performance via series of attempts [12]. Secondly, since the nonlinear SLP has many traits of universality [13], we believe that such model would provide opportunity to formulate various general statements.

Environmental Changes – Environmental changes are interpreted as important social, economic and political events such as launch of new and very effective technologies, economic or political crises, changes in legislation, revolutions, wars etc. All these events influence and complicate the economic and political life of the country. From this point onwards, such changes are defined as "environmental changes". Environmental changes were modeled by changing the pattern recognition tasks that the artificial agents have to solve.

Time Restrictions – In real life, an agent has a limited time to adapt to changing situation and to solve the pattern recognition task correctly. If the agent fails to do his task, it perishes and removes from the population. The death of an agent could suggest a bankruptcy, a lost of election, a negative decision of the trial process etc. A new agent or offspring will then take his place in the society.

Criminality – Earlier, we stated that criminality causes economic, moral, financial and psychological damages. As such, it can be interpreted as a social noise because it obscures social order. It prevents the agents from learning the correct ways to solve problems and consequently causes a dysfunctional effect. Similarly, injecting noise in training signals can very much affect the signals and disturbs the adaptation of agents. In this work, such criminality effect was imitated by noise injections, which reshuffles α fraction of desired outputs in the training data implying that we are training the agents with partially incorrect directives (α is individual to each agent).

The Agent Training to Solve a Sequence of Tasks – The nature of the behavior of a society, which is a complex system, is such that it is difficult to fully understand the evolving characteristics that exist within it. Consequently, it is impossible to program an intelligent agent with all the information it will require about the actions of the system at the beginning. The agent must therefore, be able to prioritize important aspects to some extent during learning and improves itself as it proceeds to find the best solution. In this context, we then address only the main issues that relate the dynamic changes of economic and political situations in the world and the countries' ability to adapt to changed situation.

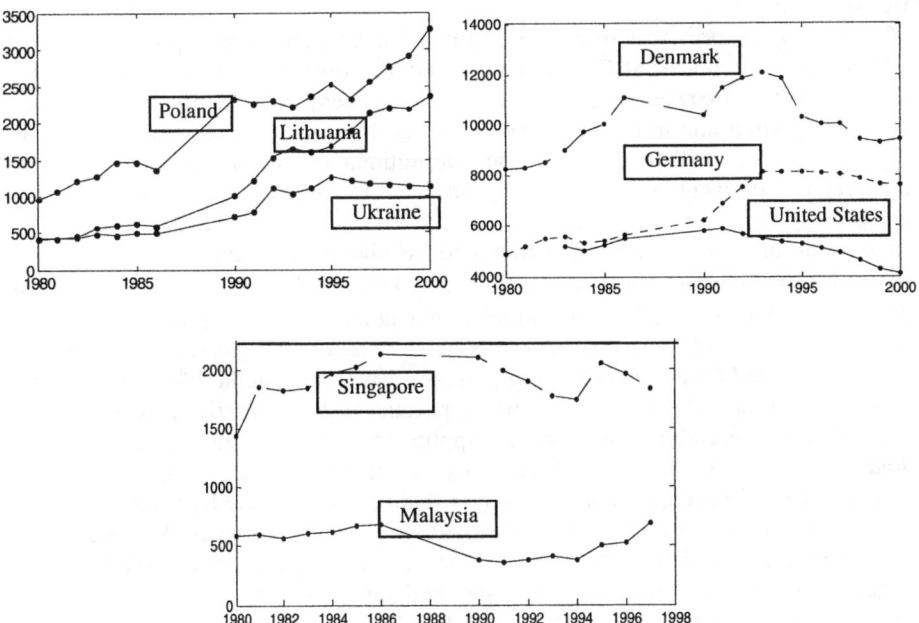

Fig. 1. Recorded crimes per 100,000 populations of selected nations

Continuity in Learning – A society consists of many agents. Each agent tries to learn proper reaction to a new situation. However, it does not mean that every agent learns only from his own experience. In a human society, "only stupid one learns by his own experience – a clever one prefers to use experience of other people". Best experience is taken over by other agents especially the new ones that just started their learning. This transfer of the best experience plays a huge role in human society and crucial for its progress. Thanks to this, society learns also as a whole and as an integrated entity [3, 4]. In our study, we imitate this aspect of the social system by allowing the appearance of new agents (newborns) and design a special way in which they will receive their initial experience.

4 The Single Layer Perceptron and Its Training Peculiarities

In our analysis of adaptation characteristics of the populations of intelligent agents, we utilize two different classification task models as introduced in [6, 14, 15]. If an agent is unable to solve a new recognition task within a limited time, it will "perish", be removed from the population and replaced by a newborn. Agent individuality is characterized by its own noise injection parameter α, which represents the noise injection intensity. When an agent failed to learn in time, it is removed and the new offspring inherits parameter α of its "mother" who is among the most successful agents at that particular moment. Therefore, at the beginning of the training process, the offspring learns from the experience of his agent mother with initial weights that start from zero.

To explain the principal trends, we utilize the simplest model possible. It is a model of the non-linear SLP commonly used to solve pattern recognition tasks. Interested readers are referred to [12, 16, 17] for a general introduction into statistical pattern recognition and artificial neural networks. In order to explain a sense of our analysis, below we will present necessary definitions and show that during training the perceptron its weights are increasing and are slowing down a speed of training process.

In our formulation, objects (situations) to be allocated to one of categories are described by feature vectors, $x = (x_1, x_1, ... x_p)$. Two 2D Gaussian classes, $N(\mu_1, \Sigma)$, $N(\mu_2, \Sigma)$, were considered in our simulation studies. The perceptron calculates a weighted sum of inputs, $sum = w_1 \times x_1 + w_2 \times x_2 + ... + w_p \times x_p + w_0$. A set of p values $w_1, w_2, ... , w_p$ is called a weight vector w and w_0 is a weight threshold value. We will use vector notation $sum = w \times x + w_0$. Very important essential of the perceptron is a transfer function. Weighted sum, sum, is supplied to nonlinear element that calculates $output$ as a non-linear function of sum. As an example one can consider sigmoid function, $output = 1/(1+\exp(-sum))$. If $sum = 0$, $output = 0.5$ (middle value). If sum is large negative, $output$ is close to 0. If sum is large positive, $output$ is close to 1. Note that a slope (incline) of function $output = f(sum)$ is the highest where $sum=0$. If sum moves toward $\pm$ *infinity*, the slope diminishes and approaches 0.

The SLP based classifier is used as a model of the "units" that compose the state: large enterprises, corporations, political, or social groups. A country is considered as a fixed number, say m, of such units (perceptrons) that differ slightly in their noise injection intensities, α_i (a fraction of incorrect training signals, i.e. criminality rate).

In order to use SLP practically one needs to know coefficients $w_0, w_1, \ldots, w_p$. To find the coefficients, we utilize training data called a training set, the vectors of the categories **A** and **B**: $x_1^{(1)}, x_2^{(1)}, \ldots, x_{N_1}^{(1)}$ from **A** and $x_1^{(1)}, x_2^{(2)}, \ldots, x_{N_2}^{(1)}$ from **B**. In perceptron training, we require that for class **A** the *output* ought to be close to a priori selected target, t_1. For another class, **B**, we have to choose another value, e.g. $t_2 = 1 - t_1$ [12, 17]. Traditional algorithm used to train SLP is back propagation, where usually a sum of squares cost function, *cost*, is minimized,

$$cost = 1/N \sum_{i=1}^{2} \sum_{j=1}^{N_i} (t_j^{(i)} - f(w' x_j^{(i)} + w_0))^2, \quad (1)$$

where w is unknown p-dimensional weight vector, w_0 is a bias term, both to be found during training process, $t_j^{(i)}$ is a desired output (a target) of the perceptron if vector $x_j^{(i)}$ is presented to its input; sign " ' " denotes transposition operation.

During training (adaptation) process new vector, $w_{(t+1)}$, is equal to the previous one, $w_{(t)}$, plus a correction term:

$$w_{(t+1)} = w_{(t)} + CT_{ij}^t, \quad (2)$$

where $CT_{ij}^t = -\eta \times (t_j^{(i)} - f(sum) \times (\partial f(sum)/\partial sum) \times (\partial sum / \partial w)$ is a correction term, η is called learning step parameter, $t_j^{(i)} - f(sum)$ is an error signal – a difference between the desired and actual outputs of the perceptron, $sum = w' x_j^{(i)} + w_0$, $\partial f(sum)/\partial sum$ is the derivative of the activation function and $(p+1)$-dimensional vector $(\partial f(sum)/\partial sum) \times (\partial sum / \partial w)$ is called a gradient.

If the weights are small, the gradient, $(\partial f(sum)/\partial sum) \times (\partial sum / \partial w)$, is large. Simple algebra shows that when the weights are large, the gradient becomes small [14, 17]. During training, the magnitudes of the weights increase progressively. As such, when the agent (the perceptron) completes its learning task, the perceptron weights are already large. Due to *these **large weights**, the agent is unable to learn a new task quickly. Such phenomenon is distinct and a novelty of our analysis.*

During each time interval, the tasks associated to each perceptron are similar. After t_{change} learning epochs, m pattern recognition tasks are substituted by new ones and the intelligent agents have to learn the new tasks rapidly. At t_{max} learning epochs, if the classification error, $P_{classif}$, exceeds threshold P_{goal}, we assume that learning was unsuccessful thus, causing the intelligent agent to perish and substituted by a "newborn".

Survived agents are given the right to produce offspring if and only if the error is small i.e. $P_{classif} < 0.4 \times P_{goal}$. Details of this can be found in [6]. A distinct and an unconventional behavior of our simulation experiments is its small classification error. Here we have a situation where after lengthy training process the magnitudes of the perceptrons' weights grow excessively large that cause the weighted sum, *arg*, to become large, too. As a result, the activation function $f(arg)$ saturates and diminishes the gradient to almost zero. Such a phenomenon implies that large weights instigate slow re-training of the perceptron model as aforementioned.

The outer world is constantly changing. In real life, both strengths and intervals between changes represent the random variables. To simplify the interpretation of the results obtained, the time intervals (a number of epochs) between two subsequent data changes were fixed *a priori* whilst the strength of the changes varied according to a certain definite rule. We changed matrix Σ mainly: $\Sigma = \begin{bmatrix} \rho & 0 \\ 0 & \beta \end{bmatrix} \begin{bmatrix} 1 & 0.98 \\ 0.98 & 1 \end{bmatrix} \begin{bmatrix} \rho & 0 \\ 0 & \beta \end{bmatrix}$, where parameter β was varying as an ordered set: β, $1/\beta$, β, $1/\beta$, ... To have the agents behavior somewhat different, small Gaussian noise was added to parameter $\rho=1$ and to the components of the mean vectors, $\mu_1 = -\mu_2 = [0.25, -0.25]$. So, the data are rotated counter clockwise and then clockwise. In order to simulate criminality we corrupt the target values in fraction α of randomly chosen training vectors in each training sweep (iteration). This means the intelligent agents are trained partially with incorrect training signals to avoid the agent from having too much information about its environment at a particular instant of time that might lead to the excessive growth of the perceptron weights.

Hence, corrupting the targets can actually prevent excessive weight growth thus avoiding the decrease in gradient and improves the agent ability to adapt to new task more rapidly. Similarly, many authors have found that noise injections are helpful in perceptron training if there is a *shortage of the training data* (see e.g. in [12, 17, 18]). In sum, the incorrect training signals have a positive influence in which *they block excessive growth of the perceptron weights* in the gradient descent training procedure. This theoretical conclusion is the core finding of our criminality analysis in changing environments.

5 Modeling of Crime Rate Dynamics

The Experimental Setup – Each country has its own laws, moral traditions, habits and institutions that control both legal and illegal entrepreneurships. Each "unit" in the country (or groups of them) is affected by all these issues. Therefore, each unit has its individual peculiarities including inner and outer illegal behaviors. Thus, we assume that during training, each intelligent agent may possess different noise intensity (parameter α_i) that affects its re-training speed and its ability to survive the environmental changes. At the start of the experiment, for all m agents (*say $m=100$*) that composed a state, we assigned different values of parameter α_i distributed uniformly in an interval of [α_{min} α_{max}]. A random character of the training data (comprising 50 training vectors from each class i.e. $N=50$) is supplied to each intelligent agent. Different values of α lead to the death of certain agents who fail to learn fast enough to satisfy the *a priori* fixed condition $P_{classif} < P_{goal}$ after the t_{max} training epochs. In the experiments, we set $t_{max} = 120$ and $P_{goal} = 0.01$. When the intelligent agent perishes, it will be replaced with a new offspring. The agent possesses its mother's noise intensity, α_i, and starts the learning process from zero initial weights. A small random variable is added to α_i each time through the mutation process. The mutations result that parameters α_i are changing in a time. In addition, we train the newborn agent with its mother's output signal until $P_{offspring} > 1.1 \times P_{mother}$. Such strategy increases the population's resistance to strong environmental changes.

Simulation Studies – Previous analysis of SLP training dynamics [14, 15, 17] revealed that in situations of good separation of the categories (i.e. small classification error), they involved long training intervals that caused excessive growth of weights magnitudes which in turn, diminish the re-adaptation ability to learn a new classification. Therefore, the interval between the environmental changes, t_{change}, contains important contextual information. In Fig. 2 and 3 we show *typical experimental results* of the simulation selected out of several dozens for frequent (t_{change}= 75) and infrequent (t_{change}=250) sequences of environmental changes. The top most plots (*a*) represent the dynamics of the power of environmental changes, β. The upper and lower part of the center plot (*b*) depicts the dynamics of a number of offspring and mature agents, respectively.

The bottom-most plot (*c*) describes the dynamics of maximal, mean and minimal fraction of incorrect training signals, α, in the population of *m* agents that is interpreted as the crime rate. The dotted lines represent the maximal and minimal values while the continuous bold line represents a dynamic change of a mean for 1500 environmental changes. In Fig. 3(*b*) *the environmental changes are infrequent*. We see that in such situation, most of the intelligent agents perish. In middle interval of strong changes, only agents with high parameter α_i values survived and were allowed to produce the newborns that inherited their mothers' traits (parameter α_i).

In Fig. 3(*c*), deaths of agents and mutations cause the level of incorrect desired outputs to increase. Contrary, in Fig. 2(*b*) we have situation where the environmental changes are *frequent*. The perceptron weights remain small due to frequent changes in the pattern recognition tasks. Small weights of the perceptron produce high gradient. Hence, the SLP learns rapidly and almost all agents survive. Therefore, in spite of large environmental changes we have small inheritance and practically no changes in the power of noise injection intensity (a middle part of Fig. 2(c)).

6 The Results and Discussions

We see that an increase in the power of the environmental changes increases the intensity of incorrect training signals, which happens when the interval, t_{change}, between environmental changes, is large. A high level of noise injection prevents weights from becoming too large and helps the intelligent agents to adapt faster to changed conditions and survive. On the contrary, if the interval, t_{change}, between environmental changes, is smaller or frequent along with an increase in the intensity of power of the environmental changes, the intensity of incorrect desired outputs will decrease. In all cases, an interval of variation of parameter α_i in the population is rather wide. This means that in population that survived, we have agents with diverse α_i, which is important for the population survival. In general, relationships between the interval between changes, the strength of the changes, optimal level of a noise and other factors of the model are very complicated. Large scale simulation studies should reveal more factors affecting the criminality and relationships between them.

Computer modeling is a useful tool in disciplines in which the empirical experimentation is difficult or impossible. In this work, the key finding is that noise injection can boost the agent ability to re-adapt when the environment changes are infrequent. Simulations show random oscillations in a noise injection intensity caused by "genetic inheritance" of the newborn agents. Possibly, such oscillations are natural peculiarity of such type of evolution. Analogously, in the human world an increase in noise injection intensity and its oscillations means that despite the danger, the increased criminality helps the society to adapt to totally new life conditions by *preventing premature ceasing of learning by the agents in the state*.

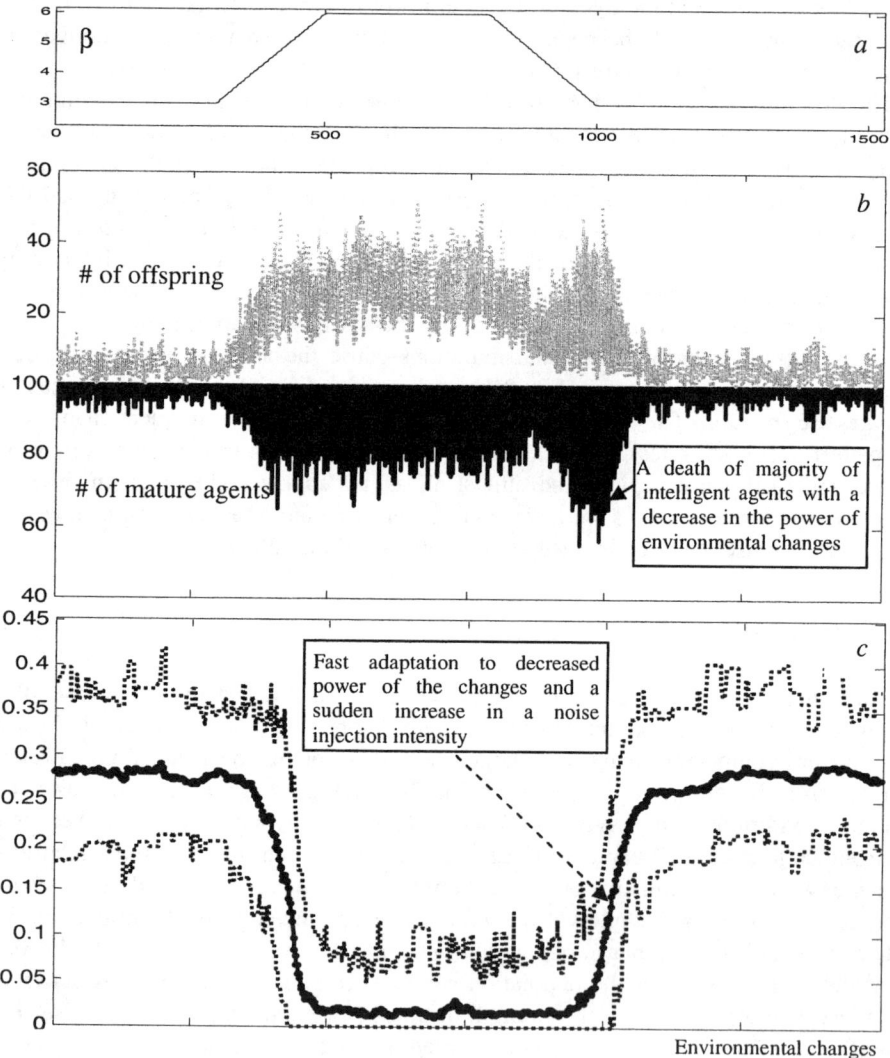

Fig. 2. Simulation result of *frequent* environmental changes with $t_{change}= 75$ epochs

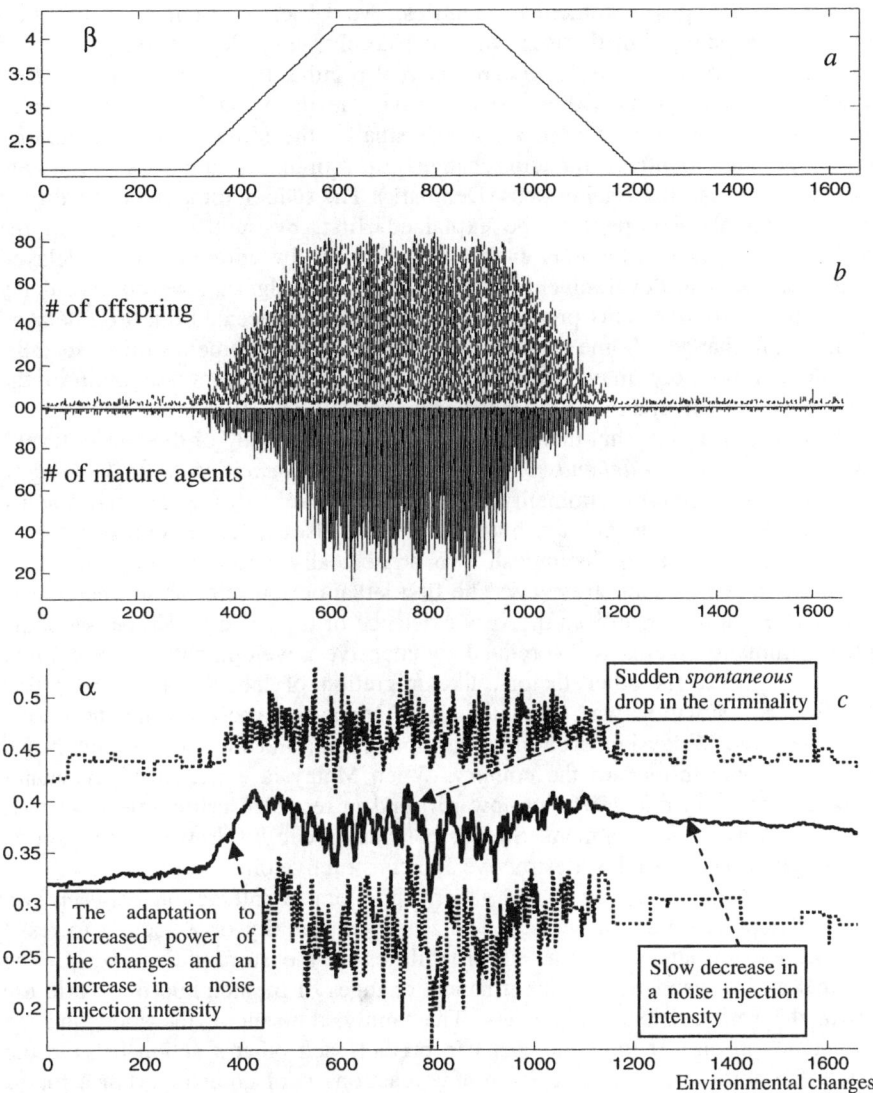

Fig. 3. Simulation result of *infrequent* environmental changes with t_{change}=250 epochs

Indeed, a great part of the modern control system of economic activities has developed as a reaction to this outburst of economic criminality in transitional period. In this sense, these states really *needed* their criminality to develop their efficient modern control systems. Thus, shown earlier the East European "sudden jump in crime rate" was not only a *consequence* of the social revolt but also *a way to re-adapt to new conditions*. Most importantly, we also observe that frequent environmental changes also played a role as partial noise injection that allows the intelligent agents to survive. In the post-communist countries, we observe coming of the new technologies, frequent political crises and numerous changes in legislation.

Random changes in the criminality are inherent peculiarity of the evolution where the population of the agents with inheritance has to adapt to changing environments. The more agents we have in population, the smaller the scale of the oscillations. Therefore, in big countries random changes in criminality are less significant (Germany, USA) as in the smaller ones (Denmark). The sudden jump in criminality in Eastern and Middle Europe may be explained, first, by sudden changes in the economic and political situations in these countries. In countries with delayed economic and political development (Ukraine, Belarus, Bulgaria), we observe delay in actual crime activities. This process might have been concealed due to a delayed tempo of social changes. Some underground capitalist development during socialist time in Poland possibly may have caused a more sudden increase both in the economic development and in the recorded crime rate (1987 – 1989).

Our study also suggests that there exists an "optimal interval" of the "noise level". We see that *this optimum depends on the power and frequency of the task changes*. Applying this conclusion to criminality analysis, we deduce that the same criminality rates can be harmful for one country but still tolerable elsewhere. These observations, in turn, provide criteria to distinguish two strategically important key situations needing quite diverse social strategies. The first situation is such that an increase in criminality rate really jeopardizes the very existence of the society. Whilst, secondly in which criminality seems to be related to intensive development of the country bearing in mind several other factors, like migration of labors force, that affect criminality. Such situation was characteristic to Malaysia where during the period from 1986-96, the Malaysian economy was flourishing, and this attracts both legal and illegal foreign labors into the country. When Malaysia experienced the major economic downturn in mid-1996, an upward trend in recorded crime rate is seen as the price of intensive development. Similar conclusion can be drawn on Singapore. Thus, criminality has two sides: destructive and constructive one.

The above remarks bring us to the general problem of interaction between two fundamental opposite sides of society. The first is the ability of a society to resist changes i.e. to stay stable despite all instabilities of the surrounding world. The stability and an independence from permanent changes of the neighboring world are crucial for the very existence of society. The family, the school, the court and the shop can function only if their member can predict each other's action [4]. On the other side, the ability of society to learn new reactions is of core importance for its ability to survive. Both society and its agents are under permanent pressure to learn new ways to react and to adapt.

Interactions between both fundamental sides of social life are highly complicated and controversial. The modern social theories tell a lot both on ways in which social systems *resist* to changes and on those in which it *accepts* the changes. Our simulations were able to discover new, interesting and important effects in this interaction. It showed the paradoxical effect of noise in the learning of intellectual agents and provided the ground to infer that the social noise (i.e. criminality) produces an analogical functional effect also in social life - *instead of disturbing the society life the criminal actions also promotes its development and adaptation.*

In comparison to real life, computer imitations allow the investigations on a wider variation of the factors that influence criminality. The imitation experiments that consider all possible values and interrelations between involved variables should be a very good starting point to reveal possible effects of the factors considered. Our imitation experiments provide good example. Experimenting with different meanings of noise level has shown that the paradoxical constructive effect in learning arises only in situations when the changes (catastrophes) are great and infrequent. Despite the fact that sociological investigations on stability and changes have a long history, the sociologists were not able to discover this effect. It is so because "verbal models" conventionally used in sociological and criminological research are not suited to discover and describe such complicated behaviors.

Consequently, the discovery of these paradoxical effects of artificially created partially incorrect training signals in learning processes suggests a new research direction. Several stages have to follow.

First, a generalized criminological tool to measure a degree of incorrect training signals has to be developed. This index should include three types of crime data.

1. Crime data that relate to various kinds of 'black market' and illegal economic activities (criminal business) – These crimes were highly characteristic for the first few independent years in countries of the East and Middle Europe. They were caused by the collapse of the planned economy and due to lack of new capitalistic legal regulation. The "explosion" and wantonness of the illegal shadow economy was so extensive that it endangered the very existence of the new independent states. However, this extremely dangerous situation gave a mighty impetus to develop modern capitalistic, legal and economic regulations.
2. One needs to include a data about crimes, which in new democratic society proved to be *especially resistant* to legal control. The criminal justice of the newly independent countries proved to be highly unsuccessful in trying to control organized criminality, production and dissemination of illegal drugs, money laundering, human trade, corruption and many others. Legal and economic tools that were quite efficient in socialist society totally failed in this new situation causing a rapid and highly dangerous increase in such criminality. Yet, it brought an acute need to develop a new efficient legislation and to modernize the institutions of criminal justice.
3. Crime data that relate to immoral activities – The great deal of the *"moral" criminality* (e.g. pornography, prostitution, sexual exploitation of children, violence within a family) should also be included in the aforesaid index. In liberalization of sex, pornography, violence, drug addiction the East and Middle European countries were much more "democratic" and permissive than the West European countries with their long democratic traditions. At the time, the newborn democracies were a real "paradise of sexual freedom" and countries of "great opportunities" to both local and international criminals. In the first years of independence, these countries became a source of criminal contagion for the rest of Europe. All this also caused mighty stimuli for intensive development of new democratic regulations in these countries.

Second, the development of the generalized index of criminality will provide an opportunity to investigate the dynamics of the criminality processes in detail, compare them with simulation data and develop more truthful mathematical models. We then state our hypotheses based on the generalized index value as the higher the index value that a country has (i) the more intensive is the development of new regulation (in situations where other things being equal) and (ii) the more successful it becomes to taming criminality.

Third, more complex models of multi-agent system training based on analysis of variety of distinct criminality factors and a variety of simulation experiment conditions should be developed. One may hope they will provide necessary contextual information and theoretical basis for real-life explanations. These models will be based on empirical data of involved variables that were received in a given country from the previous stage. The essence of these new models should be the interaction between distractive and promoting role of criminality in moment of sudden great environmental changes in the society. These models while solving really complicated tasks such as forecasting and optimization of criminality and its effect upon social development will also be able to be a useful tool.

7 Conclusions

The simulation studies of the multi-agent training systems with partially incorrect training signals suggest that a three-stage research will be a good starting point to initiate a new research direction. This will result in the paradigm shift from general statistical investigations of the criminal activities in the societies to the examination of reasons of birth of the different types of crimes by means of computer simulations. Extensive simulations studies will help create diversity of imitational patterns and will reveal the contextual factors, which affect criminality in differing circumstances. Knowledge of these can be used in our future work to understand and perhaps, comprehend criminality in our society.

A lot of work has to be done in order to validate the models, which include the initiative to develop the generalized index of criminality, to find the relationships between "the environmental changes" in social life and the parameter changes of the computer model. We also need to improve the interpretation of the simulation results when a number of distinct social factors are acting simultaneously.

Especially interesting would be an investigation of differences between "totalitarian" societies (based upon its resistance to changes) and "democratic", "pluralistic" ones (based upon intensive learning of new adaptation ways). It can be hypothesized that criminality (a noise injection to training signals) plays quite different roles in each type of societies. In addition, we could hypothesize that both types of the societies are especially contrasting in their role of social development endeavor to deter criminality in their progress. These will be the challenging topics in further trainable multi-agent systems research.

References

[1] Hazans. M. How parallel markets fuelled chronic shortage in Soviet official sector. *Baltic Journal of Economics* . 1 (2): 3-58, 1999.
[2] Pumputis A. About Needs and Systems of Human Rights. *Jurisprudencija.* Mykolas Romeris Univ. Publ. Vilnius, 15(7): 61-65, 2000.
[3] Cohen L. Felson M. So*cial change and crime rate trends: A routine activity approach. American Sociological Review*, 44: 588-608, 1986.
[4] Parsons T. *Societies: Evolutionary and Comparative Perspectives.* NY: Free Press, 1966.
[5] Justickis V. *Criminology,* Vol. 1. Law University of Lithuania Publ., Vilnius, 2001.
[6] Raudys S. Survival of intelligent agents in changing environments. *Lecture Notes in Artificial Intelligence*, Springer-Verlag, 3070: 109-117, 2004.
[7] Henricksen K., Indulska J., Rakotonirainy A. Modeling context information in pervasivecomputing systems. *Lecture Notes in Computer Science,* Springer, 2414: 167–180, 2002.
[8] United Nations. *National Accounts, Statistics: Analysis of main aggregates and detailed tables, 2000,* United Nations Publ., New York, 2002.
[9] Etorf H., Spengler , H. *Crime in Europe. Causes and Consequences.* Springer Verlag, New York, 2002.
[10] *Crime and Criminal Justice in Europe and North America 1995-1997. Report on the Sixth United Nations Survey on Crime Trends and Criminal Justice Systems* (Edited by K. Aromaa, S. Leppä, S.i Nevala and N. Ollus) Helsinki: European Institute for Crime Prevention and Control, affiliated with the United Nations (HEUNI), 2003.
[11] *Criminal Victimization in eleven Industrialized Countries. Key findings from the 1996 International Crime Victims Survey.* The Hague: Ministry of Justice, WODC.
[12] Haykin S. *Neural Networks: A comprehensive foundation.* 2nd edition. Prentice-Hall, Englewood Cliffs, NJ, 1999.
[13] Raudys S. On the universality of the single-layer perceptron model. *Neural Networks and Soft Computing* (L. Rutkowski ed.) Physica-Verlag (Springer), New York, pp. 79-86.
[14] Raudys S. An adaptation model for simulation of aging process. *Int. J. Modern Physics C*, 13(8):1075-1086, 2002.
[15] Raudys S. and Justickis V. Yerkes-Dodson law in agents' training. *Lecture Notes in Artificial Intelligence,* Springer-Verlag. Vol. 2902: 54-58, 2003.
[16] Fukunaga K. *Introduction to Statistical Pattern Recognition.* 2nd ed. Academic Press, New York, 1990.
[17] Raudys, S. *Statistical and Neural Classifiers: An integrated approach to design.* Springer Verlag, NY, 2001.
[18] Skurichina M., Raudys S., Duin R.P.W. K-nearest neighbors directed noise injection in multilayer perceptron training, *IEEE Trans. on Neural Networks*, 11: 504–511, 2000.

Minimality and Non-determinism in Multi-context Systems

Floris Roelofsen[1] and Luciano Serafini[2]

[1] Institute for Logic, Language, and Computation,
Amsterdam, Netherlands
[2] Instituto per la Ricerca Scientifica e Tecnologica,
Trento, Italy

Abstract. Multi-context systems can be used to represent contextual information and inter-contextual information flow. We show that the local model semantics of a multi-context system is completely determined by the information that is obtained when simulating the information flow specified by the system, in such a way that a *minimal* amount of information is deduced at each step of the simulation.

The multi-context system framework implicitly presupposes that information flow is *deterministic*. In many natural situations, this is not a valid assumption. We propose an extension of the framework to account for non-determinism and provide an algorithm to efficiently compute the meaning of non-deterministic systems.

1 Introduction

The representation of contextual information and inter-contextual information flow has been formalized in several ways. Most notable are the propositional logic of context developed by McCarthy, Buvač and Mason [5, 6], and the multi-context systems devised by Giunchiglia and Serafini [3, 4], which later have been associated with the local model semantics introduced by Giunchiglia and Ghidini [2]. The two approaches have been compared by Serafini and Bouquet [9] from a technical viewpoint, and by Benerecetti et.al. [1] from a conceptual perspective.

A multi-context system describes the information available in a number of contexts (i.e., to a number of people / agents / databases, etc.) and specifies the information flow between those contexts. The local model semantics defines a system to entail a certain piece of information in a certain context, if and only if that piece of information is acquired in that context, independently of how the information flow described by the system is accomplished.

The first contribution of this paper is based on the observation that the local model semantics of a multi-context system is completely determined by the information that is obtained when simulating the information flow specified by the system, in such a way that a *minimal* amount of information is deduced at each step of the simulation. We define an operator which suitably implements such a simulation, and thus determines the information entailed by the system. This operator constitutes a first constructive account of the local model semantics.

The second contribution of this paper is based on the observation that, in its original formulation, the multi-context system framework implicitly rests on the assumption that information flow is *deterministic*. In many situations, this is not a suitable assumption. In a multi-agent scenario, for example, upon establishing a certain piece of information, an agent may decide to pass this information on to either one of a group of other agents. His choice as to which agent he will inform could be made non-deterministically. Another typical situation in which information flow is inherently non-deterministic is when the information channels between different contexts are subject to temporary failure or unavailability. Consider the case of online repositories. If information is obtained in one repository, the protocol may be to pass this information on to any one of a number of associated "mirror repositories": if the communication channel with one of these is defective or temporarily unavailable, another one is tried, until at least one successful communication is established.

The local model semantics is easily adapted to account for non-deterministic systems. However, if a system describes a non-deterministic information flow, then the minimal information entailed by the system cannot be determined unequivocally. We provide a way to generate from a non-deterministic system a number of deterministic systems, the semantics of which can be determined contructively, and which, together, completely determine the semantics of the original non-deterministic system.

We proceed, in section 2, with a brief review of multi-context system syntax and local model semantics. Minimality and non-determinism are discussed in section 3 and 4, respectively. We conclude, in section 5, with a concise recapitulation of our main observations and results.

2 Preliminaries

A simple illustration of the main intuitions underlying the multi-context system framework is provided by the situation depicted in figure 1. Two agents, Mr.1 and Mr.2, are looking at a box from different angles. The box is called magic, because neither Mr.1 nor Mr.2 can make out its depth. As some sections of the box are out of sight, both agents have partial information about the box. To express this information, Mr.1 only uses proposition letters l (there is a ball on the left) and r (there is a ball on the right), while Mr.2 also uses a third proposition letter c (there is a ball in the center).

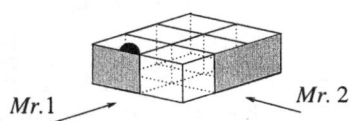

Fig. 1. A magic box

In general, we consider a set of contexts I, and a language L_i for each context $i \in I$. Henceforward, we assume I and $\{L_i\}_{i \in I}$ to be fixed, unless specified otherwise. Moreover, for the purpose of this paper we assume each L_i to be built over a finite set of proposition letters, using standard propositional connectives.

To state that the information expressed by a formula $\varphi \in L_i$ is established in context i we use so-called *labeled formulas* of the form $i : \varphi$ (if no ambiguity arises, we simply refer to labeled formulas as formulas, and we even use capital letters F, G, and H to denote labeled formulas, if the context label is irrelevant). A *rule* r is an expression of the form:

$$F \leftarrow G_1 \wedge \ldots \wedge G_n \qquad (1)$$

where F and all G's are labeled formulas; F is called the consequence of r and is denoted by $cons(r)$; all G's are called premises of r and together make up the set $prem(r)$. Rules without premises are called *facts*. Rules with at least one premiss are called *bridge rules*. A *multi-context system* (system hereafter) is a finite set of rules. A fact describes information that is established in a certain context, independent of which information is obtained in other contexts. A bridge rule specifies which information is established in one context, if other pieces of information are obtained in different contexts. So a system can be seen as a specification of contextual information available a priori plus an inter-contextual information flow.

Example 1. The situation in figure 1 can be modeled by the following system S:

$$\begin{aligned} 1 &: \neg r & \leftarrow \\ 2 &: l & \leftarrow \\ 1 &: l \vee r & \leftarrow 2 : l \vee c \vee r \\ 2 &: l \vee c \vee r & \leftarrow 1 : l \vee r \end{aligned}$$

Mr.1 knows that there is no ball on the right, Mr.2 knows that there is a ball on the left, and if any agent gets to knows that there is a ball in the box, then he will inform the other agent about it.

A classical interpretations m of language L_i is called a *local model* of context i. A set of local models is called a *local information state*. Intuitively, every local model in a local information state represents a "possible state of affairs". If a local information state contains exactly one local model, then it represents complete information. If it contains more than one local model, then it represents partial information: more than one state of affairs is considered possible. A *distributed information state* is a set of local information states, one for each context. In conformity with the literature, we will refer to distributed information states as *chains*.

Example 2. The situation in figure 1, in which Mr.1 knows that there is no ball on the right but does not know whether there is a ball on the left, is represented by a chain whose first component $\{\{l, \neg r\}, \{\neg l, \neg r\}\}$ contains two local models. As such, the chain reflects Mr.1's uncertainty about the left section of the box.

A chain c *satisfies* a labeled formula $i : \varphi$ (denoted $c \models i : \varphi$) if and only if all local models in its ith component classically satisfy φ. A rule r is *applicable* with respect to a chain c if and only if c satisfies every premiss of r. Notice that facts are applicable with respect to any chain. A chain c *complies with* a rule r, if and only if, whenever r is applicable with respect to c, then c satisfies r's consequence. We call c a *solution chain* of a system S if and only if it complies with every rule in S. A formula F is *true* in S (denoted $S \models F$) if and only if every solution chain of S satisfies F.

For convenience, we introduce some auxiliary terminology and notation. Let **C** denote the set of all chains. Notice that, as each L_i is assumed to be built over a finite set of proposition letters, **C** is assumed to be finite as well. Let $c^\perp$ denote the chain containing every local model of every context ($c^\perp$ does not satisfy any non-tautological expression); let $c^\top$ denote the chain containing no local models at all ($c^\top$ satisfies all expressions). If C is a set of chains, then the component-wise union (intersection) of C is the chain, whose i^{th} component consists of all local models that are in the ith component of some (every) chain in C. If c and c' are chains, then $c \setminus c'$ denotes the chain, whose i^{th} component consists of all local models that are in c_i but not in c'_i. Finally, let us sometimes say that a local model m is (not) in c, when we actually mean that m is (not) in some (any) component c_i of c.

3 Minimality

We order chains according to the amount of information they convey. Intuitively, the more local models a chain component contains, the more possibilities it permits, so the less informative it is. Formally, we say that c is *less informative* than c' ($c \preceq c'$), if for every i we have $c_i \supseteq c'_i$. If, moreover, for at least one i we have $c_i \supset c'_i$, then we say that c is *strictly less informative* than c' ($c \prec c'$).

Lemma 1. *Let C be a set of chains. Let c^u (c^i) denote the component-wise union (intersection) of all chains in C. Then c^u (c^i) is less (more) informative than any chain in C.*

Proof. We proof the *union* part. Let c' be a chain in C. Then for every i, every local model m in c'_i is also in c^u_i. So $c^u_i \supseteq c'_i$, and thus $c^u \preceq c'$. □

Lemma 2. $(\mathbf{C}, \preceq)$ *forms a complete lattice.*

Proof. We should prove that every finite subset of **C** has both a greatest lower bound and a least upper bound in **C**. Let C be a subset of **C** (note that **C** is finite, so C must be finite as well). Let c^u (c^i) denote the component-wise union (intersection) of all chains in C. Then, by lemma 1, c^u is a lower bound of C. Now consider a chain c', such that $c^u \prec c'$. For this to be the case, there must be a local model m, which is in c^u but not in c'. But then m must also be in some chain c^m in C, which makes $c' \preceq c^m$ impossible. So c' cannot be a lower bound of C, which implies that c^u is the greatest lower bound of C. Analogously, it can be shown that c^i is least upper bound of C. □

Note that $c^\perp$ is strictly less informative than any other chain, whereas $c^\top$ is strictly more informative than any other chain. If $c \preceq c'$ we say that c' is an *extension* of c. So, intuitively, extending c corresponds to *adding* information to it. More technically, to extend c is to *remove* local models from it. We say that c is *minimal* among a set of chains C, if c is in C and no other chain c' in C is strictly less informative than c. In particular, we say that c is a *minimal solution chain* of a system S, if it is minimal among the set of all solution chains of S.

Lemma 3. *Let c and c' be two chains, such that $c \preceq c'$. Then any formula that is satisfied by c is also satisfied by c'.*

Proof. Suppose $c \models i : \phi$. Then, per definition, $m \models \phi$ for every $m \in c_i$. As c'_i is contained in c_i, we also have $m' \models \phi$ for every $m' \in c'_i$. So $c' \models i : \phi$. □

Lemma 4. *Let C be a set of chains and let c^u denote the component-wise union of all chains in C. Then a formula is satisfied by c^u if and only if it is satisfied by every chain in C.*

Proof.
($\Rightarrow$) Follows directly from lemma 1 and lemma 3.
($\Leftarrow$) Suppose all chains in C satisfy a formula $i : \phi$. Then all local models in the ith component of every chain in C must satisfy ϕ. These are exactly the local models that make up the ith component of c^u. So c^u must also satisfy $i : \phi$. □

Lemma 5. *Let S be a system. Then the set of all solution chains of S is closed under component-wise union. That is, if C is a set of solution chains of S, then the component-wise union c^u of all chains in C is again a solution chain of S.*

Proof. Let C be a set of solution chains of S. Let c^u be the component-wise union of C. Let r be an arbitrary rule in S. Then all c' in C comply with r. Suppose, towards a contradiction, that c^u does not comply with r, i.e., c^u satisfies all of r's premises, but does not satisfy r's consequence. By lemma 4 all c' in C satisfy all of r's premises, and therefore, by assumption, they all satisfy r's consequence as well. But then, again by lemma 4, c^u must also satisfy r's consequence, which contradicts the assumption that c^u does not comply with r. So c^u must comply with r, and as r was arbitrary, c^u must be a solution chain of S. □

Theorem 1. *Every system S has a unique minimal solution chain c_S.*

Proof. Every system has at least one solution chain, namely $c^\top$. Now, let S be a system and let C_S be the set of all its solution chains. Then, by lemma 5, the component-wise union c_S of C_S is itself in C_S. Moreover, by lemma 1, c_S is less informative than any other chain in C_S. So c_S is minimal among C_S and, moreover, any chain c' in C_S which is minimal among C_S, must be equal to c_S. In other words, c_S is the unique minimal solution chain of S. □

Theorem 2. *The meaning of a system S is completely determined by its unique minimal solution chain c_S. For any formula F we have:*

$$S \models F \quad \Leftrightarrow \quad c_S \models F$$

Proof. Let S be a system and let F be a formula. Then F is true in S if and only if F is satisfied by all solution chains of S. By lemma 4, this is the case if and only if F is satisfied by the component-wise union of all solution chains of S. From the proof of theorem (1) this union constitutes the minimal solution chain c_S of S. □

Theorem (1) and (2) are extremely useful, because they establish that, to answer queries about a system S, it is no longer necessary to compute all solution chains of S; we only need to consider the system's minimal solution chain c_S.

3.1 Computing the Minimal Solution Chain

Recall that a system S can be thought of as a specification of inter-contextual information flow. It turns out that the minimal solution chain of S can be characterized as the $\preceq$-least fixpoint of an operator $\mathbf{T}_S$, which, intuitively, simulates the information flow specified by S.

Let $S^*(c)$ denote the set of rules in S, which are applicable w.r.t. c. Then:

$$\mathbf{T}_S(c) = c \setminus \{m \mid \exists r \in S^*(c) : m \not\models cons(r)\} \tag{2}$$

For every rule r in S that is applicable w.r.t. c, $\mathbf{T}_S$ removes from c all local models that do not satisfy $cons(r)$. Intuitively, this corresponds to augmenting c with the information expressed by $cons(r)$. In this sense, $\mathbf{T}_S$ simulates the information flow described by S. Clearly, $\mathbf{T}_S(c)$ is obtained from c only by *removing* local models from it. As a result, $\mathbf{T}_S(c)$ is always more informative than c.

Lemma 6. *For every chain c and every system S: $c \preceq \mathbf{T}_S(c)$.* □

We now prove that, starting with the least informative chain $c^\perp$, $\mathbf{T}_S$ will reach its $\preceq$-least fixpoint after finitely many iterations, and that this $\preceq$-least fixpoint coincides with the minimal solution chain of S. The first result is typically established using Tarski's fixpoint theorem [10]. In order to apply this theorem, we first need to show that $\mathbf{T}_S$ is monotone and continuous with respect to $\preceq$.

Lemma 7. *$\mathbf{T}_S$ is monotone with respect to $\preceq$.*

Proof. Let c and c' be any two chains such that $c \preceq c'$. We need to prove that $\mathbf{T}_S(c) \preceq \mathbf{T}_S(c')$. Suppose, towards a contradiction that this is not the case. Then there is a local model m that belongs to $\mathbf{T}_S(c')$ but not to $\mathbf{T}_S(c)$. Clearly, m must already be present in c', and therefore also in c. From the fact that m has been removed from c by $\mathbf{T}_S$ it follows that there must be a rule r in S such that c satisfies $prem(r)$, whereas m does not satisfy $cons(r)$. But then, by lemma 3, c' must also satisfy $prem(r)$, so $\mathbf{T}_S$ should have removed m from c' as well. We conclude that $\mathbf{T}_S(c) \preceq \mathbf{T}_S(c')$. So $\mathbf{T}_S$ is monotone with respect to $\preceq$. □

Lemma 8. T_S *is continuous with respect to* $\preceq$.

Proof. Let $c^0 \preceq c^1 \preceq c^2 \preceq \ldots$ be an infinite sequence of chains, each of which contains more information than all preceding ones. We need to prove that $\mathbf{T}_S(\bigcup_{n=0}^{\infty} c^n) = \bigcup_{n=0}^{\infty} \mathbf{T}_S(c^n)$. As **C** is finite, $\{c^0, c^1, c^2, \ldots\}$ must have a maximum c^m in **C**. So $\mathbf{T}_S(\bigcup_{n=0}^{\infty} c^n) = \mathbf{T}_S(c^m) = \bigcup_{n=0}^{\infty} \mathbf{T}_S(c^n)$. □

Theorem 3. T_S *has a $\preceq$-least fixpoint, which is obtained after a finite number of consecutive applications of* T_S *to* $c^{\perp}$.

Proof. Follows from lemmas 2, 7, and 8 by Tarski's fixpoint theorem [10]. □

Lemma 9. *Let c be a chain and let S be a system. Then c is a fixpoint of* T_S *if and only if c is a solution chain of S.*

Proof. A chain c is a fixpoint of $\mathbf{T}_S$ if and only if for every rule r in S, c satisfies $cons(r)$ whenever c satisfies $prem(r)$. This is the case if and only if c is a solution chain of S. □

Theorem 4. *Let S be a system. Then the minimal solution chain c_S of S coincides with the $\preceq$-least fixpoint of* T_S.

Proof. Follows directly from lemma 9. □

From theorems 3 and 4 we conclude that the minimal solution chain c_S of a system S is obtained by a finite number of applications of $\mathbf{T}_S$ to the least informative chain $c^{\perp}$. But we can even prove a slightly stronger result:

Theorem 5. *Let S be a system and let $|S|$ denote the number of bridge rules in S. Then the minimal solution chain c_S of S is obtained by at most $|S| + 1$ consecutive applications of* T_S *to* $c^{\perp}$.

Proof. Let c be a chain and let S be a system. Notice that $\mathbf{T}_S(c)$ is a fixpoint of $\mathbf{T}_S$ if and only if $S^*(\mathbf{T}_S(c))$ coincides with $S^*(c)$. Lemmas 3 and 6 imply that, in any case, $S^*(\mathbf{T}_S(c)) \supseteq S^*(c)$. In other words, during each iteration of $\mathbf{T}_S$ some (possibly zero) rules are added to S^*. In the case that S^* remains unaltered, $\mathbf{T}_S$ must have reached a fixpoint. Now we observe that during the first application of $\mathbf{T}_S$ (to $c^{\perp}$) all facts in S are added to S^*. Clearly, after that, $\mathbf{T}_S$ can be applied at most $|S|$ times before a fixpoint is reached. □

In fact, a slightly more involved, but essentially equivalent procedure was introduced for rather different reasons in [7]. This procedure was shown to have worst-case time complexity $O(|S|^2 \times 2^M)$, where M is the maximum number of propositional variables in either one of the contexts involved in S. The greater part of a typical computation is taken up by propositional reasoning within individual contexts, which itself requires exponential time in the worst case.

Example 3. Consider the system S given in example 1. Applying $\mathbf{T}_S$ to $c^{\perp}$ establishes the facts given by the first two rules of the system. But then Mr.2

knows that there is a ball in the box, so the next application of $\mathbf{T}_S$ simulates the information flow specified by the third rule of the system: Mr.2 informs Mr.1 of the presence of the ball. The resulting chain is left unaltered by any further application of $\mathbf{T}_S$, and therefore constitutes the minimal solution chain of S. The fact that this chain satisfies the formula $1 : l$ reflects, as desired, that Mr.1 has come to know that there is a ball in the left section of the box.

4 Non-determinism

The original formulation of multi-context systems implicitly rests on the assumption that information flow is *deterministic*. However, there are many natural situations in which information flow is inherently non-deterministic.

Example 4. Adriano is on holiday after having submitted his final school exams. He has promised to call his father or his mother in case his teacher lets him know that he has passed his exams. This situation can be modeled by a system S consisting of the following rule:

$$m : p \text{ or } f : p \leftarrow a : p$$

Notice that Adriano may be conceived of as an agent in a multi-agent system, who non-deterministically decides which other agents to inform when acquiring novel information. Alternatively, Adriano's parents may be conceived of as mirror repositories of information about Adriano's well-being (assuming that they tell each other everything they come to know about Adriano). Typical telephonic connections may be broken or temporarily unavailable. Analogous to the situation sketched in the introduction, Adriano will try to reach his parents, until at least one of them is informed.

In general, we would like to consider systems in which rules r are of the form:

$$F_1 \text{ or } \ldots \text{ or } F_m \leftarrow G_1 \wedge \ldots \wedge G_n \tag{3}$$

where all F's and G's are labeled formulas; all F's are called consequences of r and together form the set $cons(r)$; and as before, all G's are called premises of r and together constitute the set $prem(r)$. A rule doesn't necessarily have any premises ($n \geq 0$), but always has at least one consequence ($m \geq 1$). We call a rule deterministic if it has only one consequence, and non-deterministic otherwise. We call finite sets of possibly non-deterministic rules *non-deterministic multi-context systems* (non-deterministic systems for short). Systems which consist of deterministic rules only, are from now on refered to as deterministic systems.

A chain c complies with a non-deterministic rule r if and only if, whenever r is applicable w.r.t. c, *at least one of* its consequences is satisfied by c. A chain is a solution chain of S if and only if it complies with all rules in S. A formula F is true in S, $S \models F$, if and only if F is satisfied by all solution chains of S.

Observation 1. *Let S be a non-deterministic system, let c' and c'' be two solution chains of S, and let c be the component-wise union of c' and c''. Then it* is not generally the case that c is again a solution chain of S. Therefore, S does not generally have a unique minimal solution chain.

Example 5. Suppose Adriano's teacher lets him know that he passed his exams. The resulting system S is given by the following rules:

$$a : p \leftarrow$$
$$m : p \text{ or } f : p \leftarrow a : p$$

This system has two minimal solution chains:

$$c^m = \{\{p\}_m, \quad \{p, \neg p\}_f, \quad \{p\}_a\}$$
$$c^f = \{\{p, \neg p\}_m, \quad \{p\}_f, \quad \{p\}_a\}$$

whose component-wise union $\{\{p, \neg p\}_m, \quad \{p, \neg p\}_f, \quad \{p\}_a\}$ is not a solution chain of S.

Theorem 6. *The meaning of a non-deterministic system S is completely determined by the set C_S of all its minimal solution chains. For any formula F we have:*

$$S \models F \quad \Leftrightarrow \quad \forall c \in C_S : c \models F$$

Proof. Let S be a non-deterministic system and let F be a formula. Then F is true in S if and only if F is satisfied by every solution chain of S. Clearly, if F is satisfied by every solution chain of S, then it must in particular be satisfied by every minimal solution chain of S. Moreover, every solution chain of S is an extension of some minimal solution chain of S, which implies, by lemma 3, that F is satisfied by all minimal solution chains of S only if F is satisfied by all solution chains of S. □

Theorem 6 establishes that the meaning of a non-deterministic system S is completely determined by the set C_S of all its minimal solution chains. We will now provide a way to compute C_S, re-using the method outlined in section 3.

4.1 Computing Minimal Solution Chains

Our approach, inspired by an idea originally developed for disjunctive databases [8], is to generate from a non-deterministic system S a number of deterministic systems $S_1, S_2, \ldots, S_n$, in such a way that the minimal solution chains of S are among the minimal solution chains of $S_1, S_2, \ldots, S_n$ (note that each S_i has a unique minimal solution chain which can be computed as outlined in section 3). Hereto, we introduce the notion of a *generated system*. Let S be a non-deterministic system and let r be a rule in S. Then we say that a deterministic rule r' is *generated by* r if and only if $cons(r') \in cons(r)$ and $prem(r') = prem(r)$. We say that a system S' is *generated by* S if and only if it is obtained from S by replacing each rule r in S by some rule r' generated by r. Notice that, indeed, a generated system is always deterministic, and that any non-deterministic system S generates at most $\prod_{r \in S} |cons(r)|$ different deterministic systems.

Example 6. The non-deterministic system from example 5 generates two deterministic systems: $\{a : p \leftarrow, \quad m : p \leftarrow a : p\}$ and $\{a : p \leftarrow, \quad f : p \leftarrow a : p\}$. The only system generated by a deterministic system is that system itself.

Lemma 10. *A chain c is a solution chain of a non-deterministic system S if and only if it is a solution chain of some system S' generated by S.*

Proof.
($\Rightarrow$) Suppose c is a solution chain of S. Then c complies with every rule in S. For every rule r in S, if c complies with r, then there must be a rule r' generated by r such that c complies with r' as well. Let S' be the system $\{r' \mid r \in S\}$. Then c is a solution chain of S'.
($\Leftarrow$) Suppose c is a solution chain of a system S' generated by S. Then c complies with every rule in S'. Every rule r in S has generated some rule r' in S', and clearly, if c complies with r' then it must also comply with r. So c is a solution chain of S. □

We call a chain c a *potential solution chain* of S if and only if c is a minimal solution chain of some system S' generated by S.

Lemma 11. *Every minimal solution chain of a system S is also a potential solution chain of S.*

Proof. Suppose c is a minimal solution chain of S. Then, by lemma 10, c is a solution chain of some system S' generated by S. Let c' be the minimal solution chain of S'. Then c must be an extension of c'. By lemma 10 c' must be a solution chain of S. But then, as c is a minimal solution chain of S, c' must be equal to c. So c is a minimal solution chain of S', and therefore a potential solution chain of S. □

Observation 2. *It is not generally the case that a potential solution chain of S is also a minimal solution chain of S.*

Example 7. Suppose Adriano's teacher also called Adriano's mother to tell her the good news. The resulting system S is given by the following rules:

$$a : p \leftarrow$$
$$m : p \leftarrow$$
$$m : p \text{ or } f : p \leftarrow a : p$$

This system has two potential solution chains:

$$c^m = \{\{p\}_m, \quad \{p, \neg p\}_f, \quad \{p\}_a\}$$
$$c^{mf} = \{\{p\}_m, \quad \{p\}_f, \quad \{p\}_a\}$$

But as c^{mf} extends c^m only the latter is a minimal solution chain of S.

We call c an *essential solution chain* of S if and only if c is minimal among all potential solution chains of S.

Theorem 7. *A chain is a minimal solution chain of S if and only if it is an essential solution chain of S.*

Proof.
($\Rightarrow$) Suppose c is a minimal solution chain of S. Then, by lemma 11, c is a potential solution chain of S. If c is minimal among all potential solution chains of S, then, per definition, it is essential. Now, towards a contradiction, suppose that c is *not* minimal among all potential solution chains of S. Then there must be another potential solution chain c' of S, such that $c' \prec c$. But, by lemma 10, c' must also be a solution chain of S, which contradicts the assumption that c is a minimal solution chain of S.
($\Leftarrow$) Suppose c is an essential solution chain of S. Furthermore, towards a contradiction, suppose that c is *not* a minimal solution chain of S. Then there must be a minimal solution chain c' of S, such that $c' \prec c$. By lemma 11, c' is a potential solution chain of S. But this contradicts the assumption that c is minimal among all potential solution chains of S. □

Theorem 7 establishes that, to compute the meaning of a non-deterministic system S it suffices to compute the meaning of all deterministic systems generated by S. This can be done re-using the method developed in section 3. Given that S generates at most $\prod_{r \in S} |cons(r)|$ different systems, and that computing the meaning of each of these systems takes at most time $O(|S|^2 \times 2^M)$, we conclude that, in the worst case, computing the meaning of S takes time $O(\prod_{r \in S} |cons(r)| \times |S|^2 \times 2^M)$.

5 Conclusions

In this paper, we investigated the multi-context system formalism as a framework for representing contextual information and inter-contextual information flow.

We observed that the semantics of a multi-context system is completely determined by the information that is obtained when simulating the information flow specified by the system, in such a way that a *minimal* amount of information is deduced at each step of the simulation. Based on this observation, we defined an operator which determines the information entailed by the system by implementing a suitable simulation of the prescribed information flow. This operator provides a first constructive account of the local model semantics.

Next we observed that the multi-context system framework implicitly rests on the assumption that information flow is *deterministic*. We sketched a number of situations, in which this is not a valid assumption. We extended the framework in order to account for non-deterministic information flow, and provided a way to express the semantics of a non-deterministic system in terms of the semantics of a number of associated, deterministic systems. This allowed us to give a constructive account of the semantics of non-deterministic systems as well.

References

1. M. Benerecetti, P. Bouquet, and C. Ghidini. Contextual reasoning distilled. *Journal of Experimental and Theoretical Artificial Intelligence*, 12(3):279–305, 2000.
2. C. Ghidini and F. Giunchiglia. Local models semantics, or contextual reasoning = locality + compatibility. *Artificial Intelligence*, 127(2):221–259, 2001.
3. F. Giunchiglia. Contextual reasoning. *Epistemologia*, XVI:345–364, 1993.
4. F. Giunchiglia and L. Serafini. Multilanguage hierarchical logics, or: how we can do without modal logics. *Artificial Intelligence*, 65(1):29–70, 1994.
5. J. McCarthy. Notes on formalizing context. In *International Joint Conference on Artificial Intelligence (IJCAI 93)*, pages 555–560, 1993.
6. J. McCarthy and S. Buvač. Formalizing context (expanded notes). In *Computing Natural Language*, volume 81 of *CSLI Lecture Notes*, pages 13–50. 1998.
7. F. Roelofsen, L. Serafini, and A. Cimatti. Many hands make light work: Localized satisfiability for multi-context systems. In *European Conference on Artificial Intelligence (ECAI 04)*, pages 58–62, 2004.
8. C. Sakama and K. Inoue. An alternative approach to the semantics of disjunctive programs and deductive databases. *Journal of Automated Reasoning*, 13:145–172, 1994.
9. L. Serafini and P. Bouquet. Comparing formal theories of context in AI. *Artificial Intelligence*, 155:41–67, 2004.
10. A. Tarski. A lattice-theoretical fixpoint theorem and its applications. *Pacific Journal of Mathematics*, 5:285–309, 1955.

'I' as a Pure Indexical and Metonymy as Language Reduction

Esther Romero[1] and Belén Soria[2]

[1] Department of Philosophy, Granada University, Campus Cartuja, 18011 Granada, Spain
phone: 34-958-246244, fax: 34-958-248981
eromero@ugr.es

[2] Department of English and German Philology, Granada University, Campus Cartuja, 18011 Granada, Spain
phone: 34-958-240644, fax: 34-958-243678
bsoria@ugr.es

Abstract. Most direct reference theorists believe that 'I' is a pure indexical. This means that when 'I' is uttered, it contributes with the speaker to what is said. But, from some conceptions of metonymy as reference transfer, if 'I' is used metonymically, it has an improper meaning and the object referred to is not the speaker. We will show that all theories of metonymy as transfer are inadequate and so they cannot determine if the metonymic use of 'I' is a counter-example to its consideration as a pure indexical or not. We argue that the appropriate conception on metonymy is to consider it as a case of language reduction and that 'I', when used metonymically, is just a part of a non-textual complete noun phrase; 'I' has the semantic value that it usually has. The metonymic use of 'I' does not risk the consideration of 'I' as a pure indexical.

1 Introduction[1]

Most direct reference theorists believe that 'I' is a pure indexical. This means, among other things, that when 'I' is uttered, it contributes with an individual to what is said and this individual can only be the speaker.[2] But if it were possible to have a meto-

---

[1] Research for this work has been supported by the project BFF2003-07141 funded by DGICYT, Spanish Government. We want to thank F. Recanati for his questions and sharp remarks on our proposal on metonymy which helped us to refine it. This paper has also benefited from extensive comments and criticisms by P. Perconti and five anonymous referees for this volume.

[2] When we use the term 'speaker', we are thinking about the technical notion of the agent as a parameter of the context of utterance. This may be the writer, the producer or the speaker in the non-technical sense. We are also aware that this technical notion of speaker is rather controversial, especially when there is a dissociation of the three roles conflated under it: animator, author, and principal (see McCawley, J.D. (1999) Participant roles, Frames, and Speech Acts. Linguistics and Philosophy **22** 595-619). When the three roles are played by different persons, the principal is the agent. We are not going to take into account all these distinctions in our paper because the topic under discussion is not dependent on them.

nymic use of 'I', then 'I', from some standard conceptions of metonymy, would have an improper meaning and would not refer to the speaker. An example used to show the metonymic use of 'I' is (1).

[Mary, giving her car key to the attendant at a parking lot, tells him:] I am parked out back (1)

When Mary uses 'I' in (1), she refers to a car, her car, and not to herself, the speaker of (1). 'I' has an improper meaning that is equivalent to the proper meaning of 'My car' or 'The car I drive', etc. This use of 'I' spoils the argument that 'I' is a pure indexical.

Our objective is to show that, due to an inadequate conception of metonymy as reference transfer, the metonymic use of 'I' is sometimes wrongly conceived as a counterexample to the claim that 'I' is a pure indexical. This general proposal is not defended for the first time. For example, Nunberg has developed an explanation of metonymy as predicate transfer with which he intends to show that the metonymic use of 'I' is just an apparent counterexample to the conception of 'I' as a pure indexical [1], [2]. Nevertheless, we argue that the metonymic use of 'I' is not a counterexample for pure indexicals from another view of metonymy in which it is not considered as a transfer of meaning, no matter how transfer is described. As far as we are concerned [3], metonymy must be understood as a case of language reduction, and if this is the case, the metonymic use of 'I' does not entail a transfer from the semantic value of 'I' to another semantic value. 'I' means what it means and it has the semantic value that it usually has: the speaker. 'I' is only a part of the noun phrase (NP). From our proposal of metonymy, the metonymic use of 'I' does not constitute a counterexample to the claim that 'I' is a pure indexical.

2 Pure Indexicals in the Theory of Direct Reference

The term 'pure indexical' is introduced by Kaplan and makes reference to a particular class of directly referential expressions that are semantically complete such as 'I', 'here', and 'now' [4]. In general, a directly referential term indicates that the truth-conditions of the utterance in which it occurs involve its referent; it indicates that the proposition expressed by the utterance is singular. When we say that an expression is a pure indexical we are not only saying that it is a directly referential term but also that it is an expression highly restricted in its possibilities of reference in virtue of its conventional meaning and without considering the speaker's intention. Indeed, when 'I' is uttered, it contributes with an individual to what is said and this individual can only be the speaker; once the context is given, the contribution of 'I' in what is said can be automatically decoded.[3]

Pure indexicals are semantically complete; their conventional meaning which governs its use fully determines a 'character', that is, a fixed function from context to

---

[3] Although this is the most popular view in direct reference theory, there are also theorists within this frame that reject the existence of pure indexicals (see Predelli, S. (1998) I am not here now. Analysis **58.2** 107-115). For a relevant discussion on an opposite view, see Corazza et al. (Corazza, E., Fish, W., Gorvet, J. (2002). Who is I? Philosophical Studies **107** 1-21). This debate, however, is not related with the metonymic use of 'I'.

content; in this case, from context to referent. The content is the semantic value that intervenes in the proposition. The linguistic meaning of a referential expression determines such a function only if the expression is a 'pure indexical'. The meaning of 'I' is a fixed function that takes us from the context of utterance to the semantic value of the word in that context.

The main difference between demonstratives and pure indexicals is that the reference of a demonstrative is determined by the speaker's intention rather than by a fixed function of external features of the context of utterance. Pure indexicals are different from demonstratives in the sense that it does not seem possible to find an example of a pure indexical so that a rational subject can think at the same time that 'α is P' and 'α is not P' and this is perfectly possible with a demonstrative. It seems that I cannot think of 'me' as of two different persons at the same time. The reason why this is so is that the mental counterpart of a pure indexical is an egocentric concept and an egocentric concept is based on a fundamental epistemic relation between the subject and the objects which fall under that concept. The egocentric concepts associated with pure indexicals have the property that the fundamental relation on which such concepts is based on uniquely determines an object, in the sense that one cannot bear that relation to different objects at the same time (see [5]).

3 The Standard Conception of Metonymy as Reference Transfer

According to the standard conception of metonymy, this is a figure of signification, a trope that exploits a figurative or transferred meaning. From this approach, (2)

$$\text{The ham sandwich is waiting for his check} \tag{2}$$

would mean figuratively and metonymically the same as (3),

$$\text{The customer of the ham sandwich is waiting for his check} \tag{3}$$

since 'the ham sandwich' is substituted by 'the customer of the ham sandwich' and it means figuratively what the latter expression means. In any case of metonymy, the expression used metonymically changes its meaning so that it can be applied to the object talked about.

As far as classical rhetoric is concerned, what characterizes metonymy is that it has a semantic referent which does not coincide with the actual referent but with which it has a relation of contiguity. That is, there is what we can call a "reference transfer", and (2) can be considered a case of referential metonymy.[4] Yet, if metonymy depends

---

[4] (2) belongs to a homogeneous group of examples, based on the referential criterion. Utterances of the sentences such as 'It won't happen while I still breathe', 'She turned pale', etc. are sometimes considered cases of metonymy. We won't take into account this type of examples because not all of them are equivalent from a conceptual nor a linguistic point of view. Even if we admitted that both the examples excluded and those included had a common cognitive basis, it can be still defended that within this group there are different types of metonymy that respond to different restrictions and that, linguistically speaking, they behave differently (see Warren, B. (1999) Aspects of Referential Metonymy. In: Panther, K., Radden, G. (eds): Metonymy in Language and Thought. John Benjamins Publishing Company, Amsterdam Philadelphia 121-135).

on the existence of a transfer of the semantic value of the NP and (1) is considered as a metonymy, 'I' acquires a transferred meaning, the meaning of the expression 'the car I drive' and thus 'I' does not refer to Mary, the speaker, but to Mary's car. 'I' cannot be considered as a pure indexical.

But, how do we know, from this proposal, when an expression is used metonymically and how is its specific meaning derived? With respect to the identification problem nothing is said and with respect to the transfer of meaning it is argued that we have to resort to a relation of contiguity between the object which the words metonymically used, 'the ham sandwich', refer to, in this case, the ham sandwich, and the object actually referred to, the customer of the ham sandwich. But this strategy does not seem to let us go very far.

In the recent past, several answers to these questions have arisen from some pragmatic approaches. For instance, metonymy is conceived, following Grice's proposals on conversational implicature [6], as an implicature that is identified because the person who utters (2) in cases such as (4)

[In a restaurant, looking at the customer of the ham sandwich, a waitress says   (4)
   to another:] The ham sandwich is waiting for his check

cannot commit herself with the truth of what is said, as it is clear that the ham sandwich cannot be waiting for his check. She says something which she believes false. With the metonymic use of language the speaker conversationally implicates something different from what is said which includes the semantic referent, the ham sandwich; the speaker conversationally implicates what is said with its paraphrase, what is said with (5).

[In a restaurant, looking at the customer of the ham sandwich, a waitress says   (5)
   to another:] The customer of the ham sandwich is waiting for his check

If metonymy works in this way, we can consider the pronoun 'I' as a pure indexical. With (1), Mary says that she is parked out back, something that she does not believe because it is obviously false, and so she conversationally implicates what is said by (6),

[Mary, giving her car key to the attendant at a parking lot, tells him:] The car   (6)
   I drive is parked out back

she conversationally implicates that her car is parked out back.

The first problem with the theory of implicature is that its identification criterion, to say something the speaker believes false, is not exclusive of metonymy and that some do not even require it. Furthermore, considering metonymy as a particularized conversational implicature is not totally correct to the extent that a particularized conversational implicature is relevant when the speaker says or makes as if to say something literally and with (1) and (4) it cannot be said or made as if to say anything literally because these utterances cannot fix any literal proposition. The sentences contained in (1) and (4) have no literal and textual compositional sense. This is so, for example in (4), simply because sandwiches are inanimate objects which cannot volitionally instigate any action. Our linguistic competence prevents us from constructing a literal proposition without previously changing the conventional meaning of some of the sentence constituents or recovering some non-textual information. In addition, how do we know that (5) is the paraphrase of (4) and that (6) is the paraphrase of (1)? Or how is the specific implicature calculated? This approach does not improve the classical one [7], [8].

Nevertheless, metonymy is not conceived as a case of conversational implicature in all the pragmatic proposals. Recanati, for example, argues that it is a context dependent process of transfer that operates locally or subpropositionally [5], [9]. The transfer of meaning is, for this author, a context dependent aspect that affects what is said and, in his first writings, this transfer affects the NP.[5] Recanati explains that the interpreter of (4) does not go from the concept of the ham sandwich to the concept of the customer of the ham sandwich after having conceived the absurd literal proposition, rather she goes from one concept to another as a result of a shift of accessibility due to the activation of the literal interpretation of the predicate that accompanies the NP 'the ham sandwich' in (4); the activation of the literal interpretation of the predicate 'is waiting for his check'. When (1) is produced, the interpreter goes from the concept of I to the concept of the car I drive as a result of a shift of accessibility that the literal interpretation of the predicate 'be parked out back' triggers. In this sense, the pronoun 'I' refers to the speaker's car, and cannot be a pure indexical.

But, although from this approach the context restricts the interpretation of referential expressions to eliminate the irrelevant interpretations, once a non-literal interpretation is activated, how do we know it is a metonymy and how can we determine its correct interpretation?

With respect to the explanation of how the metonymic meaning is achieved, there have been different proposals. The cognitive linguistics proposal stands out since Lakoff and Johnson [10]. In cases such as (7)

The Times hasn't arrived at the press conference room yet. (The reporter from   (7)
the Times)

an entity is used to refer to another by the relation between institution-person responsible. The examples of metonymy are framed in metonymic concepts that are systematic: OBJECT USED FOR USER in (2), INSTITUTION FOR PEOPLE RESPONSIBLE in (7), and so forth. From this point of view, 'I' in (1) gets a transferred meaning because it stands for something else, the speaker's car. So the pronoun 'I' cannot be a pure indexical.

Although the relations that cognitive linguists establish when talking about metonymy are relations with an important linguistic relevance, these relations are also present in the semantics of phrases and clauses without them having to be always cases of metonymy. Thus, to characterize metonymy is not to specify the types of contiguity relations, but to point out that they are stands-for relations, and this is the only thing that all the metonymic examples have in common in the cognitive linguists proposals. In a stands-for relation the concept used to stand for the other gets a transferred meaning. The problem is to know which relation is the one exploited in every case. The interlocutor recognizes it because s/he understands the utterance but not because he knows what metonymic concept he has to appeal to so that one piece of information can be substituted by another. To specify the types of contiguity relations does not improve the classical approach with respect to the identification and interpretation of metonymic utterances.

---

[5] Nowadays, Recanati (Recanati, F. (2004) Literal Meaning. Cambridge University Press, Cambridge) sides with the position defended by Nunberg (see Sect. 4).

To sum up, from the proposals of metonymy as reference transfer that locate its process of interpretation at the level of what is said, we cannot defend both that the metonymic use of 'I' is a case of reference transfer and that 'I' is a pure indexical. Only the theory of implicature can. Nevertheless, all these versions of the standard conception of metonymy have several serious problems with the identification criteria of metonymy and with the description of the metonymic mechanism for its interpretation and, thus, we can consider them neither a threat to the claim that 'I' is a pure indexical nor a support to it.

Only if we solve these problems, can we get a proposal on metonymy from which to determine if the metonymic use of 'I' is a counterexample to pure indexicals theory or not. One attempt can be found in Nunberg's writings. However, his approach is directed towards saving the pure indexicals theory more than towards solving the problems that emerge from the conception of metonymy as reference transfer.

4 Metonymy as Predicate Transfer

Nunberg intends to show that the problem raised by (1) for pure indexicals theory depends on an inadequate conception of metonymy by which the transfer is always located at the referential level. He defends that, although the conceptual process of metonymy is always the same, that is, it is the same rhetoric figure, from a linguistic point of view, we have to distinguish between two different types of metonymic transfer [1], [2].

First of all, he speaks about deferred ostension or reference transfer as a process that allows a demonstrative or indexical to refer to an object that corresponds in a certain way to the contextual element picked out by the semantic character of the expression or by a demonstration. So, if we consider (8),

[The manager at a parking lot, handing a car key to the attendant who is in (8)
charge of bringing the customer's car, tells him:] This is parked out back

we can say that, as the property of being parked out back is predicated of a vehicle, 'this' in (8) refers to a car although the demonstration points to a key.

Nunberg considers (8) as an example of metonymy, in particular as deferred ostension, because in (8) the manager does not use the subject to refer to the key that he is holding, but to the car that the key goes with. Nunberg is assuming that the relation between 'this' and the car in (8) goes through the index, the key. But the demonstratum of 'this' is not the key. This is supported by the fact that, as Nunberg himself recognizes, we cannot get the kind of deferred reading involved in (8) when we use the description in place of which this pronoun is supposed to be. He shows this by saying that (9)

*The key I'm holding is parked out back (9)

is not available.

There are two tests that validate the analysis of (8) as deferred ostension: the number of the demonstrative is determined by the intended referent, the car, and in languages where demonstratives and adjectives are marked for grammatical gender, the gender will be the same as the gender of the intended referent. So, in Spanish (8) would be (8').

[The manager at a parking lot, handing a car key (fem.) to the attendant who is (8')
in charge of bringing the customer's car, tells him:] Éste está aparcado detrás
(This (masc.) is parked (masc.) out back)

The analysis of the demonstrative in (8) as a case of deferred ostension is used by Nunberg to test if other examples of metonymy in which proper names, descriptions, or other indexical expressions are involved can also be considered as instances of the same kind. But, the gender test that validates the analysis for the demonstrative in (8) does not serve to validate the analysis of (1) in terms of deferred ostension. In Spanish, (1) would be (1'),

[Mary, giving her car key to the attendant at a parking lot, tells him:] Yo estoy (1')
aparcada detrás (I am parked (fem.) out back)

and in (1') the adjective 'aparcada' is feminine while 'car' is masculine. So, the subject of (1) refers to the speaker, not to the car, and the transfer involves the predicate, not the indexical. The predicate 'parked out back' contributes with a property of persons, the property they possess in virtue of the locations of their cars. In this case, like in all cases of predicate transfer, the name of a property that applies to something in one domain can be used as the name of a property that applies to things in another domain, provided the two properties correspond in a certain way. Predicate transfer is indifferent to how the bearer of this new or derived property is referred to by an indexical or description or whatever. For example, the parking lot manager could say to the attendant (10)

The man with the cigar (Mr. McDowell, etc.) is parked out back (10)

to predicate of a person the property of having a car which is parked out back.[6] In (8), we can say that the property of being parked out back is predicated of a vehicle while in (1) and in (10) it is used to predicate something more complex, the property of having a car which is parked out back. Thus (1), according to Nunberg, must be interpreted as (1'').

[Mary, giving her car key to the attendant at a parking lot, tells him:] I have a (1'')
car that is parked out back

As far as we are concerned, the first problem in Nunberg's proposal arises in the very distinction between the two types of metonymic transfer. This distinction is dependent on the analysis of (8). But this is quite a dubious one, in the sense that, although there is a case of non-verbal ostension in (8), we do not think that there is a metonymic use of 'this'.

In (8) there is not a case of deferred ostension from the index of 'this', the key, to the referent, the car, simply because the index of 'this' is not the key but the car. Indeed, there is concord both in number and gender with the intended referent, the car, and a lack of concord, at least in Spanish, with the key. In Spanish, we cannot say '*este (masc.) llave (fem.)' nor '*Éste (masc.) es la llave (fem.) que va con el coche

---

[6] Nevertheless, in the case of (4), where there is a definite description too, he explains predicate transfer in another way. It is not the description as a whole, but the predicate 'ham sandwich' in the description which has a derived, non-literal value; it is a case of predicate transfer but located at the NP. We cannot quite grasp why (10) and (4) are not explained in the same way.

que está aparcado detrás'. A pronoun can never have as its index something with which it does not concord, and the pronoun 'this' which is marked for number [+SINGULAR] (in Spanish also for gender [+MALE]) concords with the car. Clearly, the index of 'this' is the car. The difficulty is how we can use the pronoun 'this' which is marked [+NEAR] in (8) to refer to a car that is not near.

From our explanation of (8), someone could argue, the problem of gender is solved but the problem of proximity arises and this would justify having the key as the index of 'this'. Yet, we want to maintain our explanation because this difficulty is overcome if we discover that it is possible to use demonstrative pronouns in a contrastive anaphoric way which dissolves the contrast between 'this' and 'that' [11]. When 'this/that' is used to refer to something already present in the verbal or non-verbal context we are using the pronouns in an anaphoric way.[7] In these cases, there is a special use of them in which the distinction between the physical relative proximity speaker-object vanishes. In a normal utterance of (11)

> She offered him another tie, but this/that was no better than the first (11)

none of the ties are pointed deictically as they are not in the situation described. The choice of the pronoun is not motivated by the relative proximity of the speaker to the ties, the object denoted or referred to. There is a contrastive use of the pronoun, and either 'this' or 'that' can be used. Something similar happens in (8).

In (8), the mutual cognitive environment of speakers and hearers allows the contrastive use of the pronoun 'this'. In the situation described for (8), 'cars' is part of the given information among the interlocutors and this allows the speaker to use the pronoun to distinguish the car in question from any of the other cars that the attendant should bring. However, the referent, the car, cannot be identified unless the interlocutor has more information from the non-linguistic context. In particular, the referent can be identified if, for example, the attendant is able to identify the cars by the keys. Let us take for granted that this is the case in (8), then, by handing the key to the attendant, the speaker is making a non-linguistic ostension that allows the identification of the car that goes with the key. But, the non-linguistic ostension is independent of the contrastive use of 'this'.

Let's take another example to clarify the idea that we can identify the referent by the use of a pronoun and by a non-linguistic ostension without understanding that the object pointed out by ostension is the index of the pronoun but part of its modifier. Let's take into account example (12).

> [Pointing to an empty chair left by a person when she went to the toilet, some- (12)
> one at the same table says:] She is very intelligent

This is a case by means of which 'she' is used to refer to her and to restrict its reference by pointing to the location she occupied in a previous non-verbal context which is cognitively relevant for the interlocutors. By pointing to that location and using

---

[7] "It is the interplay of new and not new that generates information in the linguistic sense. Hence the information unit is a structure made up of two functions, the New and the Given. [...] by its nature the Given is likely to be 'phoric' – referring to something already present in the verbal or non-verbal context; and one way of phoricity is through ellipsis, a grammatical form in which certain features are not realized in the structure." (Halliday, M.A.K. (1985: 275) An Introduction to Functional Grammar. Edward Arnold, London, Melbourne, Auckland).

'she', the speaker makes reference to a female person previously present in the non-verbal context. But, in this situation, we would not like to say that the index of the pronoun 'she' is the location. 'She' does not refer to a place, nor does this expression have to undergo a process of transfer from location to female third person singular. 'She' both means what 'she' means and refers to a female person. The non-verbal ostensive stimulus helps to pick out the referent of 'she' but no transfer from the properties of the location to the properties of the woman is needed. It is a feature typical of indexicals that we have to resort to non-verbal contextual information (either ostensive or non-ostensive) in order for a pronoun to be saturated. But when we identify the referent of a demonstrative or indexical with a demonstration (that points to an object which is not the referent of the demonstrative), we do not have a case of metonymic use of this demonstrative. 'She' in (12) is not used metonymically[8], unless we defend that every indexical that has an associated demonstration whose demonstratum does not coincide with the referent of the indexical but identifies it is a case of metonymy. If this is so, we have lost all the intuitions about metonymy as a figure of speech.

As we pointed out, in the traditional conception, the trait of metonymy is to refer semantically to an object that allows the recovery of the object we are really talking about, but neither 'this' nor 'she' refer semantically to the key or the empty chair. In (8) and in (12), it is the non-verbal demonstration that points to the key or to the empty chair but the pronouns refer to the car and the woman respectively, although the demonstrations specify something relevant about the referents, to wit, the key that goes with the car in (8) and the location she occupied before in (12). This is similar (although a description is used instead of a pronoun) to what happens in (13),

[Pointing to table 4, a waiter says to the owner of the restaurant:] The customer is waiting for his check at the cash register (13)

and there is not a metonymic use of language here.

(8) parallels (12), none of them are metonymies. But if (8) is not a case of metonymy, then the fact that (1) and (4) do not share the same syntactic conditions as (8) does not mean that they are two different types of metonymic transfer: they are two different types of phenomena. In this sense, the argument in favor of explaining some metonymies as predicate transfer vanishes.

Still, we might maintain that some metonymies are cases of predicate transfer but we have to recognize that this position is counterintuitive. It is curious that metonymy should depend on the use of certain noun phrases, of expressions which can be used to refer or denote, and that the change should be produced either in the meaning of the predicates for nouns and personal pronouns or in the properties of descriptions. In addition, Nunberg's proposal does not overcome the problems raised in the standard theory of metonymy as reference transfer: he does not offer identification criteria for metonymy and does not describe the metonymic mechanism for metonymic interpretation. So, Nunberg's support to the claim that 'I' is a pure indexical is not motivated.

---

[8] Indeed, we would recognize a metonymic use of 'she' in the same context specified in (12), if someone said 'She is parked out back'. But, following Nunberg, this new example should be analysed like (1), as a case of predicate transfer rather than deferred ostension.

5 Metonymy as a Case of Language Reduction. Detecting and Interpreting Referential Metonymy

If we pay attention to examples such as (1) and (4) and their traditional readings in (6) and (5), we see that the substituting element is always an abbreviated part of the substituted one which is wider. 'I' is included in 'the car I drive'; 'the ham sandwich' is included in 'the customer of the ham sandwich'.[9]

We conceive ellipsis as the omission of one or more necessary elements in a grammatical structure, specific elements that can be easily understood by the interlocutor in context (linguistic and extralinguistic) and, thus, can be explicitly recovered [12], [13]. Example (14)

[In a hairdresser's, a hairdresser tells another:] The blonde (lady) is waiting (14)
for her check

is a case of ellipsis at a phrasal level. The head of the NP (in brackets), which is the obligatory element in a phrase structure, is missing.

Let us consider examples (1) and (4) again, both referential metonymies, and let us point out the elided elements by means of (1''') and (4').

[Mary, giving her car key to the attendant at a parking lot, tells him:] (The car) (1''')
I (drive) am/*is parked out back

[In a restaurant, looking at the customer of the ham sandwich, a waitress says (4')
to another:] (The customer of) the ham sandwich is waiting for his check

In these cases, there is an ellipsis at a phrasal level. In particular, the elided elements are the head of a complex NP (that includes a modifier also complex) and its determiner together with one or several elements of the modifier that go with it (prepositions, relative pronouns, ...).

What examples (1''') and (4'), examples of metonymy, have in common with (14) is that they all need a non-textual element to be interpreted. In all cases there is ungrammaticality. But, what do they not share? Cases of ellipsis such as (14) are syntactically incomplete; there is an obligatory element with an obligatory grammatical category missing in the phrase structure, thus the syntactic slot is left empty. In (1''') and (4'), there is also ungrammaticality but this time it is revealed just semantically. When ellipsis occurs at a phrasal level it might be the case that apparently there is a complete sentence structure as in (1''') and (4'). In metonymy, although the obligatory element is also missing, the syntactic function is apparently filled by an element which does not really correspond to that function. The head of the modifier takes over

---

[9] As far as Quirk et al. (Quirk, R., Greenbaum, S., Leech G., Svartvik, J. (1985: 858) A Comprenhensive Grammar of the English Language. Longman, London) are concerned, reduction is "a grammatical principle by which the structure of a sentence is abbreviated". For them, there are two types of reduction: ellipsis and substitution by means of pro-forms. The motivation for the use of this principle is that, other things being equal, the users of a language will follow the maxim "reduce as much as you can". This preference for reduction is not justified only by a preference for economy but also because reduction contributes to clarity. By reducing given information, attention is focalized over new information.

the head of the phrase. There is an ellipsis by means of which only the noteworthy part of a restrictive modifier remains.

The use of a restrictive modifier is appropriate in a context of discourse in which there is more than one entity which can be included in the semantic value of the head of the NP and a modifier is required to distinguish the entity being referred to from other entities of this type. The task of the modifier 'of the ham sandwich' is to identify one entity from this set. It is possible, then, that the set of entities which can form a part of the semantic value of the head constitutes a piece of 'given information' (see Footnote 7). In a restaurant, it is well known that waiters' work consists in serving customers, thus, 'customers' will be given information for the waiters in a restaurant when talking about the goods and services that they have to offer them. When waiters are communicating at work, it is essential for them to pick out the specific customer in order to get him attended. In this case, a restrictive modification may be required as in 'the customer of the ham sandwich'. As 'the customer of' is given information in 'the customer of the ham sandwich', we can omit it without any problem of recovery, making a metonymic use of the NP 'the ham sandwich' as it is the case in (4). 'The ham sandwich' can be used to function as the syntactic head of the complex NP through a process of language reduction, provided that some syntactic adjustments take place. In particular, the rules of grammar do not allow the syntactic mismatch that the reduction produces in certain occasions. In the reduction from (5) to (4) there is no problem of concord between subject and verb phrase. But in the reduction from (6) to (1) several syntactic adjustments are required. We cannot have '*I is parked out back' as a result of it. The reduction produces a contextual abnormality which sometimes requires a semantic anomaly but this cannot deprive the sentence from a syntactic well-formedness. Thus, we have to change from '*I is parked out back' to 'I am parked out back'.[10] The abnormality produced in (1) entails a breach of a semantic restriction, such as the need to have a subject with the features [-HUMAN, +VEHICLE] if we consider the predicate, since the subject has to be an entity that can be parked. The lack of agreement between 'the car' and 'am parked' is but a part of the semantic abnormality that arises when the NP 'I' [+HUMAN] functions as the apparent syntactic (though not semantic) subject of the predicate 'be parked'.

The fact that the syntactic function is filled by an element which does not correspond to this function generates, from the point of view of the identification, a contextual abnormality. Indeed, the metonymic interpretation is triggered because, unlike other utterances, metonymic utterances are identified when the speaker perceives a contextual abnormality and a veiled restricted nominal element. In general, the contextual oddity or abnormality we refer to here must be understood as the use of an expression in an unusual linguistic or extra-linguistic context. If we understand the contextual abnormality in this way, we can distinguish between two modes of appearance

---

[10] Similarly, in cases of tough movement such as 'they are easy to teach', which comes from 'to teach them is easy', some syntactic adjustments are needed in order to prevent the ungrammatical expression: '*them is easy to teach'. In particular, the objective form 'them' changes into the subjective 'they', and the verb form 'is' changes into 'are' so that subject-verb agreement is maintained. For a detailed analysis of the correspondences between thematic structures and metonymy, see Romero and Soria (Romero, E., Soria, B. (forthcoming) Metonymy as a Syntactic Strategy in Assigning Informational Prominence within the Noun Phrase. XXIII AESLA, Palma de Mallorca).

of this abnormality in metonymy: mode (a), an oddity between the terms uttered, and mode (b), an oddity between the occurrence of an expression in the actual unusual context and the implicit context associated to a normal use of this expression [14].

Mode (a) can be illustrated by example (4). In the metonymic utterance (4) the normal interpretation of the predicate, 'is waiting for his check', is incompatible with the normal interpretation of the NP 'the ham sandwich' functioning as its subject. It is also a type of abnormality found in (1). Nevertheless, the rules of grammar do not allow this abnormality to affect the syntactic well-formedness of the sentence, thus, as we have just said, the metonymic reduction is possible only if several syntactic adjustments are produced when needed.

Mode (b) can be exemplified by (15). In the metonymic utterance (15)

[In a restaurant, a waiter asks a waitress, who had just served a steaming ham (15) sandwich to a customer who is shivering, why she is closing the window, and she answers:] The ham sandwich is cold

the abnormality is presented by the confrontation between the semantic value of 'the ham sandwich' in a possible usual context and the actual and unusual use of the expression in this specific situation in which both interlocutors know that the ham sandwich is warm and it is clear that when the speaker uses the expression 'the ham sandwich', he cannot be speaking merely about the ham sandwich in this context.

The abnormality is not a sufficient identification condition for metonymy as there are other examples in which it can be detected, for example, in metaphors, nonsense, fiction and so on. To identify metonymy we also have to detect a veiled restricted nominal element: the NP used abnormally is identified as the restrictive modifier of an implicit nominal element.

The contextual abnormality in (1) shows that we are not just talking about the speaker, and the context, Mary speaking to the attendant at a parking lot, gives the information that cars are the set of entities which can form a part of the semantic value of the head and that they constitute a piece of given information. (1) includes the expression 'I' that refers to the speaker. When we take into account contextual and not linguistic information alone, we realize that the object referred to by this singular term, the speaker, is not the type of object that the utterance of the explicit singular term must refer to or denote.[11] Given the utterance (1), there is no doubt that we are talking about a car. 'I' is just part of the description that serves to pick out the referent. The head of the NP must be 'car'. We detect that, in spite of being the notional head of the NP that functions as subject of the sentence, the topic talked about, a car, does not realize the syntactic function of the explicit NP. This syntactic function is realized by the expression 'I'. The slot left empty by the elided head of the NP is filled by this piece of new information 'I'. 'I' is recognized as part of the modifier restricting 'car'. It points to the specific car not because this expression acquires a new and transferred meaning but because the missing elements that this expression restricts are recovered. The metonymic use makes the hearer recover some non-explicit but re-

---

[11] Nevertheless, here there is not a referential use of a singular term whose meaning does not describe the object even when both the speaker and the interpreter believe it does. This is not one of the cases of referential uses of definite descriptions by Donnellan (Donnellan, K. (1966) Reference and Definite Descriptions. Philosophical Review **75** 281-304).

quired sub-propositional and sub-phrasal element in order to have the intended proposition, but once we recover what is non-explicit ('The car ... drive') both terms ('the car I drive' and 'I') can refer to their respective normal referents.

Metonymy is a non-textual use of language in which there is at least one non-explicit sub-phrasal constituent, a veiled restricted nominal element. So, the ellipsis that characterizes metonymy can be understood in theory of meaning as a mandatory pragmatic process that operates at a subpropositional level. To interpret (1), the NP 'I' must be completed and it becomes a more complex NP: 'The car I drive'. Now we can reconstruct what is literally said with the utterance, although what is said is not textually said. What is literally said is that the car I drive is parked out back.

6 Conclusion

In the most extended theories of metonymy, it is argued that, if used metonymically, singular terms change their meanings. From these theories, 'I' cannot be a pure indexical because in its metonymic use the meaning changes and it refers to an object different from the speaker. Our paper shows that these proposals have too many problems.

Nunberg's account of metonymy as predicate transfer seems to offer an alternative view that can explain the type of transfer that characterizes some metonymies, included the metonymic use of 'I', without risking the proposal of 'I' as a pure indexical. But the distinction between deferred ostension and predicate transfer as two different ways of explaining metonymies is not right, and the predicate transfer explanation of metonymy is not motivated.

Our proposal accommodates the data we have been discussing, by introducing the conception of metonymy as a case of language reduction. The term used metonymically, we argue, is just a part of the modifier of the singular term that really refers or denotes. To interpret (1), the NP 'I' must be completed and it becomes a more complex NP, 'The car I drive'. 'I' means what it means and it has the semantic value that it usually has, the speaker. So, if we consider that somebody who utters a sentence including a metonymic use of 'I' is prompting the hearer to understand a non-textual meaning of this utterance, we can argue that the metonymic use of 'I' does not risk the consideration of 'I' as a pure indexical.

References

1. Nunberg, G.: Indexicality and Deixis. Linguistics and Philosophy **16** (1993) 1-43
2. Nunberg, G.: Transfers of Meaning. Journal of Semantics **12** (1995) 109-132
3. Romero, E., Soria, B.: La Metonimia Referencial. Theoría **17/3** (2002) 435-455
4. Kaplan, D.: Demonstratives. An Essay on the Semantics, Logic, Metaphysics, and Epistemology of Demonstratives and other Indexicals. In: Almog, H., Wettstein, H., Perry, J. (eds): Themes from Kaplan, Oxford University Press, New York (1989) 481-563
5. Recanati, F.: Direct Reference: From Language to Thought. Basil Blackwell, Oxford (1993)
6. Grice, H. P.: Logic and Conversation. In: Cole, P., Morgan, J. (eds.): Syntax and Semantics 3: Speech Acts, Academic Press, New York (1975) 41-58

7. Romero, E., Soria, B.: Comunicación y Metáfora. Análisis Filosófico **XXII/2** (2003) 167-192
8. Romero, E., Soria, B.: A View of Novel Metaphor in the Light of Recanati's Proposals. In: Frápolli, M.J. (ed.): Saying, Meaning and Refering. Essays on François Recanati's Philosophy of Language, Palgrave Studies in Pragmatics, Language and Cognition, London (forthcoming)
9. Recanati, F.: The Alleged Priority of Literal Interpretation. Cognitive Science **19** (1995) 207-232
10. Lakoff, G., Johnson, M.: Metaphors We Live by. Univ. of Chicago Press, Chicago (1980)
11. Huddlestone, R.: Introduction of the Grammar of English. Cambridge University Press, Cambridge (1984)
12. Quirk, R., Greenbaum, S., Leech G., Svartvik, J.: A Comprenhensive Grammar of the English Language. Longman, London (1985)
13. Burton-Roberts, N.: Analysing Sentences. Longman, London (1986)
14. Romero, E., Soria, B.: Novel Metaphor and Novel Metonymy as Primary Pragmatic Processes. In: Garrido, P. (ed.): Grammar and Discourse: Interactions (forthcoming)

Granularity as a Parameter of Context

Hedda R. Schmidtke

Department for Informatics, University of Hamburg,
Vogt-Kölln-Str. 30, D-22527 Hamburg, Germany
schmidtke@informatik.uni-hamburg.de

Abstract. Spatial and temporal granularity can be understood as parameters of context restricting the set of accessible objects in a context. Starting from the idea that this selection process depends to a large extent on the relation between the grain-size of the context and the local extension of the objects, the granularity of a context is in this article formalised as a class of possible sizes in the context. This formalisation is shown to be in accordance to well-known mathematical foundations on perceptual classification. An example for the case of temporal granularity illustrates how the introduction of new elements into a context may result in a more or less smooth shifting of the granularity leading to a classification of four different types of change of granularity. The results can be applied in a wide range of fields, e.g. in research on contextual reasoning and natural language understanding.

1 Introduction

A central motivation for research on contextual reasoning is that even highly complex scenarios become tractable if we can separate the currently relevant from the irrelevant details: if an intelligent system can be enabled to keep the number of objects to be handled constantly small, it may even use algorithms of high complexity without losing efficiency. Granularity as described by Hobbs [13] can be conceived as a parameter of context that restricts the available objects and the categories and properties for characterising objects.

Existing approaches to spatial and temporal granularity can be classified as partitioning approaches or grain-size approaches. Partitioning approaches to temporal and spatial granularity map temporal and spatial individuals to a certain granularity based on a partitioning of space and time, with a hierarchy of containment providing discrete levels of granularity forming preferably a tree structure. Partitioning approaches to spatial granularity like the stratified rough sets [5] or the granular partitions [4] assign a partitioning of space to each level of granularity, with partitions of a coarser level containing several partitions of a finer level. The surface of the globe for instance can be partitioned into continents on the coarsest level, into states on a finer level, and so forth. Although this may be an adequate structuring for political and legal borders, it is well known that many spatial common sense concepts cannot be modelled using regions with crisp boundaries, which is especially problematic in the geographic domain [8, 10]. To mirror this, partitioning approaches to granularity have to be build upon a theory of vagueness like the rough set theory [20] or the theory of fuzzy granularity [27].

Partitioning approaches in the temporal domain like the modal logic approaches of the calendar logics [19] and the layered metric temporal logics [7] and also the approach of Bettini, Jajodia, and Wang [3] on time granularities in databases, data mining, and temporal reasoning are derived from calendar systems which provide a natural partitioning of time based on the grains of years, months, days, etc., thus including a notion of grain-size. However, the crisp boundaries between partitions are as problematic for partitioning approaches in the temporal domain as in the spatial [7, 18]. Partitioning approaches tend to have good inferential properties, but their applicability depends on a suitable partitioning of the domain being given in advance. A spatial partitioning assigning the postal districts of Paris to the same level of granularity as the continent Australia, for instance, would be less plausible than the above mentioned partitioning, but cannot be ruled out based on the hierarchy of containment alone.

Starting point of the grain-size approach is the hypothesis that spatial and temporal granularity are mainly determined by the concept of extension in a context, the central notion being the difference between irrelevant and relevant extensions in a context. A level of granularity under this perspective can be conceived as a coarse class of sizes constraining which objects have to be represented. In a spatial context, e.g., very small objects are often completely irrelevant and can therefore be neglected, or – in the case of non-spatial relevance – they can be represented simplified, for instance, to a single pair of coordinates; very large objects, on the other hand, either cover the whole area currently under consideration or are only partially represented. In the first case they can be treated as a background of the context, but their exact spatial location will not be relevant; in the latter case, only the specific relevant parts need to be represented.

The grain-size approach to granularity is founded on the analogy between the representation of objects and eventualities[1] in spatial and temporal granular contexts and depictions in a photography (for a discussion of the metaphor see Galton [9]), which is also employed in other investigations on the phenomenon of granularity such as [21]. A photography having a certain resolution and a limited extent is similar to a granular context-dependent representation having a limited maximal extent and limited details. If we need more details we have to zoom in, presumably eliminating objects in the periphery from the context. If we need an overview, on the other hand, we can zoom out, possibly sacrificing the smaller details. Results from research on spatial cognition and natural language understanding [15, 12] suggest that human beings use a similar strategy to structure complex spatial information into granular representations suitable for efficient communication and reasoning.

---

[1] The notion of eventuality is used as a generic term subsuming all types of temporal objects, like events, processes and states (cf. [17]).

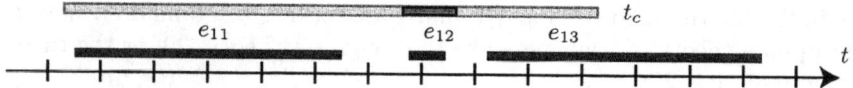

Fig. 1. A process e_1 consisting of three sub-processes e_{11}, e_{12}, and e_{13} in a temporal context given by the temporal interval t_c (*light grey*) and a grain-size illustrated by an example interval (*dark grey*)

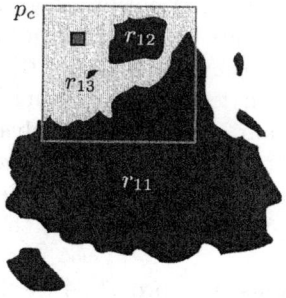

Fig. 2. A scattered region r_1 consisting of r_{11}, r_{12}, r_{13}, and three other regions in a granular spatial context given by a rectangular region p_c (*light grey*) and a grain-size illustrated by an example grain-sized rectangle (*dark grey*)

Accordingly, granularity will be characterised in this article as a parameter of context which is determined by a grain-size and an area covering the scope of the context:

The temporal/spatial scope of a context is the area currently under consideration in a given context and determines the maximal extent an object may have.
The grain-size determines the current minimal extension. A smaller object is either not relevant in the context at all or can be represented in a simplified manner.

In a grain-size approach the objects do not need to fit into cells of a partition. In Fig. 1 for instance, the scope of the granular temporal context is given by the temporal interval t_c and a grain-size. Fig. 1 shows a process e_1 consisting of three sub-processes e_{11}, e_{12}, and e_{13}. The process e_{11} would be accessible in the context: it is fully contained in t_c and larger than a grain; e_{13} is only partially accessible in the context since those parts outside of t_c can only be accessed by shifting the scope of the context; e_{12} being smaller than a grain has no relevant extension in the context. It is therefore either not represented at all or has to be represented as a punctual event. Its extension is accessible only on a finer level of granularity. The process e_1 consisting of e_{11}, e_{12}, and e_{13} is too large to be represented as a whole in the context of t_c; a context of a coarser granularity would have to be chosen to make it accessible.

The grain-size approach can also be applied in the spatial domain, provided that a class of basic extended regions is given that can serve as the grains of space. In Fig. 2, the granular spatial context is given by a rectangular region p_c, which could be, e.g., the region currently displayed in an application window, and a grain-size. The regions r_1 (consisting of r_{11}, r_{12}, and r_{13}) and r_{11} are only partially accessible in the context. The region r_{13} being smaller than a grain could be represented as a punctual object, e.g., by a grain containing it.

Structure of the Article. The aim of this article is to present a formalisation of the common structure underlying spatial and temporal, granular contexts building on the notions of grain-size and maximal scope of the context. The formalisation is founded solely upon comparisons of extensions. Therefore neither a quantitative treatment of size nor a metric are necessary, which ensures that the approach can be used in qualitative spatial and temporal reasoning. Building upon a strict order of durations of intervals in the temporal domain (Section 2), a temporal level of granularity γ is formalised as a class of intervals having an extension compatible with a context of a given maximal scope (Section 3). The conceptualisation is sufficiently general to be transferable to extensions of objects in the spatial domain. The levels of granularity can be compared with a semi-order $\prec$ (finer) and an intransitive indistinguishability relation $\approx$ (indifferent). Intransitive indistinguishability is a well-known cognitive phenomenon of classification under uncertainty [24], e.g. in perceptual classification of lengths or colours.

In Section 4 the concepts are exemplified for the domain of temporal granularity with a very simple example of a temporal granular structure. This structure is used in Section 5 in a dynamic scenario in which new objects are successively added to a context. The example illustrates, how the addition of new objects may lead to a change of the granularity of a context, and shows that the modelling can be used to differentiate between cases of more or less smooth shifts of granularity.

2 Extension

To build a theory of granularity based on the notions of the grain-size and extension in a context, the foundational notion of extension has to be studied. The levels of granularity and the extensions of objects and the durations of eventualities are ordered by ordering relations: On a finer level of granularity, a representation of higher detail – showing more and especially also smaller details of an object but perhaps not the whole object – is appropriate. This relation between extension and granularity can be encoded by building the order on granularities on top of an order on extensions of certain temporally or spatially basic entities that can represent both the grains and the maximal scope of contexts in the respective domains. Convex, closed temporal intervals can fill this role for the domain of time (for a formal specification of a temporal interval structure cf. Allen [1]). Temporal intervals have a definite extension: the duration. They can therefore be ordered using a relation < (shorter duration). Two eventualities e_1 and e_2 can then be compared for their duration by comparing intervals t_1 and t_2 containing the eventualities.[2] In contrast to the closed, convex, temporal intervals t, eventualities e are not restricted regarding their inner structure: they may consist of several unconnected parts, be located in open intervals, etc. For simplicity, I assume in the following that any eventuality e can be assigned a unique smallest interval t containing e using a function $temp$.

We can now state the above mentioned constraint that, if an eventuality e_1 is represented only in contexts of a finer granularity than an eventuality e_2, the duration of e_1 has to be shorter than that of e_2. In contrast to this, the reverse does not hold:

e_A: The test took $50\mu s$ on the system A.
e_B: The test took $80\mu s$ on the system B.
e_C: The test took one day on the system C.

In the example, $\text{temp}(e_A) < \text{temp}(e_B)$ holds, but the relative differences between the durations are close enough, in order not to demand a change of granularity. The two durations can in addition be described using the same unit: microseconds. The eventuality e_C, in contrast, has to be on a coarser granularity than e_A and e_B, since it has a much longer duration, and cannot be adequately described linguistically on the basis of microseconds. Replacing the term *one day* in the description of e_C with $86,400,000,000\mu s$ would not entail the same meaning, since the second expression suggests a higher precision than the first [16].

I will in the following present only the formalisations for the case of temporal intervals and temporal granularity, but all considerations equally apply to appropriate spatial structures, as will be illustrated. The formalisation is founded upon an equivalence relation of equal extension $\equiv$ between temporal intervals (t, t_1, t'): $\equiv$ is a transitive (A≡1), symmetric (A≡2), and reflexive relation(A≡3).[3]

$$\forall t_1, t_2, t_3 : t_1 \equiv t_2 \wedge t_2 \equiv t_3 \rightarrow t_1 \equiv t_3 \qquad (\text{A}\equiv 1)$$

$$\forall t_1, t_2 : t_1 \equiv t_2 \rightarrow t_2 \equiv t_1 \qquad (\text{A}\equiv 2)$$

$$\forall t : t \equiv t \qquad (\text{A}\equiv 3)$$

The equivalence classes of $\equiv$ are the possible sizes of intervals, i.e. their durations. The relation < (*shorter*) for comparison of durations is characterised as a strict order on temporal intervals: < is transitive (A<1) and asymmetric (A<2); only intervals having different extension according to $\equiv$ can be compared with < (A<3); the equivalence classes of $\equiv$ are totally ordered by < (A<4).

---

[2] The relation < is not to be confused with the relation *before*. $t_1 < t_2$ does not depend on the relative position of the intervals in time.

[3] For better readability, the formulae are abbreviated by saving brackets: The scope of the quantifier is to be read as maximal, i.e. until a bracket closes that was opened before the quantifier. The following order of precedence applies: $\neg, \wedge, \vee, \rightarrow, \leftrightarrow, \stackrel{\text{def}}{\Leftrightarrow}$.

$$\forall t_1, t_2, t_3 : t_1 < t_2 \wedge t_2 < t_3 \rightarrow t_1 < t_3 \quad \text{(A<1)}$$
$$\forall t_1, t_2 : t_1 < t_2 \rightarrow \neg t_2 < t_1 \quad \text{(A<2)}$$
$$\forall t_1, t_2 : t_1 < t_2 \rightarrow \neg t_1 \equiv t_2 \quad \text{(A<3)}$$
$$\forall t_1, t_2 : t_1 < t_2 \vee t_2 < t_1 \vee t_1 \equiv t_2 \quad \text{(A<4)}$$

The relation $\leq$ can be derived accordingly:

$$t_1 \leq t_2 \stackrel{\text{def}}{\Leftrightarrow} t_1 < t_2 \vee t_1 \equiv t_2 \quad \text{(D1)}$$

Spatial Extension. Extension in the domain of space is harder to formalise than in the one-dimensional time, since higher-dimensional objects can have different extensions in different directions. Apart from the maximal extension of an object, e.g. the height of a tower or the length of a pen, also its secondary extension may be important: a river and a brook, for instance, especially differ in the secondary extension. In addition, the width of a river varies along its course. Consequently, the secondary extension has to be determined locally. A suitable axiomatic characterisation of extension for the domain of two-dimensional space supporting the above axioms can be found in [22]. The approach formalises certain basic entities, called *places*, as regions having the same extension in every direction. The places p can therefore be used to provide a concept of spatial grains and the maximal spatial scope of a context in two-dimensional space, and can – like the closed convex temporal intervals t – be compared for their extension using a relation $<$. Among the models for the geometric formalisation of places in [22] are the circles of a given metric, such as, e.g., the rectangular regions used in Fig. 2. Like the eventualities in the temporal domain, the spatial objects, such as for instance r_1 in Fig. 2, do not need to be restricted regarding their inner structure.

3 Levels of Granularity

In contrast to the relations $<$ and $\equiv$, which represent exact comparison of the durations of two temporal intervals, the formalisation of granularity is based on a weaker structure. A level of granularity is characterised as a range of sizes. A finer (coarser) temporal granularity is compatible only with intervals of a shorter (longer) duration. Granularities encompassing a common size will be termed indifferent. A granularity γ compatible with two intervals of the lengths 23h and 26h would also be compatible with any interval of the lengths 23h, 24h, 25h (including, but not restricted to, all days of the calendar in every time zone), but also intervals of the lengths 23h30min, 23h31min4sec, etc.

A level of granularity γ shall be determined by a grain-size and the size of the maximal scope of contexts having this level of granularity. Accordingly, the basic relations for specifying levels of granularity are the relations *grain-size* (grn) and *context-size* (ctx) between levels of granularity (γ, γ_1, γ') and temporal intervals. The relation $\text{grn}(\gamma, t)$ ($\text{ctx}(\gamma, t)$) holds if t is a possible grain (possible maximal scope of a context) on γ. The axioms (A1)–(A4) suffice to specify levels of granularity as will be shown below (4)–(7).

Axiom (A1) ensures that each temporal interval can serve both as a grain and as a maximal scope of a context. Axiom (A2) secures that every granularity γ has both a temporal interval serving as a grain and one serving as a maximal scope, with the scope being larger than the grain. Levels of granularity being specified to formalise the notions of refinement and coarsening of representations shall themselves not be subject to refinement and coarsening (A3): If two levels γ, γ' differ in grain-size or context-size with respect to $<$, the respective other size has to be in the same relation, i.e.: if a grain of γ is smaller than a grain of γ', than any maximal scope of γ is also smaller than any maximal scope of γ', and vice versa. This axiom ensures that the relation $\prec$ (finer) defined below (D3) is in accordance to the relation $<$ between intervals, thus securing that granularities can indeed be used as coarse classes of sizes constraining the entities accessible in a granular context (cf. [22] for a less restrictive approach allowing arbitrary ranges of sizes).

The relation of compatibility ($\sim$) can now be defined (D2): A temporal interval t is compatible with a level of granularity γ iff it is at least as large as the grain-size and at most as large as the context-size. Compatibility is used below to define the accessibility of an eventuality in a context (D5). Two levels of granularity that have exactly the same compatible intervals shall be identical (A4).

$$\forall t \exists \gamma_1, \gamma_2 : \mathrm{grn}(\gamma_1, t) \wedge \mathrm{ctx}(\gamma_2, t) \tag{A1}$$

$$\forall \gamma \exists t_1, t_2 : \mathrm{grn}(\gamma, t_1) \wedge \mathrm{ctx}(\gamma, t_2) \wedge t_1 < t_2 \tag{A2}$$

$$\forall \gamma, t_1, t_1', t_2, t_2', \gamma' : \mathrm{grn}(\gamma, t_1) \wedge \mathrm{grn}(\gamma', t_1') \wedge \mathrm{ctx}(\gamma, t_2) \wedge \mathrm{ctx}(\gamma', t_2') \tag{A3}$$
$$\wedge \, (t_1 < t_1' \vee t_2 < t_2') \rightarrow t_1 < t_1' \wedge t_2 < t_2'$$

$$t \sim \gamma \overset{\mathrm{def}}{\Leftrightarrow} \exists t_1, t_2 : \mathrm{grn}(\gamma, t_1) \wedge t_1 \leq t \wedge \mathrm{ctx}(\gamma, t_2) \wedge t \leq t_2 \tag{D2}$$

$$\forall \gamma_1, \gamma_2 : \gamma_1 = \gamma_2 \leftrightarrow \forall t : t \sim \gamma_1 \leftrightarrow t \sim \gamma_2 \tag{A4}$$

Given a granularity γ, it follows that the relation grn – and ctx accordingly – holds exactly for the intervals of one size, i.e. for a certain equivalence class of $\equiv$ (1). Consequently, the levels of granularity γ are uniquely determined by both grain-size and context-size (2).

$$\forall \gamma, t, t' : \mathrm{grn}(\gamma, t) \rightarrow [t \equiv t' \leftrightarrow \mathrm{grn}(\gamma, t')] \tag{1}$$

$$\forall \gamma, \gamma', t, t' : \mathrm{grn}(\gamma, t) \wedge \mathrm{grn}(\gamma', t) \rightarrow \gamma = \gamma' \tag{2}$$

A granularity γ shall be termed finer than ($\prec$) a granularity γ' iff all intervals compatible with γ are smaller than any interval compatible with γ' (D3). It follows that γ is finer than γ' iff any maximal scope of γ is smaller than any grain of γ' (3). Two granularities are called indifferent ($\approx$) iff there are intervals compatible with both (D4).

$$\gamma \prec \gamma' \overset{\text{def}}{\Leftrightarrow} \forall t, t' : t \sim \gamma \land t' \sim \gamma' \rightarrow t < t' \tag{D3}$$

$$\forall \gamma, \gamma', t, t' : \text{ctx}(\gamma, t) \land \text{grn}(\gamma', t') \rightarrow [\gamma \prec \gamma' \leftrightarrow t < t'] \tag{3}$$

$$\gamma_1 \approx \gamma_2 \overset{\text{def}}{\Leftrightarrow} \exists t : t \sim \gamma_1 \land t \sim \gamma_2 \tag{D4}$$

It can be shown that $\prec$ has all properties of a semi-order, which is a relation encountered in the mathematical foundations of classification under uncertainty, e.g. in perceptual classification of lengths or colours [24]: $\prec$ is irreflexive (4). If two pairs of granularities are ordered, then the finer of the first pair is finer than the coarser of the second pair, or the finer of the second pair is finer than the coarser of the first pair (5). If three granularities are ordered by $\prec$, any fourth granularity is coarser than the finest of the three, or finer than the coarsest (6). Two granularities are indifferent ($\approx$) iff they are not ordered by $\prec$ (7). Two granularities are identical if they are indifferent to the same granularities (8).

$$\forall \gamma : \neg \gamma \prec \gamma \tag{4}$$

$$\forall \gamma_1, \gamma_1', \gamma_2, \gamma_2' : \gamma_1 \prec \gamma_1' \land \gamma_2 \prec \gamma_2' \rightarrow \gamma_1 \prec \gamma_2' \lor \gamma_2 \prec \gamma_1' \tag{5}$$

$$\forall \gamma_1, \gamma_2, \gamma_3, \gamma : \gamma_1 \prec \gamma_2 \land \gamma_2 \prec \gamma_3 \rightarrow \gamma_1 \prec \gamma \lor \gamma \prec \gamma_3 \tag{6}$$

$$\forall \gamma, \gamma' : \gamma \approx \gamma' \leftrightarrow \neg(\gamma \prec \gamma' \lor \gamma' \prec \gamma) \tag{7}$$

$$\forall \gamma_1, \gamma_2 : (\forall \gamma : \gamma_1 \approx \gamma \leftrightarrow \gamma_2 \approx \gamma) \rightarrow \gamma_1 = \gamma_2 \tag{8}$$

Properties of the semi-order are transitivity and asymmetry [24]. The relation $\approx$ is not transitive, i.e.: γ_1 being indifferent to γ_2 and γ_2 being indifferent to γ_3 does not entail that γ_1 is indifferent to γ_3. The given characterisation thus structures the domain of sizes without necessarily partitioning it.

The relation of compatibility provides a means to formally express criteria of accessibility in a granular, temporal context. The eventuality e is accessible in a context t_c iff the smallest interval containing e (temp(e)) is part of t_c and compatible with the level of granularity γ for which t_c determines the context-size:

$$\text{access}(t_c, e) \overset{\text{def}}{\Leftrightarrow} \text{temp}(e) \sqsubseteq t_c \land \exists \gamma : \text{ctx}(\gamma, t_c) \land \text{temp}(e) \sim \gamma \tag{D5}$$

The relation *part* ($\sqsubseteq$) is supposed here as given by the underlying axiomatisation of temporal intervals.

4 A Simple Model

The advantages and disadvantages of the presented conception of granularity can be illustrated with a very simple example of a structure that complies with the axioms. The sketched structure is a model of the axiomatisation and demonstrates its consistency, but does not claim cognitive adequacy. The structure is obtained by supposing a fixed ratio $1 : n$ between each context-size and its corresponding grain-size. This structure would, for instance, result from the simple model of a representational system in which the maximal scope of a context can be covered by exactly n grains like in the data format of a raster. Nevertheless, the structure does not imply an underlying partitioning of time, rather the raster would select certain intervals of the grain-size.

In the example, the factor $n = 10$ is chosen: A granularity γ having grain-size x is assigned a context-size of $x * 10$, with $d(t)$ indicating the duration of t in seconds:

$$\forall \gamma, t, t' : \mathrm{grn}(\gamma, t) \rightarrow [\mathrm{ctx}(\gamma, t') \leftrightarrow d(t') = d(t) * 10]$$

We thus obtain for instance the following levels of granularity, which are differentiated with an index giving their grain-size.

$$\mathrm{grn}(\gamma_s, t) \stackrel{\mathrm{def}}{\Leftrightarrow} d(t) = 1s \qquad \mathrm{ctx}(\gamma_s, t) \stackrel{\mathrm{def}}{\Leftrightarrow} d(t) = 10s$$
$$\mathrm{grn}(\gamma_{min}, t) \stackrel{\mathrm{def}}{\Leftrightarrow} d(t) = 60s \qquad \mathrm{ctx}(\gamma_{min}, t) \stackrel{\mathrm{def}}{\Leftrightarrow} d(t) = 600s$$
$$\mathrm{grn}(\gamma_h, t) \stackrel{\mathrm{def}}{\Leftrightarrow} d(t) = 3600s \qquad \mathrm{ctx}(\gamma_h, t) \stackrel{\mathrm{def}}{\Leftrightarrow} d(t) = 36000s$$

Fig. 3 illustrates the example structure. The process e_1 consisting of the three partial processes e_{11}, e_{12}, and e_{13} is partially accessible in a temporal context determined by the interval t_c. t_c is a maximal scope of a context at the level of granularity γ_{10s}. e_{13} and e_1 are only partially accessible in the context. The grain-intervals of γ_{10s} have the duration $10s$. The process e_{12} having a shorter duration of 7s could therefore be termed a punctual eventuality. Its extension is only accessible by choosing a granularity with a smaller grain-size, such as γ_{7s} for which it would be a grain, but also the granularities γ_{6s} and γ_s are

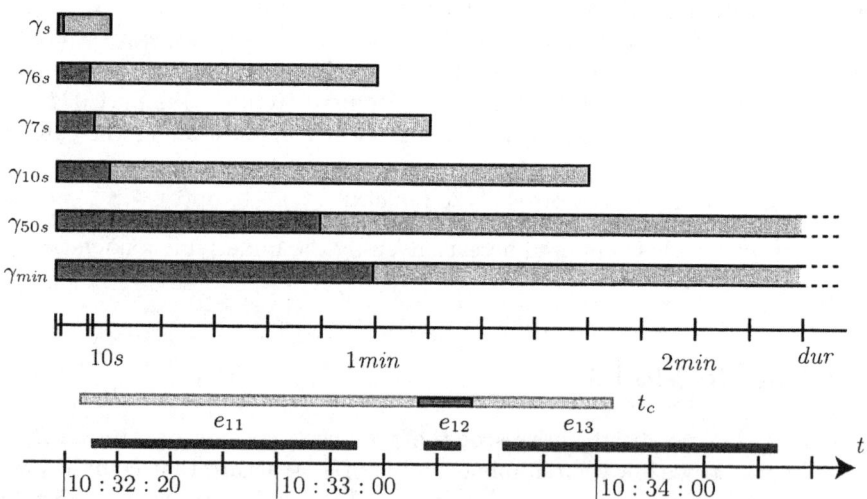

Fig. 3. Example for granularities in the temporal domain. Below: the processes e_{11} (10:32:25–10:33:15), e_{12} (10:33:27–10:33:34), and e_{13} (10:33:42–10:34:32) in relation to a temporal axis with 10s unit; above: different granularities γ represented by grain-size (*dark grey*) and context-size (*light grey*).

compatible with the duration of e_{12}. In any refinement, also the maximal scope of the context has to be narrowed, since t_c cannot be the scope of a context for any of the granularities encompassing intervals of the duration 7s.

The process e_{11} is located in an interval of duration 50s, which is in this simple structure neither compatible with γ_s nor with γ_{min}, but, e.g., with γ_{50s} having a grain-size of 50s. γ_{50s} encompassing intervals of duration 50s–500s, (i.e. 50s–8min20s) is indifferent to γ_{min}, coarser than γ_s and finer than γ_h:

$$\gamma_{50s} \approx \gamma_{min} \qquad \gamma_{7s} \approx \gamma_s \qquad \gamma_{5s} \approx \gamma_s$$
$$\gamma_s \prec \gamma_{50s} \qquad \gamma_{7s} \approx \gamma_{min} \qquad \gamma_{5s} \prec \gamma_{min}$$
$$\gamma_{50s} \prec \gamma_h$$

The granularity γ_{7s} is indifferent from both γ_s and γ_{min}. The granularity γ_{5s}, in contrast, which encompasses nearly the same range of sizes is not indifferent to γ_{min}. The finest granularity indifferent with γ_{min} is in this example exactly γ_{6s}, which has a context-size of 1min. This exact border may be seen as a disadvantage of the approach, since exact transitions are a general consequence of the axiomatisation and cannot be avoided by choosing a cognitively more plausible model. This problem is closely related to the problem of higher order vagueness (see Fine in [6]): the granularities representing vague classifications of the sizes of certain entities have themselves crisp boundaries. The vagueness of the boundaries could be modelled with a further relation of intransitive indistinguishability for modelling perceptual indifference, but the resulting entities of second order vagueness would themselves have crisp boundaries demanding a formalisation of third order vagueness, and so forth.

5 Change of Granularity

In the following, two examples are given to illustrate that the axiomatisation of granularity introduced above actually supports necessary properties of a context-dependent concept of granularity respecting notions of vagueness. It is discussed how the granularity has to be changed if a new object is introduced into a context. The discussion focuses on necessary changes of granularity; the relative position of the objects with respect to the scope of the context is neglected.

Suppose an application in which processes e_1–e_4 with durations

$$d(\text{temp}(e_1)) = 30s, d(\text{temp}(e_2)) = 20s, d(\text{temp}(e_3)) = 80s, d(\text{temp}(e_4)) = 1,5s$$

are successively added to the context. With the relation of indifference, changes of granularity can be classified as more or less smooth. The granularity in the context can then be determined in the following way: First a level of granularity has to be chosen encompassing $\text{temp}(e_1)$. A possible choice is γ_{30s} having $\text{temp}(e_1)$ as a grain. In the second step e_2 is added which is located in an interval that is not compatible with γ_{30s}, so γ_{20s} which is still compatible with $\text{temp}(e_1)$ is selected. The context in this case is not changed relevantly, since

$\gamma_{20s} \approx \gamma_{30s}$ holds. Adding e_3 in the third step does not change the context at all, since temp(e_3) is compatible with γ_{20s}. A change of granularity, or to be more exact a refinement, is necessary as soon as e_4 is added. No granularity is compatible with both temp(e_4) and any of the intervals of the previous steps. The coarsest granularity $\gamma_{1,5s}$ with grn($\gamma_{1,5s}$, temp(e_4)) encompasses elements up to a duration of $15s$.

A smoother transition is exemplified in the following sequence:

$$d(\text{temp}(e_1)) = 1s, d(\text{temp}(e_2)) = 5s, d(\text{temp}(e_3)) = 30s, d(\text{temp}(e_4)) = 1min.$$

Analogously to the above example, the first chosen granularity would be γ_s having temp(e_1) as a grain-size. The interval containing e_2 added in the second step is still compatible with γ_s. The third element e_3 demands a shift of granularity: γ_{5s} is suitable to both encompass e_2 and e_3, but e_1 is no longer accessible. The same considerations apply, when e_4 is added, whose interval again is not compatible with γ_{5s}. The coarsest granularity compatible with the intervals of e_4 and e_3 is γ_{30s}; the finest, is γ_{6s}. For γ_{30s} the relation $\gamma_{30s} \approx \gamma_{5s}$ holds but also $\gamma_s \prec \gamma_{30s}$. Here, the intransitivity of indifference has to be observed: In comparison with the last step the change of granularity is smooth, but in comparison with the first step, a clear coarsening of granularity has occurred. If we choose γ_{6s}, on the other hand, the relations $\gamma_{6s} \approx \gamma_s$ and also $\gamma_{6s} \approx \gamma_{5s}$ hold, so a clear coarsening concerning the whole series would not occur. A clear break over the whole series is only necessary, if an element is to be added that has a much longer duration. A process e_5 with $d(\text{temp}(e_5)) = 4min$ has γ_{24s} as its finest compatible granularity. $\gamma_s \prec \gamma_{24s}$ holds, and a change over the whole series has occurred, although each granularity in this alignment is indifferent to its direct predecessors and successors.[4]

The examples show how granularity can be employed as a parameter of context, and illustrate dynamic changes of granularity that can be classified for their smoothness according to the following gradation:

- a slight shift of granularity, so that a further entity can be added to the entities already encompassed,
- a slight shift resulting in a previous element being removed, but with the new level of granularity being still indifferent to all previous granularities,
- a shift of granularity, after which the resulting granularity is coarser or finer than a previous granularity,
- a shift of granularity, with the resulting granularity being coarser or finer than any previous granularity.

The different possible grades illustrate that the presented formalisation of granularity has indeed properties of a context-dependent concept of granularity that

---

[4] This reminds of the so called sorites-vagueness. An overview of this phenomenon is given by Varzi [26] and Hyde [14]. For the role of intransitivity cf. van Deemter [25] and Halpern [11]. A detailed discussion is beyond the scope of this article.

also reflects phenomena of vagueness. This is a clear advantage over the discrete levels of granularity employed in partitioning approaches to granularity, if a cognitively realistic axiomatisation is intended.

6 Conclusion and Outlook

The approach presented in this article can be used with arbitrary axiomatisations of temporal interval structures, e.g. with dense or discrete, bounded or unbounded, linear or branching time. A combination with an underlying partitioning of time is therefore also possible. However, the advantage of the formalisation is that a partitioning is not necessary. An underlying partitioning severely restricts the possible locations of temporal objects in two ways. First, only intervals of certain durations can be temporal locations for eventualities; second, these intervals have to be anchored in a fixed raster. This is an even greater restriction in the domain of space, for which no inherent partitioning comparable to the calendar system exists. The presented approach to temporal and spatial granularity omits these restrictions, allowing for a wider conceptualisation of granular, spatial and temporal contexts. The approach has been shown to be sufficiently general to handle both temporal and spatial contexts.

The axiomatisation formalises the basic notions of a grain-size approach to temporal and spatial granularity based on a cognitively realistic concept of classes of sizes. Open questions concern the construction of a cognitively adequate model of the axiomatisation. It should be empirically investigated how human beings handle granularity in spatial and temporal contexts, so that a cognitively more adequate model could be found. Another issue, which has to be resolved, is whether there is one cognitively adequate structure or a variety of structures depending on other parameters of context.

The notion of accessibility in a context was defined based on the maximal extension of an interval. For a more detailed treatment also the inner structure of an object and – for partially represented objects – also the local extension in the context have to be considered. How the representation of a spatial object can be controlled based on the local extension of the object in a context, and how simple objects of a coarser level can be constructed from objects of a finer level depending on granularity has been studied in [22] and [23], respectively.

Future work will include research on the inferential properties of the formalisation in reasoning about and within spatial and temporal contexts. The conceptualisation should therefor be embedded into a general framework for contextual reasoning [2].

Acknowledgements

I am grateful to Christopher Habel and four anonymous reviewers for comments on earlier drafts of this article.

References

1. J. Allen. Towards a general theory of action and time. *Artificial Intelligence*, 23:123–154, 1984.
2. M. Benerecetti, P. Bouquet, and C. Ghidini. Contextual reasoning distilled. *Journal of Experimental and Theoretical Artificial Intelligence*, 12(3):279–305, 2000.
3. C. Bettini, S. Jajodia, and X. S. Wang. *Time granularities in databases, data mining, and temporal reasoning.* Springer, 2000.
4. T. Bittner and B. Smith. A theory of granular partitions. In M. Duckham, M. F. Goodchild, and M. F. Worboys, editors, *Foundations of Geographic Information Science*, pages 117–151. Taylor & Francis, London, New York, 2003.
5. T. Bittner and J. G. Stell. Stratified rough sets and vagueness. In W. Kuhn, M. Worboys, and S. Timpf, editors, *Spatial Information Theory: Foundations of Geographic Information Science*, pages 270–286, Berlin, 2003. Springer.
6. K. Fine. Vagueness, truth, and logic. *Synthese*, 30:265–300, 1975.
7. M. Franceschet and A. Montanari. Branching within time: an expressively complete and elementarily decidable temporal logic for time granularity. *Research on Language and Computation*, 1(3-4):229–263, 2003.
8. A. Frank. The prevalence of objects with sharp boundaries in GIS. In P. Burrough and A. Frank, editors, *Geographic Objects with Indeterminate Boundaries*, pages 29–40. Taylor & Francis, London, 1996.
9. A. Galton. *Qualitative Spatial Change*. Oxford University Press, 2000.
10. M. Goodchild. A geographer looks at spatial information theory. In D. Montello, editor, *Spatial Information Theory: Foundations of Geographic Information Science*, pages 1–13, Berlin, 2001. Springer.
11. J. Y. Halpern. Intransitivity and vagueness. In D. Dubois, C. A. Welty, and M.-A. Williams, editors, *Principles of Knowledge Representation and Reasoning: Proc. 9th Intl. Conf. (KR 2004)*, pages 121–129. AAAI Press, 2004.
12. A. Herskovits. Schematization. In P. Olivier and K.-P. Gapp, editors, *Representation and Processing of Spatial Expressions*, pages 149–162. Erlbaum, Mahwah, NJ, 1998.
13. J. Hobbs. Granularity. In *Proceedings of IJCAI-85*, pages 432–435, 1985.
14. D. Hyde. Sorites paradox. In E. N. Zalta, editor, *The Stanford Encyclopedia of Philosophy*. CSLI (internet publication), Stanford, Ca., Fall 2002.
15. S. Kosslyn. *Image and Mind*. The MIT Press, Cambridge, MA, 1980.
16. M. Krifka. Be brief and vague! and how bidirectional optimality theory allows for verbosity and precision. In D. Restle and D. Zaefferer, editors, *Sounds and Systems. Studies in Structure and Change. A Festschrift for Theo Vennemann*, pages 439–458. Mouton de Gruyter, Berlin, 2002.
17. A. Mourelatos. Events, processes, and states. *Linguistics and Philosophy*, 2:415–434, 1978.
18. H. Ohlbach. Calendrical calculations with time partitionings and fuzzy time intervals. In H. Ohlbach and S. Schaffert, editors, *Principles and Practice of Semantic Web Reasoning: Second International Workshop*, pages 118–133, Berlin, 2004.
19. H. J. Ohlbach and D. M. Gabbay. Calendar logic. *Journal of Applied Non-Classical Logics*, 8(4), 1998.
20. Z. Pawlak. *Rough Sets*. Kluwer, Dordrecht, 1994.
21. F. Reitsma and T. Bittner. Scale in object and process ontologies. In W. Kuhn, M. Worboys, and S. Timpf, editors, *Spatial Information Theory: Foundations of Geographic Information Science*, pages 13–27, Berlin, 2003. Springer.

22. H. Schmidtke. A geometry for places: Representing extension and extended objects. In W. Kuhn, M. Worboys, and S. Timpf, editors, *Spatial Information Theory: Foundations of Geographic Information Science*, pages 235–252, Berlin, 2003. Springer.
23. H. Schmidtke. Aggregations and constituents: geometric specification of multi-granular objects. *Journal of Visual Languages and Computing*, to appear 2005.
24. P. Suppes and J. Zinnes. Basic measurement theory. In R. Luce, R. Bush, and E. Galanter, editors, *Handbook of Mathematical Psychology*. John Wiley & Sons, New York, 1963.
25. K. van Deemter. The sorites fallacy and the context-dependence of vague predicates. In M. Kanazawa, C. Pinon, and H. de Swart, editors, *Quantifiers, Deduction, and Context*, pages 59–86, Stanford, Ca., 1995. CSLI Publications.
26. A. Varzi. Vagueness. In L. Nadel, editor, *Encyclopedia of Cognitive Science*, pages 459–464. Macmillan and Nature Publishing Group, London, 2003.
27. L. Zadeh. Fuzzy sets and information granularity. In M. Gupta, R. Ragade, and R. Yager, editors, *Advances in Fuzzy Set Theory and Applications*, pages 3–18. North-Holland, Amsterdam, 1979.

A Proof: $\prec$ Is a Semi-order

The irreflexivity of $\prec$ (4) follows directly from $<$ being irreflexive as a consequence of (A<2).

Proof of (5): Suppose four granularities satisfying $\gamma_1 \prec \gamma_1' \wedge \gamma_2 \prec \gamma_2'$. The context-size of γ_1 can now be smaller, of the same size, or larger than the context-size of γ_2 (A<4). In the first two cases $\gamma_1 \prec \gamma_2'$ follows from the transitivity of $<$ (A<1), (A3), and (D3). In the third case $\gamma_2 \prec \gamma_1'$ follows accordingly. □

Proof of (6): Given three granularities satisfying $\gamma_1 \prec \gamma_2 \wedge \gamma_2 \prec \gamma_3$, any additional granularity γ must have a grain-size smaller, of the same size, or larger than the grain-size of γ_2. In the first and the second case $\gamma \prec \gamma_3$ follows from the transitivity of $<$ (A<1), (A3), and (D3). In the third case $\gamma_1 \prec \gamma_3$ follows accordingly. □

Proof of (7): $\rightarrow$: Follows directly from the definitions of $\prec$ (D3) and $\approx$ (D4).
$\leftarrow$: Suppose there are intervals t_1, t_2 compatible with γ and intervals t_1', t_2' compatible with γ' with $\neg t_1 < t_1'$ and $\neg t_2' < t_2$. According to (A<4) $t_1 \equiv t_1'$ or $t_1' < t_1$ holds, and analogously $t_2 \equiv t_2'$ or $t_2 < t_2'$. In the case of $t_1 \equiv t_1'$, t_1 is also compatible with γ', and (D4) is fulfilled; correspondingly, in the case of $t_2 \equiv t_2'$. So only the remaining case, in which $t_1' < t_1$ and $t_2 < t_2'$ hold, has to be considered: The two pairs t_1 and t_2 and t_1' and t_2' can then be compared. Because of (A<4), (A<3) and the transitivity of $<$ (A<1), any ordering of the four intervals t_1', t_1, t_2, t_2' under the restrictions $t_1' < t_1$ and $t_2 < t_2'$ has one of the intervals of γ' between intervals of γ or intervals of γ between those of γ'. Suppose w.l.o.g. the ordering is $t_1' < t_1 \leq t_2 < t_2'$. By (D2) and (A<4) it can be inferred that t_1 must be compatible with γ', and therefore $\gamma \approx \gamma'$ (D4). □

Th. (8) follows from (D4) and (A4).

Identifying the Interaction Context in CSCLE

Sandra de A. Siebra[1,2], Ana Carolina Salgado[1],
Patrícia A. Tedesco[1], and Patrick Brézillon[3]

[1] Centro de Informática – UFPE, Caixa Postal 7851, Cidade Universitária,
CEP 50732-970 Recife – PE - Brasil
`{sas, acs, pcart}@cin.ufpe.br`
[2] Faculdade Integrada do Recife (FIR) – Av. Abdias de Carvalho, 1678 – Madalena,
CEP: 50720-635 Recife – PE – Brasil
`{sas}@fir.br`
[3] LIP6, University Paris 6, 8 rue du Capitaine Scott, 75015 Paris, France
`Patrick.Brezillon@lip6.fr`

Abstract. Collaborative learning can be motivated through environments that provide tools for communication. However, most available environments do not provide support neither for the students' reflection and self-regulation processes, nor to the teachers' activities. This support could be provided through the analysis of the interactions occurred within the CSCLE and stored in a group memory. The main weakness of the approaches up to now is to neglect the context where the interactions occur and the inadequate mechanism of the interactions' persistence. In this light, we propose the creation of a Learning Interaction Memory (LIM) to store the learning interactions. It will be modelled in a multidimensional structure and included in its model contextual elements to enrich and qualify the stored knowledge. By including contextual elements in the LIM, it will be possible to adapt dynamically its information to the concrete current situation, in order to provide more useful just-in-time feedback.

1 Introduction

Collaborative learning [16] is a kind of social activity involving a community of learners and teachers, where members share and acquire knowledge. As Vygotsky [31] pointed out, "in a collaborative scenario, students interchange their ideas for coordinating when they work for reaching common goals. When dilemmas arise, the discussion process involves them in learning". When the learners work in groups they reflect upon their ideas (and those of their colleagues), explain their opinions, consider and discuss those of others, and as a result, they learn. In this way, each learner acquires individual knowledge from the collaborative interaction. In fact, interacting with our peers gives us a forum to discuss our ideas, to take a stand on our views, to reflect about and to elaborate on them. Consequently, the study of collaborative interactions is a fundamental element for the evaluation of the learning and teaching processes, and for the provision of support by the system to its users.

In this state of affairs, a basic requirement to support this study of collaborative interactions is to provide interaction mechanisms with persistence. Without persistence, interaction is ephemeral and cannot be shared afterwards with people who were not involved at the time it occurred [2]. With persistence, the interactions will constitute a group memory (GM) mechanism, which make past information about the interactions readily and selectively available when required. According to Paiva [22], when dealing with a CSCLE (Computer Supported Collaborative Learning Environment), we need to keep track of the group interactions, in order to follow the changes of social roles, beliefs, conflicts, misconceptions, and views of the task within the group.

The GM should be modelled in such way that stored interactions can be viewed from different perspectives (e.g. information can be easily crossed or filtered) and could be presented selectively, according to users' needs (i.e. depending on their context, users could access different information). Indeed, the GM model should facilitate the interaction analysis as means to provide access to the informations and construction and generation of customized reports to teachers and students. Moreover, to qualify each interaction and to fully understand many actions or events, it is important to take the context [6] where each interaction occurs into consideration. For example, understanding the action of "opening a window" depends on what is referred: a real window or a window on a graphical user interface [7].

In this light, this paper proposes the creation of a kind of GM called Learning Interaction Memory (LIM) to store the learning interactions occurred in CSCLE.

This LIM will be modelled in a multidimensional structure (implemented using a data warehouse [20]) and analysed via On-Line Analytical Processing (OLAP) [21]. Thus, the interactions can be explored in different dimensions and levels of detail, and specific feedback can be provided to both teachers and students. The LIM includes in its model contextual informations to enrich the stored knowledge. As means to help to identify the relevant contextual elements in CSCLE, this paper uses a conceptual framework for collaborative learning [29].

Through the analysis of the information stored in the LIM (that includes contextual information), it will be possible to characterize interactions for a better understanding of the collaborative learning process and to give support to teachers in their activities (e.g. guiding and evaluating the student) and to students in the reflection and self-evaluation processes.

The remainder of this document is structured as follows: Section 2 discusses about the reasons to create a GM. Section 3 presents the reasons for contextualizing the information in the GM. Section 4 presents our proposal to the LIM. Section 5 describes a conceptual framework for analysing the use of context in the learning environments. Finally, section 6 presents our conclusions and further works.

2 Why to Create a Group Memory?

In CSCLE, learning is promoted by interaction among peers, and/or between peers and a more experienced collaborator (e.g. the teacher or a tutor student). Thus, learning takes place through interactions and what is learned can be used when the learner tries to solve a similar problem independently. Therefore, collaboration enables knowledge sharing.

Nowadays, there are several proposals of CSCLE (e.g. [18][19]) that provide a wide variety of interaction mechanisms such as forums, blackboards, chats, email and videoconferences. However, one of the main problems with the use of interaction mechanisms in CSCLE is the frequently inappropriate match between the interaction mechanism available and the persistency required [2]. Group members may find it difficult to recall and justify their decisions when using interaction mechanisms with low or no persistence. Important information may be lost or need to be reproduced several times in order to achieve the desirable level of common knowledge [2].

Although many CSCLE (e.g. [14][18][19][30]) provide a way to store previously sent or received messages (e.g. by using sequential log files, normally organized in temporal order), what it is really needed is a common space to store the information in order to comfortably refer to it and add new contributions. This common space is the GM. Thus, the GM is the record of the complete group interaction process. It is a common organized memory that corresponds to the discussion database. It is the result of a process of accumulating data generated by group members during discussions in synchronous and asynchronous tools.

This dialogue history is viewed as an important resource in collaborative dialogue since it provides a common reference to previous activity that may encourage reflection and more effective collaboration [13]. This kind of information would help teachers to track the students' evolution process. Besides, the analysis of the group memory would enable users to reuse historical information to solve future problems, reminding participants of previous ideas (encouraging elaboration on them) and possibly serving as an agenda for further work.

3 Why to Contextualize the Information in the Group Memory?

Learning always takes place in dynamic environments, characterised by a collection of relevant conditions and surrounding influences that make a situation unique and comprehensive called context [6]. Each attribute of the context (e.g. location, user knowledge level, task name) is called contextual element. Situations with apparently the same context can differ from each other in some aspect. This diversity and unpredictability of the aspects are factors that influence the identification and representation of contextual elements related to group interactions [4].

The issue of context has been an important area of research in recent years, although, there is no consensus yet about what context really means, what its implications are and how it can be generalised [3]. Several domains have already elaborated their own working definition of context. In Artificial Intelligence, the context constraints the problem solving [6]. In a Human-Machine Interaction, context is a set of information that could be used to define and interpret a situation in which agents interact [7]. In the context-aware applications community, the context consists of a set of information for characterising the situation, where humans, applications and the immediate environment interact [15]. Bazire et. al. [1] show that all the definitions found on the web can be assembled around six questions: Who? What? When? How? Where? And Why?

There are two main research streams in context: (1) a community interested in making context explicit in the representation of the human reasoning [32] to support systems [23] through logics [5]; and (2) a community, interested in context-aware applications, has recently emerged [12][17]. Such applications take into account contextual information directly accessible from physical sensors in their reasoning.

In both communities, very few researchers are interested by different contexts at different granularity as in a collaborative community [3][8]. In CSCL, one must deal with several contexts at different granularity, such as the context of the group (why this group is constituted), the individual contexts of the members (e.g. their technical origins) and the context of the project (e.g. the artifact to build) [11].

In the learning process the more details the system can provide about user's interactions, the more it can support their reflection and knowledge construction. These details are the contextual informations. For example, a user can accept different recommendations depending on whether it is Monday or Friday (close to the weekend), whether he/she is at the office or at home, whether it is raining or it is terrible hot, whether he/she feels good or bad. However, although context is a relevant aspect in the learning process, to the best of our knowledge, there are no CSCLE explicitly using the concept of context in their development.

Furthermore, in conversation, context plays a fundamental role in disambiguating utterances: in many cases only the context can provide the correct cues to give the right interpretation to a sentence. In situations where geographically separated individuals have to collaborate (especially if they are interacting asynchronously), technological support for understanding and storing the contextual elements involved (for example, location and users' goals) is very important. This identification of contextual elements can help to clarify users' utterances, as well as to repair misinterpretations. Moreover, by knowing the context, teachers and systems can decide better on which is the adequate feedback to the learner. In fact, context provides the semantic enrichment of the GM information and allows the CSCLE to better support user reflection.

Other point to consider is that the informations generated through the analysis of the data in the GM (considering their context) could be used to improve and support awareness [24] in CSCLE. Awareness is an understanding of the activities of others, which provides a context for your own activity. It is used to facilitate effective group communication and coordination.

4 The Learning Interaction Memory

The LIM is a group memory where all the interactions occurred within CSCLE's communication tools can be stored. For the purpose of facilitating the execution of analytical analysis, the LIM will be modelled in a multidimensional structure [21], implemented using a data warehouse [20]. Indeed, using multidimensional modelling the LIM's informations can be viewed from different perspectives (e.g. information can be easily crossed or filtered) and could be presented selectively, according to users' needs (i.e. depending on their context, users could access different information).

Other point is if the LIM is modeled in a multidimensional way, analytical queries can be applied in it using OLAP technology [20]. OLAP queries can be used as much by the users as by the CSCLE to support students and teachers. In this way, it will be possible to answer questions such as: which kind of knowledge has been shared within the environment? Which members are participating more actively? Which are the most frequent problems that have been found during the learning processes via the environment? Which topics are being more difficult to students? Which students are not motivated? Which students have already faced the problem Y, considering the context X?

The LIM considers contextual knowledge (such as access time, access local, user level and role that users are playing during the interaction) in its modeling to adapt dynamically its information to the concrete situation based in past facts. All this knowledge is not a part of the actions to execute or the events that occur, but will constrain the execution of an action or event interpretation without intervening in it explicitly [6]. Making explicit and using context in collaborative learning is a way to improve the conditions in which each student participates in the group.

Thus, to build the LIM's multidimensional structure, we had to proceed with an analysis of which contextual elements will be considered relevant in a CSCLE, as a means to characterize and to better understand each interaction in particular.

In order to represent and cluster the relevant contextual elements present on learning interactions we have adapted the generic conceptual framework proposed in [27]. The resulting framework is presented in the next section.

5 A Conceptual Framework for Learning Context

The question of context is central to interaction analysis. It is the fairly straightforward observation that analyses look beyond simply the interaction between students. They look at the context in which that interaction emerges – the social, cultural and organisational factors that affect interaction, and on which the user will draw in making decisions about actions to take and in interpreting the system's and/or user's response. Consequently, identifying the contextual elements relevant to characterise the interaction is very important to enrich and qualify the information storage in the LIM.

Rosa et al. [27] proposed a conceptual framework for analysing the use of context in groupware. Their framework is a generic classification of contextual elements for groupware, so it neither covers the particularities of a specific domain nor applies to a particular category of groupware environments like CSCLE. Thus, we have adapted the framework (see [29] for details) according to the specificities of CSCLE and the context hierarchy, as an attempt to identify the contextual elements. In this resulting framework (see Table 1), contextual information is clustered in the five following categories:

1. *Information about scheduled tasks (task context)* - in CSCLE several tasks are possible (for example, to study a lesson, do exercises, discuss about a subject);
2. *Information about people and groups (Individual and Group contexts)* - The knowledge about the characteristics of individuals and the group as a whole is a resource that can be used by teachers to encourage interaction and collaboration [26].

   - *Group Context* - In CSCLE, users join other users (building a group), generally, to discuss about a given problem or subject in order to construct

their knowledge. Thus is important have in the LIM some knowledge about the group to understand the evolution of the individuals in CSCLE;
- *Individual Context* - CSCLE generally have a user model to store individual users' characteristics (static characteristics such as personal data and dynamic characteristics such as knowledge level). Such models are kept for both the students and teachers, and help CSCLE to cater for their individual needs. Thus, the attributes of this category can be obtained from the users' models and they help to characterise the user, as well as let other users better understand her/his doubts, difficulties and actions in the CSCLE;

Table 1. Conceptual Framework for Learning Context

Information Type	Associated Contexts	Contextual Elements Considered	
Scheduled Tasks	Task	• Name • Description • Goals • Deadlines • Estimated Effort	• Activities • Constraints • Workflow • Value • Kind of activity
Group Members	Group	• Name • Members • Roles • Abilities • Previous Experience	• Aim • Course • Initial Group Structure • Course
	Individual	• Nickname • Name • Abilities • Academic Education • Previous Experience • Location • Interests • Native Language	• Age Group • Knowledge Level • Preferences • Learning Speed • Email • Access Frequency • Main Access Turn • Average Session Length
Relationship between people and tasks	Interaction (Synchronous)	• Group in-charge • Related Task • Gesture Awareness • Presence Awareness • User Participation Style • Concluded Activities	• Messages Exchanged (including: author, addressee, date, time, discussion subject, message subject, abstraction used in the argumentation model, related message)
	Interaction (Asynchronous)	• Group in-charge • Related Task, • Related Messages • Location • Presence Awareness	• Messages Exchanged (including: author, addressee, date, time, discussion subject, message subject, abstraction used, related message)
	Planning	• Rules • Coordination Procedures • Course Planning	• Aim of the Course • Pedagogical Strategies • Responsibilities
Setting	Environment	• Connection quality • Organizational Structure	• Standard Procedures • Tools used
Completed Tasks	Historical	• Task Name • Working period]	• Activities • Contextual Elements used to carry out the task

3. *Information about the relationship between people and tasks (Interaction and Planning Contexts)* - In CSCLE it is important to know who is doing what, i.e. what the task's execution plan and what is being discussed into the environment. Indeed the interaction analysis is important for discovering more about the student (e.g. her/his difficulties or doubts). This type of information is represented here.

   - *Interaction Context* - In this specification, we took into account the fact that synchronous and asynchronous interactions may alternate in group-based learning. Tasks are carried out groupwise, though this does not imply that group members are always working synchronously;
   - *Planning Context* - It consists of information about the course execution plan (generally present in the pedagogical model of CSCLE). The Planning Context could be implemented using the idea of proceduralised context and contextual graphs presented in [9][10]. The interest of contextual graphs is to account for all the variants used by the learners that lead also to the same solution. In learning, an interesting side effect of this approach is to identify clearly when a learner goes towards a dead-end way before the learner actually reaches it.

4. *Information about the environment where the interaction takes place (Environment Context)* - It consists of information that characterises the environment where the interaction takes place and influences task completion and

5. *Information about tasks and activities already concluded (Historical Context)* - The information in this category tries to characterise the interactions that have already occurred. Its goal is to provide background information about the experiences learned either from the same group or from similar tasks performed by other groups. In this category, all the aforementioned contextual information generated and used (proceduralised context) is stored for future retrieval. This is exactly the goal of the Learning Interaction Memory (LIM). It is the repository of the "group memory" (including contextual elements). In this way a situation can be reconstructed with the context in which it occurred. Whereas the LIM provides historical context, students and/or teachers can access past incidents. This can also be used to share the latest news, seek advice, compare notes, etc. Thus, it might be a source of reflection for both the teacher and the student.

Our framework caters for one aspect that Artificial Intelligence in Education had not considered until fairly recently: the needs of a group. In order to better motivate users, and to improve their interaction quality, it is fundamental to understand the weaknesses and strengths of a group. The historical information included in our framework allow teachers and students to reflect upon past interactions, and to learn from their (or other's) past performances or errors.

Finally, by using this conceptual framework as a fundamental building block, we intended to include the concept of context into the LIM's multidimensional model. This enables the semantic enrichment of the information contained in the model and allows the CSCLE to better support user reflection.

5.1 Levels and Dimensions of the Context

To model the LIM, we need to identify different types of contexts at different levels, trying to reach all the elements related to CSCLE. These contexts have not the same granularity [10] and make difficult a simple representation of the contextual cues in our model. Thus, our framework organise context in different levels such as individual context, group context, task context, environment context and interaction context, from the more general to the more specific (see Fig. 1).

As mentioned before, all these contexts are not at the same level of generality. For example, the task's context is more general than the group's context because it is the context that everybody knows about what to have to be done within the CSCLE. The context of group contains more general contextual information than the context of each individual. This also does not imply that a context at one level is a subpart of a more general context [10]. Indeed, a focus at the level of an individual is related to the individual context, but also to the group context in which is the individual and the task context which the group is working in. At a given situation that takes place inside the CSCLE, its context can be understood at different levels. While executing a task (for example, develop a software program), the context of a student (individual context) can be explained in terms of the perspective of the group, the context determines that this student A is, for example, building the module X that will have to be integrated to the code being constructed by the students B and C.

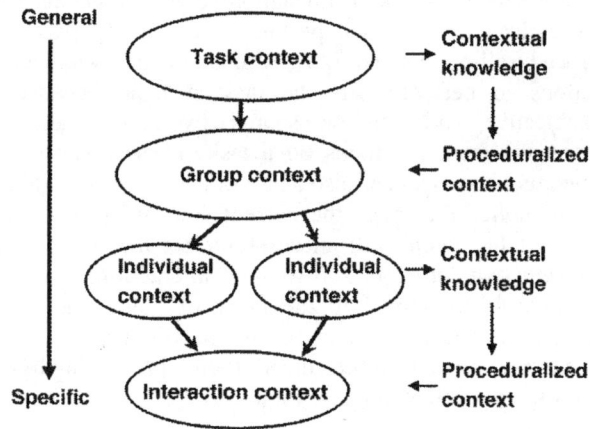

Fig. 1. Context Levels and the dynamic dimension of the context

Brézillon and Pomerol [25] proposed a classification for differentiating the contextual elements related to task performing. The set of contextual elements that are relevant to the task realization and can be mobilized to understand a given situated problem is called *contextual knowledge*. Each person uses a large amount of *contextual knowledge*, different from one person to another, to picture a situation. The knowledge that is shared by all people involved but is not used to perform a task is called *external knowledge*. During the execution of a task, a portion of the contextual knowledge is actually employed. This portion is called *proceduralised context*.

The proceduralised context is a part of the contextual knowledge which is invoked, structured and situated according to a given focus (e.g. the task at hand or the interaction). Thus, context has a dynamic dimension intertwined with its static dimension. This dynamics comes from the fact that along the learning process, a part of the *contextual knowledge* is assembled, organized, structured in a *proceduralised context* [25] that is built and used at one step of the process, and after stored in the group memory as contextual knowledge. For example, a teacher establishes tasks (task context) and a group of students contextualize these tasks to develop efficient practices. Put differently, each group of students develops their own practice, tailoring the task in order to take into account the current group context (e.g. background, ability, other group components), which is particular and specific (see Fig. 1). Two groups having the same task to complete will build two different proceduralised contexts. First, because they have two different bodies of contextual knowledge (group context). Second, because they also have different interpretations of the task according to their different body of contextual knowledge. Other example is the interaction context contains proceduralised context pieces in the focus of attention of the two students (see Fig. 1). These pieces of knowledge are extracted from the contextual knowledge (in the individual context) of each student (e.g. what is the student level or what is the experience of the student) and this proceduralised context will have an influence on the discussion. Because in the discussion is important to know details about the users interacting.

This supposes some mechanisms of translation between contexts at different levels of the context hierarchy. In fact, an important issue is the passage from contextual knowledge to proceduralised context [25]. The contextual knowledge needs some further specifications to perfectly fits the task at hand because the contextual knowledge is subjective and can be shared by many individuals [8]. This proceduralization results from the focus on a task. These precisions and speciation brought to the contextual knowledge is also a part of the proceduralization process [8].

It is important to notice that once the current step of the focus is completed (a proceduralised context has been built successfully for the focus), the focus moves towards the next step (e.g. other task or a new interaction). At this moment, the proceduralised context is integrated in the body of contextual knowledge of the person. Additionally, a piece of contextual knowledge can become external knowledge (becomes not relevant to the current focus) [9]. In this light, before model the LIM (in a multidimensional structure), we need to:

1. Identify the relevant pieces of contextual knowledge;
2. Organise the contextual pieces in a hierarchical structure from the more general to the most specific (see Fig. 1), and
3. Define rules to transform contextual knowledge at one level into a proceduralised context at the specific level below. For example, in order to execute the task "to give an example for illustrating a new concept", the pieces of contextual knowledge could be: the student has a scientific background, the previous lesson was on "Techniques for organising Dialogs" and the new concept is Argumentation Models. A proceduralised context could be an extension of an example given to that student at the previous lesson. For example, "Argumentation model is an way to organise dialogs. The use of an argumentation model implies classifying all elements of discussions in pre-defined abstractions of the model (e.g. Question and Issue) and connecting them via a set of pre-defined relations (e.g. provokes and generates)". One way to implement this rules is to have as many production rules as contextual elements. An example is:

```
RULE-6

IF previous lesson = "Techniques for organising
dialogs" and new concept = "Argumentation Model" and
student background = "scientific"

THEN Execute action "Extend the example given to that
student at the previous lesson"
```

6 Conclusions and Future Work

In this paper, we have proposed a Learning Interaction Memory to store the interactions (in a multidimensional structure) occurred in CSCLE. The LIM considers contextual elements in its modelling to adapt dynamically its information to the concrete situation based in past facts. In order to identify these contextual elements in CSCLE, we will use a conceptual framework for collaborative learning. And, considering the collaborative learning environment where we distinguish different types of context at different levels, we also take into account the movement from one context to another one. This is ensured by a proceduralization of a part of the knowledge from the first context to the second context. By including contextual elements in our modelling, we are able to semantically enrich the support provided to participants.

It should be noted that the LIM, proposed in this paper, is a component of a larger project, carried out by the authors, which aims at storing and analysing collaborative interactions in order to provide good quality just-in-time feedback for both students and teachers. The full-length description of the project has been done in [28].

Thus, our future research will concentrate on study how the contextual-knowledge pieces can be combined to produce proceduralised context. Subsequently, we intended to model the LIM in a multidimensional structure and prototyping this model using data warehouse technologies [20][21].

References

1. Bazire, M.; Brézillon, P.; Tijus, Ch. Eléments intervenants dans le contexte d'une activité finalisée. Ed. J.M.C. Bastien. *Actes des Deuxièmes Journées d'Etude en Psychologie Ergonomique* – EPIQUE'2003, INRIA (2003) 281-286.
2. Borges, M. R. S.; Pino, J.A. Requirements for Shared Memory in CSCW applications. *Proceedings of 10th Annual Workshop On Information Technologies and Systems (WITS'2000)*, Brisbane, Australia (2000) 211-216.
3. Borges, M.R.S.; Brézillon, P.; Pino, J.A.; Pomerol, J.C. Bringing Context to CSCW. In: Proceedings of the 8th International Conference on Computer Supported Cooperative Work in Design. CSCWD´2004, Xiamen, China (2004).
4. Borges, M.R.S.; Meire, A.P.; Pino, J.A. An interface for supporting versioning in a cooperative editor. *Proceedings of the 10th International Conference on Human - Computer Interaction*, v. 2, Lawrence Erlbaum Associates Publishers, Crete, Greece (2003) 849-853.
5. Bouquet, P.; Ghidini, C.; Giunchiglia, F.; Blanzieri, E. Theories and uses of context in knowledge representation and reasoning. Akman V., Bazzanella C. (eds.) In: *Journal of Pragmatics*. 35(3) Elsevier Science (2003) 455-484.

6. Brézillon P. Context in problem solving: A survey. *The Knowledge Engineering Review*, v. 14, n.1 (1999) 1-34.
7. Brézillon P. Making context explicit in communicating objects. C. Kintzig, G. Poulain, G. Privat and P-N. Favennec (Eds.), *Communicating Objects,* Hermes Science Editions, Lavoisier (2002).
8. Brézillon P.; Pomerol J. Ch. Is context a kind of collective tacit knowledge? M. Jacovi and A. Ribak (Eds.). *European CSCW 2001 - Workshop on Managing Tacit Knowledge.* Bonn, Germany (2001) 23-29.
9. Brézillon, P. Context dynamic and explanation in contextual graphs. *Modeling and Using Context (CONTEXT-03),* P. Blackburn, C. Ghidini, R.M. Turner and F. Giunchiglia (Eds.). LNAI 2680, Springer Verlag (2003) 94-106.
10. Brézillon, P. Representation of procedures and practices in contextual graphs. *The Knowledge Engineering Review,* 18(2) (2003) 147-174.
11. Brézillon, P.; Borges, M.R.S.; Pino, J.A.; Pomerol, J.C. Context-based Awareness in Group Work. *Proceedings of the 17th International Flairs Conference*, Miami Beach, Fl, USA (2004).
12. Byun, H.E.; Cheverst, K. Exploiting User Models and Context-Awareness to Support Personal Daily Activities. In: Proceedings of Workshop on User Modelling in Context-Aware Applications. 8th International Conference on User Modelling UM2001. Sonthofen (Germany) (2001).
13. Collins, A; Brown, J.S. The computer as a tool for learning through reflection. H. Mandl and A. Lesgold (Eds). *Learning Issues for Intelligent Tutoring Systems,* Springer-Verlag, New York (1988) 1-18.
14. Constantino-González, M.; Suthers, D. Coaching Collaboration by Comparing Solutions and Tracking Participation. Eds. P. Dillenbourg, A. Eurelings & K. Hakkarainen. Proceedings EuroCSCL 2001, pp. 173-180. Maastricht, The Netherlands, 2001.
15. Dey, A. K., Salber, D. Abowd, G.D. A Conceptual Framework and a Toolkit for Supporting the Rapid Prototyping of Context-Aware Applications. *Context-Aware Computing, Human-Computer Interaction (HCI) Journal,* v. 16 (2001) 97-166.
16. Dillenbourg P. What do you mean by collaborative leraning? Ed. P. Dillenbourg. *Collaborative-learning: Cognitive and Computational Approaches,* Oxford: Elsevier (1999) 1-19.
17. Dourish, P.; Bellotti, V. Awareness and Co-ordination in Shared Workspaces. J. Turner and R. Kraut (Eds.) *Proceedings of the conference of the Computer Supported Cooperative Work (CSCW'92).* ACM Press, Toronto, Ontario (1992) 107-114.
18. Eleuterio M., Barthès JP, Bortolozzi F. Mediating collective discussions using an intelligent argumentation-based framework. *Proceedings of Computer Supported Cooperative Learning (CSCL 2002),* Boulder, Colorado, USA (2002).
19. Fuks H.; Gerosa, M.A.; Lucena, C.J.P. Using the AulaNet Learning Environment to Implement Collaborative Learning via Internet. In: Innovations 2003 - World Innovations in Engineering Education and Research, iNEER, USA (2003).
20. Inmon, W.H. Building the data warehouse (2nd ed.). New York: John Wiley & Sons (1996).
21. Kimball, R.; Reeves, L.; Ross, M.; Thomthwaite, W. The data warehouse lifecycle toolkit: tools and techniques for designing, developing and deploying data warehouses. New York: John Wiley & Sons (1998).
22. Paiva, A. Learner Modelling for Collaborative Learning Environments. B. Du Boulay and R. Mizoguchi (Eds). *Proceedings of the 8th World Conference on Artificial Intelligence in Education* (AIED'97), IOS Press (1997) 215-222.

23. Pasquier, L.; Brézillon, P.; Pomerol J.-Ch. Context and decision graphs in incident management on a subway line. CONTEXT 1999. In: *Lecture Notes in Artificial Intelligence*, N. 1688, Springer Verlag (1999) 499-502.
24. Pinheiro, M. K.; Lima, J. V.; Borges, M. R. S. "A framework for awareness support in groupware systems", Computers in Industry, Netherlands, Vol. 52, n. 1 (2003) 47-57.
25. Pomerol J.-Ch.; Brézillon P. Dynamics between contextual knowledge and proceduralized context. *Proceedings of the 2nd International and Interdisciplinary Conference on Modeling and Using Context*. Lecture Notes in Artificial Intelligence, n. 1688, Springer Verlag (1999) 284-295.
26. Rittenbruch, M. ATMOSPHERE: A Framework for Contextual Awareness. *International Journal of Human-Computer Interaction*, v. 14, n. 2 (2002) 159-180.
27. Rosa, M. G. P.; Borges, M. R. S.; Santoro, F. M. A Conceptual Framework for Analyzing the Use of Context in Groupware. In: Proceedings of International Workshop on Groupware, Autrans, France, *Lecture Notes in Computer Science*, Berlin, Germany, v. 2806 (2003) 300-313.
28. Siebra, S. A; Salgado, A.C.; Tedesco, P. A. Structuring Participants' Interactions in Collaborative Learning Environments. Proceedings of WebMedia/LA-Web 2004 - WCSCW, Ribeirão Preto, São Paulo, Brasil. (2004) 181- 187.
29. Siebra, S. A; Salgado, A.C.; Tedesco, P. A.; Brézillon, P. A Context-based Analytical Environment for CSCLE. Internal Report, CIN – UFPE, Recife – PE. Available in: http://www.cin.ufpe.br/~sas/reports (2004).
30. Soller, A.; Wiebe, J. ; Lesgold, A. A Machine Learning Approach to Assessing Knowledge Sharing During Collaborative Learning Activities. Proceedings of Computer Support for Collaborative Learning. pp. 128-137, Boulder, CO, 2002.
31. Vygostki, L. Mind and society: The development of higher psychological processes. Cambridge, MA:Harvard University Press (1978).
32. Young, R. A. Demonstratives, Reference, and Perception. CONTEXT 2003. In: *Series Lecture Notes in Computer Science*, Springer Verlag (2003) 383-396.

Operational Decision Support: Context-Based Approach and Technological Framework

Alexander Smirnov, Michael Pashkin, Nikolai Chilov,
and Tatiana Levashova

St. Petersburg Institute for Informatics and Automation
of the Russian Academy of Sciences,
39, 14th line, St.Petersburg, 199178, Russia
{smir, michael, nick, oleg}@mail.iias.spb.su

Abstract. The paper presents a context-based approach to operational decision support. The approach focuses on modelling and solving a problem considering changes in the dynamic environment. The problem is modelled by abstract and operational contexts integrating information provided by information sources and domain knowledge. The approach involves ontology management operations for knowledge integration, context management techniques for information organisation, and object-oriented constraint network mechanisms for problem definition and solving. The approach is tested as an adaptive service for on-the-fly portable hospital configuration.

1 Introduction

The goal of intelligent support to operational decision making is to assess the relevance of information & knowledge to a decision and to gain insight in seeking and evaluating possible decision alternatives. Operational decision making faces problems of management and sharing of huge amount of knowledge, personalization of decision making, availability of up-to-date and accurate information provided by the dynamic environment.

The paper presents an approach to operational decision support as a follow-up of KSNet-approach [1]. KSNet-approach was an effort in knowledge logistics as part of knowledge management activities. Knowledge logistics deals with integration and transfer of the right knowledge from distributed sources to the right person within the right context in the right time to the right purpose [1, 2, 3]. The presented approach inherits main aspects of KSNet-approach such as ontology model for knowledge representation, ontology-based knowledge integration, constraint-based problem formalization, problem solving by a constraint solver. KSNet-approach was tested on a number of case studies in areas of e-business [2] and health service logistics [3].

The case study of health service logistics showed that in some cases to find the right solution for the problem it is not enough to have knowledge how to express and how to solve the problem. The problem may need some actual information, e.g. current location of an object, time, weather conditions, etc. The question of having real-time information is very important since operational decisions must rely on timely, accurate, directly usable, and easily obtainable information provided by the dynamic environment. The task of an easy aggregation of information and data from different sources is a central theme of the Semantic Web. To handle this task context mechanism is used [4]. Context is defined as any information that can be used to characterize the situation of an entity where an entity is a person, place, or object that is considered relevant to the interaction between a user and an application, including the user and applications themselves [5].

Since the main goal of decision making is to find a solution for a problem the idea behind the approach consists in creation of a knowledge-based model of the user problem and to solve it considering dynamic nature of the environment. Decision makers, operators, and other participants involved in the decision making are meant as users. The concept "problem" is used for both a task at hand (the user problem) to be solved or a current situation to be described.

The rest of the paper is organized as follows. An approach overview is given in Section 2. User problem modelling by means of object-oriented constraint networks is discussed in Section 3. Section 4 presents a technological framework of context-based decision making. A testbed of the approach is presented in Section 5.

2 Approach Overview

The final goal of decision making is to find a solution for the problem at hand. Dynamic nature of the environment induces to application of context as a model allowing to take into account changes in the environment. The approach aims at modelling the user problem and solving it. Because of this, integration of information and knowledge in context in a problem-oriented manner is proposed. Two types of context are used: abstract and operational. *Abstract context* is a knowledge-based problem model uniting knowledge and information relevant to the problem. *Operational context* is an instantiation of the abstract context with data provided by information sources.

Operational decision making within the approach is considered consisting of two stages: a preliminary stage and a decision making stage. The *preliminary stage* is responsible for creation of semantic models for the components of a decision support system (sec. 2.1), accumulating domain knowledge, and linkage of domain knowledge to its environment. The *decision making stage* deals with integration of knowledge and information relevant to the problem, problem modelling and solving.

2.1 Preliminary Stage

Within the approach ontology representation is used for the explicit description of the semantics for the main components of a decision support system. The choice of ontologies was determined by several reasons: ontology model is believed to be a means to overcome the problem of semantic heterogeneity; ontologies provide reusable domain knowledge; knowledge represented by ontologies is sharable and understandable for both humans and computers. The last point of knowledge understandability plays a critical role in the proposed approach since it aims at problem solving. The problem has to be explicitly specified in order to the computer be able to solve it.

The following components are defined: *information sources*, *domain knowledge*, and *users*. The *common component representation* allows unification of information and knowledge provided by the components, enables to handle the information sources in a same manner, and simplifies integration of the information and knowledge.

The approach orients to an availability of domain knowledge. The knowledge has to be collected before it can be used in problem solving. Knowledge collecting includes phases of knowledge representation and integration. Due to ontology model heterogeneous knowledge being collected is represented in a uniform way.

In order to obtain actual information from the environment the ontologies are linked to information sources that keep track of environment changes.

2.2 Decision Making Stage

The starting point for the decision making stage is the user request containing a formulation of the user problem in a user presented view. Based on the request the knowledge relevant to it is searched for within the collected knowledge, extracted, and integrated. Ontology-driven knowledge integration enables to involve methods of consistency checking for the integrated knowledge and to create a consistent context. To operate on the extraction of relevant knowledge, its integration and consistency checking *ontology management* techniques are used.

The consistent context is considered as an *abstract context* that is an ontology-based problem model supplied with links to information sources that will provide data needed for the given problem. The linked information sources instantiate the abstract context producing the *operational context* that is the problem model along with problem data (Fig. 1). Changes in the environment result in changes in the operational context. Obtaining information and its organisation in contexts are *context management* issues.

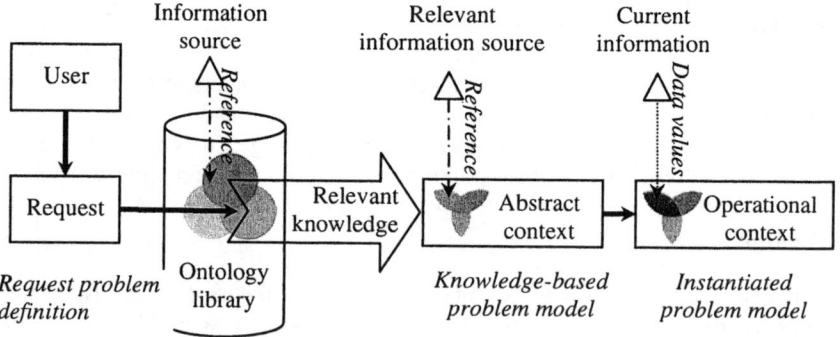

Fig. 1. User problem modelling

In order to the problem be solved automatically it has to be represented in a way suitable for its processing. To express the problem in a natural way and to take into account various constraints that the environment imposes the mechanism of object-oriented constraint networks (OOCN) [6] is proposed to be employed. Constraints can provide the expressive power of the full first-order logics [7] that is tended to be used as the key logics for ontology formalization. The other logics used for this purpose are an extension or reduction of the first-order logics. The problem expressed by a set of constraints is to be solved by specialized constraint solvers as *constraint satisfaction problem* (CSP).

Fig. 2 demonstrates a developed integrated framework for operational decision support. The framework is organised in three integrated levels. Problem level deals with modelling and solving the user problem formulated by the request. Technological level offers technologies for the problem level achievement. The bottom level proposes technical resources for the implementation. Below the main elements of the framework are discussed.

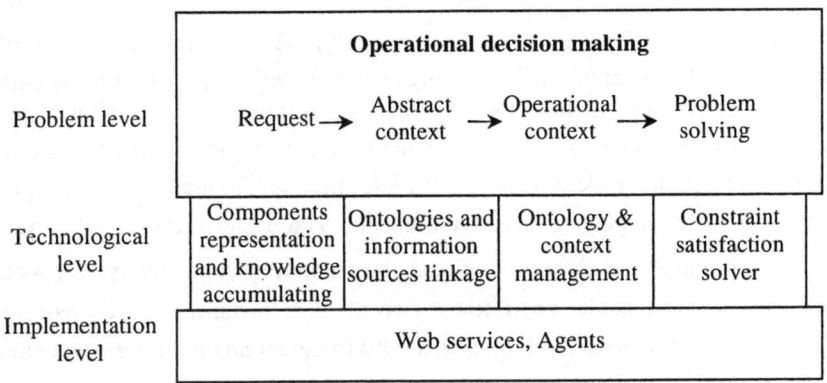

Fig. 2. Integrated framework for operational decision support

3 Object-Oriented Constraint Networks as Problem Model

CSP consists of three parts: a set of variables; a set of possible values for each variable (its domain); and a set of constraints restricting the values that the variables can simultaneously take.

Current ontology representation languages and ontology modelling tools provide with great number of axioms to knowledge description. No one constraint solver has mechanisms to completely support this variety of axioms [8]. To express the problem by a set of constraints that would be compatible on the one hand, with ontology model, and, on the other hand, with internal solver representations from the variety of axioms, the most typical constraints occurring in ontology specifications or supporting by constraint solvers are identified.

Typical ontology modelling primitives are classes, relations, functions, and axioms. The most commonly encountered relations are relations modelling class taxonomy (is-a relation), class hierarchy (part-of relation), inheritance, range and cardinality restrictions. Most used axioms are disjointness and equality. A correspondence between the primitives of ontology modelling and OOCN is shown in Table 1.

Table 1. Primitives of ontology model and OOCN

Ontology Model	OOCN
Class	Object
Attribute	Variable
Attribute domain (range)	Domain
Axioms and relations	Constraints

The following representation of the problem (*P*) by the formalism of OOCN is proposed:

$$P = (O, A, D, C) \qquad (1)$$

where O – a set of *object classes* ("*classes*"); A – a set of class attributes ("*attributes*"); D – a set of attribute domains ("*domains*"); C – a set of *constraints*. The set of constraints includes six types of constraints for modelling relations encountered in ontologies and constraint networks: C_1 – (class, attribute, domain) relation used to model triple of classes, attributes pertinent to them, and restrictions on the attribute value ranges; C_2 – taxonomical ("is-a") and hierarchical ("part-of") relations used to model class taxonomy and class hierarchy respectively; C_3 – classes compatibility used to model condition if two or more instances can be parts of the same class; C_4 – associative relationships used to model any relations and axioms of external ontologies neglected by the internal formalism; C_5 – attribute cardinality restriction used to define how many values the attribute can have; C_6 – functional relations used to model functions and equations. Constraints of the types defined can be mapped onto solver language constructs easily.

The representation proposed is considered as the internal knowledge representation for the operational decision support. Ontologies used for contexts creation are supposed to be modelled by the internal representation (1). Reusable ontologies represented in other formalisms have to be translated into the internal representation before operations on context creation.

4 Technological Framework

4.1 Components Representation and Knowledge Storage

The components of a decision support system are modelled as follows: *domain knowledge* is modelled by ontology model; semantics of *information sources* is described by *information source capabilities* model; *users* are modelled by *user profile* model.

Ontology library serves as a repository for the collected knowledge. It is responsible for common knowledge representation. It provides a vocabulary for the knowledge representation and supports the internal knowledge representation formalism (OOCN formalism) for specification of the ontologies it has in. Components of the ontology library are multiple ontologies of two types: domain ontology and tasks & methods ontology. Domain ontology represents conceptual knowledge about the domain, tasks & methods ontology formalises tasks identified for the domain and hierarchies of problem solving methods.

Every task within the tasks & methods ontology is considered as a goal. The method hierarchy is a decomposition of the task; it provides a sequence of methods (taking into account alternative ones) to achieve the goal. The tasks and methods are represented by classes. A signature of a class in tasks & methods ontology specifies the set of arguments each method takes by the set of class attributes, the type of arguments and the type of the result the method returns by attribute domains. Domain and tasks & methods ontologies are interrelated by functional constraints (Fig. 3) showing that certain method uses the attribute of the domain ontology as an input argument or that certain attribute of the domain ontology is calculated by the method (i.e., this attribute is an output argument for the method). Methods represented by classes become entities independent to the representation language and can be treated as first-order objects. They can be added to and removed from the ontology easily using basic primitives of the ontology description language, at the same time not changing the method specifications. The tasks & methods ontology transforms bidirectional constraints into unidirectional methods.

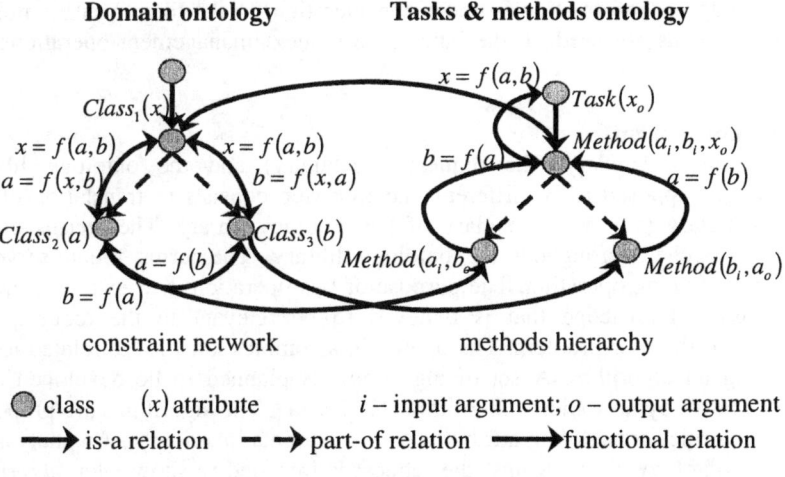

Fig. 3. Knowledge representation

4.2 Environment Coupling

Due to information sources and users are represented by the same formalism as the ontologies the relation of domain knowledge to the environment is indicated through associative constraints between attributes of domain ontology and attributes of the representations for information sources and users. The indication means that the attribute of the ontology class gets values provided by the information source or user (Fig. 4).

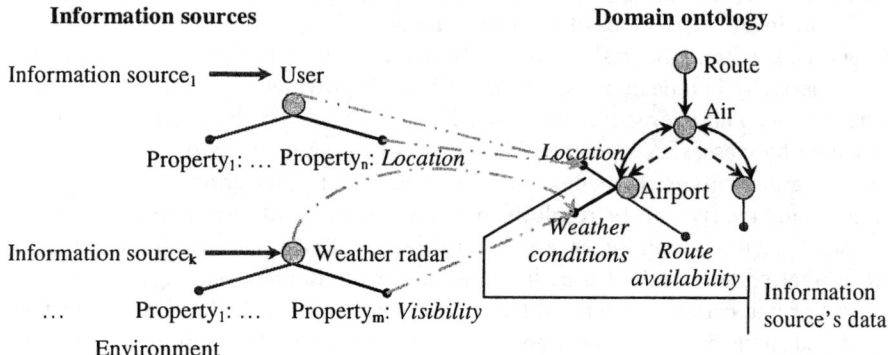

Fig. 4. Links between domain knowledge and information sources

4.3 Management of Ontologies and Contexts

Ontology management or context management techniques are involved depending on either knowledge or information is under consideration. In the former case ontology management means are used, in the latter case context management operations are applied.

Ontology Management
Since the user vocabulary (the request vocabulary) and the ontology library vocabulary are supposed to be different, the first step consists in translation of the request vocabulary into the vocabulary of the ontology library. Then terms of the request are searched throughout the ontology library. The terms found serve as "seeds" for the slicing operation. The purpose of this operation is to extract from the ontology library knowledge that is believed to be relevant to the request, and consequently to the user problem. The operation assembles knowledge related to the "seeds" using an algorithm. A set of algorithms is planned to be developed as a revision of known by now slicing algorithms [9, 10, 11]. The algorithms are proposed to be based on rules considering attributes and constraints inheritance. An analysis of relevance of ontology slices against the request is intended to show what algorithm generates best results.

Generally, if knowledge collected in the ontology library is represented by knowledge representation formalisms differing from the formalism of OOCN then ontology slices can be formed with reasoning rules provided by the formalisms the source ontology specifies. After that, the slices should be represented by the internal formalisms with possible loss of the source axiomatic. So far, the question of development of "real" semantically rich ontologies has been an open issue. Moreover, development of techniques for slices definition depending on ontology representation formalism needs certain efforts, which is not the main purpose of the approach. Because of these reasons it is decided that the source ontology will be represented by means of the internal formalisms. Ontologies represented in other formalisms are to be imported into the internal formalism before slicing.

The slices are merged resulting in a piece of knowledge describing or relevant to the problem (Fig. 5, slicing "2"). In order to the figure not to be too busy, in the ontology slice functional dependencies between attributes are emulated by a functional relation between appropriate classes. The automatic merging operation is believed to be achievable by reasons of the common vocabulary and formalism used for the internal knowledge representation, simple representation formalism excluding complex interdependencies that can have effect on semantic ambiguity, and a certain area of interests (application domain) ensuring collecting reliable knowledge providing by experts having fundamental understanding of the domain. A merging algorithm is to be developed based on an analysis of semiautomatic [e.g., 12, 13, 14] and automatic [e.g., 15, 16, 17, 18, 19] ontology merging algorithms.

Due to relations between ontologies and the environment, assigned in the ontology library, the merged knowledge is connected to information sources and users that will provide data. In this sense users are considered as information sources. Below, information sources will refer to users as well.

Context Management

Context management aims at obtaining, organisation, and integration of information relevant to the problem. The information is provided by information sources the merged knowledge refers to.

If an information source supports a complex data structure a slice from the information source representation is formed including the structure elements needed for the problem solving (Fig. 5, slicing "3"). Associative constraints between attributes of the information source and the knowledge describing the problem show what data are needed. If an information source supports a simple data structure the slice is the representation of the information source.

The slices of information sources and ontology-based problem description are merged. Since CSP operates over domains the task here is to develop methods allowing to merge domains in a more reasonable way. The difficulty is to define whether domain knowledge representation or information source representation is more reliable. The viable alternative will lead to a more accurate problem model and in some cases of reduced domains, will reduce search space. The result of merging checked on the consistency is considered to be the abstract context. Consistency is proposed to be checked by an automatic tool provided for F-logic [20]. Checked context is a problem model that has an interpretation for all predicates and functions included.

To be sure that the abstract context is a complete problem model a sufficiency of this context to the request and to the problem is expected to be assessed. The sufficiency criteria are proposed as follows: every meaningful request term is represented in the context; every input argument of a method is associated with an attribute of a domain ontology class providing values for this argument; for every attribute that is a function calculated by a method an association between this attribute and a method output argument exists. The list of the criteria probably will be extended after some practical achievements in abstract context composition and a more careful analysis of the resulting abstract contexts. An insufficient abstract context leads to refinements of the context and ontologies through searching and integration of lacking knowledge from external knowledge sources including humans.

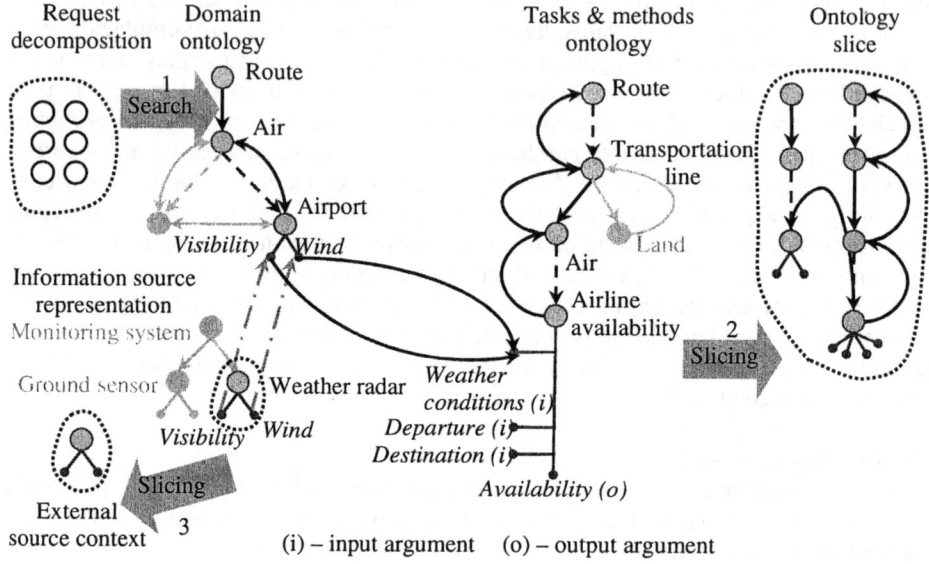

Fig. 5. Organisation of relevant information and knowledge

The information sources the abstract context is connected to instantiate it through a redefinition of domains. The abstract context with fully or partially redefined domains is an operational context (Fig. 6). At the same time, the operational context is OOCN to be processed as CSP.

Since operational decision making deals with repeatable decisions the contexts are retained. To operate them the context versioning mechanism is intended to be used.

4.4 Problem Definition and Solving

Three main forms of CSP are distinguished [7]:

1. The Decision CSP. Given a network, decide whether the solution set of the network is non-empty;
2. The Exemplification CSP. Given a network, return some tuple from the solution set if the solution set is non-empty, or return nil otherwise;
3. The Enumeration CSP. Given a network, return the solution set.

If all the domains in an operational context have been redefined the operational context is considered as a situation description. This case corresponds to CSP of the 1st type. If an operational context contains not redefined domains, it means that a solution is expected. In that case CSP falls under the 2nd or the 3rd form.

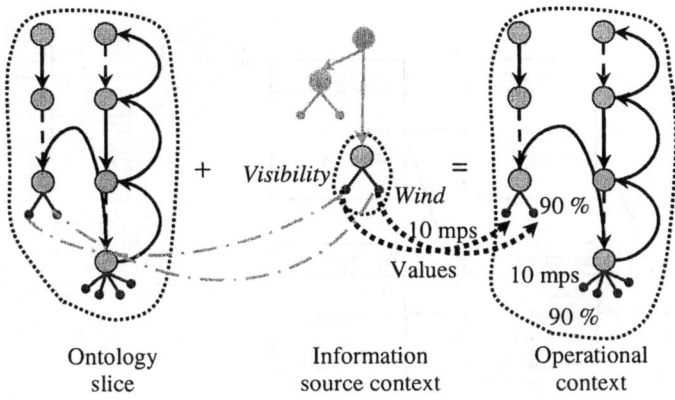

Fig. 6. Operational context producing

5 Implementation and Case Study

As a case study for experimentation with the system KSNet a problem of on-the-fly portable hospital configuration in the hypothetical Binni region [21] has been considered. As a result of the analysis of the problem a number of modules (subproblems) were defined (Fig. 7). The following notation is used in the figure: bold label denotes the common problem, underlined labels denote subproblems and italic labels denote example parameters common for two or more subproblems. Based on these subproblems the abstract contexts are identified.

Simplified examples of the identified abstract contexts on the basis of a domain ontology is presented in Fig. 8. Rectangles denote classes with attributes, solid lines denote associative relationships. Part "a" illustrates abstract context for "Resource Allocation" subtask, part "b" illustrates abstract context for "Hospital Allocation" subtask, and part "c" illustrates abstract context for "Routing" subtask. Abstract contexts include references to information sources with actualized information. These references are not shown in Fig. 8.

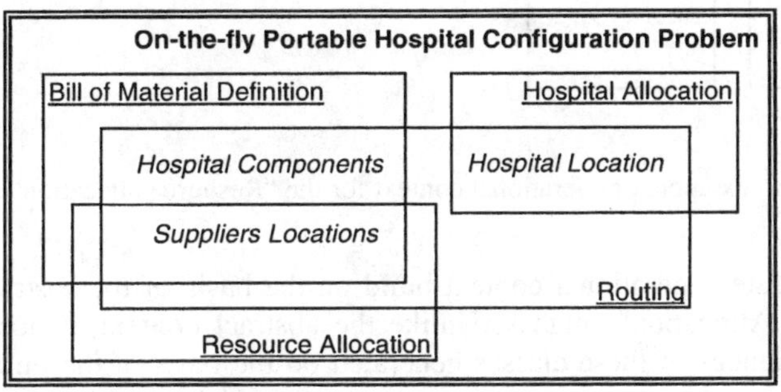

Fig. 7. On-the-fly portable hospital configuration problem and its subproblems

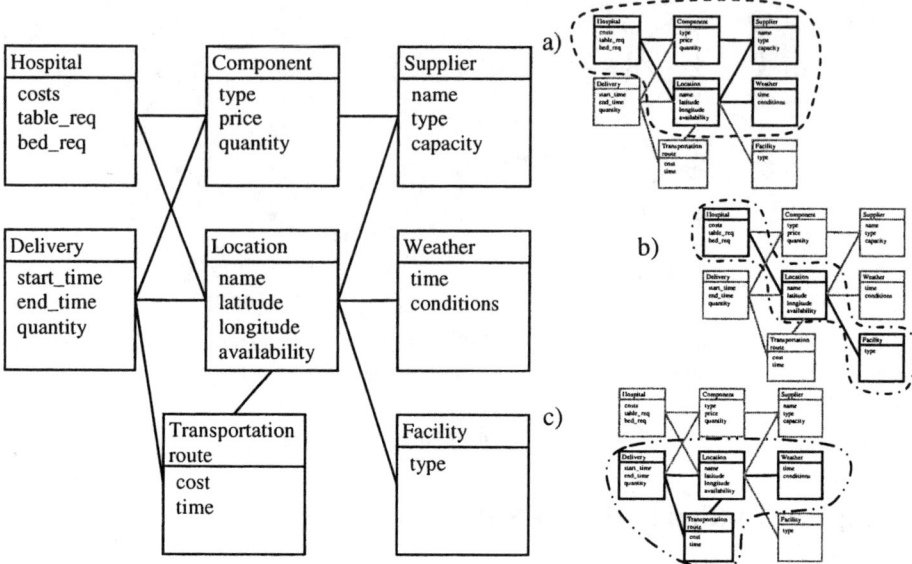

Fig. 8. Examples of abstract contexts ("a", "b", and "c") based on the same domain ontology (magnified on the left)

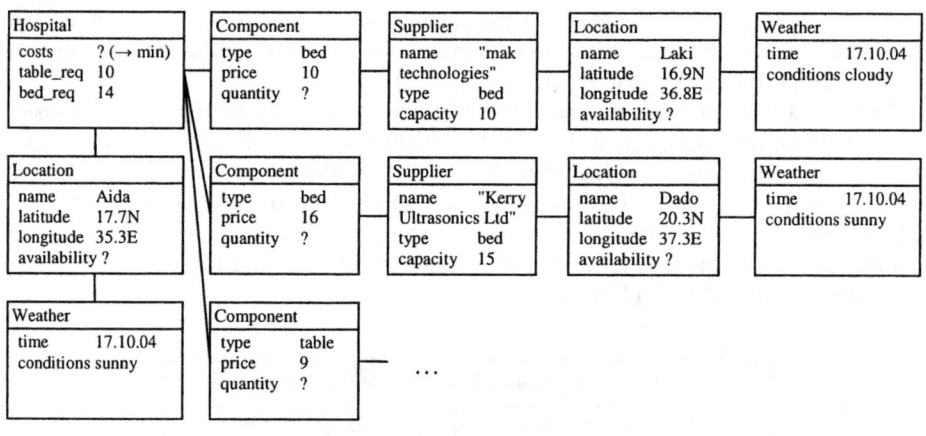

Fig. 9. An example of operational context for the "Resource allocation" subtask

Fig. 9 illustrates operational context build on the basis of the abstract context for the "Resource Allocation" subtask. Unlike the abstract context, it does not contain classes but instances of these classes generated on the basis of the information from knowledge sources (databases, sensors, experts, etc.). Values with question marks are to be calculated based on the functional constraints or (in case of the hospital costs) are the goal functions for evaluating the solution feasibility.

As it can be seen due to the chosen knowledge representation formalism of object-oriented constraint networks, the operational context is a formalized constraint satisfaction task. Since the tasks to be solved are presented in the same formalism but may differ in parameters and in structure, an on-the-fly problem modification and solving based on adaptive software modules are proposed. For this purpose the described approach implements "adaptive services" that can modify themselves when solving a particular problem. The problem is described by the operational context. Upon receiving the request the service loads this context and generates an executable module for its solving on-the-fly (Fig. 10). ILOG constraint satisfaction technology was chosen as a constraint solver for the implementation of the approach.

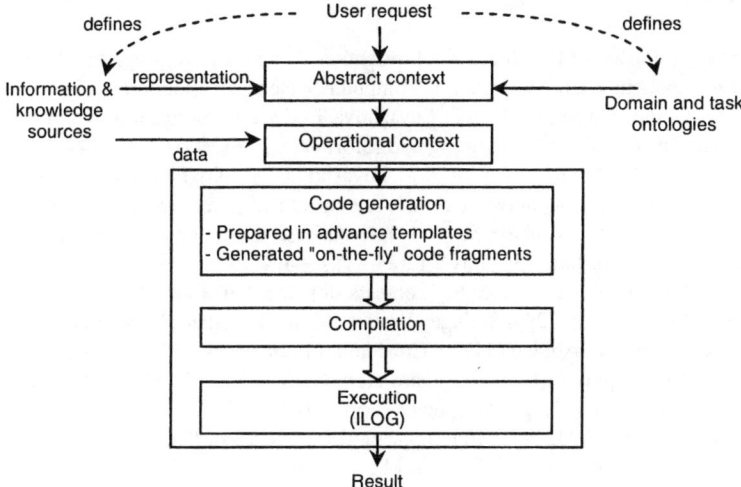

Fig. 10. The concept of the on-the-fly compilation mechanism

6 Conclusion

The paper describes a context-based approach to operational decision support as a follow-up of KSNet-approach developed earlier. The approach aims at modelling and solving the decision maker problem. The problem is represented as ontology-based contexts integrating information and knowledge relevant to it. Constraint-based problem model is used for problem solving.

The approach is based on common knowledge representation for the main components of a decision support system: domain knowledge, information sources, and the decision maker. It incorporates ontology management techniques for integration of knowledge relevant to the decision maker problem, context management methods for integration of information relevant to the problem and for organisation of the contexts, and object-oriented constraint networks mechanism for problem solving.

The approach testing is illustrated through an adaptive service for on-the-fly portable hospital configuration.

Acknowledgement

The paper is due to the research carried out as a part of Partner Project # 1993P funded by EOARD of the USAF, project # 16.2.44 of the research program "Mathematical Modelling and Intelligent Systems", the project # 1.9 of the research program "Fundamental Basics of Information Technologies and Computer Systems" of the Russian Academy of Sciences, the project funded by grant # 05-01-00151 of the Russian Foundation for Basic Research, and the CRDF partner project with US ONR and US AFRL & EOARD.

References

1. Smirnov A., Pashkin M., Chilov N., Levashova T. Knowledge Logistics in Information Grid Environment. Future Generation Computer Systems, 20 (2004) 61—79.
2. Smirnov A., Pashkin M., Chilov N., Levashova T. KSNet-Approach to Knowledge Fusion from Distributed Sources. Computing and Informatics, 22 (2003) 105—142.
3. Smirnov A., Pashkin M., Chilov N., Levashova T., Krizhanovsky A. Fusion-based knowledge logistics in network-centric environment: intelligent support of OOTW operations. Proceedings of the Seventh International Conference on Information Fusion, Stockholm, Sweden, June 28 – July 1 (2004) 487—494.
4. Guha R., McCool R., and Fikes R. Contexts for the Semantic Web. Lecture Notes in Computer Science, Vol. 3298. Springer-Verlag, Berlin Heidelberg New York (2004) 32—46.
5. Dey A.K., Salber D., Abowd G.D. A Conceptual Framework and a Toolkit for Supporting the Rapid Prototyping of Context-Aware Applications. In: T.P. Moran, P. Dourish (eds.): Context-Aware Computing, A Special Triple Issue of Human-Computer Interaction, Lawrence-Erlbaum, Vol. 16 (2001) http://www.cc.gatech.edu/fce/ctk/pubs/HCIJ16.pdf.
6. Smirnov A., Levashova T., Pashkin M., Chilov N. Ontology-oriented Multiagent Approach to Building of Systems for Knowledge Integration from Distributed Sources. Information Technologies and Computing Systems, No. 1 (2002) 62—81 (in Russian)
7. Bowen J. Constraint Processing Offers Improved Expressiveness and Inference for Interactive Expert Systems. International Workshop on Constraint Solving and Constraint Logic Programming (2002) 93—108.
8. Caseau Y. Abstract Interpretation of Constraints over an Order-Sorted Domain. Proceedings of International Logic Programming Symposium, San Diego, USA (1991) 435—454.
9. Chaudhri V.K., Lowrance J.D., Stickel M.E., Thomere J.F., Wadlinger R.J. Ontology Construction Toolkit. Technical Note Ontology, AI Center. Report, January 2000. SRI Project No. 1633. 85 p.
10. Levashova T.V., Pashkin M.P., Shilov N.G., Smirnov A.V. Ontology Management. Journal of Computer and System Sciences International. Part II, 42 (5) (2003). 744--756.
11. Swartout B., Patil R., Knight K., Russ T. Toward Distributed Use of Large-Scale Ontologies. Tenth Knowledge Acquisition for Knowledge-Based Systems Workshop (KAW'96), Banff, Canada, (1996) URL: http://www.isi.edu/isd/banff_paper/Banff_final_web/Banff_96_final_2.html.
12. McGuinnes D.L., Fikes R., Rice J., Wilder S. An Environment for Merging and Testing Large Ontologies. Proceedings of the Seventh International Conference on Principles of Knowledge Representation and Reasoning, Breckenridge (2000) 483—493.

13. Noy N.F., Musen M.A. PROMPT: Algorithm and Tool for Automated Ontology Merging and Alignment. Proceedings of the Seventeenth National Conference on Artificial Intelligence, Austin (2000) 450—455.
14. Noy N.F., Musen M.A. Anchor-PROMPT: Using Non-Local Context for Semantic Matching. Proceedings of the IJCAI01 Workshop on Ontologies and Information Sharing (2001)
15. Stumme G., Maedche A. FCA-Merge: Bottom-Up Merging of Ontologies. Proceedings of the 7th International Joint Conference on Artificial Intelligence, Seattle, USA, August 1-6 2001, San Francisco/CA. Morgan Kauffmann (2001)
16. Kalfoglou Y., Schorlemmer M. IF-Map: An Ontology-Mapping Method Based on Information-Flow Theory. Journal of Data Semantics In: Lecture Notes in Computer Science, Vol. 2800. Springer (2003) 98—127.
17. Kokla M. and Kavouras M. Fusion of Top Level and Geographic Domain Ontologies Based on Context Formation and Complementarity. International Journal of Geographical Information Science, 15 (7) (2001) 679—687.
18. Doan A., Madhavan J., Domingos P., Haleny A. Learning to Map between Ontologies on the Semantic Web. In: 11th International World Wide Web conference, Honolulu (2002)
19. Lammari N., Metais E. Building and Maintaining Ontologies: a Set of Algorithms. Data & Knowledge Engineering, 48 (2) (2004) 155—176.
20. Decker S., Erdmann M., Fensel D., Studer R. Ontobroker: Ontology Based Access to Distributed and Semi-Structured Information. In: R. Meersman et al. (eds.): Semantic Issues in Multimedia Systems. Proceedings of DS-8. Kluwer Academic Publisher, Boston (1999) 351—369.
21. Rathmell R.A. A Coalition Force Scenario "Binni – Gateway to the Golden Bowl of Africa. In: A. Tate (ed.): Proceedings on the International Workshop on Knowledge-Based Planning for Coalition Forces. Edinburgh, Scotland (1999) 115—125.

Threat Assessment Technology Development

Alan N. Steinberg

CUBRC, Inc., Buffalo, NY, USA
alan.steinberg@comcast.net

Abstract. A concept for performing threat analysis is developed. The goal is to establish a systematic approach for predicting, detecting and characterizing threat activity; allowing automation of some of these functions. The proposed approach explicitly addresses the fundamental problems of (a) sparse and ambiguous indicators of potential or actualized threat activity buried in massive background data; and (b) uncertainty in threat capabilities, intent and opportunities. Threats are modeled in terms of potential and actualized relationships between perpetrators (threatening entities) and targets (threatened entities). Threats may be intentional or unintentional (e.g. potential natural disasters or human error). Intentional threats can also have unintended consequences. Attack hypotheses are adaptively generated, evaluated and refined as the understanding of the situation evolves. This effort builds upon advances in Situation, Ontology and Estimation theory.

1 Definition of Terms

Estimation or prediction of states and events in the world is the general province of data fusion. Considerable progress has been made in developing and automating data fusion techniques applicable to such problems as signal/feature extraction (classified as "level 0" problems in the standard JDL Data Fusion model[1-4]) and object recognition, location and tracking (level 1 problems). There has been less success in developing systematic approaches to the higher levels of data fusion, and specifically in Situation Assessment (level 2) and Impact Assessment (level 3).

Per the JDL Data Fusion Model, Threat assessment is a level 3 data fusion process. Indeed, the original model version [1] used 'Threat Assessment' as the general name for level 3, indicative of the importance of that topic. In a subsequent revision [2], the concept of level 3 has been broadened to that of Impact Assessment, defined as

> *the process of estimation and prediction of effects on situations of planned or estimated/predicted actions by the participants; to include interactions between action plans of multiple players (e.g. assessing susceptibilities and vulnerabilities to estimated/predicted threat actions given one's own planned actions).*[1]

---

[1] Discussions concerning further revisions to the JDL model continue within the U.S. Data and Information Fusion Group, which developed and maintains the model. A recent study[3,4] suggests redefining the data fusion levels as follows.

Generally speaking, threat assessment can include some or all of the following functions:

- Threat Event Prediction: Determining likely threat events ("attacks"): who, what, where, when, why, how;
- Indications and Warning: Recognition that an attack is imminent or under way;
- Threat entity detection & characterization: determining entities' identity, attributes, composition, location/track, activity capability, intent;
- Attack Assessment:
  - Responsible country/ organization/ individual;
  - Intended target(s);
  - Intended effect (e.g. physical, political, economic, psychological effects);
  - Threat capability (e.g. weapon and delivery system characteristics);
  - Force composition, coordination & tactics (goal and plan decomposition);
- Consequence Assessment: Estimation and prediction of event outcome states and their cost/utility.

In comparison to the lower-level data fusion problems, the relative difficulty of the higher-level Situation and Impact/Threat Assessment problems can be attributed to the following three factors:

1. *Weak spatio-temporal constraints on relevant evidence*: Evidence relevant to a level-1 estimation problem (e.g. target recognition or tracking) can be assumed to be contained within a small spatio-temporal volume, generally limited by kinematic or thermodynamic constraints. In contrast, many situation and threat assessment problems can involve evidence that is wide-spread in space and time, with no readily defined constraints;
2. *Weak ontological constraints on relevant evidence*: The types of evidence relevant to threat assessment problems can be very diverse and can contribute to inferences in unexpected ways. This is why much of intelligence analysis – like detective work – is opportunistic, *ad hoc* and difficult to codify in a systematic methodology. Rather, the methodology in threat assessment is second-order: not to discover instantiations of pre-scripted threat scenarios, but (i) to discover patterns of

---

- Level 0: Signal/Feature Assessment – estimation and prediction of signal or feature states;
- Level 1: Entity Assessment – estimation and prediction of entity attributive states (i.e. of entities considered as individuals);
- Level 2: Situation Assessment – estimation and prediction of the structures of parts of reality (i.e. of relations among entities & their effects on entities);
- Level 3: Impact Assessment – estimation and prediction of the utility/cost of outcome states;
- Level 4: Performance Assessment – estimation and prediction of a system's performance.

Under this proposed scheme, a set of Resource Management levels, formal duals of the Data Fusion levels, encompasses such functions as signal/signature management, individual resource management, coordinated resource management, goal management and system engineering.

information from which hypotheses concerning potential threat situations and can be constructed, including threat situations that have not been previously experienced or anticipated and (ii) to nominate searches for data that could either confirm or refute such hypotheses;
3. *Poorly-modeled causality*: Threat assessment often involves inference of human intent and of human behavior. Such inference is basic not only to predicting future events (e.g. attack indications and warning) but also in understanding past and current activity. Needless to say, our models of human intent, planning and execution are far less complete and far more fragile than the physical models used in target recognition or target tracking.

In the present discussion, we are particularly concerned with *intentional* threats, rather than such unintentional threats as from natural disasters. That is to say, we are particularly interested in predicting, detecting and characterizing cases in which an agent intends to do harm to something or somebody. This, of course, does not assume that intended threat actions necessarily occur as intended or at all, or that they yield the intended consequences.

Furthermore, we need to recognize threat assessment as part of the more general problems of assessing intentionality and of characterizing, recognizing and predicting intentional acts in general.

Because of these three factors and this focus, the threat assessment process will need to be adaptive and opportunistic. That is to say, it cannot expect to have a pre-defined set of threat scenarios available for a simple pattern-matching recognition algorithm (e.g. using graph-matching). Rather, situation and event hypotheses will need to be generated, evaluated and refined as the understanding of the situation evolves.

2 Threat Assessment System Architecture

The proposed threat assessment architecture (Figure 1) involves the following processes:

1. Data Collection: sensor management and data mining to obtain reports of real-world entities, relationships and events. Reports can be considered to have the form of "Infons" $(r,x_1,...,x_n,h,k,p)$, for attributes/relations r, entities x_i, locations h, times k and probabilities p;[2]
2. Hypothesis Generation: building candidate threat hypothesis consistent with the available data. Such hypotheses will be instantiations of situations and events per the threat ontology (described on Section 3);[3]

---

[2] Readers familiar with Situation Logic as developed by Barwise and Perry [5], Devlin [6], *et al*, will note our expansion to a probabilistic model in place of the original bipolar truth-value model. This extension allows consistent representation of factual, conditional, hypothetical and estimated information.

[3] Hypotheses can be represented as sets of Infons $(r,x_1,...,x_n,h,k,p)$. Level 1 data fusion hypotheses (e.g. those concerning target identification and tracking) typically involve estimating or predicting 1-place relationships (or, possibly, n-place relations with n-1 bound or parameterized variables). Level 2 and 3 fusion hypotheses generally concern multi-place relationships.

3. Hypothesis Evaluation: ascribing a likelihood score to each candidate hypothesis and requesting additional data expected to either support or refute the hypothesis;
4. Hypothesis Selection: selecting among candidate hypotheses on the basis of global likelihood;
5. Alerting: Providing indications of current and predicted threat situations and threat events for Event Prediction, Indications & Warning, Threat characterization, Attack Assessment and Consequence Assessment;
6. Model Management: building and refining threat models. Model management is typically performed as an off-line task, involving abductive and inductive processes for pattern explanation and generalization.

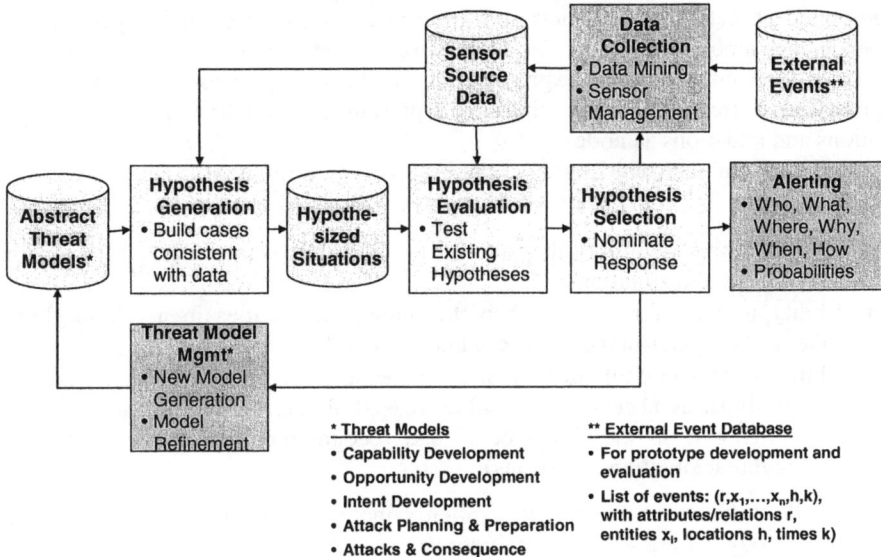

Fig. 1. Threat assessment functional flow. The adaptive process is shown by which hypotheses concerning threat situations and threatened events are successively generated, evaluated and selected for response or for further refinement. A secondary feed-back loop refines the models whereby threat situations and events are recognized or predicted

In general, Hypothesis Generation is expected to be the most complex and difficult of the processes in on-line threat analysis, because of the above-listed three factors:

1. Weak spatio-temporal constraints on relevant evidence;
2. Weak ontological constraints on relevant evidence;
3. Dominance of relatively poorly-modeled causal processes (specifically, human and group behavior vice simple physical processes).

This is a major contrast with level 1 data fusion, in which Hypothesis Generation is relatively straight-forward. It is expected that threat assessment in general – and particularly threat hypothesis generation – will remain an intensely manual process for some time. We anticipate that the present study will contribute to the development of automated decision aids for this purpose and, eventually, to full automation of more and more of the threat hypothesis generation process.

Adaptive data collection – seeking evidence to support or refute threat hypothesis – will also evolve to greater levels of automation, largely driven by anticipated advances in data mining and collection management.

3 Ontology of Entities, Situations and Inferences

Our method of representing threat situations and threat events expands on the common method of representing situations via first-order acyclic directional graphs (ADGs) and recognizing situations by graph matching. In such methods, situations are represented as networks in which nodes represent entity states and edges represent dependencies among the states. The most common approach uses Bayesian networks, in which dependencies are expressed ad conditional probabilities. The present approach also uses ADGs, but explicitly represents n-place first- and second-order relations and situations as nodes.

Based on Combinatorial Logic (Curry-Feys [7]) and Situation Logic (Barwise and Perry [5], Devlin [6], *et al*), the method provides the added benefits of

- Explicit representation of the roles of multiple entities in complex relationships, interactions and situations;
- Ability to reason in cases when the number of entities in a relationship or situation is indeterminate, as is common in intelligence threat analysis;
- Ability to reason explicitly about attributes and relationships of attributes and relationships, as necessary for inductive and abductive machine learning (e.g. generalizing from individual cases and recognizing higher-order similarities among threat situations or events).

We follow the above sources in explicitly including attributes and relationships into our ontology. Doing so is at least a convenience, allowing us to reason about such abstractions without the definitional baggage or extensional issues in reductionist formulations; e.g. in which uncertainties in the truth of a proposition involving multiple entities are represented as distributions of multi-target states $X = \{x_1,...,x_n\}$.

It is additionally difficult to argue that a multi-target state of the sort of interest in Situation/Threat Assessment can always be inferred from a set of single-target states $X = \{x_1,...,x_n\}$. It does appear feasible, however, to restrict our ontology of relationships to those of finite and determinate order; i.e. any given relation maps from a space of specific finite dimensionality $R:X^{(n)} \to Y$.

A Threatening Situation is one in which there is some likelihood of certain types of potential actions (e.g. attacks) by some agent (which may or may not be a person or human agency) against threatened entities (which often include people or their possessions). Indicators of Threatening Situations are the Capability, Opportunity and Intent of agents to carry out such actions.

Figure 2 shows the elements of a representative threat situation hypothesis and some important inference paths. The three sub-elements of the threat situation – an entity's capability, opportunity and intent to affect one or more entities – are each decomposed into sub-elements. The Threat Assessment process (a) generates, evaluates and selects hypotheses concerning entities' capability, intent and opportunity to carry out an attack and (b) provides indications, warnings and characterizations of attacks that occur.

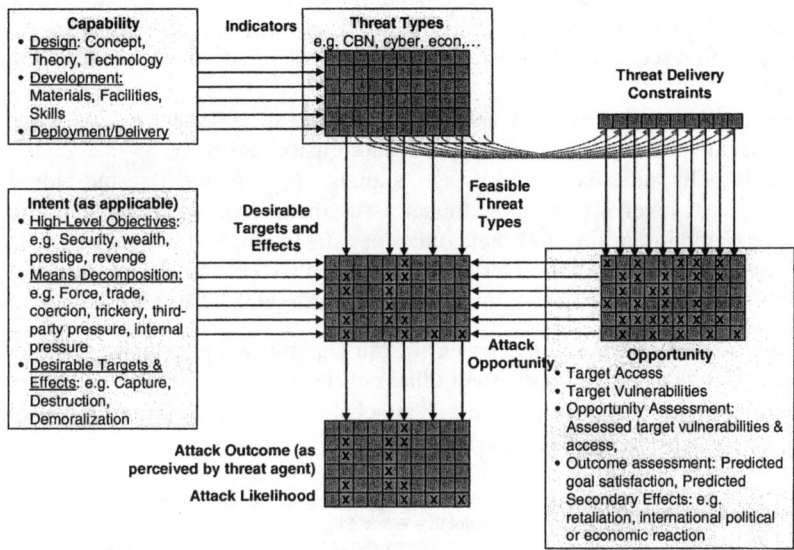

Fig. 2. Elements of a Threat Situation Hypothesis, in which the three principle components – capability, opportunity and intent – are decomposed into elements that can be subject of information discovery or inference

The threat entity's *capability* to carry out a specific type of threat activity depends on its capability to design, develop and deploy or deliver the resources (e.g. weapon system) used in that activity. The hypothesis generation process searches for indicators of such capability and generates a list of feasible threat types.

Intent is inferred by means of goals decomposition, based on decision models for individual agents and for agent conclaves.

The postulated threat type constrains a threat entity's *opportunity* to carry out an attack against particular targets (e.g. to deploy or deliver a type of weapon as needed to be effective against the target). Other constraints are determined by the target's accessibility and vulnerability, and by the threat entity's assessment of opportunities and outcomes.

Capability, opportunity and intent all figure in inferring attack-target pairing. The Threat Assessment Process evaluates threat situation hypothesis on the basis of likelihood, in terms of the threat entity's *expected perceived net pay-off*. Be it noted that this is an estimation, not of the actual outcome of a postulated attack, but an estimation of the *threat entity's estimation* of the outcome.

The system's ontology provides a basis for inferencing; capturing useful relationships among entities that can be recognized and refined by the intelligence analysis process. An ontology can capture a diversity of inference bases, to include logical, semantic, physical, and conventional contingencies (e.g. societal legal or customary contingencies). The represented relationships can be conditional and/or probabilistic. Such a representational framework is necessary for achieving interoperability and shared understanding; especially for higher-level inferencing (as in situation and threat assessment) as well as for inferencing across fusion levels (relating signals and features to individuals, individuals to situations and situations to potential outcomes). Using the IDEF5 method, entities are distinguished as Kinds, Individuals, Referents, First-Order and Second-Order Attributes and Relations, and Processes.

This ontology will permit an inferencing engine to generate, evaluate and select hypotheses at various fusion levels. In data fusion level 0, 1 and 2, these are, respectively, hypotheses concerning signals (or features), individuals and relationships. A level 3 (threat or impact assessment) hypothesis concerns potential interactions among entities and their outcomes. In Intentional Threat Assessment we are, of course, concerned with interactions in which adverse outcomes are intended.

Figure 3 shows the high-level Intentional Threat model. Threat hypotheses include

- inference of *threat situations*; i.e. of the capability, opportunities and intent of one entity x to (adversely) affect other entities y;[4]
- prediction of *threatened events* ("attacks"); i.e. interactions whereby entities intend to adversely affect others.

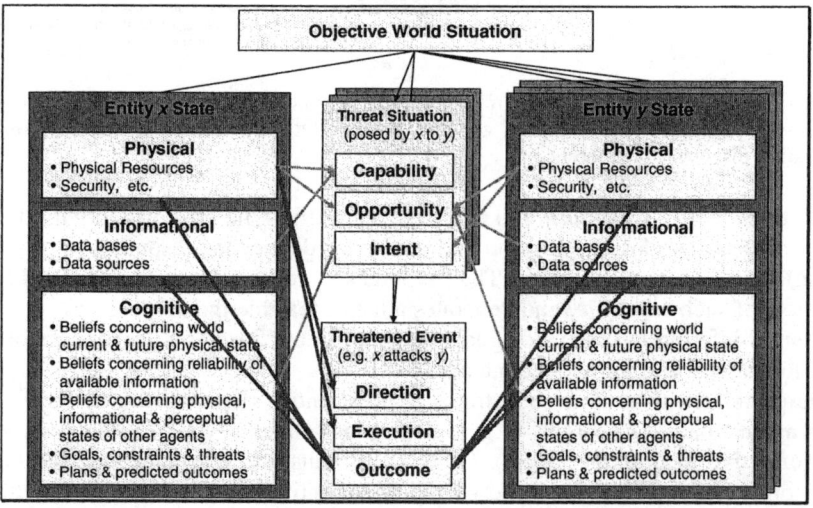

Fig. 3. Generic model of an intentional threat, including threatening entity (x), threatened entities (y), threat situation and threatened event. *Colored arrows* indicate principal factors that influence the states of situations events or the involved entities. These, then, become key inference paths in estimating or predicting the states of such elements

---

[4] For simplicity of exposition, we will ignore cases of self-threat; where $x=y$. Also, given our focus on intentional threats, in the subsequent discussion, we assume the presence of agents who intend to do harm to somebody or something. Once again, this does not assume that intended threat actions necessarily occur as intended or with the intended consequences.

A threat situation is one which Interactions of particular types are likely. These interactions (loosely "attacks") can be characterized in terms of their Direction (i.e. command and control), Execution (i.e. actual, as opposed to planned or commanded, actions) and Outcome (which, of course, may vary from intended outcomes).

Threat situations and threatened events are inferred on the basis of the attributes and relationships of the entities involved. As indicated by green arrows in the figure, estimates of physical, informational and perceptual states of such entities are used to infer actualized and potential relationships among entities; and specifically to infer threat situations, in terms of one entity's capability, opportunity and intent to interact with (usually other) entities. The red arrows indicate information used to characterize a predicted or occurring threat event, in terms of its direction (i.e. planning, command and control), execution and outcomes.

4 Formal Foundations

Following Devlin [6], we define a situation as a structured part of reality that is discriminated by some agent. Depending on the way information is used; virtually any entity may be treated either as an individual or as a situation. For example, an automobile may be discussed and reasoned about as a single individual or as an assembly of constituent parts. The differentiation of parts is also subject to choices: we may disassemble the automobile into a handful of major assemblies (engine, frame, body ...) or into a large number of widgets or into a huge assembly of atoms.

Accordingly, the number of entities in a situation can be undecided. That is to say, the same situation can have an indeterminate number of entities, depending on the interests and focus of attention of agents reasoning about or experiencing the situation.

As illustrated in Figure 4, this broad definition permits a uniform method of inference across such diverse problems as model-based target recognition, computer vision and scene understanding, force structure analysis, course-of-action analysis, situation dynamic analysis and threat/impact assessment.

Situation or threat assessment – whether implemented by people, an automatic process or some combination thereof – requires the capability to make inferences of the following types:

- Determining criteria for recognizing relationships, to include logical, semantic, causal and conventional (e.g. moral, legal, cultural) expectations. This requires a validated, comprehensive ontology; but one in which the uncertainties in the inference are captured (e.g. in terms of conditional probabilities);
- Determining criteria for recognizing situations. This requires some formal method of situation semantics and situation logic;
- Determining criteria for contextually-conditioned estimation of target states and relationships. This requires some form of inferential calculus, in which uncertainties in the ontology, in sensor/source reports and in the inference process are systematically represented and manipulated.

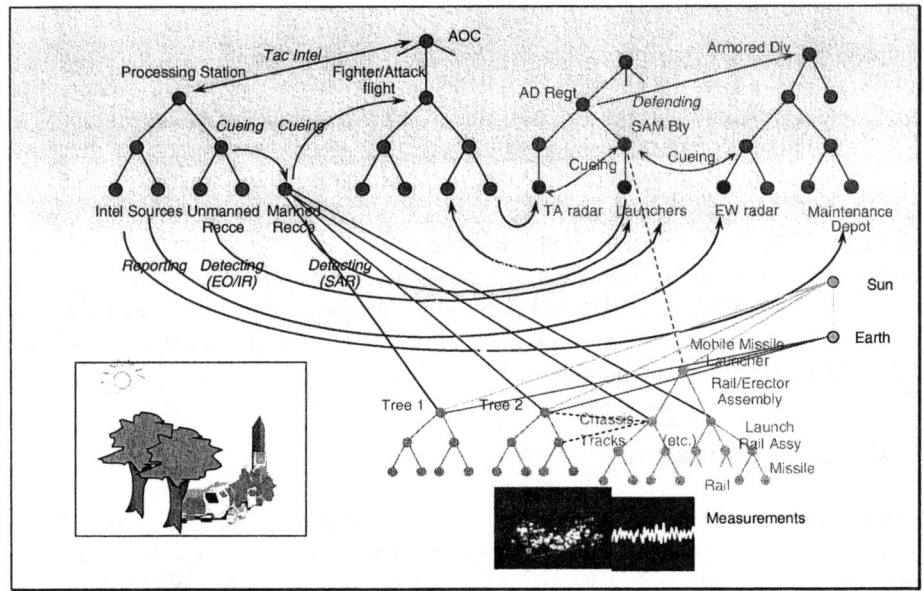

Fig. 4. Example of a situation/threat hypothesis, in which individual sensor measurements (*lower right*) and extracted scene features (*lower left*) are analyzed and extended to infer diverse physical, institutional, semantic and other relationships, allowing further inferencing of the states or related entities. Relationships among elements of a threat organization (*red nodes*), threatened entities and their defenders (*blue nodes*) are depicted by edges. Also components of selected entities (*orange and green nodes*) and their relationship to other scene (*yellow nodes*) entities are depicted

Broadly speaking, relationships to be inferred and exploited in situation assessment can include:

- Logical/Semantic (e.g. definitional, analytic, taxonomic)
- Physical (e.g. spatio-temporal, causal, nomic)
- Functional (e.g. structural or organizational role)
- Conventional (e.g. ownerships, legal and other societal conventions)
- Cognitive (e.g. sensing, perceiving, believing, fearing)

Relationships of particular interest in threat assessment include the following types:

- Relationships among objects in a threat organization (organization role/subordination, communications, type similarity, deployment, kinetic interaction, financial and information exchange, etc.);
- Relationships among blue defending and defended assets (sensor spatio-temporal, measurement calibration, confidence, communication, coordination, etc.);
- Relationships between sensors and sensed entities (intervisibility, assignment/ cueing, sensing, data association, countermeasures);

- Relationships between red and blue entities (detecting, tracking, targeting, etc.);
- Relationships between entities of interest and other entities (terrain features; solar & atmospheric effects; weapon launch and impact points; etc.).

It should be noted that most relationships are not directly observed, but inferred from the (attributive) states of entities, their context and from other relationships. In a related application, indirect inference of this sort is essential to model-based automatic target recognition (ATR), in which the spatial and spectral relationships among components are inferred and these, in turn, used to infer the state of the constituted entity. Scene understanding extends such analysis to infer such relationships as occlusion, illumination, shadowing, and causal relations such as radiative heating and terrain disturbance. These can be extended to infer functional relations among components (e.g. "the left tank tread is moved by this drive wheel") and organizational relations ("these tanks are following the lead of that one"). Additional relationships can include informational and perceptual states (x detects y, x is tracking y, x perceives the state of y to be z, x predicts at time t_1 that y will transition to state z at time t_2, etc.).

Other applications in which relationships are inferred from attributive entity states are in target state transition prediction and tracking (e.g. using Kalman filters or Markov random fields), force structure analysis, link and network analysis, etc.

As in model-based ATR, inferring relationships presumes a model, having a form such as those shown in the figure. Here dependencies among relational and attributive states are expressed by conditional probabilities. Building and managing such predictive models can be relatively straightforward, as in model-based ATR applications (in which the types of relations are few and well understood, deriving from the mechanical design of targets of various types – here there problem is usually not in the fidelity of the models but in the size of the ambiguity set – i.e. in the number and similarity of competing models). At the other extreme are the relatively poorly modeled domains of human social activity. This, however, is never really random, only very complex. The problem is in encoding the structure in the variability of the human-dominated problems that are of most interest in Situation Assessment in general and most particularly in Threat Assessment.

As in Level 1 inferences (i.e. with 1-place predicates), we can write production rules based on logical, semantic, causal, or material (etc.) relationships among predicates of any length.

Written in terms of conditional probabilities, with density functions f, n-place predicates $P^{(n)}, Q^{(n)}, R^{(n)}$, and situations S), characteristic inference patterns include the following:

- Situational Inferences
  $P^{(n)}(X^{(n)}) \Rightarrow S$

- Level 1 Inferences
  $f(P^{(1)}(x)|Q^{(1)}(x),S)$
  (e.g. single target likelihood functions or Markov transition densities)

- Level 2 Inference Examples
  $f(P^{(2)}(x,y)|Q^{(1)}(x),R^{(1)}(y),S)$ (Level 1→2 deduction)
  $f(\exists y[P^{(2)}(x,y)]|Q^{(1)}(x),S)$ (Level 1→2 induction)

$f(P^{(1)}(x)] | Q^{(2)}(x,y), S)$ (Level 2 →1 deduction)
$f(P^{(2)}(x,y)] | Q^{(2)}(x,y), S)$ (Level 2 →2 deduction).[5]

Some situations can be crisply defined; e.g., a chess game, of which the constituent entities and their relevant attributes and relationships are explicitly bounded. Other situations may have fuzzy boundaries. Fuzziness is present both in situation types (e.g. the concepts economic recession or naval battle) and of real-world situations (e.g. the 1930s, the Battle of Trafalgar). Both can naturally be characterized via fuzzy membership functions. A fuzzy membership function f can be equated to a continuous distribution on infon polarities: If $\sigma = (R, x_1, ..., x_n, h, k, p)$ and $s \models \sigma$, then $f_s(\sigma) = p$.

References

1. White, F.: A Model for Data Fusion. In: Proceedings of the First National Symposium on Sensor Fusion, vol. 2. GACIAC, IIT Research Institute, Chicago (1988), 143-158
2. Steinberg, A., Bowman, C., White, F.: Revisions to the JDL data fusion model. In: Joint NATO/IRIS Conference Proceedings, Quebec (1998). Reprinted in: Sensor Fusion: Architectures, Algorithms, and Applications, Proceedings of the SPIE, Vol. 3719 (1999). Reprinted in: Hall, D., Llinas, J. (eds.) Handbook of Multisensor Data Fusion. CRC Press, London (2001), 2.1-2.19
3. Steinberg, A., Bowman, C.: Rethinking the JDL data fusion model. In: Proceedings of the 2004 MSS National Symposium on Sensor and Data Fusion (2004), vol 1
4. Llinas, J., Bowman, C., Rogova, G., Steinberg, A., Waltz, E., White, F.: Revisiting the JDL Data Fusion Model II. In: Proceedings, Seventh International Conference on Information Fusion, Stockholm (2004) 1218-1230
5. Barwise, J., Perry, J: Situations and Attitudes. Bradford Books, MIT Press (1983)
6. Devlin, K.: Logic and Information. Press Syndicate of the University of Cambridge (1991)
7. Curry, H., Feys, R.: Combinatory Logic, volume 1, North-Holland Publishing Company, Amsterdam (1974)

---

[5] Level 2 →2 deduction can include, for example, estimating multi-target likelihood functions or multi-target Markov transition densities.

Making Contextual Intensional Logic Nonmonotonic

Richmond H. Thomason

Philosophy Department, University of Michigan,
Ann Arbor, MI 48109-2110, USA

Abstract. This paper motivates and presents a nonmonotonic version of Contextual Intensional Logic, a type-theoretic logic intended as a general formalism for reasoning about context. In developing this logic, it is necessary to think through interactions between nonmonotonic and intensional logic that are interesting in their own right. The paper concludes with an extended example how nonmonotonic lifting rules can be deployed in inter-contextual reasoning.

1 Introduction

For several years, I have advocated a type-theoretic approach to the logical formalization of context. A presentation of Contextual Intensional Logic (CIL), is available in [Thomason, 2003], along with references to the earlier work.

It has often been claimed that "lifting rules"—rules that transfer information from one context to another—need, in general, to be nonmonotonic. See, for instance, [McCarthy, 1993, Guha, 1991]. Examples of fully developed nonmonotonic logics of context are hard to find, perhaps because the logical issues are complex, and because apparently we do not yet have techniques for implementing a nonmonotonic logic of context efficiently.

Nevertheless, examples of cross-contextual information transfer exhibit the properties that provide the best sort of motivation for a nonmonotonic logic; in particular, they provide examples of the qualification problem. When problems of this sort arise in a domain, this provides convincing evidence that it will be difficult or impossible to create large-scale formalizations of the domain without using a nonmonotonic logic.[1]

Suppose, for instance, that we are dealing with a domain in which communicating processors are distributed throughout a building, some of them equipped with sensors of various sorts. Each processor can be identified with a context in this application. When information from sensors is imported from one processor to another, reliability is always a consideration; and a nonmonotonic logic is a very natural way to formalize the role of reliability. We could, for instance, begin with a rule

---

[1] This, in effect, is the argument in [Guha, 1991] for the use of nonmonotonic lifting rules in CYC.

(1.1) $\forall c \forall c' \forall p[[\neg Ab1(c,c',p) \wedge [c]p] \rightarrow [c']p)]$

saying that if processor c is not abnormal for c' with respect to the proposition p, then p can be imported from c to c'. (Here, [c] and [c'] are modalities recording the beliefs of the two contexts.) Nonmonotonicity is provided by a semantics that minimizes the extension of abnormality predicates, including $Ab1$.

An abnormality theory can then provide conditions under which instances of $Ab1$ may fail. A rule might say, for instance, that if c is scheduled for repair then all propositions believed by c are abnormal:

(1.2) $\forall c \forall c' \forall p[[\mathit{ScheduledForRepair}(c) \wedge [c]p \wedge \neg c = c'] \rightarrow Ab1(c,c',p)]$

If we want to provide for the possibility that some exceptional scheduled repairs for c will not indicate that transmissions from c are unreliable, we would make Rule (1.2) nonmonotonic, and add the further qualifications to the abnormality theory.

We could elaborate the example by considering how a processor without a temperature sensor could import indirect evidence about its local temperature. We might begin with a default to the effect that temperature can be imported from a processor in the same room. This might need to be qualified if, for instance, one processor is near a heater and the other is not.

These examples have the open-ended flavor that characterizes instances of the qualification problem, and provide a strong motivation for developing a logic of context that supports nonmonotonic lifting rules, even if workable implementations of these logics are not immediately available.

2 Motivation and Syntax of CIL

With this section, I'll begin a brief review the formalism of Contextual Intensional Logic. In this paper, I will refer to the static version of the logic only. There are interactions between nonmonotonicity and dynamic operators,[2] and these are sufficiently complicated to be beyond the scope of this paper.

CIL is motivated by the desire to base a logic of context on type-theoretic logics modeled after Church's formalization of higher-order logic in [Church, 1940]. Since there is a modal element to the logic of context, the immediate ancestor of CIL is Montague's Intensional Logic, [Montague, 1970, Gallin, 1975].

The types of CIL are defined as follows.

1. e is a type.
2. t is a type.
3. i is a type.
4. If σ and τ are types, so are $\langle \sigma, \tau \rangle$ and $\langle w, \tau \rangle$.

(2.1) **CIL Types**

---

[2] In particular, context-changing dynamic operators need to be made nonmonotonic.

whose interpretation depends on context. The type $\langle w, \tau \rangle$ represents functions from possible worlds to objects of type τ.

A CIL language $\mathcal{L}$ is a function from the set of CIL types to nonempty sets of expressions; where τ is a type, $\mathcal{L}(\tau)$ is the set of basic constants of type τ. For each type τ, there is a denumerable set $Var(\tau)$ of variables of type τ. $\mathcal{L}(\tau) \cup Var(\tau)$ is the set of basic expressions of type τ. We require that if $\sigma \neq \tau$ then $\mathcal{L}(\sigma)$ and $\mathcal{L}(\tau)$ are disjoint.

The following definition extends $\mathcal{L}$ to a function $\mathcal{L}^*$ taking each type into the set of (basic or complex) expressions that type.

Basic expressions: $\mathcal{L}(\tau) \cup Var(\tau) \subseteq L^*(\tau)$.

Identity: If ζ and $\xi \in \mathcal{L}^*(\tau)$, then $\zeta = \xi \in L^*(t)$.

Functional application: If $\zeta \in \mathcal{L}^*(\langle \sigma, \tau \rangle)$ and $\xi \in \mathcal{L}^*(\sigma)$, then $\zeta(\xi) \in \mathcal{L}^*(\tau)$.

Lambda abstraction: If $\zeta \in \mathcal{L}^*(\tau)$, then $\lambda x_\sigma \zeta \in \mathcal{L}^*(\langle \sigma, \tau \rangle)$.

Intension: If $\zeta \in \mathcal{L}^*(\tau)$, then $^\wedge \zeta \in \mathcal{L}^*(\langle w, \tau \rangle)$.

Extension: If $\zeta \in \mathcal{L}^*(\langle w, \tau \rangle)$, then $^\vee \zeta \in \mathcal{L}^*(\tau)$.

Character formation: If $\zeta \in \mathcal{L}^*(\tau)$ then $^\cap \zeta \in \mathcal{L}^*(\langle i, \tau \rangle)$.

Content determination: If $\zeta \in \mathcal{L}^*(\langle i, \tau \rangle)$ then $^\cup \zeta \in \mathcal{L}^*(\tau)$.

3 Domains and Models of CIL

3.1 Domains

Let D be a function taking the basic types e, w, and i into nonempty sets, or *domains*, $D(e)$, $D(w)$, and $D(i)$. D is extended as follows to a function assigning domains to arbitrary types. The values $\top$ and $\bot$ are arbitrary different individuals, taken to stand for truth and falsity; these two elements belong only to the domain $D(t)$.

(3.1.1) $D(t) = \{\top, \bot\}$.
(3.1.2) $D(\langle \sigma, \tau \rangle) = D(\tau)^{D(\sigma)}$.

3.2 Models of CIL

We continue the review of CIL with a sketch of the semantics.

A (static) model $\mathcal{M}$ of $\mathcal{L}$ on a domain function D is an assignment of a member $[\![x_\tau]\!]_\mathcal{M}$ of $D(\tau)$ to each variable x_τ in $Var(\tau)$, and of a member $[\![\xi]\!]_{\mathcal{M},i,w}$ of $D(\tau)$ to each basic constant in $\mathcal{L}(\tau)$, for each $i \in D(i)$ and $w \in D(w)$.

Models assign a value $[\![\xi]\!]_{\mathcal{M},i,w} \in \tau$ to each expression $\xi \in \mathcal{L}^*(\tau)$, for each $i \in D(i)$ and $w \in D(w)$, subject to the following rules of semantic interpretation.

 0. For $x_\tau \in Var(\tau)$, $[\![x_\tau]\!]_{\mathcal{M},i,w} = [\![x_\tau]\!]_\mathcal{M}$.
 1. $[\![\zeta = \xi]\!]_{\mathcal{M},w} = \top$ if $[\![\zeta]\!]_{\mathcal{M},i,w} = [\![\xi]\!]_{\mathcal{M},i,w}$. Otherwise $[\![\zeta = \xi]\!]_{\mathcal{M},w} = \bot$.
 2. $[\![\zeta(\xi)]\!]_{\mathcal{M},i,w} = [\![\zeta]\!]_{\mathcal{M},i,w}([\![(\xi)]\!]_{\mathcal{M},i,w})$.

3. Where $\zeta \in \mathcal{L}^*(\tau)$ and $x_\sigma \in Var(\sigma)$, $[\![\lambda x_\sigma \zeta]\!]_{\mathcal{M},\text{i},\text{w}} =$ the function f from $D(\sigma)$ to $D(\tau)$ such that $f(d) = [\![\zeta]\!]_{\mathcal{M}^{d/x},\text{i},\text{w}}$.
4. Where $\zeta \in \mathcal{L}^*(\tau)$, $[\![^\wedge\zeta]\!]_{\mathcal{M},\text{i},\text{w}} =$ the function f from $D(w)$ to $D(\tau)$ such that $f(w') = [\![\zeta]\!]_{\mathcal{M},\text{i},w'}$.
5. Where $\zeta \in \mathcal{L}^*(\langle w, \tau\rangle)$, $[\![^\vee\zeta]\!]_{\mathcal{M},\text{i},\text{w}} = [\![\zeta]\!]_{\mathcal{M},\text{i},\text{w}}(w)$.
6. Where $\zeta \in \mathcal{L}^*(\tau)$, $[\![^\cap\zeta]\!]_{\mathcal{M},\text{i},\text{w}} =$ the function f from $D(i)$ to $D(\tau)$ such that $f(i') = [\![\zeta]\!]_{\mathcal{M},i',\text{w}}$.
7. Where $\zeta \in \mathcal{L}^*(\langle i, \tau\rangle)$, $[\![^\cup\zeta]\!]_{\mathcal{M},\text{i},\text{w}} = [\![\zeta]\!]_{\mathcal{M},\text{i},\text{w}}(i)$.

(3.2.1) **Semantic Rules for CIL**

In CIL, contexts are split into two components: (1) a modal component, which has type $\tau_M = \langle\langle w, t\rangle, \langle w, t\rangle\rangle$ and (2) an indexical component, which has type i. The first component is a function that inputs a proposition and outputs a proposition; given a proposition p, this function returns the proposition that p is known in the context. The second component is an index. Indices represent decontextualization policies; think of an index as providing a decontextualized value for each context-sensitive lexical item in the logic's vocabulary.

4 Interactions Between Modality and Nonmonotonicity

As far as I know, there are no extended studies in the literature of the sort of nonmonotonicity that we will need. In epistemic approaches to nonmonotonicity, e.g. in autoepistemic logic, modal logics provide a basis for nonmonotonicity. But this is not the interaction between nonmonotonicity and modality that we are interested in here, where we want to have a mechanism for nonmonotonic reasoning that we can apply to epistemic operators.

Two papers, [Lin, 1988] and [Thomason, 2000], discuss nonmonotonic reasoning about epistemic operators. Like the present paper, both of these papers use a circumscriptive framework. But these papers consider special cases that from the present standpoint are too restricted to be of much use; [Lin, 1988], inspired by [Shoham, 1988], considers the case in which a single agent's ignorance is maximized, and [Thomason, 2000] concentrates on how to achieve something like mutuality by using defaults. In discussing how to design a nonmonotonic modal logic for reasoning about intercontextual information exchanges, we will have to make a fresh start.

Let's concentrate for the moment on a simple IL language $\mathcal{L}_1$. $\mathcal{L}_1$ is like a propositional modal logic with one modal operator, [B] (belief). But it also has higher-order formulas, such as $\forall p_{\langle w,t\rangle}[[B](p) \to {}^\vee p]$, which do not belong to propositional modal logic.

We will formalize belief as a simple predicate of propositions; this means that [B] will have type $\langle\langle w,t\rangle, t\rangle$. To define the vocabulary of $\mathcal{L}_1$, we make the following declarations.

$\mathcal{L}_1(\langle e\rangle) = \{a\}$
$\mathcal{L}_1(\langle e,t\rangle) = \{Ab\}$

$\mathcal{L}_1(\langle\langle w,t\rangle,t\rangle) = \{\text{[B]}\}$
$\mathcal{L}_1(\tau) = \emptyset$ for all τ other than e, $\langle e,t\rangle$, and $\langle\langle w,t\rangle,\langle w,t\rangle\rangle$

That is, $\mathcal{L}_1$ has one name a of an individual, one 1-place predicate Ab, and one modal operator. We ignore contextual effects for the moment, by working with a domain function D that only provides one index: $D(i) = \{i_0\}$.

We will adopt a preferential semantics approach to the semantics of nonmonotonicity. In the case of $\mathcal{L}_1$, this means at least that a model $\mathcal{M}$ is preferred to a model $\mathcal{M}'$ if the extension that $\mathcal{M}$ gives to Ab is a proper subset of the extension that $\mathcal{M}'$ gives to Ab. But $\mathcal{M}$ assigns an extension to Ab in each world in $\mathcal{M}(w)$—which world is this?

There is something to reflect about here: familiar approaches to circumscription minimize the extensions of abnormality predicates, where these extensions are conceived as sets of tuples of individuals. In generalizing to the intensional case, we need to consider what to do about the dependence of these extensions on worlds.

We could reproduce the familiar approaches by fixing the world argument. We associate a designated world (the "actual world") $\mathcal{M}_w$ with each model $\mathcal{M}$. And we prefer a model $\mathcal{M}$ of $\mathcal{L}_1$ to $\mathcal{M}'$ if

$$\{d \,/\, [f(d)](w) = \top\} \subset \{d \,/\, [f'(d)](w) = \top\},$$

where f is $[\![Ab]\!]_{\mathcal{M},i_0,w}$ (the function associated with Ab by $\mathcal{M}$ at $\langle i,w\rangle$) and f' is $[\![Ab]\!]_{\mathcal{M}',i_0,w}$.

This simple solution does incorporate nonmonotonicity into the logic, but does this in a highly local way that precludes appropriate interactions with modal operators. Consider, for instance, the fact that this approach confines defaults to just one world, the actual world. Therefore, this approach would allow no natural way for the logic to impose a constraint to the effect that all defaults are believed, without making believers omniscient.

Let's look at this in more detail. In $\mathcal{L}_1$, the constraint would amount to

(4.1) $\forall p_{\langle w,t\rangle}[\Box^\wedge[\neg Ab(a) \to {}^\vee p] \to \text{[B]}(p)]$.

Here, $\Box$ is the absolute necessity operator (which takes a proposition p into the proposition that p is true in all worlds). $\Box$ can be defined in IL: in fact, $\Box$ is $\lambda p_{\langle w,t\rangle}{}^\wedge[p = {}^\wedge[a = a]]$.[3] Now, take a model $\mathcal{M}$ on D, where $D(e)$ consists of just one individual; $D(e) = \{d\}$. Let a denote d. Let $Ab(a)$ be false in $\mathcal{M}$ in the actual world w_0 but true in the all other worlds; we can assume that $\mathcal{M}$ is maximally preferred, since preference is determined only by the behavior of Ab in the actual world. Suppose that [B] is a DS5[4] modal operator, and denotes

---

[4] See any textbook on modal logic for information about deontic S5. For instance, see [Cresswell and Hughes, 1996][p. 58], where this modal logic is called "K+E." Or see [Fagin et al., 1995][pp. 57–61], where the logic is called "KD45."

[3] In displaying formulas of CIL and NMCIL, we follow the practice of flagging only the first occurrence of a variable with its type.

nonomniscient belief. Where [B] denotes in $\mathcal{M}$ the function f,[5] this means (1) that there is a function K from $D(w)$ to subsets of $D(w)$ such that $f(g)(w) = \top$ iff $g(w') = \top$ for all $w' \in K(w)$, and (2) that $w_1 \in K(w_0)$ for some $w_1 \neq w_0$.

Now, define a proposition p in this model as follows: $p(w) = \top$ iff $w \neq w_0$. Then $f(p)(w_0) = \bot$, because $p(w_1) = \bot$ and $w_1 \in K(w_0)$. But, on the other hand, $\Box(\char`\^[\neg Ab(a) \to p])$ is true in $\mathcal{M}$, where p is assigned the proposition p. So (4.1) is falsified. This example shows that (4.1) is invalidated by natural minimal models, when minimization is localized to the actual world, unless the agent is omniscient.

Similarly, if we were to introduce a counterfactual conditional $>$, along the lines of [Stalnaker, 1968] or [Lewis, 1973], into a language for CIL with only one abnormality $Ab(a)$, it would be natural to require that if a formula ϕ does not entail $Ab(a)$, and $\neg Ab(a)$ is in fact true, then $\phi > \neg Ab(a)$ is also true. That is, counterfactuals should not introduce gratuitous violations of defaults. We could not obtain this result if minimization were confined to a single world.

Our solution to this problem, which we present formally in the next section, minimizes a model over a set of worlds related to the actual world w_0. We do not require that w_0 be a member of this set of worlds. In NMCIL, then, defaults are formalized as epistemic constraints; and, for instance, ϕ is not a consequence of a theory like $\{\neg Ab(a) \to \phi\}$ that intuitively yields ϕ as a default consequence, unless we confine ourselves to models in which the actual world is related to itself. I do not see this as a disadvantage; taking this view of things and adhering to it simplifies some aspects of the logic, and clears up conceptual confusions that are apt to arise in working with nonmonotonic formalisms.

Embedding nonmonotonic logic in Intensional Logic in this way also solves a minor problem that afflicts extensional formalizations of circumscription. It is natural to think of propositions as true by default, but circumscription doesn't apply to sentences (IL expressions of type t) as such, but only applies indirectly to sentences that involve a predicate and arguments. It is hard to see how a sentence like 'It is not snowing' could be true by default on the circumscriptive approach.

The disadvantage to this approach is that it introduces the need to characterize the relation that chooses the worlds to be minimized. In practical formalization, it is more difficult to formalize worlds and relations over them than familiar individuals and relations over them. This is certainly an added complication, but I don't think it is an insuperable difficulty. In cases where formalizing the worlds explicitly is problematic, all worlds can be minimized. In cases where something can actually be said about worlds, I think the approach I describe here may actually be superior.

Some readers may wonder why I globalize circumscription with respect to worlds but not to indices. The main reason for this is merely the desire to keep things as simple as possible. There are strong analogies between the two cases, and probably applications can be envisaged in which it would be useful

---

[5] More precisely, where $[\![[B]]\!]_{\mathcal{M},i,w} = f$ for all $i \in D(i)$ and $w \in D(w)$.

to circumscribe over a set of closely related indices. However, there seems to be no need for this in the cases that arise most naturally.

4.1 The Language of NMCIL

The only difference between the language of CIL and NMCIL, its nonmonotonic variant, is that certain basic predicates are classified in NMCIL as abnormality predicates. We begin with a definition of the predicational types of the language; the idea is that a type denoting an n-place function to truth-values is predicational.

> *Definition of predicational types:*
> A type τ of NMCIL is n-*place predicational* if there are types $\sigma_1, \ldots, \sigma_n$, for $n \geq 0$, such that $\tau = \langle \sigma_1, \langle \ldots, \langle \sigma_n, t \rangle \ldots \rangle \rangle$. τ is *predicational* if for some $n \geq 0$, τ is n-place predicational.

Note that according to this definition, the primitive type t is predicational. We are treating sentences as implicit predicates of possible worlds.

Derivatively, an expression of NMCIL is predicational if its type is predicational, and is n-place predicational (for $n \geq 0$) if it has an n-place predicational type. An induction on n shows that if $\langle \sigma_1, \langle \ldots, \langle \sigma_n, t \rangle \ldots \rangle \rangle = \langle \sigma'_1, \langle \ldots, \langle \sigma'_n, t \rangle \ldots \rangle \rangle$ then $\sigma_i = \sigma'_i$ for $1 \leq i \leq n$. Therefore, if ζ is m-place predicational and n-place predicational then $m = n$.

A NMCIL language $\mathcal{L}$ is a pair $\langle \mathcal{L}^1, \mathcal{L}^2 \rangle$ of functions. The first function, as in CIL, takes the set of CIL types to nonempty sets of expressions. As before, where τ is a type, $\mathcal{L}^1(\tau)$ is the set of basic constants of type τ. The second function takes each type into a set $\mathcal{L}^2(\tau)$ of abnormality concepts of that type. $\mathcal{L}^2(\tau) = \emptyset$ unless the type τ is predicational. Of course, we require that $\mathcal{L}^1(\tau)$ and $\mathcal{L}^2(\tau)$ are disjoint, for all types τ.

> *Definition of $\mathcal{L}_{Ab}$:*
> $\mathcal{L}_{Ab}$, the set of abnormality predicates of $\mathcal{L}$, is $\cup \{\mathcal{L}^2(\tau) \,/\, \tau \in \textit{Types}\}$.

The set $\mathcal{L}(\tau)$ of basic expressions of type τ is $\mathcal{L}^1(\tau) \cup \mathcal{L}^2(\tau)$. From here on, the syntax of NMCIL is the same as that of CIL.

In effect, the only syntactic difference between CIL and NMCIL is that certain basic expressions of predicational type are designated as abnormality predicates in NMCIL.

4.2 Model Theory of NMCIL

A model $\mathcal{M}$ for NMCIL on a domain function D is defined as for CIL—that is, as in Section 3—except for two additions. (1) With each model $\mathcal{M}$ on the domain function D a *world of evaluation*[6] $\mathcal{M}_w \in D(w)$ is associated. (2) Also,

---

[6] I avoid the term "actual world" here because the world of evaluation does not play the same role in the interpretation of NMCIL that it does in two-dimensional modal logic.

each model $\mathcal{M}$ involves a nearness relation. This is a two-place relation $\mathcal{M}_K$ over the set $D(w)$ of worlds.

In the following series of definitions, we introduce a preference relation over NMCIL models. First we introduce a relation of congruence over models: two models are congruent if they differ only in ways that are allowed to vary in the course of circumscription.

We assume in this definition that a designated index $\mathcal{M}_i$ is associated with each model $\mathcal{M}$, as well as a designated world $\mathcal{M}_w$.

Definition of $\mathcal{M} \cong_{\mathcal{L}_V} \mathcal{M}'$:
Let $\mathcal{L} = \langle \mathcal{L}^1, \mathcal{L}^2 \rangle$ be a language for NMCIL. Let $\mathcal{L}_V$ be a set of basic vocabulary items of $\mathcal{L}$. Let $\mathcal{M}$ and $\mathcal{M}'$ be models of $\mathcal{L}$ on the same domain function D, and with the same world of evaluation, designated index, and nearness relation. That is, $\mathcal{M}_w = \mathcal{M}'_w$, $\mathcal{M}_i = \mathcal{M}'_i$, and $\mathcal{M}_K = \mathcal{M}'_K$.

Then $\mathcal{M} \cong_{\mathcal{L}_V} \mathcal{M}'$ iff $[\![\zeta]\!]_{\mathcal{M},i,w}$ and $[\![\zeta]\!]_{\mathcal{M}',i,w}$ can differ only when the index i is $\mathcal{M}_i$, the world w is such that $\mathcal{M}_w \mathcal{M}_K w$, and ζ is an abnormality predicate of $\mathcal{L}$ or is in $\mathcal{L}_V$.

I.e., $\mathcal{M} \cong_{\mathcal{L}_V} \mathcal{M}'$ iff (1) for all variables ζ, $[\![\zeta]\!]_{\mathcal{M},i_0,w} = [\![\zeta]\!]_{\mathcal{M}',i_0,w}$, where $i_0 = \mathcal{M}_i$, and (2) for all basic expressions ζ of $\mathcal{L}$ that are not variables, if $\zeta \notin \mathcal{L}_{Ab} \cup \mathcal{L}_V$ or $i \neq \mathcal{M}_i$ or not $\mathcal{M}_w \mathcal{M}_K w$ then $[\![\zeta]\!]_{\mathcal{M},i,w} = [\![\zeta]\!]_{\mathcal{M}',i,w}$.

We now define inclusion for expressions of predicational type. We regard each such expression as involving an extra argument for worlds, which is inherited from the world parameter of the satisfaction relation. This argument figures in the inclusion.

Definition of $\mathcal{M} \preceq_{w,i,\zeta} \mathcal{M}'$:
Let D be a domain function, and let $\zeta \in \mathcal{L}^2(\tau)$, where τ is an n-place predicational type: $\tau = \langle \sigma_1, \langle \ldots, \langle \sigma_n, t \rangle \rangle \ldots \rangle$. Let $w \in D(w)$, let $i \in D(i)$, and let $\mathcal{M}$ and $\mathcal{M}'$ be NMCIL models on D, with $\mathcal{M}_K = \mathcal{M}'_K$. Let $K = \mathcal{M}_K$.

Then $\mathcal{M} \preceq_{w,i,\zeta} \mathcal{M}'$ iff for all $d_1 \in D(\sigma_1)$, ..., $d_n \in D(\sigma_n)$, and all w' such that $w K w'$, if $[\![\zeta]\!]_{\mathcal{M},w',i}(d_1)\ldots(d_n) = \top$ then $[\![\zeta]\!]_{\mathcal{M}',w',i}(d_1)\ldots(d_n) = \top$.

Second, we define the preference relation over models. This corresponds to the more or less standard case of parallel circumscription in which a designated set of constants is allowed to vary along with the abnormality predicates; see, for instance, [Lifschitz, 1994]. The only difference here is that the circumscription is performed simultaneously over a set of worlds.

Definition of $\mathcal{M} \leq_{\mathcal{L}_V,i} \mathcal{M}'$:
Let $\mathcal{M}$ and $\mathcal{M}'$ be models of a NMCIL language $\mathcal{L}$ on the same domain function D, and with the same actual world and nearness relation: $\mathcal{M}_w = \mathcal{M}'_w$ and $\mathcal{M}_K = \mathcal{M}'_K$. Let $w_0 = \mathcal{M}_w$, and let $\mathcal{M} \cong_{\mathcal{L}_V} \mathcal{M}'$, where $\mathcal{L}_V$ is a set of basic vocabulary items of $\mathcal{L}$. Let $i \in D(i)$.

Then $\mathcal{M} \leq_{\mathcal{L}_V,i} \mathcal{M}'$ iff for all $\zeta \in \mathcal{L}_{Ab}$, $\mathcal{M} \preceq_{w,i,\zeta} \mathcal{M}'$ for all w such that $w_0 K w$.

Finally, nonmonotonic logical consequence is defined as consequence restricted to minimal models of the premises.

Definition of $\leq_{\mathcal{L}_V,i}$-minimality for T and of $\|\hspace{-2pt}\sim_{\mathcal{L}_V}$:
Let T be a set of formulas of a NMCIL language $\mathcal{L}$, let $\mathcal{M}$ be a model of T on a domain function D, and let $i \in D(i)$.
$\mathcal{M}$ is $\leq_{\mathcal{L}_V,i}$-minimal for T iff for all models $\mathcal{M}'$ of T such that $\mathcal{M}' \leq_{\mathcal{L}_V,i} \mathcal{M}$, $\mathcal{M} = \mathcal{M}'$.
Where $\phi \in \mathcal{L}^*(t)$, $\mathcal{M} \models \phi$ iff $[\![\phi]\!]_{\mathcal{M},i_0,w_0} = \top$, where $w_0 = \mathcal{M}_w$ and $i_0 = \mathcal{M}_i$.
$T \|\hspace{-2pt}\sim_{\mathcal{L}_V} \phi$ iff $\mathcal{M} \models \phi$ for all models $\mathcal{M}$ that are $\leq_{\mathcal{L}_V,i}$-minimal for T, where $i = \mathcal{M}_i$.

This concludes our presentation of the syntax and model theory of NMCIL. The following two sections illustrate how NMCIL can be used as a tool for formalization.

5 A Belief Transfer Example

We begin with a simple example that suppresses the indexical aspects of NMCIL. Imagine that an agent's beliefs are divided into three modules. Two of these, m_1 and m_2, are specialized information sources. The third module, m_3, is a synthesizer, responsible for combining information from m_1 and m_2. The language $\mathcal{L}_2$ for this application has four basic expressions of type e: a_1, a_2, b_1, and b_2. Think of the first two constants as names of switches and the second two as names of lights. (The idea is that the switch denoted by a_i is connected to the light denoted by b_i.) The language has one 1-place predicate, ON. This predicate has type $\langle e, t \rangle$; it denotes the property of being on, for lights and switches. Thus, if a_1 denotes $light_1$ and b_1 denotes $switch_1$, $\text{ON}(a_1)$ means that $light_1$ is on, and $\text{ON}(b_1)$ means that $switch_1$ is on.

Finally, $\mathcal{L}_2$ has three modal operators, [1], [2], and [3]; one for each of the three modules. In general, to allow for iteration of modalities, we would formalize such modal operators as constants of type $\langle\langle w,t\rangle\langle w,t\rangle\rangle$. In the context of this paper, however, there is no need to iterate modal operators, so I will simplify things by treating all such modal operators as constants of type $\langle\langle w,t\rangle, t\rangle$.

The background theory for this example has, we assume, axioms ensuring that [1], [2], and [3] are DS5 modal operators. For instance,

$$\forall p_{\langle w,t\rangle} \forall q_{\langle w,t\rangle} [1]^{\wedge}[{}^\vee p \wedge {}^\vee q] \leftrightarrow [[1]p \wedge [1]q]$$

might be such an axiom. We won't go into those details here.

We are interested in domain functions D which provide only one index: $D(i) = \{i_0\}$. In this case, there are no variations in how the modules assign contents to characters, so that we can neglect such effects in this example.

We wish to model a belief transfer scheme according to which m_3 gathers beliefs from m_1 unless it believes they are incompatible with m_2, and from m_2 unless it believes they are incompatible with m_1. For this, we need two abnormality predicates of type $\langle\langle w,t\rangle,t\rangle$. (This is the type of 1-place predicates of propositions.) $Ab1$ characterizes propositions that are abnormal in that there is reason preventing their transfer from m_1 to m_3. $Ab2$ characterizes propositions that are abnormal in that there is reason preventing their transfer from m_2 to m_3.

The following axioms formulize belief transfer. Axiom (5.1) says that normally, m_3's beliefs transfer to m_1. Axiom (5.2) says that m_3's beliefs that m_1 believes to be incompatible with some belief of m_2's are abnormal. Axioms (5.3) and (5.4) are the corresponding axioms for transfer from m_2 to m_3.

(5.1) $\forall p_{\langle w,t\rangle}[[\texttt{[1]}p \land \neg Ab1(p)] \to \texttt{[3]}p]$
(5.2) $\forall p_{\langle w,t\rangle}[[\texttt{[1]}p \land \exists q_{\langle w,t\rangle}[\texttt{[2]}q \land \texttt{[3]}^\wedge[^\vee q \to \neg^\vee p]]] \to Ab1(p)]$
(5.3) $\forall p_{\langle w,t\rangle}[[\texttt{[2]}p \land \neg Ab2(p)] \to \texttt{[3]}p]$
(5.4) $\forall p_{\langle w,t\rangle}[[\texttt{[2]}p \land \exists q_{\langle w,t\rangle}[\texttt{[1]}q \land \texttt{[3]}^\wedge[^\vee q \to \neg^\vee p]]] \to Ab1(p)]$

To round out the transfer axioms, we add another axiom, according to which m_3 believes that the states of the switches are coupled with the states of the corresponding lights.

(5.5) $\texttt{[3]}^\wedge[[\text{On}(a_1) \leftrightarrow \text{On}(b_1)] \land [\text{On}(a_2) \leftrightarrow \text{On}(b_2)]]$

We'll suppose that the task of m_1 is to monitor switches, and that of m_2 is to monitor lights. Then a set of *switch data axioms* is any subset of

$$\{\texttt{[1]}^\wedge\text{On}(t) \,/\, t \in \{a_1,a_2\}\} \cup \{\texttt{[1]}^\wedge\neg\text{On}(t) \,/\, t \in \{a_1,a_2\}\},$$

and a set of *light data axioms* is any subset of

$$\{\texttt{[2]}^\wedge\text{On}(t) \,/\, t \in \{b_1,b_2\}\} \cup \{\texttt{[2]}^\wedge\neg\text{On}(t) \,/\, t \in \{b_1,b_2\}\}.$$

Such a set represents a combination of judgments of a module about the environment it is intended to observe. Inconsistent combinations are allowed. A *light-switch theory* is a set $T = T_0 \cup T_1 \cup \{\text{Axiom (5.5)}\} \cup T_2 \cup T_3$, where:

T_0 is a theory of the modal operators,
$T_1 = \{\text{Axiom (5.1)}, \text{Axiom (5.2)}, \text{Axiom (5.3)}, \text{Axiom (5.4)}\}$,
T_2 is a set of switch data axioms, and
T_3 is a set of light data axioms.

We circumscribe this theory by allowing the abnormality predicates and the modal operator $\texttt{[3]}$ to vary, while everything else is held constant. And we circumscribe over the whole set of worlds.[7] The idea is to determine what m_3

---

[7] That is, the relation $\mathcal{M}_K$ relates $\mathcal{M}_w$ to every world.

will believe, given what the observing modules believe, when abnormalities are minimized.

It can be shown that the nonmonotonic consequences of a light-switch theory $T = T_0 \cup T_1 \cup \{\text{Axiom (5.5)}\} \cup T_2 \cup T_3$ are the monotonic consequences of $T_0 \cup \{\text{Axiom (5.5)}\} \cup T_2 \cup T_3 \cup T_4$, where T_4 says that m_3 believes the uncontradicted literals in the observing contexts.

For instance, if $T_2 = \{\texttt{[1]}\char`\^\neg\text{ON}(a_1)\}$ and $T_3 = \{\texttt{[2]}\char`\^\neg\text{ON}(b_1), \texttt{[2]}\char`\^\neg\text{ON}(b_2)\}$, then $T_4 = \{\texttt{[3]}\char`\^\neg\text{ON}(b_2)\}$.

Circumscription and abnormality theories are general-purpose logical mechanisms, and it may well be that defaults are used locally by the information-gathering modules m_1 and m_2 in drawing conclusions. That factor was not introduced into this simple example, but it could be added by making the connection between the states of the switches and the states of the lights they control nonmonotonic.

6 A Temporal Example

We illustrate the more general case in which the content of expressions varies from context to context with a temporal example. Here there are three information sources m_1, m_2, and m_3—which we can simply think of as documents recording information—and a fourth module m_4 that integrates information from each of the three sources. In this application, a date consists of a week and a day. There are three weeks, denoted by individual constants WEEK$_1$, WEEK$_2$, and WEEK$_3$. The days are denoted by individual constants MONDAY, etc. The documents record the weather, but are so limited in scope that they can only talk about whether the weather is clear.

Ordinarily, we might use a 2-place predicate for the weather records, saying, for instance, CLEAR(WEEK$_2$)(TUESDAY), but we allow the documents to be elliptical. There are in fact three predicates for weather recording: CLEAR0 is a 0-place predicate indicating that it is clear on the current day, CLEAR1 is a 1-place predicate indicating, when attached to a name of a day, that it is clear on that day of the current week, and CLEAR2 is a fully decontextualized 2-place predicate indicating that it is clear by giving the date explicitly: e.g., CLEAR2(WEEK$_2$)(TUESDAY).

The observing modules m_1, m_2, and m_3 are allowed to be elliptical; the synthesizing module m_4 gathers information from the observers in decontextualized form. Each module is assigned a date. In this example, we assume the following assignment.

m_1: ⟨Friday, Week$_1$⟩
m_2: ⟨Wednesday, Week$_2$⟩
m_3: ⟨Monday, Week$_3$⟩
m_4: ⟨Sunday, Week$_3$⟩

These assignments are known to the synthesizing module, m_4.

A sentence whose content depends on an index, like CLEAR0, corresponds to the type $\langle i, \langle w, t\rangle\rangle$; functions of this type input an index, and output a proposition. The function corresponding to CLEAR0, for instance, would output the proposition that is true in worlds where the weather is clear on Friday of Week$_1$ in the index for m$_1$. Our formalization of this example uses quantifiers of this type in the rules for information transfer.

We introduce constants of type i standing for the indices of each module. (Remember, an index represents a complete policy for disambiguating and decontextualizing expressions.) The constants c_1, c_2, and c_3 represent the indices of m$_1$, m$_2$, and m$_3$, respectively.

The following axioms about change of content begin with an axiom saying that CLEAR2 produces a constant character when applied to its arguments. Axiom (6.1) uses an O operator, which is a sort of necessity operator for indices. This operator, which has type $\langle\langle i, t\rangle, t\rangle$, can be defined out of logical primitives as follows.

$$O = \lambda x_{\langle i,t\rangle}[x = {}^{\cap}\top].$$

The second axiom defines the predicate DAY, which ranges over days of the week. The remaining axioms in this group are *decontextualization axioms*: they provide rules for decontextualizing elliptical characters. Axiom (6.3) gives the decontextualization rule for CLEAR0, Axiom (6.4) gives the rule for CLEAR1.

(6.1) $\forall x_e \forall y_e \exists p_{\langle w,t\rangle} O[{}^{\cup}p = {}^{\wedge}\text{CLEAR}(x)(y)]$

(6.2) $\forall x_e [\text{DAY}(x) \leftrightarrow [x = \text{MONDAY} \vee x = \text{TUESDAY} \vee \ldots \vee x = \text{SUNDAY}]$

(6.3) $\forall x_i [{}^{\cap\wedge}\text{CLEAR0}](x) = p \leftrightarrow$
  $[x = c_1 \wedge p = {}^{\wedge}\text{CLEAR}(\text{WEEK}_1)(\text{FRIDAY})]$
  $\vee [x = c_2 \wedge p = {}^{\wedge}\text{CLEAR}(\text{WEEK}_2)(\text{WEDNESDAY})]$
  $\vee [x = c_3 \wedge p = {}^{\wedge}\text{CLEAR}(\text{WEEK}_3)(\text{MONDAY})]$
  $\vee [x = c_4 \wedge p = {}^{\wedge}\text{CLEAR}(\text{WEEK}_3)(\text{SUNDAY})]\,]$

(6.4) $\forall x_i \forall y_e [{}^{\cap\wedge}\text{CLEAR1}](x)(y) = p \leftrightarrow$
  $[\text{DAY}(y) \wedge [x = c_1 \wedge p = {}^{\wedge}\text{CLEAR}(\text{WEEK}_1)(y)]$
  $\vee [x = c_2 \wedge p = {}^{\wedge}\text{CLEAR}(\text{WEEK}_2)(y)]$
  $\vee [x = c_3 \wedge p = {}^{\wedge}\text{CLEAR}(\text{WEEK}_3)(y)]$
  $\vee [x = c_4 \wedge p = {}^{\wedge}\text{CLEAR}(\text{WEEK}_3)(y)]\,]\,]$

The detection of inconsistencies in information transfer takes place at the level of content, after decontextualization has taken place. Therefore, the transfer rules for this example are exactly analogous to those of the previous example. For instance, the axioms for transfer of beliefs from m1 to m4 are like Axioms (5.1) and (5.2).

(6.5) $\forall p_{\langle w,t\rangle}[[\,\texttt{[1]}p \wedge \neg Ab1(p)] \rightarrow \texttt{[4]}p]$

(6.6) $\forall p_{\langle w,t\rangle}[[\,\texttt{[1]}p \wedge \exists q_{\langle w,t\rangle} \exists r_{\langle w,t\rangle}[\texttt{[2]}q \wedge \texttt{[3]}r \wedge \texttt{[4]}^{\wedge}[[{}^{\vee}q \wedge {}^{\vee}r] \rightarrow \neg{}^{\vee}p]]]$
  $\rightarrow Ab1(p)]$

This axiomatization style is not only fairly modular, but it allows the beliefs of the various modules to be formalized, using *ist*, in a way that matches the form in which they (presumably) appear in the knowledge sources themselves. We can use, for instance, an axiom like $ist(c_1)([1])(\text{CLEAR})$ to say that module m_1 believes that it is clear.

7 Conclusion

Although many authors who have advocated a general logic of context have pointed out the need for nonmonotonicity, nonmonotonic logics of context have not been much explored. In other papers, I advocated using an intensional type theory modeled on Montague's Intensional Logic as a general framework for the logic of context. This paper was devoted to showing how to make this logic nonmonotonic, providing the basic semantic definitions, and to showing how it might be used to formalize some simple examples.

I used model preference relations, and in particular circumscriptive versions of model preference, because these techniques seem to generalize most readily from First Order Logic to other sorts of logics. If a logic has a well defined model theory, it should be possible to introduce nonmonotonicity by introducing a preference relation over the models. However, if you take this approach, it is important to think through interactions between these relations and the components of the models, and it is vital to make sure that in the contemplated applications there is a natural way to distinguish the aspects of models that are to vary.

I hope that this paper has shown at least that nonmonotonicity can be introduced in a straightforward way, and that circumscription can be naturally applied at least in simple examples of belief transfer. This paper itself does not attempt to go beyond the proof of concept stage; there is much work to be done on the logic and its applications.

NMCIL is intended as a contribution to the theory of context. At present, there is a large gap between the theory and the desired applications. I hope that future work will begin to close this gap by providing ways, informed by the logical theory, of efficiently implementing special cases. The fact that this has happened in other areas of nonmonotonic logic—and especially, in planning formalisms—is encouraging. But I do not expect this to be easily accomplished, or to happen quickly.

The pieces of a general logic of context that I advocated in earlier papers on CIL, such as [Thomason, 1999], are not yet entirely integrated. I have shown how to make CIL dynamic, providing a way to interpret formulas by tracing a trajectory through a series of contexts, using local beliefs and contextualizations along the way. In this paper, I have shown how to make CIL nonmonotonic. It remains to combine the two. This integration, I hope, would provide a more flexible and more plausible framework for integrating nonmonotonic reasoning into the logic of context.

References

[Church, 1940]Alonzo Church. A formulation of the simple theory of types. *Journal of Symbolic Logic*, 5(1):56–68, 1940.

[Cresswell and Hughes, 1996]Max J. Cresswell and G.E. Hughes. *A New Introduction to Modal Logic*. Routledge, London, 1996.

[Fagin et al., 1995]Ronald Fagin, Joseph Y. Halpern, Yoram Moses, and Moshe Y. Vardi. *Reasoning about Knowledge*. The MIT Press, Cambridge, Massachusetts, 1995.

[Gallin, 1975]Daniel Gallin. *Intensional and Higher-Order Logic*. North-Holland Publishing Company, Amsterdam, 1975.

[Guha, 1991]Ramanathan V. Guha. Contexts: a formalization and some applications. Technical Report STAN-CS-91-1399, Stanford Computer Science Department, Stanford, California, 1991.

[Lewis, 1973]David K. Lewis. *Counterfactuals*. Harvard University Press, Cambridge, Massachusetts, 1973.

[Lifschitz, 1994]Vladimir Lifschitz. Circumscription. In Dov Gabbay, Christopher J. Hogger, and J. A. Robinson, editors, *Handbook of Logic in Artificial Intelligence and Logic Programming, Volume 3: Nonmonotonic Reasoning and Uncertain Reasoning*, pages 298–352. Oxford University Press, 1994.

[Lin, 1988]Fangzhen Lin. Circumscription in a modal logic. In Moshe Y. Vardi, editor, *Proceedings of the Second Conference on Theoretical Aspects of Reasoning about Knowledge*, pages 113–127, San Francisco, 1988. Morgan Kaufmann.

[McCarthy, 1993]John McCarthy. Notes on formalizing contexts. In Ruzena Bajcsy, editor, *Proceedings of the Thirteenth International Joint Conference on Artificial Intelligence*, pages 555–560, San Mateo, California, 1993. Morgan Kaufmann.

[Montague, 1970]Richard Montague. Pragmatics and intensional logic. *Synthese*, 22:68–94, 1970. Reprinted in *Formal Philosophy*, by Richard Montague, Yale University Press, New Haven, CT, 1974, pp. 119–147.

[Shoham, 1988]Yoav Shoham. *Reasoning about Change: Time and Causation From the Standpoint of Artificial Intelligence*. The MIT Press, Cambridge, Massachusetts, 1988.

[Stalnaker, 1968]Robert C. Stalnaker. A theory of conditionals. In Nicholas Rescher, editor, *Studies in Logical Theory*, pages 98–112. Basil Blackwell Publishers, Oxford, 1968.

[Thomason, 1999]Richmond H. Thomason. Type theoretic foundations for context, part 1: Contexts as complex type-theoretic objects. In Paolo Bouquet, Luigi Serafini, Patrick Brézillon, Massimo Benerecetti, and Francesca Castellani, editors, *Modeling and Using Contexts: Proceedings of the Second International and Interdisciplinary Conference, CONTEXT'99*, pages 352–374. Springer-Verlag, Berlin, 1999.

[Thomason, 2000]Richmond H. Thomason. Modeling the beliefs of other agents. In Jack Minker, editor, *Logic-Based Artificial Intelligence*, pages 375–473. Kluwer Academic Publishers, Dordrecht, 2000.

[Thomason, 2003]Richmond H. Thomason. Dynamic contextual intensional logic: Logical foundations and an application. In Patrick Blackburn, Chiara Ghidini, and Roy M. Turner, editors, *Modeling and Using Context: Fourth International and Interdisciplinary Conference*, pages 328–341. Springer-Verlag, 2003.

Modeling Context as Statistical Dependence

Sriharsha Veeramachaneni[1], Prateek Sarkar[2], and George Nagy[3]

[1] SRA Division, ITC-IRST Povo, TN 38057, Italy
sriharsha@itc.it
[2] Palo Alto Research Center, 3333 Coyote Hill Road,
Palo Alto, CA 94304, USA
psarkar@parc.com
[3] ECSE Dept. Rensselaer Polytechnic Institute,
110 Eighth Street, Troy, NY 12180, USA
nagy@ecse.rpi.edu

Abstract. Theories of context in logic enable reasoning and deduction in contexts represented as formal objects. Such theories are not readily applicable to systems that learn by induction from a set of examples. Probabilistic graphical models already provide the tools to exploit context represented as statistical dependences, thereby providing a unified methodology to incorporate context information in learning and inference. Drawing on a case study from optical character recognition, we present the various types of dependences that can occur in pattern classification problems and how such dependences can be exploited to increase classification accuracy. Learning under different conditions require differing amounts and kinds of samples and different trade-offs between modeling error due to overly strict independence assumptions and estimation error of models that are too elaborate for the size of the available training set. With a series of examples based on frames of two patterns we show how each kind of dependence can be represented using graphical models and present examples from other disciplines where the particular dependence frequently occurs.

1 Introduction

A recognition problem often pertains to the interpretation of a collection of observations in a scene. A simple approach to the solution is to interpret/classify each observation independently of others. In reality observations and their interpretations are often interdependent. This interdependence is manifested as high likelihood of occurrence of some combinations of observations and interpretations, and relative rarity or improbability of some others. Modeling the interdependence of observations and interpretations can improve the accuracy of recognition. Therefore it can be argued that scene understanding (whatever that means for the application) can be better performed by processing each object in the *context* of the others in the scene rather than independently [25]. However, it is not practical in most application domains to respect arbitrary interdependence of observations and interpretations in formulating a solution to the recognition problem. Choosing a good dependence model can improve the efficacy of learning (e.g., convergence, robustness) from available data, as well as the efficiency (e.g., accuracy, speed) of interpretation. Optical Character Recognition (OCR) is one application area where extensive work has been done in modeling and exploiting the relationships between the objects (in this case images of characters) to be classified and their intended interpretations (class labels) [14].

The recognition of a symbol, signal or object can be done in isolation by using a classification rule that has been learnt from labeled samples available for 'training' by operating on the observable data ('features') about each object (or pattern). Higher classification accuracy can be achieved by the recognition of groups of related objects, taking into account the relationship between them in addition to their individual characteristics. The cause of the relationship may be multiple observations of the same characteristic ('feature dependence'), temporal or spatial contiguity (i.e., co-articulation, ligatures, alignment), constraints imposed by a message ('language context'), or a common source ('style'). The additional information that is derived from the co-occurring objects is often called 'context.'

In logical approaches to AI, contexts are modeled as abstract mathematical entities whose values determine the values of other logical entities [12, 24]. In a setting where we have to induce general truths from thousands of noisy examples, each characterized by tens or hundreds of attributes, a statistical approach seems convenient. Moreover for the interpretation of a collection of objects, the context sometimes cannot - and often need not - be explicitly articulated as long as it is understood that one exists. In such scenarios the dependence among entities or more precisely, between the representations of the objects, can be modeled by dependent parameterized statistical distributions. Multivariable statistics provide a natural way to induce the effect of context on the interpretation of objects without the need for its explicit representation.

A close look reveals several different types of relationships between features, labels, and sources. From a statistical perspective, the joint distribution of all the variables specifies the problem completely. Because of the large number of variables, the joint distribution may be difficult to estimate accurately, and the underlying relationships are obscured rather than revealed by a model that is too elaborate for the application. We can overcome this problem by bringing to bear prior knowledge about the problem at hand to avoid modeling relationships that can be neglected.

Directed graphical models (of which Bayesian Networks are a special case) offer a way to represent even complex models by avoiding the specification of conditional independence relations. Recent advances in graphical models provide rules for efficient computation of both the joint distributions (learning), and of the conditional distributions required for classification (inference). We show how such models provide a systematic representation and efficient computational tools for the classification of groups of patterns under diverse contextual assumptions.

We present the equations and graphical models for a variety of contextual classification methods that have already found application in practice. We study the simple problem of classifying a pair of objects (patterns) illustrating with equations and examples how different types of context can be modeled. We also comment on the implications of context on the acceptable sample size of the training set, and give examples of past and future applications of broadly defined context.

Section 2 provides a brief introduction to probabilistic graphical models and a short summary of previous work. In Section 3, we show how differing assumptions on the statistical dependence affect underlying joint distributions and where such assumptions are justified. Section 4 discusses methods to learn (i.e., estimate the parameters for) the various models. Section 5 gives some examples of the expected dependence in diverse applications. Many of these are based on OCR (Optical Character Recognition) and ASR (Automatic Speech Recognition) because some of the more complex models were developed there. In the final section we demonstrate how quickly the number of parameters grows with larger frames (i.e., more patterns per frame), more features, and more classes, and discuss the trade-offs between model complexity and sample size.

2 Probabilistic Graphical Models

In probabilistic analysis with many random variables all the marginal and conditional probabilities on a subset of the variables can be computed from the joint distribution if it is known. However often in practice the joint distribution has to be estimated from (relatively) sparse data, and it is essential that we impose restrictions on the nature of the joint distributions to make this feasible.

On the other hand if all the variables are mutually independent, then the marginal distribution on each variable is simpler to estimate, and the joint distribution is simply the product of the marginals. Under such assumptions, no variable conveys information about any other variable. Useful probabilistic models often lie between these two extremes. Graphical models have emerged as an interesting class of models that constrain (or simplify) the joint density via conditional independence relations among variables which are represented by nodes in the model. The set of edges embody information about conditional independence among the variables. The graphical model formalism, drawing from graph theory and probability theory, provides a unifying framework for classical and new models such as mixture models, factor analysis, hidden Markov models, Kalman filters, Bayesian networks, Markov random fields, Ising models, and conditional random fields. [11, 13, 7] provide a good overview of graphical models.

Bayesian networks, pioneered by Pearl [15], are a class of directed graphical models where the joint distribution of random variables are completely specified by functions (or tables) stored in the nodes, which represent the distribution of the node variable when conditioned on its parents. Edge direction in these networks can be interpreted as 'causation' providing a basis for the design of the graph structure in any problem domain [16]. We recommend [15] for an excellent philosophical and mathematical foundation for Bayesian networks. There are many web tutorials for a more urgent introduction to the standard notations and diagram conventions.

Once the underlying graphical model and its parameters are known, most applications involve inference, i.e., the computation of a conditional or marginal distribution from the known joint distribution, such as P(character-label = A|bitmap = $observation$). The graphical structure of the model leads to efficient inference algorithms, such as variable elimination [1], local message passing (for directed acyclic graphs) [15], or the junction tree algorithm when there are cycles in the graph. Heckerman [10] has written an excellent tutorial on exact inference. Often exact inference (exact reduction of the joint probability function) is computationally infeasible, requiring the use of approximate algorithms such as variational methods, Monte Carlo sampling methods and loopy belief propagation [11].

There is also extensive literature on learning the parameters of a graphical model. Jordan [11] provides a good review of previous work. In situations where the graphical structure is known a priori, the parameters of the distributions at each node are estimated either from fully observable data (direct maximum likelihood estimates) or from partially observed data (maximum likelihood estimates through Expectation-Maximization or gradient ascent). Learning the model structure from data is also an active research area.

3   Modeling Context as Statistical Dependence

We consider the following pattern classification problem. The patterns to be classified arrive as ordered sets $z = (x_1, x_2)$ of two patterns each. Each pattern has a class label y which is one of two classes A or B. Consequently the ordered-pair pattern z has a class label (y_1, y_2). Each pattern is composed of two features $x_i = (u_i, v_i)$ that are used for classification. Given the observed pair (x_1, x_2) the objective is to classify it by assigning to it the appropriate pair label (y_1, y_2). We introduce the notion of a frame $F = (x_1, y_1, x_2, y_2) = (u_1, v_1, y_1, u_2, v_2, y_2)$ which is the ordered set of patterns and the corresponding class labels.

$F = (x_1, y_1, x_2, y_2)$ (each frame consists of two patterns and their labels)

$x_i = (u_i, v_i)$ (each pattern has a label and two features)

When the joint probability distribution over all possible frames $P(x_1, y_1, x_2, y_2)$ is completely known, the classifier with the highest accuracy (the maximum a posteriori classifier) can be constructed that assigns the class $(y_1^\star, y_2^\star)$ to the frame (x_1, x_2) where

$$(y_1^\star, y_2^\star) = \operatorname*{argmax}_{(y_1, y_2)} P(x_1, y_1, x_2, y_2) \tag{1}$$

3.1 Complete Dependence

One way to exploit context during learning and inference is to make no assumptions of statistical independence a priori but to learn, under the most general model, the absence of such dependence relations from the training data. This method, however, often results in an inaccurate model because of the enormous amount of training data required to learn it accurately. It is therefore necessary to make the appropriate independence assumptions justified by the amount of data available as well as by prior knowledge about the problem.

3.2 No Dependence

The simplest independence model can be obtained by assuming mutual independence of the variables except for the dependence of each feature on the label of the corresponding pattern (without which classification would be impossible). Then the joint probability can be written as

$$P(x_1, y_1, x_2, y_2) = P(u_1, v_1, y_1, u_2, v_2, y_2)$$
$$= P(u_1|y_1)P(v_1|y_1)P(u_2|y_2)P(v_2|y_2)P(y_1)P(y_2) \quad (2)$$

The probabilistic graphical model depicting the assumption of no dependence is shown in Figure 1. The way to read the graph is to note that given the class label (y_i) of the pattern the pattern features (u_i, v_i) are statistically independent. The assumption of no dependence leads to the so-called *naive Bayes* classifier

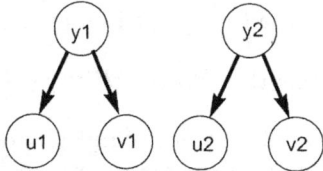

Fig. 1. Graphical model - No dependence

which often performs surprisingly well, especially in situations where there is little training data available.

3.3 Intra-pattern Class-Conditional Feature Dependence

Here we drop the assumption of independence between the features u_i and v_i given the class-label y_i and therefore the complete joint distribution $P(u_i, v_i|y_i)$ has to be specified. Such dependence can be incorporated into the graphical model as shown in Figure 2. Such intra-pattern dependence is generally well understood and is discussed in most texts for binary and Gaussian variables [6, 22, 23]. The covariance matrix represents only the second-order (pairwise) dependence between the feature variables but suffices to completely specify the

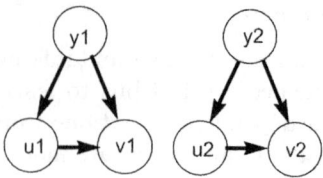

Fig. 2. Graphical model for intra-pattern class-conditional feature dependence

interdependence of Gaussian random variables regardless of dimensionality. It is possible to construct examples where taking correlation into account leads to zero error rate even when classification based only on an independence assumption is no better than a random choice.

When there are a large number of features, it may be desirable to exploit only the dependence of highly correlated features. Partial independence methods are based on dependence chains or dependence trees of features [3, 4, 6].

A different way of modeling dependence is by means of mixture distributions, which are most often used in the multidimensional case for Gaussian distributions [6, 19]. Even if the features are class-conditionally independent within each component distribution, they may be dependent in the class-conditional mixture distribution.

3.4 Inter-pattern Class Dependence

Inter-pattern class dependence is often called *linguistic context* in OCR. When such dependence is modeled the frame distribution is given by

$$P(x_1, y_1, x_2, y_2) = P(x_1|y_1)P(x_2|y_2)P(y_1, y_2) \qquad (3)$$

That is, the joint class probability is not completely described by the marginal probabilities. Figure 3 shows an example of two sequences of patterns to be recognized by a word recognizer. Although, the first three patterns are identical in both the words, knowing that the labels for the last two patterns are either 'ks' or 'AD' makes either the label 'boo' or '600' more likely for the first three patterns. The graphical representation of inter-pattern class dependence is shown in Figure 4.

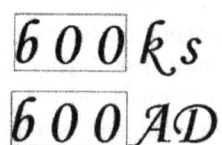
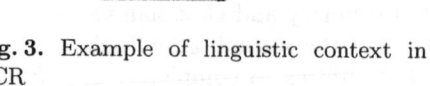

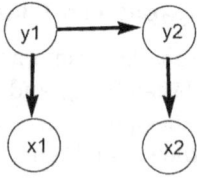

Fig. 3. Example of linguistic context in OCR

Fig. 4. Graphical model for inter-pattern class dependence

3.5 Inter-pattern Class-Feature Dependence

In some classification problems, features of a particular pattern depend on the co-occurring classes (but not necessarily on the features of co-occurring patterns). An example of such context is shown in Figure 5. The vertical location of the apostrophe depends on whether the preceding letter was a 'T' or an 'n', but not on the particular rendering of the preceding letter. That is, the features of the apostrophe are independent of the preceding letter given its label. The graphical model that captures this kind of context is shown in Figure 6.

Such dependence can be modeled graphically as shown in Figure 6 and the joint frame distribution is given by

$$P(x_1, y_1, x_2, y_2) = P(x_1|y_1, y_2)P(x_2|y_1, y_2)P(y_1)P(y_2) \qquad (4)$$

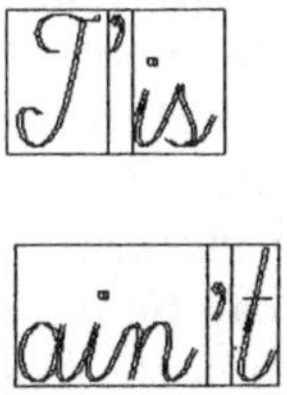

Fig. 5. Example of inter-pattern class-feature dependence

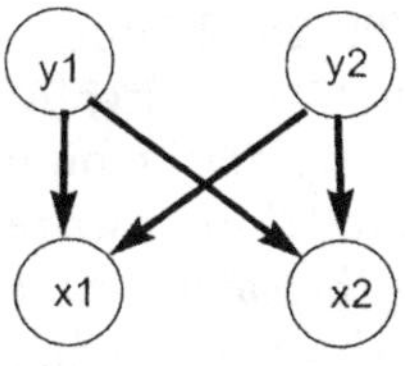

Fig. 6. Graphical model for inter-pattern class-feature dependence

3.6 Inter-pattern Feature Dependence

(a) There are several reasons for the dependence between features of patterns in a field. In speech the articulation of a pattern, and in cursive handwriting the rendering of a pattern often determine the features of the next pattern. For example, in Figure 7 the shape of the letter 'a' depends on the shape of the preceding letter. This type of dependence can be modeled as shown in Figure 8.
(b) Another cause for such dependence is the *isogeny* (commonality of origin or source) of pattern groups. For example, handwriting or typeface consistency

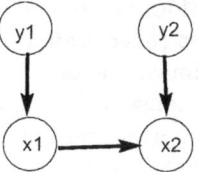

Fig. 7. Example of ligatures in OCR

Fig. 8. Graphical model for inter-pattern feature dependence (ligatures, co-articulation etc.)

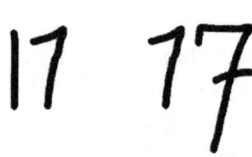

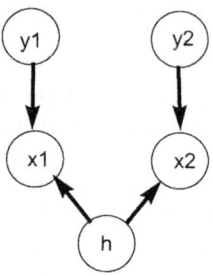

Fig. 9. Example of inter-pattern feature context due to styles

Fig. 10. Graphical model for inter-pattern feature dependence (style.)

can lead to such a dependence in recognition problems with multiple writers or typefaces [20, 26]. Figure 9 shows the handwriting of two hypothetical writers A and B. The '7' of writer A is identical to the '1' of writer B. In such a scenario, no classifier operating on individual patterns can discriminate between '7' from writer A and '1' from writer B. However, when the test patterns appear as pairs written by the same writer, then the above confusion can be resolved using the context of the co-occurring pattern, if the other pattern in the field is a '1' from writer A or a '7' from writer B. Of course when the identity of the writer is known during classification such context is irrelevant, but this information is often lacking. Such a situation can be modeled as shown in Figure 10 where the variable h represents the hidden writer identity. The probabilistic graphical model for such dependence and the corresponding joint distributions is given by

$$P(x_1, y_1, x_2, y_2) = \sum_h P(x_1|y_1, h)P(x_2|y_2, h)P(y_1)P(y_2)P(h) \qquad (5)$$

4 Parameterization and Learning

In the previous section we showed how the various varieties of context can be visualized with probabilistic graphical models (*qualitative specification*). Now, in order to completely specify the models we need to choose the appropriate parameterization, i.e., to choose the actual families of distributions for the various dependences (*quantitative specification*). Once the model is completely specified we can estimate the parameters from the available training data (i.e., learning) in order to construct the maximum a posteriori classifier (cf. Equation 1).

In most pattern recognition applications, labeled training samples are expensive and scarce. Therefore one always faces a small-sample estimation problem in deriving the model parameters [18]. The number of parameters to be estimated increases rapidly with the number of variables that are considered dependent. Therefore most models avoid considering dependence beyond the second order (pairwise correlation). Even second-order models require estimating a number of parameters proportional to the square of the number required under independence assumptions.

Although there exist general techniques for learning the parameters for probabilistic graphical models, we discuss some specific techniques that are used for estimation of parameters for the models presented above. Since we cannot go into the intricacies of statistical parameter estimation, we indicate the major families of techniques for learning for the pattern pair recognition problem considered in the previous section and give references to more detailed sources.

4.1 Independent Variables

When the class-conditional feature distributions are modeled by a Bernoulli distribution, the necessary parameters are obtained by averaging feature values over the training samples. For Gaussian features, the sum of the squares of the feature values are also needed to obtain estimates of the class-conditional variances.

4.2 Intra-pattern Feature Dependence

The parameters of second-order models are the means and covariance matrices (or, equivalently, correlation coefficients) of the feature variables. The elements of the covariance matrix for two variables are the average values of the product of the variables minus the product of their individual averages over the training samples. The sample variances and covariances are usually scaled by $n/(n-1)$, where n is the number of samples, to obtain an unbiased estimate. The expected value of an unbiased estimator over different training sets is equal to the value of the estimated parameter [2]. When the training sample size is small the variance in the estimate of the covariance matrix can be reduced by increasing its bias by a smoothing technique called *regularization* [8].

Estimation is more difficult for mixture models if the mixture components are not explicitly indicated in the training sample. For Gaussian mixtures, algorithms and computer code based on Expectation Maximization [5, 19] are available. This is an iterative two-phase process. In the first phase, all the patterns are assigned probabilistic labels (called hidden variables) that indicate their affinity to each mixture component. In the second phase, the parameters of each mixture component are estimated from the samples weighted according to the coefficients determined in the first phase. This process converges to the population parameters of the mixture model under broad conditions.

4.3 Class-Label Dependence

The transition probabilities of a first-order Markov model can be estimated directly from the sample bigram frequencies. In our example, it would suffice to count the number of occurrences of AA and BB in the training set if we assume the Markov chain to be in steady state. Some care is necessary for estimating the probabilities of very low or (rarely) very high frequency samples. *Laplace smoothing* is often performed to obtain robust estimates.

For the lexical approach, it is necessary to estimate the probabilities of all possible frame-labels. For long strings, the above caveat for low-frequency samples is even more important.

4.4 Inter-pattern Feature Dependence

For Markov models, we can follow the same route as in 4.3. If the features are class-conditionally independent, then we need to count only the frequencies of n-grams of corresponding features. If they are dependent, we must count n-grams of feature vectors.

When the styles are labeled in the training set, the feature probabilities for each style can be estimated as in 4.1 and 4.2. If not style-labeled, we can use Expectation Maximization for certain families of feature distributions. The mixture components for the class-conditional features are estimated from all the frames that include any pattern with the label for which we are estimating the features. Only the style variables are hidden.

We have not mentioned HMM (Hidden Markov Models) because they are most useful for unsegmented patterns of unknown length or duration, like cursive writing and speech, rather than discrete patterns. HMM is based on a hidden (unobservable) Markov state sequence. Only some transitions are permitted: in handwriting and speech, the network topology is usually restricted to left-to-right with self transitions. Features are generated in each state according to predetermined but unknown class-conditional probability distributions. Methods have been developed to estimate the (hidden) transition probabilities and the state-dependent feature probabilities from unsegmented sample sequences [17]. Most of these methods are specialized formulations of Expectation Maximization. The corresponding frame classification is based on Dynamic Programming (the Viterbi algorithm) to determine the most likely sequence of labels in the frame. Other algorithms such as the A^* search and Stack search are also used.

Table 1. Number of parameters to be estimated for various dependence models when the class-conditional feature distribution is assumed Gaussian

Dependence	Number of parameters
None	CF means, CF variances
	$C - 1$ class-probabilities
Intra-pattern feature dependence only	CF means, $CF(F+1)/2$ covariances
	$C - 1$ class-probabilities
Class-label dependence only	$C^L - 1$ frame-label probabilities
	CF means, CF variances
Inter-pattern feature dependence (style only)	$K - 1$ style probabilities
	KCF means, KCF variances
	$C - 1$ class probabilities
C-Number of classes; F-Number of features; K-Number of styles; L-Frame length	

Table 1 shows how the number of parameters to be estimated depends on the number of patterns per frame, the number of classes, and the number of features under the various assumptions on the dependence structure. Increasing the complexity of the models increases the number of samples necessary for accurate estimation of the model parameters. Consequently better modeling of dependence (or the joint distributions) does not necessarily imply an improvement in classification accuracy. The choice of model complexity must be guided by considerations of relative gain in performance, and the sample size necessary to train the model. We need to carefully choose the appropriate dependence model given the size and type of training data available and from prior knowledge about the problem.

5 Some Applications

5.1 Intra-pattern Feature Dependence

It is difficult to find features that are truly uncorrelated. For instance, the width and height of blob patterns and alphabetic symbols, and even their geometric moments, tend to be correlated. So are the short-term frequency components of phonemes, the height and trunk-diameter of trees, and the multispectral reflection coefficients of vegetation.

An interesting example of feature dependence is induced during digitization by the random phase of the spatial sampling grid relative to a blob pattern [9]. The pixels on each edge are positively correlated (they appear or disappear together), while pixels on opposite edges are negatively correlated [21].

Sometimes principal components (also called Karhunen Loeve expansions or Hotelling transform) are extracted in an attempt to obtain orthogonal (uncorrelated) features, but this leads only to features that are overall, rather than class-conditionally, independent [6].

5.2 Inter-pattern Class Dependence

The best known examples of context are those among letters (in English 'u' is more likely given the previous letter is 'q' than if the previous letter is 'i'), among words ('sunny day' is more likely than 'sunny night'), and among phonemes in speech. However, one is also likely to find high positive correlation among the classes (species) of adjacent trees, crops, or malignant cells. These correlations are universally exploited in OCR and ASR, but only seldom in remote sensing and in biomedical image analysis.

5.3 Inter-pattern Feature Dependence

Markov type dependence occurs in ligatures in handwriting and as co-articulation in speech. Both are due to obvious physical constraints. In remote sensing, adjacent pixels are often correlated simply because the instantaneous field of view

Fig. 11. The 'Checkerboard shadow illusion' created by E. H. Adelson. (http://web.mit.edu/persci/people/adelson/checkershadow_illusion.html. Used with permission.) The tiles labeled 'A' (1, 2) and 'B' (3, 3) have exactly the same grey level

of the sensor (its point spread function) is larger than the spatial sampling interval. In forestry, one might expect that the height-dependent shadow cast by each tree, regardless of its species, affects the height of its neighbors.

Style dependence may occur because of external but unknown factors. In printed matter, it is reasonable to assume that each document has a dominant typeface. A single hand-written form, or a single telephone conversation, exhibits some homogeneity due to the writer or speaker. Trees within the same stand share the same weather and soil conditions. Cells from some organ of a given individual are more similar than cells from different organs or different individuals.

Figure 11 shows an optical illusion where the human brain is incapable of ignoring context in labeling the tiles as either 'dark' or 'light'. Pixel-features of a tile are dependent on the pixel-features of neighboring tiles due to shared illumination. In this example the dependence can be factored as: given the color-class ('dark' or 'light'), and stylistic factors (illumination, i.e., presence or absence of shadow) the features of neighboring tiles are independent.

5.4 Complete Dependence

Although the situation where everything depends on everything else may be quite prevalent it is often practical to make restrictive assumptions on the dependence structure, thereby increasing the small-sample robustness at the expense of increased modeling bias leading to higher accuracy.

6 Discussion and Conclusion

We have examined how different varieties of contextual information as statistical dependences can be modeled by means of directed probabilistic graphical models. The most commonly modeled types of dependence are those between features of the same pattern and linguistic label context. Dependence between nearby patterns can be modeled by Markov chains, which can be extended to Markov fields in image processing. Dependence between patterns within a frame, regardless of their position, can be modeled using styles. We have provided references to some common techniques used for learning the parameters for the various models.

We have seen that even for a simplified case of frames with only two patterns there are a number of varieties of contextual information that can be modeled and exploited. For frames with more patterns this number is larger. Although usually several types of context occur simultaneously, in most problems we expect that a few types dominate the others. The choice of a good dependence (or context) model involves art and careful empirical case analysis. We believe that rapid advances in computer speed and storage provide an opportunity to model and exploit increasingly complex manifestations of context for more accurate pattern classification.

References

1. M. Aji, S and R. J. McEliece. The generalized distributive law. *IEEE. Trans. on Information Theory*, 46(2):325–343, 2000.
2. H. D. Brunk. *An Introduction to Mathematical Statistics*. Ginn&Co., Boston, 1960.
3. C. K. Chow. A recognition method using neighbor dependence. *IRE Trans. Elec. Comp.*, EC-11:683–690, 1966.
4. C. K. Chow and C. N. Liu. Approximating discrete probability distributions with dependence trees. *IEEE Trans. Info. Theory*, IT-14:462–467, 1968.
5. A. P. Dempster, N. M. Laird, and D. B. Rubin. Maximum likelihood from incomplete data via the EM algorithm. *J. Royal Statistical Society*, Series B(39):1–38, 1977.
6. R. Duda and P. Hart. *Pattern Classification and Scene Analysis*. Wiley, 1973.
7. B. J. Frey. *Graphical Models for Machine Learning and Digital Communication*. MIT Press, 1998.
8. H. Friedman. Regularized discriminant analysis. *Journal of American Statistical Association*, 84(405):166–175, 1989.
9. D. I. Havelock. The topology of locales and its effect on position uncertainty. *IEEE Trans. PAMI*, 13(4):380–386, 1991.
10. D. Heckerman. A tutorial on learning with bayesian networks. Technical report, Microsoft Research, Redmond, Washington, 1995.
11. M. I. Jordan, Z. Ghahramani, T. Jaakkola, and L. K. Saul. An introduction to variational methods for graphical models. *Machine Learning*, 37(2):183–233, 1999.
12. J. L. McCarthy. Notes on formalizing context. In *IJCAI*, pages 555–562, 1993.
13. K. Murphy. An introduction to graphical models. Technical report, Intel Research, 2001.

14. G. Nagy. Teaching a computer to read. In *Proceedings of the Eleventh International Conference on Pattern Recognition*, volume 2, pages 225–229, 1992.
15. J. Pearl. *Probabilistic Reasoning in Intelligent Systems: Networks of Plausible Inference.* Morgan Kaufmann, 1988.
16. J. Pearl. *Causality: Models, Reasoning and Inference.* Cambridge University Press, 2000.
17. L. R. Rabiner. A tutorial on hidden markov models and selected applications in speech recognition. In *Proc. of the IEEE*, volume 2, pages 257–285, 1989.
18. S. J. Raudys and A. K. Jain. Small sample effects in statistical pattern recognition: Recommendations for practitioners. *IEEE Trans. PAMI*, 13(3):252–263, 1991.
19. R. A. Redner and H. F. Walker. Mixture densities, maximum likelihood, and the EM algorithm. *SIAM Review*, 26(2):195–235, 1984.
20. P. Sarkar and G. Nagy. Style consistent classification of isogenous patterns. *IEEE Trans. PAMI*, 27(1):14–22, 2005.
21. P. Sarkar, G. Nagy, J. Zhou, and D. Lopresti. Spatial sampling of printed patterns. *IEEE Trans. PAMI*, 20(3):344–351, 1998.
22. R. Schalkoff. *Pattern Recognition.* Wiley, 1991.
23. J. Schurmann. *Pattern Classification.* Wiley, 1996.
24. L. Serafini and P. Bouquet. Comparing formal theories of context in AI. *Artif. Intell.*, 155(1-2):41–67, 2004.
25. G. T. Toussaint. The use of context in pattern recognition. *Pattern Recognition*, 10(3):189–204, 1978.
26. S. Veeramachaneni and G. Nagy. Style context with second-order statistics. *IEEE Trans. PAMI*, 27(1):88–98, 2005.

Robust Utilization of Context
in Word Sense Disambiguation

Xiaojie Wang

School of Information Engineering,
Beijing University of Posts and Telecommunications,
Beijing, 100876, China
xjwang@bupt.edu.cn

Abstract. Context is the only means to identify the sense of a polysemous word. All algorithms for word sense disambiguation make use of information within a context window of the target word. What is the best window size for word sense disambiguation has been long a problem. Different contexts generally give different results even for a same algorithm. In this paper, we exploit an algorithm which is more robust with the varying of different context used. This method aims to lower the uncertainty brought by classifiers using different context window sizes and make more robust utilization of context while perform well. Experiments show our approach outperforms some other algorithms on both robustness and performance.

1 Introduction

Word sense disambiguation (WSD) has long been a central issue in Natural Language Processing (NLP). In many NLP tasks, such as Machine Translation, Information Retrieval etc., WSD plays a very important role in improving the quality of systems. Many different algorithms have been used for this task, including some machine learning algorithms, such as Naïve Bayesian model, decision trees, example based learners and some combinations of different algorithms[1][2][3][4]. Since context is the only means to identify the sense of a polysemous word, all algorithms for sense disambiguation make use of the context of the target word to provide information.

Ide in [5] identified three kinds of contexts: micro-context, topical context and domain. In practice, a context window (l, r), which includes l words to the left and r words to the right of the target word, is predetermined by human or chosen automatically according to a performance criterion. Only information in the context window is then used for training classifiers and disambiguating new occurrences. Since Weaver[6] hoped we could find a minimum value of the window size, which leads to the correct choice of sense for the target word, some works have been done on finding the best windows size for WSD. Yarowsky[7] argued the optimal value is sensitive to the type of ambiguity. In English, semantic or topic-based ambiguities warrant a larger window (from 20 to 50), while more local syntactic ambiguities warrant a smaller window (3 or 4). Leacock et al. [8] showed the local context (in a small window size) is superior to topical context as an indicator of word sense when using a statistical classifier. Yarowsky[9] suggested that different algorithms prefer different window sizes. Followed by these works, it is clear that different window sizes might cause different sense selection for an occurrence of the target word even when a same algorithm is used. Yarowsky[9] gave a detail investigation on how the performances change with different window sizes for several different algorithms and several different kinds of words.

In stead of finding an optimal window size, we here try to investigate another side of the coin, that is, robust utilization of context. We want to find a way which can get a more robust result with the varying of windows sizes. We will firstly construct a sequence of Naïve Bayesian classifiers along a sequence of orderly varying sized windows of context, and make sense selection for both training samples and test samples using these classifiers. We thus get a trajectory of sense selection for each sample, and then use the sense trajectory based k-nearest-neighbors to make final decision for test samples.

This method is motivated by an observation that there is an unavoidable uncertainty when a classifier is used to make sense selection for occurrences of ambiguous words by using different context windows. Our approach aims to alleviate this uncertainty and thus make more robust utilization of context while perform well. Experiments show our approach outperform some other algorithms on both robustness and performance.

The remainder of this paper is organized as follows: Section 2 gives the motivation of our approach, describes the uncertainty in sense selection brought by classifiers themselves. In section 3 we present the decision trajectory based approach. We then implement some experiments in section 4, and give some evaluations and discussions in section 5. Finally, we draw some conclusions.

2 Sense Selection Varying with Different Window Sizes

Even for human, different window sizes might cause different sense selections for a same occurrence of an ambiguous word. Here I give a simple example in Chinese, English native speakers can find such kind of example in English. Considering word "看" (It has two different senses: "read" and "think") in senesce S1.

S1: 我/ 看/ 这/ 本/ 书/ 值得/ 一/ 读/.
   I think this piece book worthy a read.

When we use a context window (1,1), there is no strong evidence in favor of either of the two senses. When we use (3,3), because the collocation of the target word with word 书 give a very strong indication for 看's sense, it is natural that we select the sense of "read" for 看. In this context window, the answer is definitely correct. But when we use window (6,6), we can find the sense selection of "read" is wrong, the sense for the target word we should select in this context is "think". This is still not the final selection. By appending more words to sentence S1, we can have another sentence where the sense of the target word returns to "read".

Here, the occurrence of the ambiguous word is the same, what make the sense selection different are not different occurrences but different context windows. The fact is that as long as we want to distinguish an occurrence of an ambiguous word, we had to choose a window size. That different contexts cause different sense selections means there is an inherent uncertainty when we make the sense selection as an external observer of an ambiguous word in context. The uncertainty is brought by observers but not the ambiguous word itself.

It is the same when we use a computer to do WSD. Yarowsky[9] gave a detail investigation on how the performances change with different window sizes for several different algorithms and several different types of words. Supposing an algorithm is an external observer, the window size is an inherent parameter which is necessary for the observer to implement an observation. Different sense selections thus can be thought as results brought by the observer itself.

The relation between the window size and the sense selection is to some extent similar with relation between a particle's position and its momentum in Heisenberg Uncertainty Principle. The sense selection has no meaning if a window size does not accompany.

By the Uncertainty Principle, when we measure the position and the momentum of a particle, we cannot measure them with zero-variance simultaneously. In Quantum Theory, the wavefunction is used to describe state of a particle. The method to deal with this problem in quantum theory suggests us an idea to deal with the similar problem in WSD.

Firstly, like that in Uncertainty Principle, let the sense of an occurrence of the target ambiguous word be the object to be observed. We think that since the existence of the uncertainty of sense selection brought by different window sizes, sense selection for the occurrence at only one context window cannot give a complete description of its sense. To grasp a complete description of its sense, it is necessary to get sense selections along a series of observation, i.e. using a sequence of context windows to get a trajectory of sense selection.

Secondly, unlike that in Uncertainty Principle, although, theoretically, we can use a trajectory of sense selections to fully describe the meaning of an ambiguous word, in most of NLP tasks, such as Machine Translation, we need a definitely sense selection to complete the tasks. So, we make final sense selection based on a trajectory of sense decision. Since the final selection is based on a decision trajectory along different window sizes, we thus think it may helpful to alleviate the uncertainty brought by difference of context windows.

In this way, our approach aims to improve robustness of classifiers in making use of context. Here the robustness means that sense selection is not sensitive to the window size. This kind of robustness is especially important to WSD system in noise or oral corpus, where there are many occasionally inserted words near the target word. Besides robustness, to achieve better performance is also necessary, if robustness is at a low level of performance, it is useless.

3 Decision Trajectory Based WSD

In our approach, we use Naïve Bayesian (NB) classifiers to construct sense decision trajectory for both training samples and test samples, and then use k-nearest-neighbors (KNN) to make final decision.

Let w be an ambiguous word, it has n different senses, $s_1 \ldots s_i \ldots s_n$. Supposing we have q training samples $S_1 \ldots S_j \ldots S_q$, where q_i samples are tagged with sense

s_i, $\sum q_i = q$. We present our approach in two stages: training stage and test stage. Figure 1 gives a skeleton of the algorithm.

Training stage:
1. To construct a operator vector C along trajectory T :
$$C = (C(p_1),...,C(p_k),...,C(p_m)).$$
$C(p_k)$ is a NB classifier learned by all the tagged data using p_k as the context window.
2. For each training sample S_j, operating C upon it to construct a decision trajectory, ω_j ($j = 1,...,q$).

Test stage:
1. For a new sample S, construct its decision trajectory ω by operating C upon it.
2. For $j = 1,...,q$, calculate $d(\omega, \omega_j)$
3. KNN-based sense choice for S.

Fig. 1. The algorithm of trajectory-based WSD

In the training stage, we first choose a sequence of context windows.
$$T_m : (p_1,... p_k ... p_m)$$
Where $p_k = (l_k, r_k)$ is a context window which includes l_k words to the left of a occurrence of word w and r_k words to the right. We call T_m a trajectory of context windows, p_k is a window point in this trajectory. For example, a trajectory ((1,1), (3,3), (5,5), (7,7), (9,9)) includes 5 points.

For each window point p_k in T_m, we construct a classier by using NB algorithm based on contexts of training samples in p_k. Let $C(p_k)$ denote the classifier, it can be thought as an operator that make sense selection by operating upon samples. With the change of the window point, we can get a operate vector as follow:
$$C = (C(p_1),...,C(p_k),...,C(p_m))$$
For a sample S_j for sense s_i, we use $C(p_k)(S_j)$ to denote using $C(p_k)$ to classify S_j, we can get a sense selection denoted by $\omega_j(p_k)$, that is to say, $C(p_k)(S_j) = \omega_j(p_k)$. We call $\omega_j(p_k)$ a point decision. If $\omega_j(p_k) = s_i$, we borrow a term to call S_j an eigen-sample of the operator $C(p_k)$, s_i is its eigenvaluve.

With the change of the window point, we have a point decision sequence for sample S_j along the window trajectory T_m, we denoted it by
$$\omega_j = (\omega_j(p_1),...,\omega_j(p_m))$$

We call it decision trajectory of sample S_j along the context windows trajectory T_m. If all elements of ω_j is s_i, i.e. $\omega_j = (s_i, \ldots, s_i)$, we call S_j an eigen-sample of operator C, ω_j is a eigen-trajectory of C.

In this way, we transfer training samples into training decision trajectories. When we get an eigen-trajectory from a sample, we may make sense selection with very high confidence. But when all the training samples are eigen-samples, it is not a good thing for new samples. We will discuss this case in section 5.2.

This end training stage of our approach, we now have a context windows trajectory T_m, a sequence of classifiers $C(p_k)$ along T_m, and a decision trajectory for each training sample. All these compose of our classifier for a new sample in test. When a new sample is given, we first calculate a decision trajectory ω for it by using C operating upon it. Let

$$\omega = (\omega(p_1), \ldots, \omega(p_m))$$

We then calculate the similarity measures between ω and ω_j $j = 1, \ldots, q$ by using (3.1).

$$Sim(\omega, \omega_j) = \frac{\sum_{i=1}^{m} \delta(\omega(p_i), \omega_j(p_i))}{m} \qquad (3.1)$$

Where $\delta(x, y) = 1$ at $x = y$, and $\delta(x, y) = 0$ at $x \neq y$. We then choose h training decision trajectory samples experimentally as ω's h nearest neighbors. Supposing there are h_i samples tagged with sense s_i, $\sum h_i = h$, $h_i \leq q_i$, we define i^* as in (3.2), where ω_{ij} denotes j th training decision trajectory sample for sense s_i. and choose s_{i^*} as the final sense selection for the new sample.

$$i^* = \arg\max_{1 \leq i \leq n} \sum_{1 \leq j \leq h_i} Sim(\omega, \omega_{ij}) \qquad (3.2)$$

When all training samples are eign-samples, the similarity measures between ω and ω_{ij} with same i are the same, (3.2) is simplified to (3.3).

$$i^* = \arg\max_{1 \leq i \leq n} |\{\omega(p_k) = i, k = 1, \ldots, m\}| \qquad (3.3)$$

Here, KNN decision is simplified to majority voting along the decision trajectory of the new sample.

4 Experiments

4.1 Experimental Data

We implement our experiments on data from Senseval-2 English lexical samples training data, which include data for 72 polysemous words (we exclude two words with only one sense, so the number of words in our experiment is 70). Since we just want to compare the robustness and performance of our algorithm with others, we ignore other features in context, only word co-occurrences in given windows are used as only source of features through all different algorithms in this paper.

4.2 Experimental Method

In order to do a comparative study, we implement not only our algorithm, but also four other related algorithms in our experiments. They fall into two types. NB[10] and KNN[11] are two components of our approach. Locally weighted NB(LWNB, [12]) and Ensemble NB(ENB [13]) are two combinational approaches. Since our aim is to compare not only the performance but also the robustness of these algorithms, we implement each algorithm in following way. We note our approach TB_KNN when (3.2) is used for final decision, and TB_VOTE when (3.3) is used for final decision.

We simply choose a sequence of context windows $p_k = (k, k)$ $k = 1,...,40$. We then construct a sequence of window trajectories.

$$T_i = (p_1, ..., p_i) \quad i = 1,...,40$$

We implement TB_KNN and TB_VOTE on each trajectory from T_1 to T_{40}. We denote these trajectories TG1. Obviously, our T_i-based decision and p_i-based decision in fact make use of same context surrounding the target word. The last point in decision trajectory T_i is just the NB decision on p_i, while others use small window size than p_i. So we also have NB point decisions (noted by P) from p_1 to p_{40}.

KNN is implemented along the same sequence of context window, from p_1 to p_{40}.

For the implementation of algorithm LWNB, we use the measure in [11] to find k nearest neighbors for each sample, and then construct a NB classifier according to [12]. This algorithm is also implemented for each context window along the sequence from p_1 to p_{40}.

ENB is implemented according to [13]. Different left and right window sizes we used are (1,2,3,4,5,6,10,20,40). Since one implementation of this algorithm make use of all these different window sizes. It cannot be implemented along above windows sequence, so there is only one implementation for this algorithm.

For each ambiguous word, we implement above experiments respectively, each experiment is a 10-fold cross-validation, at each time 90% of the data are used as training data, 5% are used as development data, and other 5% are used as test data. We use development data to select parameters for each algorithm, such as the number of features in each algorithm, and the nearest neighbor numbers h in TB_KNN, KNN and LWNB.

4.3 Experimental Results

We give the results curves for English word 'art' in Figure 1, and for word 'carry' in Figure 2. In all the figures, x-axis is the context window points, y-axis is F-measure, and different marker style is for different algorithms. Results curves for other target words generally have similar shapes and relative positions.

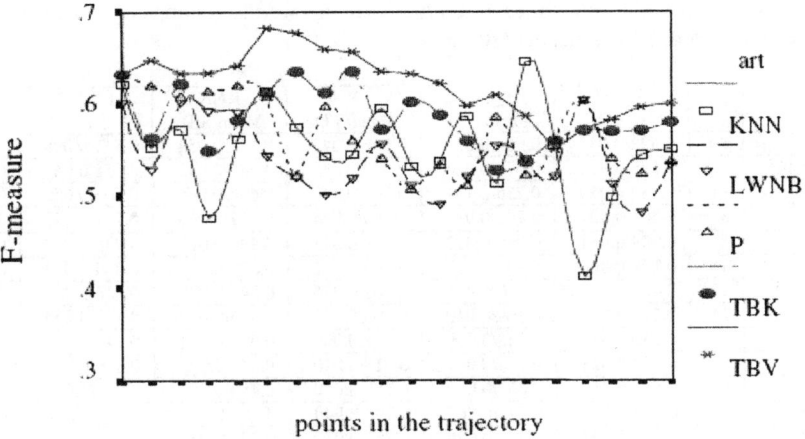

Fig. 1. F-measure curves of word "art" for different algorithms

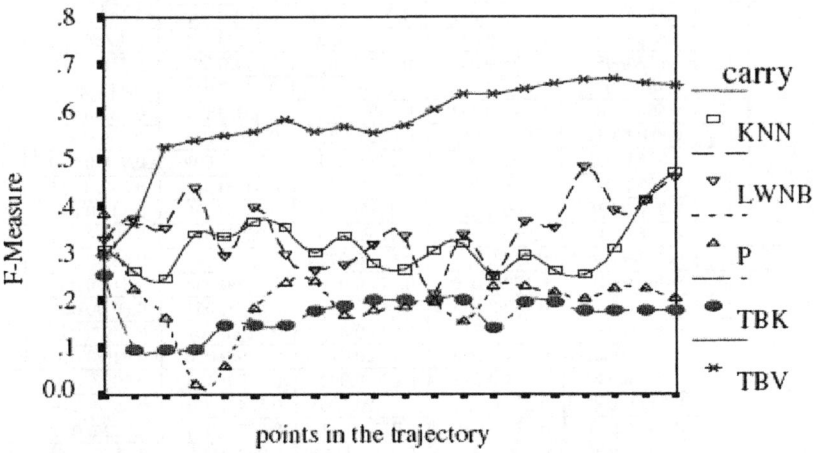

Fig. 2. F-measure curves of word "carry" for different algorithms

We list a summary for SENSEVAL-2 English words in Table 1. In experiments, we computer three values: Mean, Maximum and Standard Deviation. Since the limitation of space, we list two of them for each word in Table 1 and report the results of comparison on all of them in section 5.1. In TB_KNN column of Table 1, there are Mean and Standard Deviation of F-measure of 40 different trajectories from T_1 to T_{40}. Results are summarized in the same way in column TB_VOTE. For column P, KNN and LWNB, two values are Mean and Standard Deviation of F-measure of 40 different points from p_1 to p_{40}. In column ENB, there is only one F-measure. We mark the best Mean and S.D. with bold in each row respectively.

Table 1. A summary of results for SENSEVAL-2 English words, volumn N gives the number of sense for each word, S.D. is standard deviation

Word	N	TB_KNN Mean/S.D.	TB_VOTE Mean/S.D.	P Mean/S.D.	KNN Mean/S.D.	LWNB Mean/S.D.	ENB
art	5	**0.622**/0.035	0.584/**0.032**	0.566/0.043	0.551/0.052	0.541/0.038	0.675
authority	7	**0.696/0.011**	0.671/0.026	0.620/0.039	0.540/0.042	0.555/0.045	0.665
bar	11	**0.759/0.019**	0.255/0.131	0.206/0.130	0.610/0.060	0.605/0.059	0.200
begin	8	**0.806/0.027**	0.792/0.029	0.749/0.028	0.731/0.043	0.731/0.043	0.784
blind	5	**0.959/0.011**	0.958/0.012	0.949/0.012	0.928/0.029	0.945/0.014	0.958
bum	4	**0.898/0.000**	0.891/0.010	0.895/0.010	0.879/0.024	0.884/0.014	0.898
call	15	**0.485**/0.031	0.130/**0.027**	0.162/0.076	0.347/0.079	0.416/0.078	0.339
carry	25	**0.574/0.037**	0.169/0.040	0.195/0.072	0.313/0.057	0.347/0.070	0.186
chair	3	**0.950/0.007**	0.948/0.009	0.947/0.007	0.932/0.021	0.940/0.010	0.947
channel	7	**0.772/0.021**	0.738/0.037	0.718/0.040	0.544/0.075	0.609/0.051	0.772
child	6	**0.822/0.013**	0.804/0.019	0.751/0.050	0.756/0.056	0.761/0.045	0.795
church	4	**0.805**/0.054	0.771/0.067	0.765/**0.040**	0.663/0.060	0.690/0.050	0.824
circuit	6	**0.817/0.032**	0.795/0.058	0.785/0.037	0.704/0.041	0.728/0.039	0.802
collaborate	2	0.998/**0.005**	0.996/0.008	0.992/0.009	0.986/0.023	**0.999**/0.005	1.000
colorless	2	**0.858/0.019**	0.842/0.024	0.836/0.028	0.815/0.037	0.845/0.022	0.846
cool	3	**0.964/0.008**	0.953/0.013	0.940/0.025	0.912/0.048	0.925/0.025	0.961
day	6	**0.817/0.011**	0.806/0.018	0.796/0.011	0.770/0.027	0.774/0.025	0.821
detention	3	**0.931/0.017**	0.905/0.021	0.855/0.051	0.847/0.070	0.888/0.035	0.919
develop	15	**0.532/0.044**	0.091/**0.027**	0.099/0.051	0.364/0.062	0.388/0.065	0.095
draw	24	**0.436/0.072**	0.022/0.038	0.027/**0.032**	0.172/0.106	0.209/0.162	0.284
dress	12	**0.623/0.056**	0.273/0.058	0.251/0.069	0.474/0.083	0.518/0.068	0.336
drift	9	**0.489/0.079**	0.416/0.105	0.459/0.079	0.314/**0.075**	0.449/0.096	0.503
drive	12	**0.688**/0.109	0.172/**0.053**	0.159/0.078	0.451/0.137	0.481/0.091	0.119
dyke	2	**0.980/0.012**	0.971/0.017	0.955/0.021	0.926/0.037	0.946/0.022	0.983
face	7	**0.926/0.003**	0.921/0.007	0.921/0.004	0.917/0.009	0.918/0.010	0.929
facility	4	**0.802/0.032**	0.774/0.075	0.777/0.047	0.647/0.047	0.697/0.035	0.777
faithful	2	**0.909/0.020**	0.906/0.024	0.885/0.036	0.837/0.030	0.841/0.030	0.879
fatigue	5	**0.924/0.010**	0.919/0.014	0.916/0.010	0.908/0.019	0.907/0.020	0.928
feeling	5	**0.777/0.020**	0.738/0.023	0.660/0.072	0.631/0.080	0.653/0.075	0.790
find	15	**0.464/0.019**	0.179/0.026	0.179/0.036	0.203/0.054	0.314/0.055	0.171
fine	3	**0.757/0.034**	0.752/0.035	0.741/0.051	0.637/0.051	0.695/0.066	0.724
fit	2	**0.900/0.027**	0.872/0.049	0.829/0.058	0.729/0.101	0.749/0.083	0.889
free	8	**0.951/0.008**	0.941/0.009	0.935/0.019	0.929/0.030	0.949/0.010	0.949
green	8	**0.781/0.020**	0.704/0.045	0.686/0.092	0.549/0.105	0.603/0.090	0.780
grip	6	**0.836/0.030**	0.821/0.033	0.710/0.106	0.624/0.070	0.681/0.049	0.834
hearth	3	**0.765/0.020**	0.731/0.026	0.717/0.051	0.637/0.074	0.712/0.057	0.731
holiday	4	**0.966/0.000**	0.966/0.000	0.964/0.009	0.962/0.009	0.966/0.003	0.966
keep	17	**0.632/0.037**	0.246/0.054	0.150/0.096	0.496/0.054	0.525/0.072	0.319
lady	3	**0.879/0.008**	0.875/0.012	0.843/0.026	0.807/0.039	0.822/0.044	0.874
leave	12	**0.489**/0.060	0.235/0.072	0.219/0.069	0.426/**0.051**	0.438/0.054	0.310
live	7	**0.751/0.020**	0.728/0.021	0.730/0.035	0.572/0.062	0.644/0.059	0.746
local	2	**0.879/0.035**	0.861/0.038	0.785/0.069	0.875/0.051	0.854/0.047	0.871
match	8	**0.566/0.040**	0.516/0.052	0.527/0.050	0.435/0.090	0.443/0.092	0.555
material	7	**0.627/0.017**	0.587/0.036	0.585/0.033	0.490/0.058	0.510/0.062	0.628
mouth	9	**0.749/0.018**	0.733/0.036	0.718/0.038	0.638/0.081	0.679/0.048	0.721
nation	4	**0.937/0.007**	0.920/0.021	0.901/0.026	0.914/0.029	0.915/0.030	0.925
natural	11	**0.734/0.028**	0.320/0.077	0.324/0.092	0.674/0.054	0.730/0.036	0.358
nature	6	**0.732/0.020**	0.682/0.040	0.635/0.071	0.571/0.091	0.614/0.067	0.723
oblique	2	0.951/0.027	0.876/0.054	0.854/0.078	**0.959**/0.022	0.951/**0.021**	0.919
play	21	**0.625/0.032**	0.063/0.035	0.057/0.047	0.353/0.058	0.506/0.055	0.093
post	9	**0.790/0.013**	0.768/0.017	0.743/0.030	0.684/0.054	0.710/0.038	0.780
pull	14	**0.461/0.017**	0.117/0.018	0.145/0.053	0.310/0.082	0.353/0.084	0.204

word	n						
replace	4	**0.644/0.026**	0.565/0.040	0.541/0.063	0.629/0.054	0.631/0.075	0.499
restraint	5	**0.724/0.019**	0.677/0.028	0.633/0.064	0.511/0.088	0.571/0.056	0.642
see	19	**0.583/0.035**	0.153/0.039	0.174/0.039	0.407/0.065	0.440/0.061	0.234
sense	7	**0.761/0.045**	0.702/0.087	0.653/0.051	0.523/0.078	0.619/0.075	0.738
serve	11	**0.679/0.021**	0.358/0.044	0.312/0.087	0.566/0.054	0.592/0.043	0.508
simple	2	**1/0**	1/0	1/0	1/0	1/0	1
spade	3	**0.920/0.038**	0.913/0.045	0.917/0.038	0.846/0.049	0.865/0.040	0.879
stress	6	**0.734/0.026**	0.670/0.028	0.641/0.065	0.682/0.054	0.699/0.058	0.753
strike	18	**0.500/0.020**	0.149/0.036	0.164/0.055	0.267/0.068	0.373/0.086	0.229
train	9	**0.657/0.021**	0.614/0.036	0.588/0.047	0.537/0.052	0.570/0.065	0.622
treat	6	**0.731/0.038**	0.680/0.113	0.678/0.080	0.598/0.067	0.642/0..055	0.795
turn	27	**0.547/0.037**	0.071/0.050	0.093/0.076	0.270/0.074	0.395/0.060	0.185
use	6	**0.802/0.013**	0.787/0.032	0.781/0.025	0.705/0.044	0.767/0.030	0.799
vital	3	**0.18/0**	0.18/0	0/0	0/0	0/0	0
wander	4	**0.921/0.012**	0.912/0.018	0.912/0.017	0.896/0.037	0.908/0.018	0.907
wash	11	**0.608/0.011**	0.529/0.024	0.530/0.025	0.526/0.027	0.547/0.037	0.475
work	19	**0.590/0.040**	0.030/0.044	0.047/0.048	0.386/0.105	0.485/0.096	0.018
yew	2	**0.999/0.003**	0.999/0.004	0.987/0.014	0.963/0.022	0.983/0.012	1

5 Evaluation

5.1 Comparison with Other Algorithms

As we have mentioned, we compare results of each algorithm on both performance and robustness. We can compare them in two ways. One is to use the context-performance curve. Performance can be compared basing on the point-wise values of F-measure along a sequence of context windows, while robustness can be reflected by the flatness of the curve. Intuitively, a flat curve is more robust than a sharp one. Another way is to compare the overall features along the context windows the algorithms use. Means and Maximums along a sequence of corresponding window trajectories can be used to reflect the performance. While Standard Deviation can be used to reflect the robustness, a sequence with small standard deviation is more robust than that with a big one.

From Figures 2 and Figures 3, we can get an intuitive impression that TB_KNN not only achieves the best performance at most of points, but also has the flattest curve shape. This means TB_KNN outperforms other algorithm on both performance and robustness. This conclusion can be detailed by comparing Mean, Standard Deviation (listed in Table 1) and Maximum of TB_KNN with their correspondences in other algorithms.

Firstly, we compare values in the TB_KNN column with their correspondences in column P. For all 70 English words in Table 1, all Means of TB_KNN are consistently better than those in P. Only 2 S.D. bigger than those in P and 8 words have low Maximum than those in P. Comparing values in the TB_KNN column with their correspondences in column KNN, we can find only 1 Mean of TB_KNN are lower than those in KNN, only 2 S.D. bigger than those in KNN, 5 word have low Maximum than those in KNN. All differences are significant at least at the level of 0.01. This means our decision trajectory based classifier is better than a NB classifier or a KNN classifier in both robustness and performance. The combination of different context along a trajectory takes advantages of both NB and KNN methods. It seems that KNN directly based on word co-occurrence features suffers deeply from the problem of data sparseness in context. While KNN based on decision trajectory can alleviate the influence of data sparseness. In our final KNN decision, sense selection is also not sensitive to the number of nearest neighbors.

Comparing values in TB_KNN column with their correspondences in column LWNB, we can also find almost all values in TB_KNN are better than their correspondences in LWNB. The differences are not so bigger than those described in above paragraph, especially when the number of training samples is relatively big. In [12], the number of training samples is large.(most of them are more than several hundreds) they used 50 local training samples to construct a NB classifier. It is always impossible in our experiments and in most WSD tasks.

All of the Maximum and even most of Means in TB_KNN column are bigger than values in ENB. Comparing with ENB, We think the trajectory based approach may make use of NB decisions in a more systematical way than selecting some classifiers for voting in ENB, and also, our approach receives benefits from the final KNN decision, which can make some exceptions under consideration.

Let us give discussion on how our trajectory-based approach makes use of information in context.

Firstly, although each NB classifier uses bag-of-words as its features, because window size for NB classifiers is extended sequentially, the decision trajectory thus reflects influences brought by context words in different positions. That is to say, changing the position of a co-occurrence word in a sentence might cause different final decision in trajectory-based approach. While in point-based approach, as long as the co-occurrence word is in the context window, a classifier based on bag-of-words features always makes the same selection no matter how to change the position of that word. From this view, trajectory-based approach in fact makes use of position information of words in context.

Secondly, because of its implicit utilization of position information of context words, it may make use of information from some decisions locally correct but wrong in a bigger context. For example, we consider sentence S1 in section 2 again.

S1:我/ 看/ 这 /本/ 书/ 值得/ 一/ 读/.
 I think this piece book worthy a read

On the one hand, as we have said, when we use context window (3,3), we select the sense of "read" for 看. Although it is a wrong sense selection for this word in this sentence (when context window is (6,6)), it is a correct selection for the local collocation (when 看 collocates with 书, its sense is "read"). By saving this information, we can not only make use of information of sense selection for the sentence, but also information for this collocation. In other words, it seems that S1 gives us two samples for different senses of the target word.

On the other hand, theoretically, we think that a polysemous word changes their probability for different sense with the change of context window is one kind of pattern for sense ambiguity, trajectory based approach seems an efficient way to grasp this kind of pattern.

5.2 Trajectory

In TB_KNN, we need to calculate sense decision trajectory for each training sample, not all of these trajectories are eigen-trajecories. In TB_VOTE, we do not calculate sense decision trajectories for training samples, all training decision trajectories are regarded as eigen-trajectory, final decision for a new sample reduces to simple majority voting along the trajectory. Comparing TB_KNN and TB_VOTE, we can find that both performance and robustness of TB_VOTE significantly fall. This means existence of non-eigen-trajectories is in fact helpful.

In above experiments, we generate a context trajectory by adding one context word each time. We further explore if a looser trajectory can get the same performance. We first exclude even points in original trajectories in above experiments to get some new trajectories. For example, by excluding even points of the original context trajectory group TG1, we get:

$$T'_j = \{p_1,..., p_{2j-1}\}, j = 1,...,20$$

Note this T'_j is different from T_j in above experiments, where T_j is:

$$T_j = \{p_1,..., p_k,..., p_j\}.$$

Table 2. Results on the looser trajectories

Word	TG2 Mean/S.D.	TG3 Mean/S.D.	Word	TG2 Mean/S.D.	TG3 Mean/S.D.
art	0.63/0.03	0.62/0.04	hearth	0.79/0.01	0.78/0.01
authority	0.69/0.01	0.67/0.01	holiday	0.97/0	0.97/0
bar	0.78/0.02	0.77/0.02	keep	0.64/0.05	0.63/0.06
begin	0.82/0.03	0.82/0.05	lady	0.88/0.01	0.87/0.01
blind	0.96/0.02	0.96/0.01	leave	0.46/0.06	0.52/0.04
bum	0.898/0	0.90/0.01	live	0.74/0.02	0.74/0.02
call	0.48/0.04	0.47/0.05	local	0.91/0.02	0.90/0.03
carry	0.52/0.10	0.53/0.15	match	0.56/0.01	0.59/0.05
chair	0.95/0.01	0.95/0.00	material	0.65/0.03	0.65/0.03
channel	0.76/0.01	0.74/0.03	mouth	0.74/0.02	0.75/0.04
child	0.83/0.02	0.83/0.02	nation	0.94/0.01	0.93/0.01
church	0.82/0.06	0.79/0.06	natural	0.72/0.06	0.72/0.08
circuit	0.80/0.04	0.79/0.04	nature	0.74/0.02	0.74/0.02
collaborate	1.00/0.01	1.00/0.01	oblique	0.96/0.02	0.95/0.01
colorless	0.87/0.02	0.85/0.01	play	0.61/0.09	0.60/0.11
cool	0.96/0.01	0.96/0.01	post	0.79/0.02	0.78/0.01
day	0.82/0.01	0.82/0.01	pull	0.43/0.09	0.46/0.093
detention	0.94/0.02	0.93/0.02	replace	0.64/0.02	0.61/0.06
develop	0.53/0.05	0.54/0.05	restraint	0.73/0.02	0.70/0.02
draw	0.40/0.10	0.31/0.10	see	0.60/0.08	0.52/0.10
dress	0.65/0.06	0.63/0.05	sense	0.75/0.06	0.71/0.07
drift	0.53/0.13	0.47/0.13	serve	0.69/0.04	0.66/0.02
drive	0.63/0.14	0.58/0.15	simple	1/0/1	1/0
dyke	0.98/0.02	0.98/0.03	spade	0.90/0.04	0.90/0.05
face	0.93/0.01	0.92/0.00	stress	0.75/0.03	0.73/0.04
facility	0.81/0.07	0.82/0.09	strike	0.40/0.13	0.39/0.16
faithful	0.90/0.01	0.87/0.01	train	0.65/0.02	0.67/0.03
fatigue	0.92/0.01	0.92/0.01	treat	0.69/0.09	0.68/0.12
feeling	0.79/0.02	0.74/0.03	turn	0.55/0.04	0.53/0.08
find	0.439/0.05	0.40/0.07	use	0.80/0.02	0.79/0.02
fine	0.69/0.03	0.71/0.04	vital	0.18/0	0.18/0
fit	0.92/0.02	0.93/0.02	wander	0.92/0.02	0.91/0.02
free	0.95/0.00	0.95/0.00	wash	0.60/0.01	0.61/0.01
green	0.77/0.01	0.76/0.02	work	0.58/0.03	0.56/0.03
grip	0.83/0.02	0.80/0.04	yew	1.00/0.00	0.99/0.01

In this way, we get 20 different trajectories TG2: $T'_1,...,T'_{20}$, T'_j includes half number of points comparing with its correspondence T_{2j} in above experiments. The longest context trajectory includes 20 points. We repeat above TB_KNN experiment along these new trajectories. We repeatedly exclude even points to generate TG3 which include at most 10 points in their trajectories. We also repeat same TB_KNN experiment on TG3. Results on TG2 and TG3 are listed in Table 2.

Comparing Mean and S.D. in Table 2 with those correspondences in TG1 in Table 1, we can find that performance of classifiers using the trajectories with small number of points do not decrease significantly. That is to say, a shorter context trajectory can also achieve good robustness and performance.

6 Conclusions

This paper presents a new algorithm for WSD. We firstly construct a sequence of NB classifiers along orderly varying sized windows of context, and get a trajectory of sense selection for each sample, then use the sense trajectory based KNN to make final decision for test samples.

This method is motivated by an observation that there is an unavoidable uncertainty when a classifier is used to make sense selection for occurrences of ambiguous words by using different context windows. Our approach aims to alleviate this uncertainty and thus make more robust utilization of context while perform well. Experiments show our approach outperform some other algorithms on both robustness and performance.

Acknowledgement

Part of this work was done while the author was at Nara Institute of Science and Technology. He appreciates Professor Yuji Matsumoto and the Matsumoto laboratory in NAIST very much for lots of help he had received.

References

1. Florian, R., Yarowsky, D.: Modeling Consensus: Classifier Combination for Word Sense Disambiguation. In *Proceedings of EMNLP'02*, Philadelphia, PA, USA. (2002) 25-32.
2. Klein, D., Toutanova, K., Ilhan, H. T., Kamvar, S. D., Manning, C. D.: Combining Heterogeneous Classifiers for Word-Sense Disambiguation. In *Workshop on Word Sense Disambiguation at ACL 40*, (2002) 74-80.
3. Kilgarriff, A., Rosenzweig, J.: Framework and results for English Senseval. Computers and the Humanities. 34(1) (2000) 15-48.
4. Mihalcea., R.: Word Sense Disambiguation Using Pattern Learning and Automatic Feature Selection, *Journal of Natural Language and Engineering*, 8(4) (2002) 343-358.
5. Ide, N., Veronis, J.: Introduction to the Special Issue on Word Sense Disambiguation: The State of the Art. *Computational Linguistics*, 24(1), (1998) 1-40.
6. Weaver, W.: Translation. Locke, William N. and Booth, A. Donald (Eds.), Machine translation of languages. John Wiley & Sons, New York, (1955) 15-23.

7. Yarowsky, D.: Decision Lists for Lexical Ambiguity Resolution: Application to Accent Restoration in Spanish and French. In *Proceedings of the 32nd ACL.* (1995). 88-95.
8. Leacock, C., Miller, G. A., Chodorow, M.: Using corpus statistics and WordNet relations for sense identification. *Computational Linguistics*, 24(1). (1998) 147-165.
9. Yarowsky, D., Florian, R.: Evaluating Sense Disambiguation Performance Across Diverse Parameter Spaces. *Journal of Natural Language Engineering*, Vol.8, No 4. 2002.
10. Manning, C. D., Schutze, H.: *Foundations of Statistical Natural Language Processing.* MIT Press. (1999).
11. Ng, H. T., Lee., H. B.: Integrating Multiple Knowledge Sources to Disambiguate Word Sense: An Exemplar-Based Approach. In *Proceedings of the Thirty-Fourth ACL.* (1996).
12. Frank, E., Hall, M., Pfahringer, B.: Locally Weighted Naïve Bayes. In *Proceedings of the Conference on Uncertainty in Artificial Intelligence.*(2003).
13. Pedersen, T.: A Simple Approach to Building Ensembles of Naive Bayesian Classifiers for Word Sense Disambiguation. In *Proceedings of the NAACL-00*, May 1-3, Seattle. (2000).

Understanding Actions: Contextual Dimensions and Heuristics

Elisabetta Zibetti[1] and Charles Tijus[2]

[1] LDCI, Université Paris 8/EPHE,
41 rue Guy Lussac 75005 Paris, France
Elisabetta.Zibetti@iedparis8.net
[2] Laboratoire Cognition & Usages,
Université Paris 8 - 2, rue de la Liberté, 93526 Saint-Denis Cedex 02, France
tijus@univ-paris8.fr

Abstract. The aim of this paper is to analyze how people build an internal representation of actions done by others by matching observable data (objects, events, spatial and temporal context) with knowledge and by making inferences about not observable (intention, goal, anticipation, and planning) data. For that purpose, we list the set of dimensions that define *"what an action is"* accompanied with their contextual aspects, and the heuristics used by the human cognitive system. Finally, we propose that processing objects properties in terms of *function-to-patient relation* between objects, and in terms of *patient-to-function* within objects, is a possible way to explain the context effects with respect to the fine-grained vs. the large-grained semantic organization of the spatial and temporal context.

1 Introduction

People do interpret what happens around them. By interpretation, we mean the process that leads to a circumstantial mental representation created in order to understand a specific situation. That is the overall process that starts from the perception of objects, events and space-time relations between the objects (bottom-up processing) and ends with the production of goals and intentions, the establishments of causal links, and the predictions of future events, in others words *"knowledge activation"* (top-down processing). This overall process leads to the observer's comprehension of a specific situation.

An important part of the building of mental representations is interpreting the physical actions[1] done by others. This is a common cognitive process based on both external data (the cognitive system picks up information from the physical world) and internal data (knowledge). The interpretation process comprehends evaluating action effects in the physical world, such as primary and secondary effects, but also establishing the corresponding agents as sources of actions, and attributing a hierarchy of goals to each of them, that is used to anticipate next actions.

---

[1] We used the term "physical actions" in contrast with "mental actions" such as remembering thinking, etc.

We can think of understanding actions as being constrained by several factors related to a number of dimensions. For instance, goals attribution in action interpretation is constrained by the knowledge about what is seen, in other words by "*ontological commitment*" [1]: what is seen is seen as being an instance of the extension of a concept [2]. According to Gruber [1] different observers of the same visual scene might have a same or a different ontological commitment. Having the same ontological commitment in understanding visually perceived actions provide to the observers a shared interpretation. Additionally, it might allow them to attribute to the agent that performed the action the same kind of perception, knowledge, and goals that the observer would have in the same situation. Oppositely, these results of the interpretation process could differ when the observers do not share the same ontological commitment.

In addition to knowledge, the most constraining of these factors is probably how much of the context is processed, with on the one hand a shared ontological commitment based on shared context and, on the other hand, a different ontological commitment, based on unshared context. Note that a different ontological commitment generally results in misunderstanding.

Let's explain this point with the two following examples. In a film of Hitchcock (example A, Young and innocent, 1937), two women walking on the beach are seeing by far (1)- "*a man which was knelt close to a lengthened body, going away*" and coming closer (2) - "*that the body on the beach is that of a young woman who has been just strangled*". The two women infer that "*the man assassinated the young woman*". They did not have the temporal contextual information that the man walked like them on the beach and that he saw a body lengthened on the beach and found a young woman strangled. Hitchcock gave the spectator information in such a way that if the two women do have a same ontological commitment, the spectator has a different one. This example is related to *temporal context*.

In a Charlie Chaplin movie (example B, The idle class, 1918), a woman tells her husband that she's leaving because he is drinking too much. The husband's back is towards the audience and, from behind, the spectator perceives him to suddenly tremble and infers that the husband is crying. Then, the husband turns around and the spectator realizes that he is shaking his next cocktail. In this example, the woman and the spectator do not share the same visual perspective. The wife sees her husband holding and trembling the shaker while the spectator only perceives the husband's body movements from behind. This example is related to *spatial context*. Those two examples raise a first fundamental question: what are the temporal and spatial contextual dimensions that contribute in action understanding?

Another crucial dimension in action understanding is the internal hierarchical level of action attribution. Hierarchical levels of action have been described, for instance by Kemke [3], as being the realization level (physical local, body movement), the semantic level (physical global, environmental effect level), the pragmatic level (motivational/intentional). An example cited by Kemke [3] is the act of moving ones finger in a certain way (realization level), which coincides with switching the light (semantic level) on which again, coincides with alarming a burglar (pragmatic level). Bottom level actions that correspond to sensory-motor actions of the realization level, such as body movements in example A: "*the man was going away*", and in example B: "*the husband was having sudden tremors*" are more likely to lead to shared ontological commitments. In opposite, upper level actions corresponding to causes, such as in example A: *"the man looking for help vs. fleeing after the murder"*, and in example B: *"the husband crying vs. shaking alcoholic beverage"* are more likely to be providers of different commitments.

It is often argued that context helps understand actions. Our point of view is that the description of "what action in context is", should explain differences in ontological commitment (as seen in the above examples) as effects of both the hierarchical level of action and the amount of context. For instance, ambiguity of simple action patterns of body gestures (e.g. *"a man which was knelt close to a lengthened body,* in example A) increases with the hierarchical level of interpreting what is seen as action (touching, hurting, strangling, killing by strangulation, killing anyway). An ambiguity degree that can be balanced by the amount of context in such a way that *"the higher you interpret the more context you need"*. In other words, the richer the context is, the easier it will be to disambiguate and comprehend a body gesture at a given level of interpretation.

The goal of this paper is twofold: to propose a rational list of action dimensions that could be used for ontological commitment and to associate to these dimensions contextual effects as a basis for action understanding heuristics. These heuristics help making inferences from the processing of the physical and relational properties of the objects surrounding the potential agents. Thus context is studied along the processing of the dimensions of what is seen as being actions.

Among the dimensions of action, we first list the dimensions of the external data, from the display of a single state to the observation of a whole action realization. Second, we list the hierarchical levels of action interpretation, that is what we call the intentional dimensions, from basic sensory motor actions to task realization. Third, we present the components of what is processed to interpret what is seen as being an action realization, from the processing of objects through *between-objects relational properties* (one object being used to act on another) and through *within-objects properties* (one object being transformed in order to further use it). These three sets of dimensions are accompanied with their contextual effects. This is on the line of problem solving theories [4] that have already defined some of the dimensions (states, state transformation, actions, goal, ans so on) that we used.

2 From State to Action: The Dimensions of External Data

2.1 States

Defining *"what is a state"* is theoretically fundamental because events and actions are generally defined as state transformation, or production [5]. As pointed out by Le Ny [6] and Baudet [5], the notion of state includes a temporal location and a state is generally defined as the set of the static properties of the external and physical objects of a given situation.

As a matter of fact, observers individualize a state like a concrete fact, an occurrence in the world: perceiving a single state of an evolving situation provides valuable clues to put forth explanatory assumptions on what s/he perceives, to guide his/her comprehension. Thus a state appears to describe (i) the spatial organization of the physical properties of the objects that compose the situation at time «t» and, interestingly as consequences, (ii) it sets up the «pre-requisites» that are to be satisfied to allow an event or an action. For example, a car parked in front of the garage gate condemns the entry. This state informs any observer about the inaccessibility of the garage. Hence, we have a contextual pre-requisite that is not satisfied for other cars to enter the garage. It follows that, processing the visual scene, as a static image, is sufficient to activate action representation when the spatial arrangements of objects is satisfying certain contextual prerequisites. For example, we found that a single image activates action representations [7]. In our study, using a lexical decision task with cross-modal semantic priming, the image displays a ball (invariant) on a plane made of two segments (variant). The ball stands on the above segment and the second one is adjacent on the right. The two segments vary in orientation (without slope, ascending slope, downward slope). After the presentation of the image, the target that follows is a verb of displacement (*Compatible Target: falls, rolls, goes down,..*), a different verb (*Incompatible Target: eats, dictates, confuses,..*) or a non-word (*nuavre, crufe, plavet,..*). We found a priming effect for the compatible verbs, modulated by the orientation of the support. It appears thus possible to activate representations of displacement actions (expressed in language by verbs) by processing the physical and relational properties of objects. Thus, the cognitive system of the observer is able to recognize an action even with no temporal information. Nevertheless, this is possible only if the spatial information provided by the spatial context satisfies the "pre-requisite" for potential action. This processing of "pre-requisites" is what we suppose to be the first spatial contextual effect in actions understanding.

2.2 Happenings

To be able to perceive an event, one needs to perceive a change of state. Note that the cognitive system cannot capture or represent a change if the current state (t+1) is not related with a previous one (t). Processing the invariants, from which to perceive the variations, makes possible to capture the dynamics of the event. It follows that "*What is a state*" is also crucial to define the current state of an evolving situation as different from the previous state.

Therefore, we call happenings the lowest kind of change of state in the world. That is simple events that do not (apparently) involve actors, purpose, or intention, but produce perceptible movement in objects as a change in time and space [8]. Examples might be rainfall, water boiling, sun rising, etc. As Nelson [8] mentions "*there are also complex events with structure but no assignable purpose or actors (e.g. thunderstorms, earthquakes). These seem naturally to fall under more structured semantic categories even if they can't be assigned a purpose*".

2.3 Events

We define an event as a temporal change of physical nature on at least an object in time and space. As for happenings, this definition is thus built only on observable data of the physical world. For example: "*object A moves towards the line horizontally....*" (a ball rolls on the street); "*the object B changes color at t+1*" (a red light turns green). Nevertheless different from happenings, this change occurs on an object in space and time. Therefore because the change concerns the physical object, it requires at least two successive states in which the object differs at least for one of its physical properties. In the most abstract construal, a change is an ordered pair of facts: the fact that is obtained prior to the change and the fact that is obtained after the change took place [9].

Another alternative theoretical approach consists in defining the event as a sequence of intervals, each one being considered as a sub-event. This approach makes it possible to describe what occurs during an event [10, 11]. This is probably the reason why in the psychological literature, events are often described as being high level cognitive descriptions of a segment of time at a given location that is perceived by an observer to have a beginning and an end.

Both approaches of events, (i) as being a passage from a state to another, and (ii) as being a sequence of intervals, appear valid. The first point of view (i.e., an event as passage from a state to another) allows us to sketch the event as a segment of time with a beginning and an end (i.e. the initial state and the goal state). The second approach (i.e., an event as sequence of intervals) allows us to better tackle the question of the identification and the organization of a continuous stream of events [12], and thus to better understand how, from the perception of events sequences, one leads to an elaborate cognitive comprehension, in terms of causes, or actions, by establishing causal links between the various "moments" that capture the changes [13, 14]. A classical example is Michotte [15] findings. Observers that see two balls moving one after the other on the same direction, relate that the first *pushing* the second even if the former did not even touch the latter. In addition, the second approach captures the temporal context effects. For example, using the Heider & Simmel [16] video animation that shows three geometrical figures interacting randomly, and asking participants to stop the video when an action was done, we found that participants provide more and more elaborate action description, as well as agents' description, when the length of the temporal context increases [13]. Thus, from the perception of simple but compelling events, people give general descriptions about the objects and the events they see, and when the temporal context provided increases, they extract elaborate actions sequences describing goals and intentions of agents.

2.4 Actions

It is generally admitted that events and actions are not synonymous of the same notion. Nevertheless definitions sometime express opposite points of view. For Zacks and Tversky [17] for example, *"actions happen objectively in the world, whereas events arise in the perception of observers"* (p.4). Opposite to those authors, our position is that events are what happen objectively in the world while actions are cognitive interpretations of an observer about perceived events. Actions are not built solely on observable data of the physical world as events do. Actions differ from events as being cognitive entities. They are more specific than events. Actors perform them as an organized activity to accomplish an objective. Physical actions include one movement or a series of movements performed by an agent in order to achieve a goal under some intentional purpose by acting on objects, or on other potential agents. Thus, seeing actions done by others implies a certain interpretation made of the presence of an agent, the attribution of intention, of goal and of a cause. In this view, the action is only a possible cohesive and coherent cognitive interpretation of an event in its context.

2.5 Summary

What is seen can be a single state (e.g. an image), or a change of state that defines an event. Processing a single state appears sufficient to interpret what is seen in terms of possible action realization when contextual prerequisites are satisfied. Many authors establish a distinction between event and action [e.g., 5, 14, 15, 18, 19], as the action implies the concept of agent, causality, goals and attribution of intentions. That means that the differences between event and action relate basically to cognitive attribution on behalf of an observer. According to Casati and Verzi [20] « *Whether or not actions are treated as events, one might be tempted to distinguish between actions proper (such as John's raising his arm) and bodily movements (such as John's arm rising), or between intentional actions (John's walk) and unintentional ones (John's falling into a hole)*".

Thus, the main question is how "*what is not seen*" that is the intentional dimension of actions, can be inferred from "*what is seen*" that is the external information that comes from the physical world. As we mentioned in the introduction section, given that differences in the interpretation arise from different ontological commitment, based on unshared temporal or spatial context, we hypothesize that context plays a major role in inferring what cannot be seen (goals, intentions, next states, and so on) from what is seen (events, objects and objects transformations) in order to arise to a coherent interpretation.

3 From Movement to Action: The Intentional Dimension of Action

A physical action interpretation is grounded on both internal data (knowledge) and spatial or/and external data that can be made of single states or made of events, both being sources for seeing action realization. Since the external data can be more or less informative, - according to the contextual information provided -, their interpretation as action can be more or less complete and being of different hierarchical levels.

The tree levels that we list below should make explicit the increment of contextual information processed by the cognitive system and the organization of this information during the interpretation of intentional actions.

3.1 Primitive Pattern Activities or Biological Movements

By pattern, we mean a low-level structured motion activity. The common notion of gesture in gesture recognition is an example, in which a specific set of hand, arm and legs movements are modeled (either explicitly or by observation) and later recognized [21].

The individual and cultural variation in gesture events or activities[2] is enormous, but there are numerous hands, arm and leg pattern gestures that, at their primitive

---

[2] Activities consist of motion patterns that are temporally periodic and possess compact spatial structure. Examples include walking, running, jumping, etc.

level, seem to be "universal". Simple patterns executed by some actors are also easily recognizable independent of context [22]. The classical examples include walking, running, jumping, grasping, throwing, etc, performed by people. For instance, isolated from any kind of contextual information, locomotion actions are better recognized than social or instrumental actions [23]. It appears that human movements are perceived via interactions between low-level motion measurements and high-level cognitive processes, although the precise nature of these interactions is not yet well understood.

3.2 Simple Body Actions

Simple body actions are primitives of the behavior of the body of an agent while performing some actions. Contrary to primitive pattern activities, simple body actions compose more complex actions. Attached one to the other, the chain of simple body action permits the agent to realize the goal at hand. This is the kind of know-how one can find in detailed instructions (in a cooking book, for instance). It seems clear that Simple body actions play an important role in recognizing and understanding more complex actions. Nevertheless, the object properties, and the context as well, are also crucial. For example, vigorously shaking the hand is a low level structured motion activity that takes place in more complex actions related to objects such as shaking someone's hand (saying hello) vs. shaking a shaker (preparing a cocktail). A low-level structured motion activity that becomes part of a more complex action with an object can be differently interpreted according to the context shared by the observers (cf. examples A and B in the introduction section).

As shown above in the *"shaking hand"* example" (saying hello vs. having a drink), observing action patterns in isolation provides ambiguous action interpretation, while the evidence for a particular action concept *pop out* immediately when the gesture is perceived in conjunction with at least an object even in an extremely poor context sharing[3]. Therefore those simple body actions need some sort of local contextual knowledge to make sense. This local knowledge comprehends basically the objects and their own physical and relational properties. So that one can have *"shaking somebody's hand"*, or *"shaking maracas"*, *"shaking a cocktail"*, etc... This kind of actions seems to be quite stable within various contexts. Nevertheless, their meaning can change or be differently specified when the surrounding scene is highly stereotyped. For example *"shaking somebody's hand"* can be *"saying goodbye"*, when the contextual scene depicts the end of a party vs. *"signing a contract"* after a business arrangement, vs. *"receiving congratulations"* in a university class room where a jury stands in front of a candidate, and so on. Those differences in the interpretation of the same body action performed in highly stereotyped contexts are related with the notions of goals attribution. Thus, the contexts effects described in the next section will be those related to the hierarchical higher level of action described by Kemke [3] as being the pragmatic level (motivational/intentional).

---

[3] As matter of fact, in language to catch the meaning of a simple body action verb, one needs to figure out the part of the body that perform the action, and more informatively, the object to which the movement is directed at.

3.3 Goal Directed Actions

Contrary to primitives pattern activities (*walking*) and to simple body actions as primitive (*shaking hands*), goal-directed actions are long time running pieces of complex actions done to reach an unseen goal (e.g., *to prepare a cocktail*). Those goal-directed actions are interpretations made on the basis of what an observer is given from the physical world and of the knowledge s/he possesses about it. In other words goal-directed actions have a richer psychological structure than simply body actions.

Therefore, the cognitive system appears to make use of some heuristics for interpreting actions as directed towards a specific goal. The first heuristic is to relate successive actions: actions have a purpose, the purpose of attaining a goal. It follows that actions that precede the attainment of an action-goal are interpreted as having been done in order to make that action-goal possible.

A second heuristic is based on goal-directed behavior: if an observer believes the agent's goal is to serve a cocktail, he will probably interpret the actions that precede this event (i.e. *pouring the alcohols into the shaker, mixing the liquids,..*) as a function of the goal. Both the probability that the agent does indeed have this goal and the probability that he will attain it should increase as the plan unfolds (i.e. *closing the shaker, shaking the beverage, pouring it in a glass and so on*).

A third heuristic comes from top-down goal decomposition. For example "*serving a cocktail*" (i.e. the action-goal) can be broken down into subordinate actions (i.e. sub-goals) such as "*pouring the alcohols into the shaker*", "*mixing them*", "*pouring it in a glass*", and so on. All those are plan sequences that "*serving a cocktail*" requires in order achieving the intended goal. In fact, studies about the recall of visually perceived actions show that memory for actions follows schema-plans [24, 25, 26]. Actions situated near the top of the action hierarchy, closer to achievement of the action-goal, are more easily memorized than those situated at the bottom of the hierarchy [25] and that an action is more easily recognized and recalled when it is followed by achievement of an action-goal than when it is not [26]. These results showed that a perceived action can be recursively decomposed into sub-goals because of the "*in order to*" criterion. This "*in order to*" link generates a hierarchical structure in which the action-goal is at the top. However, another interpretation of those results is that actions goals bring implicitly all the temporal and environmental[4] contextual information that allows the realization of the goal, and therefore are better recognized and memorized. This, according to the lines of our hypothesis is a primary context effect.

A fourth heuristic comes from bottom-up processing and is mainly based on processing the temporal context: inferring goal-directed actions show another kind of hierarchical structure, less related to knowledge and more associated to perceptually-driven mechanism that automatically segments the motion stream into meaningful actions. Numbers of authors [12, 14, 27] have demonstrated that a continuous stream of events is indeed organized into discrete perceptual units of action. When observers are asked to segment continuous activity into either coarse or fine natural units, they generally coincide around breakpoints [10]. In particular when observers are asked to

---

[4] In terms of satisfaction of contextual prerequisites.

segment human goal directed actions into coarse units (i.e. the largest unit that seems natural), their description of events include specifics information about objects and only general information about actions. In contrast when asked to segment the action into fine units (i.e. the smallest units that seems natural), they give more specific information about actions and more general information about objects [14]. This indicates that understanding high hierarchical level actions, when linking action to action in large sequences, is based on fully processing the objects. This is not the case when interpreting low hierarchical level action.

3.4 Summary

We distinguished three types of actions. Primitive pattern activities (*walking, running,*) performed by agents are directly perceptible and do not necessarily require interaction with other objects to be interpreted as actions. Simple body actions are incomplete and ambiguous if perceived in isolation of any kind of local contextual information (*to push -a button-*). Both primitive pattern activities and simple body movements enter in the composition of goal directed actions, but they are not sufficient to allow the observer to interpret to which purpose they are performed if they are not structured around a temporal context and the satisfaction of contextual prerequisites.

We have seen also some of the heuristics that the human cognitive system uses to compose these primitives in compound actions. However, a simple concatenation of primitives does not provide the meaning of what is done. It is necessary to organize them in a hierarchical goal-sub-goal structure that provides the context for interpretation.

4 From Objects to Objects: The Functional Dimensions of Objects

We have seen that physical external data, such as states, events, simple patterns of activities or body movements, provide, with respect to the level we address, contextual information. Nevertheless in action interpretations those elements do not seem sufficient to provide the super-ordinate goals of the agent.

Generally speaking, when an observer describes an agent performing an action in a specific context, he/she infers the goal of the agent on the basis of the opportunities offered by the actual environment, and on the basis of the agent's properties. That is, according to the physical, functional and relational properties of the objects that compose the environmental context and according to the temporal contextual changes.

From now, we still don't have a clear understanding of the factors that constrain the interpretation of action and how the interpretation process takes into account the context in order to disambiguate events and build a coherent action understanding. However, we can analyze the specific role that is played by the functional and patient properties of objects that compose the visual scene. Hereafter, we will focus on the role of physical object properties as a key contextual dimension for understanding actions.

4.1 Agent, Functional and Patient Properties of the Objects

We defined the action as the cognitive interpretation of a change done by at least one agent that transforms at least one object of its environment. The agent, which is the entity that performs the action, is generally a self-animated entity, such as a person or an animal. The objects being changed are physical objects and include persons, animals and unanimated objects. Thus, when a transformation of objects is seen, it might be an event, with no agent (the ball is rolling), or it might be the effect of an action done by an agent (John is pushing the ball). In the second case, the main question is how an observer attributes to an agent the perceived changes. A debate exists about how one chooses an entity as being the agent of the action [e.g.; 28]. The identification of the "source" of the movement that accompanied action is indeed fundamental, because it leads to one interpretation of action, rather than another, accordingly to the attribution of goals supposed to be those of the agent. It follows that the first condition for attributing to somebody the role of agent is its belonging to the category of intentional beings. The second condition is the capture of the relational properties between the properties of the agent and those of the objects being changed. Because if an object has been changed, this means that some of its properties, that we named *"patient properties"* have been matched with some of the properties of the agents, that we name *"functional properties"*. For instance in this view, in the example *"John [A] pushing the ball [B]"*, the changed properties (i.e. *patient properties*) is the place of the ball, and Johns' hand (that is used to push the ball) is the *functional property*. So the function of the change of state of the ball (in stopped vs. rolling) is attributed to John who is then the agent of the action. Nevertheless, for our rational two things need to be pointed out:

(i) If [B] were a large and solid truck, the causal attribution would be different: the observer would be looking on the visual scene, for a more plausible agent or for a tool that eventually could help John pushing the truck. In others words, the matching between the patient and the functional properties, provide the direction of the actions and then the determination of the agent role. In other words who or what caused the change.

(ii) The ball could have fallen down even if John was not there: this is a co-occurrence of two facts (the ball falling and John's movement) that happens to be linked by the observer because of the spatial proximity of [A] and [B] in the scene and of the temporal causal relation between the moving of [A] and the falling down of [B]. These are contextual relational properties in the sense that because a human observer is not an ideal observer, they are expression of a possible point of view. The position of the observer determines what is seen. Such contextual manipulation of the position of the observer is exactly what sight gags are based on.

Now the main question we raise here is how does an observer process the contextual information in order to build a coherent action interpretation?

People use acting objects on patient objects in many of the everyday actions they perform (cutting the meat -with a knife-, filling the bowl –with milk-, cutting a paper -with scissor-, wipe the glass -with a paper towel-, and so on). These acting unanimated objects can be tools (the knife) but also objects (paper towel) that because of their functional properties are used to obtain an active effect on a patient object (to cut the meat, to wipe the glass). To some extent, objects (and not only tools) can be considered as possessing the same properties of the agents if they produce an effect. In both cases (animated, unanimated or tools), an object can be used as a functional acting object on a patient object if its functional properties (the cutting property of the blade, the absorption drying property of the paper towel) match with the patient properties of the patient object (the tenderness of the meat, the wetness of the glass).

Thus, the distribution of functional and patient properties among the objects of a scene (the spatial context) appears to be an important provider of inferences about action and heuristic for the interpretation.

We will explain this first heuristic with the following example. Seeing for instance the set of following objects, "*John, salt box, meat, plate, hand, knife, stove, fork*", the set of oriented application of functional properties give a set of transformations related to basic actions and to a main complex action, which is here having a dinner: *Johns' hand to meat, meat to stove, salt to meat, meat to plate, hand to fork, hand to knife, fork to meat, knife to meat, and so on.* The order of use of these "*function-to-patient*" relations comes from knowledge (*cooking the meat before cutting it in order to eat it*), but many of the basic actions come from matching objects properties in order to find which one can act on another one. We argue that action interpretation and action anticipation come from processing objects properties with the heuristic of matching objects properties in a *function-to-patient oriented relation*.

This *function-to-patient oriented relation* allowed the observer to caught the organization of the contextual information and then use it to build his/her interpretation. This heuristic can be applied due to the structural properties of the objects. In fact, objects have parts that are assembled in a certain manner to provide a structure: the handle and the cutting blade of the knife are assembled in a way the hand can act on the handle (patient property) and the cutting blade (functional property) can act on an object that can be cut. In others words the parts of the structured objects allowed inferences about actions when almost two objects are in interaction or are seen as having possible interaction accordingly to the opportunity or constraints provided by the context.

A second important heuristic for action interpretation is related to the concept of *affordances* [29]. Affordances are relations between perception and action, in others words the "seeing" by an observer about how to act on an object. This "seeing" might be seeing the match between the functional properties the observer is having and the patient properties of the object (i.e. the paper towel example).

A third, but causal heuristic is as follows: "*if someone acts on something (*a patient object*), this is done in order to use it for doing something (*functional object*)*: taking the knife (patient object) is for cutting (functional object). This is a very powerful heuristic, since objects are "*calling*" each other. Thus an observer can organize the perception of an event in a specific context around the relationships established between the objects of the scene.

A fourth heuristic is based on the set of successive actions that are inferred from objects transformation, using the *patient-to-function* transformation. Each of the past actions can be either a prerequisite of the successive actions, or a subordinate action[5] of the next action. Thus the last anticipated transformation provides the super-ordinate

---

[5] Determination of preceding transformations as being prerequisites or subordinate actions is not under the scope of this paper.

goal that subsumes the preceding sets of actions. The objects *"Salt box, meat, plate, hand, knife, stove, fork"* will provide *"having a dinner"* because the last transformation is cutting the meat for eating.

A fifth heuristic is a heuristic that is strongly related to the concept of ontological commitment, that is *"if someone is given a goal, then he or she intends to reach that goal"*.

One can argue that the world is not so smart that the set of seen objects sufficiently provides the necessary *function-to-patient* matches to create a chain of actions and determine a possible goal, - we agree of course -, nevertheless the context effects captured as *function-to-patient oriented relations* allows two processes. The first is the process of differentiating the objects that participate in action from the objects that don't. The second process, which is to schedule a succession of actions by ordering the use of objects, leads the observer to a coherent interpretation. If processing elementary actions can provide a number of heuristics for interpreting events as actions, we have shown that processing objects can also provide a number of heuristics based on what is perceived.

5 Discussion and Conclusion

In order to analyze the different components that take part in interpreting action done by others, from visual perception of single states or from events, it was nevertheless necessary to briefly define what we mean by action, what we mean by context, and attempt to see how the interpretation process makes use of context. Therefore we proposed that processing objects properties in terms of *function-to-patient relation* between objects, and in terms of *patient-to-function* within objects, is a possible way to explain the context effects with respect to the fine-grained vs. the large-grained semantic organization of the spatial and temporal context. Thus, there were different levels of events to be distinguished, and these include at least the following: state, happenings, events, human body patterns, and goal-directed actions performed by active human agents.

Generally speaking, it can be said that observing and understanding a situation means integrating successive events into a coherent whole, because when a series of events is not perceived as a succession of actions with causal links, it just gives rise to a series of change of state descriptions (*x moved right, moved left*, etc.). How causality is established in order to account for the perceived changes is a huge epistemological question that we will not treat here. However, if the temporal context is often related with the causality concept[6], according to our hypothesis it is also related with *effectivity*. This is a notion we introduce in order to underline the

---

[6] Two of the meanings of causality are related to actions understanding and context. The first is about perceptive causality established, on the ratios of space-temporal contingency, between two distinct elements [e.g., 15]. This first definition makes basically reference to the whole physical conditions and parameters that the physical stimulus must respect so that a relation of causality can be perceived between two or several events. The second meaning of causality is about intentional causality, which can be attached to the problems of goal and intentions attribution, in others words to the "psychological causes" inferred by the observers and attributed to agents [e.g., 16].

difference between cause and effects in action understanding. In the literature, causality is often used for both causality and *effectivity*. The effects from a psychological view can be assimilated to the notion of expectation and anticipation. Therefore, in a very restricted formula, we can say that from the point of view of the observer, the temporal context is made of the previous perceived actions that cause the current on-line action. This temporal context of the current action helps establishing *effectivity (what will be the results of what is happening right now given what has already been done)* as well as expectations about future actions that have to be anticipated. Given the heuristic that *"if someone acts on something, this is done in order to use it for doing something else"*, the larger the temporal context and the greater the number of transformations that have been operated on objects are, the more constrained *effectivity* and anticipation of what could happen is.

In short, because physical actions that have been done by people in the physical world have mainly the purpose and the effect of transforming objects or situations, we reasoned that objects should play a crucial role at the time of interpreting actions. Furthermore, that it is starting from their properties that the context effects make sense [30].

Acknowledgments. The authors are grateful to Rob Pratt for reviewing of this paper.

References

1. T.R. Gruber. A translation approach to portable ontologies. *Knowledge Acquisition, 5(2),* 199-220, (1993).
2. B. Bachimont. Engagement sémantique et engagement ontologique: conception et réalisation d'ontologies en Ingénierie des connaissances. In J. Charlet, M. Zacklad, G. Kassel & D. Bourigault (Eds.), Ingénierie des connaissances, évolutions récentes et nouveaux défis. Paris: Eyrolles (2000).
3. C. Kemke. *About the Ontology of Actions*, Technical Report MCCS-01-328, Computing Research Laboratory, New Mexico State University, (http://www.cs.umanitoba.ca/~ckemke/), (2001).
4. A. Newell & H. A. Simon. Human problem solving. Englewood Cliffs, N.J., Prentice Hall (1972).
5. S. Baudet. Représentation d'état, d'événement et d'action. *Cognition et Langages, 100,* 45-64, (1990)
6. J.F. Le Ny. *La sémantique psychologique*. Paris: Presses Universitaires de France, (1979).
7. E. Zibetti, F.S. Beltran & C.A. Tijus. Le rôle des propriétés de la situation dans l'activation de la représentation d'action: étude à partir d'une tâche de décision lexicale. *Journée "Sémantique du verbe"* - Université Paris 8, 18 November (2004).
8. R. Nelson. NSF/DARPA Workshop Perception of Action, *Final Report*. Prepared by: A. Bobick MIT Media Lab, (1997).
9. G.H. Von Wright. *Norm and Action. A Logical Inquiry*, London: Routledge and Kegan Paul (1963).
10. R. Langacker. *Foundations of Cognitive Grammar*, vol. 1. Stanford, Ca.: Stanford University Press, (1987).
11. D. McDermott. A temporal Logic for reasonning about process and plans. *Cognitive Science, 6,* 101-155, (1982).

12. D. Newtson. Attribution and the unit of perception of ongoing behavior. *Journal of Personality and Social Psychology, 28,* 28-38, (1973).
13. E. Zibetti, E. Hamilton & C.A. Tijus. The role of Context in Interpreting Perceived Events as Action. In V. Akman, P. Bouquet, R.Thomason, R. Young (Eds.). *Lectures Notes in Artificial Intelligence, vol. 1688, Modeling and Using Context,* New-York: Springer. 431-441, (1999).
14. J.M. Zacks, B. Tversky, & G. Iyer. Perceiving, remembering, and communicating structure in events. *Journal of Experimental Psychology-General, 130,* 29–58, (2001).
15. A.E. Michotte. *The Perception of Causality.* New York: Basic Books, (1963).
16. F. Heider, & M. Simmel. An experimental study of apparent behavior. *American Journal of Psychology, 57,* 243-259., (1994).
17. J.M. Zacks & B. Tversky. Event structure in perception and conception. *Psychological Bulletin, 127,* 3-21, (2001).
18. G.A. Miller & P.N. Johnson-Laird. *Language and perception.* Cambridge, Ma.: Harvard University Press, (1976).
19. T.A. Van Dijk. Semantic macro-structured and knowledge frame in discourse comprehension. In Just & Carpenter (Eds.), *Cognitive processes in comprehension,* Hillsdale, New Jersey, (1977).
20. R. Casati & A. Varzi. "Events", *The Stanford Encyclopedia of Philosophy,* Edward N. Zalta (ed.), *(Fall* 2002).
21. J.M. Rehg. NSF/DARPA Workshop Perception of Action, *Final Report.* Prepared by: A. Bobick MIT Media Lab, (1997).
22. G. Johannson. Visual perception of biological motion and model of its analysis. *Perception & Psychophysics, 14,* 201-211, (1973).
23. W.H. Dittrich. Action categories and perception of biological motion. *Perception, 22,* 15-22, (1993).
24. C.E. Cohen & E.B. Ebbesen. Observational goals and schema activation: A theoretical framework for behavior perception. *Journal of Experimental Social Psychology 15,* 305-329, (1979).
25. E. Lichtenstein, & W.F. Brewer. Memory of goal-directed events. *Cognitive Psychology, 12,* 412-445, (1980).
26. W.F. Brewer, & D.A. Dupree. Use of plan schemata in the recall and recognition of goal-directed actions. *Journal of Experimental Psychology: Learning, Memory and Cognition, 9,*117-129, (1983).
27. D. Newtson & G. Engquist. The perceptual organization of ongoing behavior. *Journal of Experimental Social Psychology, 12,* 436-450, (1976).
28. R. Gelman & E. Spelke. The development of thoughts about animate and inanimate objects: Implications for research on social cognition. In J.H. Flavell & L. Ross (Eds.), *Social Cognitive Development Frontiers and Possible futures.* Cambridge, Ma.: Cambridge University Press. 43-66, (1981).
29. J.J. Gibson. The theory of affordance. In R.E. Shaw & J. Brandsford (Eds.) *Perceiving, Acting, and Knowing.* Hillsdale, New Jersey: Erlbaum, (1997).
30. L. Ganet, P. Brézillon & C. Tijus. Explanation as Contextual Categorization. Modeling adn Using Context. 4th Interdisciplinary *Conference CONTEXT 2003, LNAI 2680.* Standford, CA, USA, Springer-Verlag Berlin Heidelberg, 142-153 (2003).

Applications of a Context-Management System

Andreas Zimmermann, Andreas Lorenz, and Marcus Specht

Fraunhofer Institute for Applied Information Technology,
Schloss Birlinghoven,
53754 Sankt Augustin, Germany
{Andreas.Zimmermann, Andreas.Lorenz,
Marcus.Specht}@fit.fraunhofer.de

Abstract. In ubiquitous computing, the user will be interrupted in performing a task only if the information is relevant to the task or highly important in the situation to justify the interrupt. The information selection and presentation therefore should be adapted to the user and his current context of use. Nowadays, uncounted Content-Management Systems provide access to a large amount of information, but without context, information is just data. This paper introduces Context-Management as a new approach for the design of context-aware systems in ubiquitous computing. As a proof of concept we illustrate three prototypical implementations of contextualized information systems in different applications domains and decompose the underlying framework into its foundational components.

1 Introduction

Today mobile devices deliver information that is adapted to the current location, network bandwidth or screen size of the user's device and commercial solutions push the mobile market to providing content and services on a variety of devices and visual displays. Future communication technology must take a step out towards new "smart" information services that use sensor fusion and multi-modal user interfaces for contextualized user interactions.

In ubiquitous environments the current context of use becomes more and more important as the computer disappears. Nevertheless in today's location based services the adaptation to location is only rarely combined with a personalization to the individual user [9]. From our point of view it is not enough to supply content or services that consider single environmental or user characteristics but to identify approaches for the integration and interpretation of different sensing components for modeling the user context more appropriately. The combination of sensor data for context acquisition and user modeling appears to be one of the key challenges for personalized information systems of the future. In this sense delivering information in the *right form* to *right place* at the *right time* has mainly two facets:

1. Personalization allows users to get information adapted to their needs and interest or other personal characteristics. It is often closely connected to the use with personal devices. User models deliver the main parameters for selecting and adapting information presentation.

2. Contextualization extends personalization so that also environmental states can be taken into account for more intelligent services. For contextualization also a variety of displays and environmental sensors become important for the user interaction, the interaction tracking, and interpretation of sensor data.

In the following paper we will introduce a toolkit for contextualizing applications and propose an infrastructure to integrate context and content management. We will demonstrate with three case studies how we have built personalized and contextualized systems based on this infrastructure.

2 A Context-Management Framework

Recent projects at the Fraunhofer Institute for Applied Information Technology (Fraunhofer FIT) developed context modeling components and personalization engines in order to make systems adaptable and adaptive to the user's behavior. Many of these components were specialized applications, tailored to one special domain or environment and rarely reusable in other or subsequent applications. Common problems that emerge during the development and the reuse of components are the strong dependency on the domain, no open and standardized interfaces, no uniform representation of context models and user profiles, and the use of different and distributed data sources and highly dynamic domains where variable properties change in short time intervals.

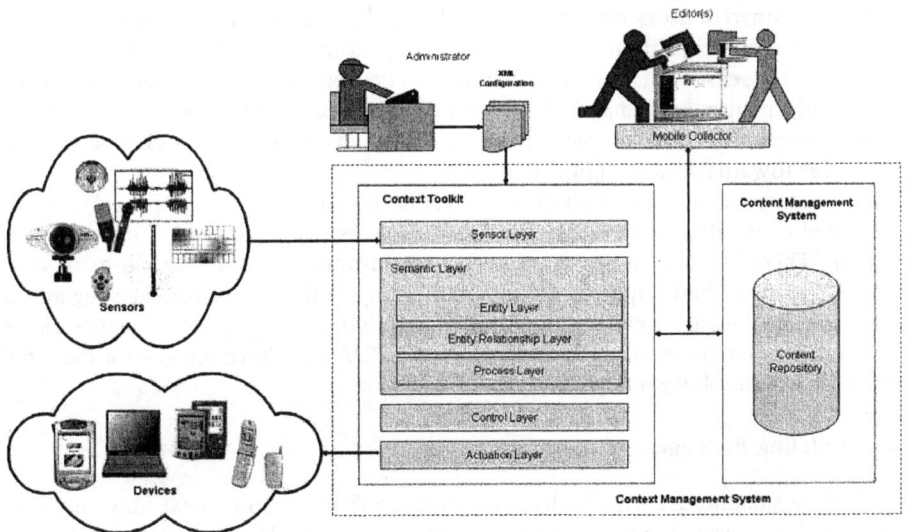

Fig. 1. The Components of a Context-Management Framework

Related work like the Context Toolkit [3, 18], the Context-Middleware [4] or the Context-Framework [7] address many issues related to context gathering, representation and supply. In fact, these approaches do not reduce the high complexity of context for inexperienced developers. The generalization from the variety of technical implementations and from our experiences in working in different application domains led us to the following definition of Context-Management as a new approach to overcome many of the existing problems:

Context-Management allows the creation, integration and administration of context-aware applications. Context-Management considers the definition of relevant context-parameters, the link between these parameters and information sources, their utilization for and the definition of the targeted adaptive behavior.

Fig. 1 illustrates the components of a Context-Management System: A *Context-Toolkit* for developers in the center, a *Content-Management System (CMS)* at the right hand side, an administrator tool for the design of the context-aware system and an editor tool for the creation of links between contents and specific context parameters they relate to. As an abstraction from our previous work [13, 22, 23], we implemented a Context-Toolkit consisting of four packages of *sensoring*, *modeling*, *controlling* and *actuating* components for the realization of context-aware applications.

2.1 Sensor Package

As information collectors, sensory mechanisms perform an observation of the users' behavior and interaction with the system, whereby the sensitivity, speed, and accuracy depends on the specific technology. Resent research in smart sensor-networks enables placing large numbers of intelligent senor-components ("smart dust") in the environment. Such smart sensors are equipped with small processors that enable for intelligent information acquisition [19]. In self-organizing networks, such as Intel's iMote approach [1], sensor technologies builds ad-hoc sensor-networks and deliver requested information on demand. At the end of the processing line, a server connected to the internet receives all data, processes the information and delivers inferred knowledge towards desired applications.

The sensory function (as well as the actuating function, cf. Paragraph 2.4) may be distributed over different devices of the mobile users [8] or embedded in their environment. Therefore, a light sensor of a PDA, an infrared sensor of an automatic door or a GPS receiver build in a car are all contributing to the sensor-package regardless to their physical location and environment. In turn, one device potentially hosts one or more sensor components of different types, e.g. a PDA may have sensors for the light-condition, a microphone and pen-input.

2.2 Modeling Package

In our understanding a user may have a context, and in turn a context may enclose a user, too. The context model captures the current situation the users act in, their preferences and interests, their social dependencies, their physical and technical environment, and many more aspects. Overall, there exist many different views on what dimensions such a model has to cover like *identity, location, time,* and *environment* [6] or *user context, computing context* and *physical context* [20]. In addition, there are several types of contexts, like *primary and secondary contexts* [2] or *static and dy-*

namic contexts [22]. To pay attention to the variety of approaches and not to limit its potentials we choose attribute-value pairs as a flexible context representation in our approach. The semantic components of this package receive and enrich the sensor data, and assign values to corresponding entity attributes. Thus, the components of the control package are supplied with an accurate image of the current situation and receive change events.

2.3 Control Package

Based on the available knowledge about the user's context model, controlling components generate sequences of commands for actuator components, in order to control the behavior of the application [11]. The commands assembled by the control components vary in their level of abstraction: simple *commands* and more complex *strategies*. The selected command or strategy can be seen as the link between these two questions: What information is taken into account for adaptation and which part or functionality of the user interface is adapted and how? Additionally, a direct communication link between those components and the domain enables answering requests from external systems or applications, and the realization of shared initiative and shared control approaches [14, 15, 16]. The Context-Toolkit offers rule templates for the definition of a hierarchically ordered set of rules as one control component.

2.4 Actuator Package

The actuators handle the connection back to the domain by mapping the decisions taken by the control package into real world actions. Actuators implements domain-dependent methods that directly render changes within the environment. Depending on the level of integration, these methods may be part of the target application. As a feedback, messages indicating the success or failure of actions are sent back.

As for sensors, actuators are specialized software-components that process small tasks like delivering one information snippet or displaying data on a particular device. Again, as a complement to the sensors package actuators may be distributed to many devices and in turn, one device may host one or more independent actuator components, e.g. like actuators for video-streaming or the adjustment of the display-brightness on one PDA.

3 Developing Context-Aware Systems with Context-Management

The framework described in the previous section forms the software basis for developers of contextualized applications. In the following we will illustrate tools that facilitate three main steps in the creation of context-aware systems: The design of the application using the authoring tool for the *Context Toolkit*, linking context information to content using the *Mobile Collector* and reproducing contextualized content using the *Content Player*.

3.1 Designing a Context-Aware Application

The Context-Toolkit developed by the Fraunhofer Institute provides ready-to-use software components that comply with the packages presented in the previous section, as well as an appropriate authoring tool. Fig. 2 shows a screenshot of the *Modeling*-panel of this tool, with which an administrator has designed the context model to be used in the application. This panel allows for the definition of domain entities and several types of context. The left hand side of Fig. 2 depicts the dynamic and the static context we added to the list of context types, as well as the attributes they consist of. The static part of the context model contains some meta-information about the contents, like their identification and the category they belong to.

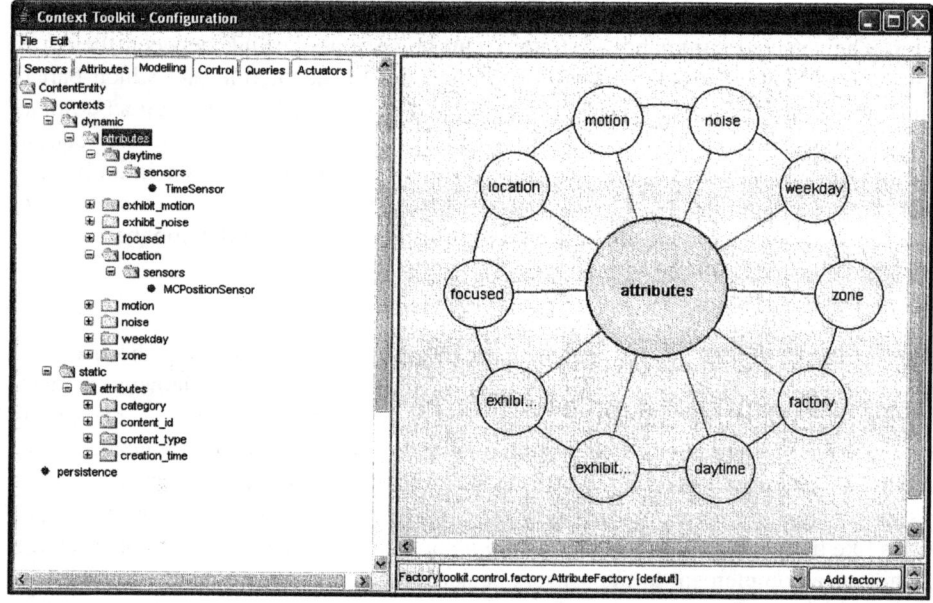

Fig. 2. The Context-Modeling Panel

Furthermore, the tool supports the allocation of sensors (information sources) to attributes (information interpreters). By creating a cross-linked network between sensors and attributes, attributes are connected with zero, one, or more sensors and in turn sensors deliver information to one or more attributes. Internally, all attributes receive events sent by sensors they are registered to as listeners. In addition, specific attributes may have no sensor as an information source and are used as containers for values determined by the control components. In Fig. 2 the attribute *daytime* depends on the *TimeSensor*, and the attribute *location* is connected with the positioning sensor *MCPositionSensor*.

The authoring tool offers a panel for the definition of rules, that determine the behavior of the targeted application. For the administration of the rules the Context-Toolkit offers rule templates for the definition of a hierarchically ordered set of rules.

Each rule is a precondition-action pair, whereat a set of actions is performed if the preconditions are met.

3.2 Annotating Content with Context Information

The annotation of content with context data is a special task that is most effectively performed on the move directly experiencing the particular context of use. In particular, our approach supports editors with a tool for "recording" or capturing context data together with content from the CMS. The *Mobile Collector* offers content providers a very efficient tool for the production of contextualized content.

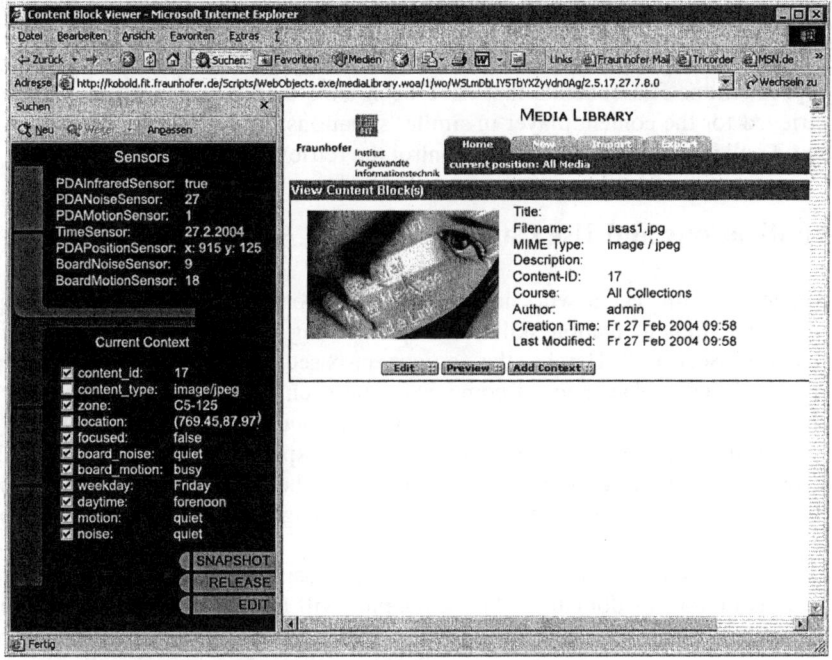

Fig. 3. A screenshot of the Mobile Collector displayed on a Tablet PC

Fig. 3 illustrates the Mobile Collector running on a Tablet PC. The right hand side shows the web front-end of a common CMS providing the author with functionality for adding, removing, searching and browsing content of different types such as images, sounds, videos or even entire HTML-pages. The browser plug-in on the left depicts the current context in the lower panel and the current sensor values in the upper panel. It reflects the context model defined by the Context Toolkit and allows the editor to capture the current context, to edit context-attribute values and to (un)select attributes that are considered (ir)relevant in this specific situation.

If the right content is selected and the context is adjusted appropriately, the editor can easily create the link between those two by just clicking on the snapshot-button and store this link to the persistent memory of the Mobile Collector. By using this context-snapshot approach content can easily be annotated with context meta-data. As an application example this is used in the European Project RAFT for creating authentic and contextualized learning content [10].

3.3 Reproducing Content

The Content Player is a modified browser that runs on a mobile device like a Personal Digital Assistant (PDA). Like for the Mobile Collector, sensors connected to the device are read out and their values are sent to the server. The server interprets the sensor values and determines the intended behavior of the targeted device. Therefore the server filters content from the CMS depending on the context the device currently is in. The server sends the adapted contents suiting the current situation to the content player, where the browser refreshes the displayed page with the new contents. The Context-Toolkit supports the determination of the appropriate behavior and the content reproduction. The contents that have been annotated with the Mobile Collector are retrieved for the content player in similar situations. The definition of rules by the Context-Toolkit enables the system to control the retrieval.

4 Applications and Use Case Studies

In comparison with recent work in this area, our work supports system developers with different implementation skills on implementing components on each level as introduced in Section 2. Hereby, the component-based approach facilitates the addition, removal and replacement of components on each level independently. This approach allows for an allocation and focusing of resources to the several layers (e.g. experts in sensoring may work independently from experts in machine learning algorithms or decision finding). Furthermore, our goal is the close integration of the work of software engineers with the work of product managers and application designers, as well as content providers.

The Context-Management System has been the basis for the realization of several applications in various domains. Three examples will be presented in this section as an illustration of how the described infrastructure was applied and how the abstract models defined earlier were instantiated. The three examples are an intelligent advertisement board, a museums guide and a treasure hunting game. All three applications have in common that they adapt their behaviour when specific context attributes change their values.

4.1 Intelligent Advertisement Board

At the CeBIT 2004 Fraunhofer FIT has presented an intelligent advertisement board as an illustrative application of the Context-Management System. The context-sensitive advertising board can be used for example at train stations, airports or in shop-windows. The board is equipped with simple and reliable sensors and reacts intelligently to the surrounding environment. It is able to respond to changing conditions like noise, trains arriving and departing, time of the day or people standing in front of the board showing interest. The content the board advertises consists of images of the categories *sports*, *food*, *shopping* and *news*, and for the times of day *morn-*

ing, *noon* and *evening*. In addition, all adverts have at least one further level of detail showing continuative information.

Regardless of the context, the system would just present the adverts randomly. With simple contextualization, the display reacts more intelligently. Fig. 4 depicts the instantiation of the model proposed in Section 2 for this specific application.

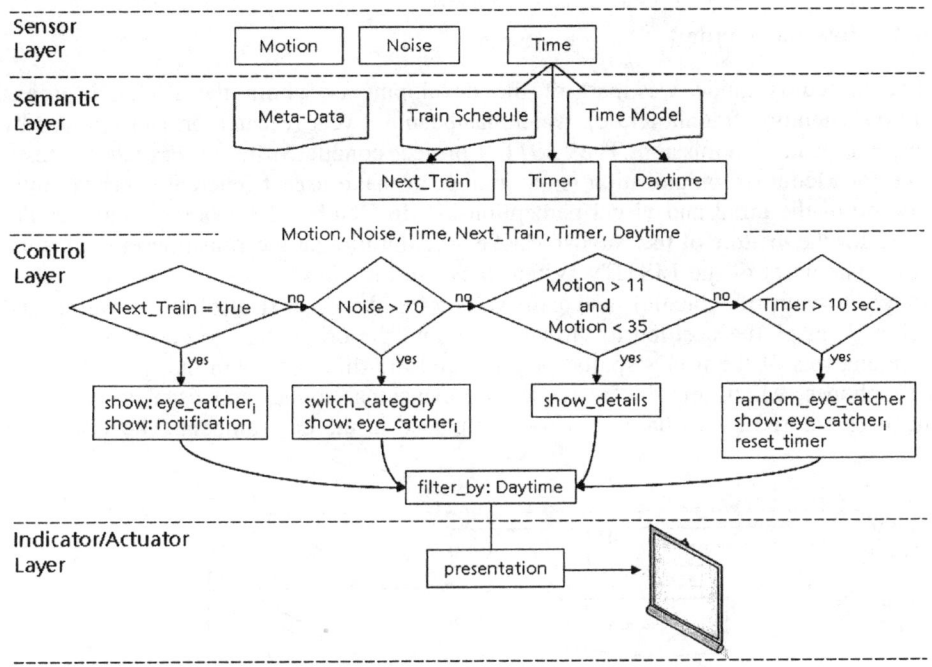

Fig. 4. Realization of an Intelligent Adervisement Board

The advertisement board is equipped with several sensors: A robust Webcam as well as a standard microphone are plugged into the system as sensors for the attributes MOTION and NOISE. Both sensors deliver values between 0 and 100 as an indicator for the degree of activity around the advertisement board. This value is determined by the camera sensor through pixel analysis of consecutive pictures and by the microphone sensor through measuring the sound intensity. The time is an additional simple but important sensor.

As mentioned in Section 3.1, the static part of the context model describes meta-information about the advertisements, in this case their identification, the category they belong to (i.e. sports, news, food, shopping) and an identifier for their more detailed successor. The attributes DAYTIME and NEXT_TRAIN derive their current values from specific models that abstract from the time.

The rules specifying the system behaviour of the intelligent advertisement board can be seen in Fig. 4, as well. By considering the time of the day, adverts for nearby restaurants are shown in the evening rather than for having breakfast. Furthermore, the board presents eye-catching and attracting pictures, if people do not show any interest. If someone's attention is attracted, the advertising board displays further and continuative information appositely to its eye-catching parent. The train schedule superposes all other context information. In case of arriving or departing trains, the advertising-board reacts with an appropriate announcement and warning, and moves its presentation style to less eye-catching, more informative notifications.

4.2 Museums Guide

The museums guide was part of the developments within the LISTEN project (http://listen.imk.fraunhofer.de), which has been a 3-year research project founded by the European Commission. The LISTEN project conducted by the Fraunhofer Institute for Media Communication is an attempt to make use of inherent everyday integration of the aural and visual perception [5]. In October 2003 this system was applied for the visitors of the August Macke art exhibition at the Kunstmuseum in Bonn [21]. The users of the LISTEN system move in the physical space wearing wireless (http://www.ar-tracking.de), and gain values for the context attributes location and focus. It maps the position to virtual zones and the orientation to object identifiers. The analysis of the user's spatial position and the time results in the speed the user headphones, which are able to render 3-dimensional sound, and listen to audio sequences emitted by virtual sound sources placed in the environment. The visitors of

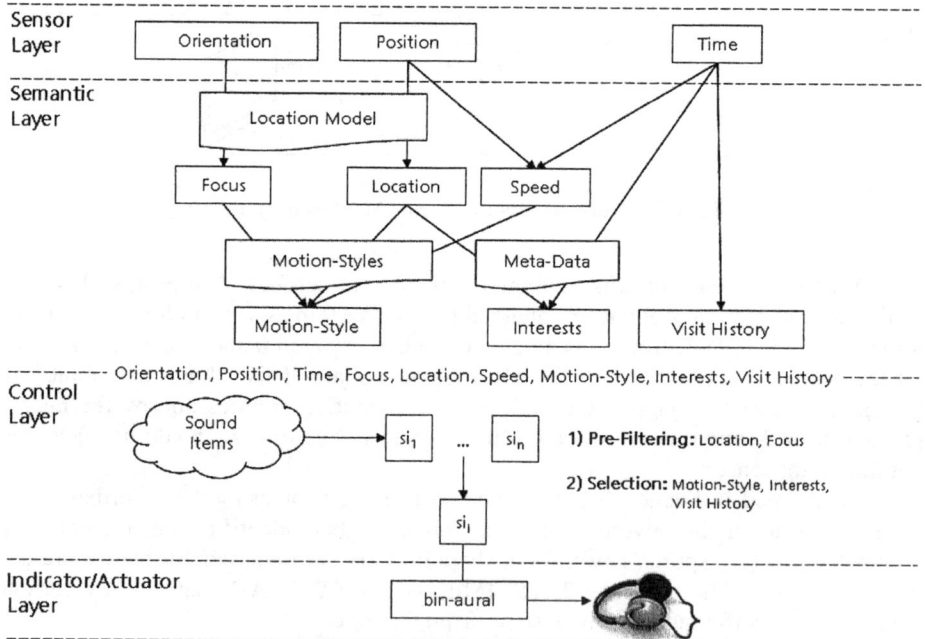

Fig. 5. Instantiation of the Model for a Museums Guide

```
<play_bench_sound>
  <precondition>
    <and>
      <equals type="symbol">
        <left_reference reference_type="attribute_value"
                        attribute_name="visitor.location"
        />
        <right_reference reference_type="value"
                         value="BENCH"
        />
      </equals>
      <equals type="symbol">
        <left_reference reference_type="attribute_value"
                        attribute_name="visitor.motion_style"
        />
        <right_reference reference_type="value"
                         value="STANDING_FOCUSED"
        />
      </equals>
    </and>
  </precondition>
  <actions>
    <play_sound>
      <parameters>
        <sound_name reference_type="value"
                    value="bench_03.aiff"
        />
      </parameters>
    </play_sound>
  </actions>
</play_bench_sound>
```

Fig. 6. Section of the Rule Set for the Museums Guide

the museum experience personalized audio information about exhibits through their headphones. While using the LISTEN system, users automatically navigate an acoustic information space designed as a complement or extension of the real space. The selection, presentation and adaptation of the content of this information space take into account the users current context.

One of the main objectives of the LISTEN project was the emphatic abandonment of any portable device for the user except the headphones. For the personalization process this means that explicit feedback is not available in our application. Thus, the user's movements are the only interface for interaction. The personalization process in LISTEN is based on the context dimensions time, position and head orientation, and on an extensive amount of annotations for sound pieces respectively visual items (i.e. paintings). Fig. 5 provides an overview of how the abstract framework model proposed earlier was instantiated for the LISTEN system.

The Location Model allows the LISTEN system to interpret the user's position and head orientation, which are delivered by a fine granular tracking system moves with. In combination with the meta-information concerning the paintings and the time allows an assumption about the users interests (e.g. the more time the visitor spends with a specific exhibit, the more s/he likes it [17]. Furthermore, we chose to employ stereotypes to define the user's motion style in a museum environment [23].

With the aid of a rule-system the system is able to accordingly adapt the scenery and to cause different sound presentations. First, a pre-filtering of sound-items is performed based on the user's current location and focus. From this list the best suited sound-item is selected referring to the users visit history, motion-style and interests. In result of the administrator's work defining the behaviour of the LISTEN application, Fig. 6 depicts an example for the rule PLAY_BENCH_SOUND as part of the rule-system. It illustrates, that the rule is divided in a precondition and an actions part. The rule causes the system to play the sound file named BENCH_03.AIFF, if the visitors location is equal to BENCH and the motion style is STANDING_FOCUSED.

Concerning the LISTEN system, the actuator layer renders the audio information in order to change the audio augmentation. Basically this denotes a certain sound-item being played from a specific sound source.

4.3 Treasure Hunting

The treasure hunting game has been developed as an example for a mobile gaming application where competing teams have to find a treasure placed somewhere in a pre-defined playfield using personal digital assistants (PDAs). While exploring the playfield the teams receive simple quizzes related to their current location on their PDAs. If the team selects the right answer to the question, a hint on the destination of the hidden treasury is shown. The more questions are answered correctly, the more information about the destination is gathered. Additionally, after answering the question correctly, the team is allowed to place a mine at this location. When another team reaches a position, where a mine has been activated, they will get an acoustic notification and loose the associated question. The maximum duration of the game is set to one hour, but if the treasury is discovered earlier, the game ends. In June 2004 a public demonstration took place when a group of 20 young scientists played the game with three competing teams consisting of 8 to 13 years old kids chasing the treasure.

The WLAN tracking technology from Ekahau (http://www.ekahau.com) was coupled with our toolkit to have access to the position of the players, the most important context information of the game. Fig. 7 shows the instantiation of the abstract model illustrated earlier for the specific domain of treasure hunting and how the information is combined with the context within the game application. To simplify the figure, all explicit interaction processes like selecting answers to puzzles or activating a mine at a certain location are left out here and only implicit user interaction with the system is shown.

The main information needed for the system are the *position* of the PDA (i.e. the position of one treasure hunting group) and the *time* since the game has been launched. Those two values are delivered by the Ekahau positioning sensor and the server clock. The modelling layer of the treasure hunting game holds three models relevant for the interpretation of the values provided by the sensors: Location Model, Quiz Locations and Mine Locations. The Location Model transforms the position of a PDA into its Location, i.e. the name of the zone the PDA is currently located. The Location Model, the locations of all quizzes and already placed mines are centrally accessible. Furthermore, the combination of the time with the location creates a Location History, which is needed for retracing the group of hunters and the locations they have already been to. The additional attribute MINE_COUNT keeps track of the number of mines already placed in the field by the group of hunters.

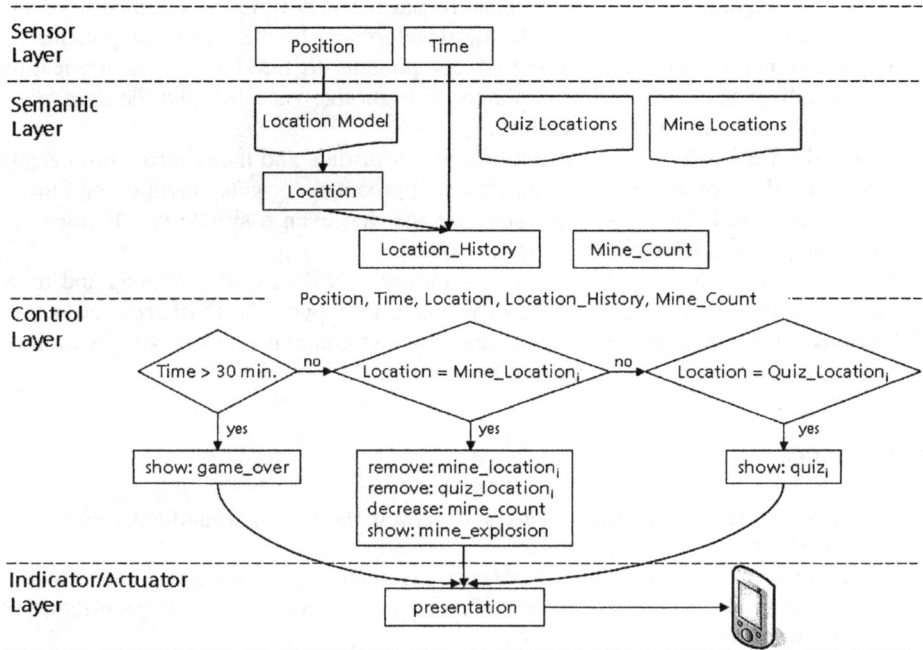

Fig. 7. A Treasure Hunting Game Realized With the Context-Management System

Based on the attribute values, the system behaviour is defined by the rules on the control layer, defining when the game ends, mines explode and quizzes are shown. The rule interpretation is always triggered, when the value of the location or the time attribute changes. In the last step (actuator layer) the information to be shown to the group of hunters is transmitted to their PDA and presented on the screen.

5 Conclusions and Future Work

The work presented gives an overview on current developments towards a Context-Management System at Fraunhofer FIT. Based on the identification of the main functions different packages of components have been introduced and prototypically implemented to support sensor data management, context abstraction, and the control of actuator output. We have described implementations of sample applications for intelligent information distribution and selection, as well as the collection and connection of context data with contents.

The demand for context-sensitive functionalities constitutes a crucial challenge for system developers as well as for product managers, application designers and system integrators. They have to handle the system heterogeneity, starting from the hardware and software protocols, including the integration within various mobile and wireless environments, considering seamless integration of services from multiple providers, and release and validation of new protocols. At the same time, managers, developers and system integrators have to react to shortened delivery time for a competitive and dynamic market. In order to integrate the development of the basic components by software engineers with the tailoring of components to build the final application logic, we deliver the framework completely configurable via XML and the user front-end for non-experts in computer science.

Especially for semantic enrichment of raw sensor-data and the control components we are currently experimenting with different approaches for generalization and inference-mechanisms. From our experiences we see that even a simple set of rules defined by end users can have very complex output.

For our future work we plan to build a variety of different applications and tools based on the described infrastructure to evaluate the approach. Furthermore, clearer definitions for the integration of static and dynamic contexts and the combination of user data and environment data have to be verified.

References

1. Culler, D.E., Mulder, H.: Smart Sensors to Network the World. Scientific American, 290 (2004) 52-59
2. Dey, A.K., Abowd, G.D.: Towards a better understanding of context and contextawareness, Technical Report GIT-GVU-99-22, College of Computing, Georgia Institute of Technology (1999)
3. Dey, A.K., Abowd, G.D., Salber, D.: A Conceptual Framework and a Toolkit for Supporting the Rapid Prototyping of Context-Aware Applications, Human-Computer Interaction Journal, 16(2-4) (2001) 7–166
4. Ebling, M.R., Hunt, G.D.H., Lei, H.: Issues for Context Services for Pervasive Computing. Workshop on Middleware for Mobile Computing, Heidelberg, Germany (2001)
5. Eckel, G.: LISTEN Augmenting Everyday Environments with Interactive Soundscapes. Proceedings of the I3 Spring Days Workshop "Moving between the physical and the digital: exploring and developing new forms of mixed reality user experience", Porto, Portugal. (2001)
6. Gross, T., Specht, M.: Awareness in Context-Aware Information Systems. Mensch&Computer - 1. Fachübergreifende Konferenz, Bad Honnef, Germany, Teubner, (2001) 173–182
7. Henricksen, K.: A Framework for Context-Aware Pervasive Computing Applications School of Information Technology and Electrical Engineering, University of Queensland, Queensland (2003)
8. Hodes, T.D., Katz, R.H.: Composeable Ad Hoc Location-Based Services for Heterogeneous Mobile Clients. ACM Wireless Networks, 5 (1999) 411–427
9. Jameson, A.: User-Adaptive Systems: An Integrative Overview. Tutorial at UM and IJCAI (1999)
10. Kravcik, M., Kaibel, A., Specht, M., Terrenghi, L: Mobile Collector for Field Trips. IEEE Journal Educational Technology & Society 7(2) (2004) 25-33

11. Lei, H., Sow, D.M., Davis, J.S., Banavar, G., Ebling, M.R.: The Design and Applications of a Context Service, ACM SIGMOBILE Mobile Computing and Communications Review 6(4) (2002) 45–55
12. Liu, J., Chu, M., Liu, J., Reich, J., Zhao, F. State-Centric Programming for Sensor-Actuator Network Systems. Pervasive Computing 2 (2003) 50-62
13. Lorenz, A.: Towards a new role of agent technology in user modelling. 12th Workshop on Adaptivity and User Modeling in Interactive Systems, Karlsruhe, Germany (2003)
14. Oppermann, R.: Adaptive User support. Hillsdale: Lawrence Erlbaum Associates (1994)
15. Oppermann, R., Thomas, C.G.:, Supporting Learning as an Iterative Process, Proceedings of the European Conference on Artificial Intelligence in Education, Lisboa, Potugal (1996)
16. Oppermann, R. and Specht, M.: A Context-Sensitive Nomadic Exhibition Guide, Second Symposium on Handheld and Ubiquituous Computing, Bristol, UK, Springer-Verlag (2000)
17. Pazzani, M.J., Billsus, D.: Learning and Revising User Profiles: The Identification of Interesting Web Sites, Machine Learning 27 (1997) 313–331
18. Salber, D., Dey, A.K., Abowd, G.D.: The Context Toolkit: Aiding the Development of Context-Enabled Applications, Proceedings of the 1999 Conference on Human Factors in Computing Systems, Pittsburgh, PA (1999) 434–441
19. Satyanarayanan, M.: Of Smart Dust and Brilliant Rocks. Pervasive Computing 2 (2003) 2-4
20. Schilit, B.N., Adams, N.I., Want, R.: Context-Aware Computing Applications. Proceedings of the Workshop on Mobile Computing Systems and Applications, Santa Cruz, CA (1994) 85–90
21. Unnützer, P.: LISTEN im Kunstmuseum Bonn, KUNSTFORUM International, 155 (2001) 469–470
22. Zimmermann, A., Lorenz, A., Specht, M.: Reasoning From Contexts. Proceedings of the 10th Workshop on Adaptivity and User modelling in Interactive Systems, Hannover, Germany (2002) 114–120
23. Zimmermann, A., Lorenz, A.: Listen: Contextualized presentation for audioaugmented environments. Proceedings of the 11th Workshop on Adaptivity and User modelling in Interactive Systems, Karlsruhe, Germany (2003) 351–357

Author Index

Abrilian, Sarkis 225
Akhras, Fabio N. 1
Arló Costa, Horacio 15
Augustinaitis, Arunas 410

Baccino, Thierry 278
Bartolini, Claudio 353
Bazire, Mary 29
Bertolotto, Michela 339
Bianchi, Claudia 41
Brézillon, Patrick 29, 55, 464
Bullot, Nicolas J. 69
Burg, Bernard 353
Byron, Donna 83

Calvi, Gianguglielmo 368
Castro, Paul 112
Chilov, Nikolai 476
Christiansen, Henning 97
Cohen, Norman H. 112
Connelly, Kay 197
Dahl, Veronica 97
De Brabanter, Philippe 126
Devillers, Laurence 225

Esbjörnsson, Mattias 140

Flanagan, John A. 155
Fortu, Ovidiu 169

Graham, Connor 382
Grinberg, Maurice 183

Harrer, Andreas 292
Hristova, Evgenia 183
Hussain, Aini 410

Justickis, Viktoras 410

Kattan, Anton 382
Khalil, Ashraf 197

Koch, Fernando 382
Kramer, Ronny 210

Lamolle, Myriam 225
Landragin Frédéric 240
Leake, David 254
Lee, Eunseok 268
Lee, Seunghwa 268
Léger, Laure 278
Levashova, Tatiana 476
Loke, Seng W. 353
Lorenz, Andreas 556

Maguitman, Ana 254
Malzahn, Nils 292
Mampilly, Thomas 83
Mancini, Maurizio 225
Martin, Jean-Claude 225
Martín-Vide, Carlos 304
Matsumoto, Naoko 316
McLoughlin, Eoin 339
Misra, Archan 112
Mitrana, Victor 304
Modsching, Marko 210
Moldovan, Dan 169

Nagy, George 326, 515
Nagy, Naomi 326

O'Sullivan, Dympna 339

Padovitz, Amir 353
Pashkin, Michael 476
Pelachaud, Catherine 225
Pezzulo, Giovanni 368
Pumputis, Alvydas 410

Rahwan, Iyad 382
Rapaport, William J. 396
Raudys, Sarunas 410
Reichherzer, Thomas 254

Roelofsen, Floris 424
Romero, Esther 436

Salgado, Ana Carolina 464
Sarkar, Prateek 515
Schmidtke, Hedda R. 450
Schneider, Edgar W. 326
Schulze, Joerg 210
Serafini, Luciano 424
Sharma, Vinay 83
Siebra, Sandra de A. 464
Smirnov, Alexander 476
Sonenberg, Liz 382
Soria, Belén 436
Specht, Marcus 556
Steinberg, Alan N. 490

Taysom, William 15
Tedesco, Patrícia A. 464
ten Hagen, Klaus 210

Thomason, Richmond H. 501
Tijus, Charles 278, 542
Tokosumi, Akifumi 316

Vassallo, Nicla 41
Veeramachaneni, Sriharsha 515

Wang, Xiaojie 529
Weilenmann, Alexandra 140
Wilson, David 339

Xu, Tianfang 83

Youn, Heeyong 268

Zaslavsky, Arkady 353
Zeini, Sam 292
Zhang, Xiaoli 326
Zibetti, Elisabetta 542
Zimmermann, Andreas 556

Lecture Notes in Artificial Intelligence (LNAI)

Vol. 3559: P. Auer, R. Meir (Eds.), Learning Theory. XI, 692 pages. 2005.

Vol. 3554: A. Dey, B. Kokinov, D. Leake, R. Turner (Eds.), Modeling and Using Context. XIV, 572 pages. 2005.

Vol. 3533: M. Ali, F. Esposito (Eds.), Innovations in Applied Artificial Intelligence. XX, 858 pages. 2005.

Vol. 3528: P.S. Szczepaniak, J. Kacprzyk, A. Niewiadomski (Eds.), Advances in Web Intelligence. XVII, 513 pages. 2005.

Vol. 3518: T.B. Ho, D. Cheung, H. Liu (Eds.), Advances in Knowledge Discovery and Data Mining. XXI, 864 pages. 2005.

Vol. 3508: P. Bresciani, P. Giorgini, B. Henderson-Sellers, G. Low, M. Winikoff (Eds.), Agent-Oriented Information Systems II. X, 227 pages. 2005.

Vol. 3505: V. Gorodetsky, J. Liu, V. A. Skormin (Eds.), Autonomous Intelligent Systems: Agents and Data Mining. XIII, 303 pages. 2005.

Vol. 3501: B. Kégl, G. Lapalme (Eds.), Advances in Artificial Intelligence. XV, 458 pages. 2005.

Vol. 3492: P. Blache, E. Stabler, J. Busquets, R. Moot (Eds.), Logical Aspects of Computational Linguistics. X, 363 pages. 2005.

Vol. 3488: M.-S. Hacid, N.V. Murray, Z.W. Raś, S. Tsumoto (Eds.), Foundations of Intelligent Systems. XIII, 700 pages. 2005.

Vol. 3476: J. Leite, A. Omicini, P. Torroni, P. Yolum (Eds.), Declarative Agent Languages and Technologies II. XII, 289 pages. 2005.

Vol. 3464: S.A. Brueckner, G.D.M. Serugendo, A. Karageorgos, R. Nagpal (Eds.), Engineering Self-Organising Systems. XIII, 299 pages. 2005.

Vol. 3452: F. Baader, A. Voronkov (Eds.), Logic for Programming, Artificial Intelligence, and Reasoning. XI, 562 pages. 2005.

Vol. 3446: T. Ishida, L. Gasser, H. Nakashima (Eds.), Massively Multi-Agent Systems I. XI, 349 pages. 2005.

Vol. 3438: H. Christiansen, P.R. Skadhauge, J. Villadsen (Eds.), Constraint Solving and Language Processing. VIII, 205 pages. 2005.

Vol. 3430: S. Tsumoto, T. Yamaguchi, M. Numao, H. Motoda (Eds.), Active Mining. XII, 349 pages. 2005.

Vol. 3419: B. Faltings, A. Petcu, F. Fages, F. Rossi (Eds.), Constraint Satisfaction and Constraint Logic Programming. X, 217 pages. 2005.

Vol. 3416: M. Böhlen, J. Gamper, W. Polasek, M.A. Wimmer (Eds.), E-Government: Towards Electronic Democracy. XIII, 311 pages. 2005.

Vol. 3415: P. Davidsson, B. Logan, K. Takadama (Eds.), Multi-Agent and Multi-Agent-Based Simulation. X, 265 pages. 2005.

Vol. 3403: B. Ganter, R. Godin (Eds.), Formal Concept Analysis. XI, 419 pages. 2005.

Vol. 3398: D.-K. Baik (Ed.), Systems Modeling and Simulation: Theory and Applications. XIV, 733 pages. 2005.

Vol. 3397: T.G. Kim (Ed.), Artificial Intelligence and Simulation. XV, 711 pages. 2005.

Vol. 3396: R.M. van Eijk, M.-P. Huget, F. Dignum (Eds.), Agent Communication. X, 261 pages. 2005.

Vol. 3394: D. Kudenko, D. Kazakov, E. Alonso (Eds.), Adaptive Agents and Multi-Agent Systems II. VIII, 313 pages. 2005.

Vol. 3392: D. Seipel, M. Hanus, U. Geske, O. Bartenstein (Eds.), Applications of Declarative Programming and Knowledge Management. X, 309 pages. 2005.

Vol. 3374: D. Weyns, H.V.D. Parunak, F. Michel (Eds.), Environments for Multi-Agent Systems. X, 279 pages. 2005.

Vol. 3371: M.W. Barley, N. Kasabov (Eds.), Intelligent Agents and Multi-Agent Systems. X, 329 pages. 2005.

Vol. 3369: V.R. Benjamins, P. Casanovas, J. Breuker, A. Gangemi (Eds.), Law and the Semantic Web. XII, 249 pages. 2005.

Vol. 3366: I. Rahwan, P. Moraitis, C. Reed (Eds.), Argumentation in Multi-Agent Systems. XII, 263 pages. 2005.

Vol. 3359: G. Grieser, Y. Tanaka (Eds.), Intuitive Human Interfaces for Organizing and Accessing Intellectual Assets. XIV, 257 pages. 2005.

Vol. 3346: R.H. Bordini, M. Dastani, J. Dix, A.E.F. Seghrouchni (Eds.), Programming Multi-Agent Systems. XIV, 249 pages. 2005.

Vol. 3345: Y. Cai (Ed.), Ambient Intelligence for Scientific Discovery. XII, 311 pages. 2005.

Vol. 3343: C. Freksa, M. Knauff, B. Krieg-Brückner, B. Nebel, T. Barkowsky (Eds.), Spatial Cognition IV. XIII, 519 pages. 2005.

Vol. 3339: G.I. Webb, X. Yu (Eds.), AI 2004: Advances in Artificial Intelligence. XXII, 1272 pages. 2004.

Vol. 3336: D. Karagiannis, U. Reimer (Eds.), Practical Aspects of Knowledge Management. X, 523 pages. 2004.

Vol. 3327: Y. Shi, W. Xu, Z. Chen (Eds.), Data Mining and Knowledge Management. XIII, 263 pages. 2005.

Vol. 3315: C. Lemaître, C.A. Reyes, J.A. González (Eds.), Advances in Artificial Intelligence – IBERAMIA 2004. XX, 987 pages. 2004.

Vol. 3303: J.A. López, E. Benfenati, W. Dubitzky (Eds.), Knowledge Exploration in Life Science Informatics. X, 249 pages. 2004.

Vol. 3301: G. Kern-Isberner, W. Rödder, F. Kulmann (Eds.), Conditionals, Information, and Inference. XII, 219 pages. 2005.

Vol. 3276: D. Nardi, M. Riedmiller, C. Sammut, J. Santos-Victor (Eds.), RoboCup 2004: Robot Soccer World Cup VIII. XVIII, 678 pages. 2005.

Vol. 3275: P. Perner (Ed.), Advances in Data Mining. VIII, 173 pages. 2004.

Vol. 3265: R.E. Frederking, K.B. Taylor (Eds.), Machine Translation: From Real Users to Research. XI, 392 pages. 2004.

Vol. 3264: G. Paliouras, Y. Sakakibara (Eds.), Grammatical Inference: Algorithms and Applications. XI, 291 pages. 2004.

Vol. 3259: J. Dix, J. Leite (Eds.), Computational Logic in Multi-Agent Systems. XII, 251 pages. 2004.

Vol. 3257: E. Motta, N.R. Shadbolt, A. Stutt, N. Gibbins (Eds.), Engineering Knowledge in the Age of the Semantic Web. XVII, 517 pages. 2004.

Vol. 3249: B. Buchberger, J.A. Campbell (Eds.), Artificial Intelligence and Symbolic Computation. X, 285 pages. 2004.

Vol. 3248: K.-Y. Su, J. Tsujii, J.-H. Lee, O.Y. Kwong (Eds.), Natural Language Processing – IJCNLP 2004. XVIII, 817 pages. 2005.

Vol. 3245: E. Suzuki, S. Arikawa (Eds.), Discovery Science. XIV, 430 pages. 2004.

Vol. 3244: S. Ben-David, J. Case, A. Maruoka (Eds.), Algorithmic Learning Theory. XIV, 505 pages. 2004.

Vol. 3238: S. Biundo, T. Frühwirth, G. Palm (Eds.), KI 2004: Advances in Artificial Intelligence. XI, 467 pages. 2004.

Vol. 3230: J.L. Vicedo, P. Martínez-Barco, R. Muñoz, M. Saiz Noeda (Eds.), Advances in Natural Language Processing. XII, 488 pages. 2004.

Vol. 3229: J.J. Alferes, J. Leite (Eds.), Logics in Artificial Intelligence. XIV, 744 pages. 2004.

Vol. 3228: M.G. Hinchey, J.L. Rash, W.F. Truszkowski, C.A. Rouff (Eds.), Formal Approaches to Agent-Based Systems. VIII, 290 pages. 2004.

Vol. 3215: M.G.. Negoita, R.J. Howlett, L.C. Jain (Eds.), Knowledge-Based Intelligent Information and Engineering Systems, Part III. LVII, 906 pages. 2004.

Vol. 3214: M.G.. Negoita, R.J. Howlett, L.C. Jain (Eds.), Knowledge-Based Intelligent Information and Engineering Systems, Part II. LVIII, 1302 pages. 2004.

Vol. 3213: M.G.. Negoita, R.J. Howlett, L.C. Jain (Eds.), Knowledge-Based Intelligent Information and Engineering Systems, Part I. LVIII, 1280 pages. 2004.

Vol. 3209: B. Berendt, A. Hotho, D. Mladenic, M. van Someren, M. Spiliopoulou, G. Stumme (Eds.), Web Mining: From Web to Semantic Web. IX, 201 pages. 2004.

Vol. 3206: P. Sojka, I. Kopecek, K. Pala (Eds.), Text, Speech and Dialogue. XIII, 667 pages. 2004.

Vol. 3202: J.-F. Boulicaut, F. Esposito, F. Giannotti, D. Pedreschi (Eds.), Knowledge Discovery in Databases: PKDD 2004. XIX, 560 pages. 2004.

Vol. 3201: J.-F. Boulicaut, F. Esposito, F. Giannotti, D. Pedreschi (Eds.), Machine Learning: ECML 2004. XVIII, 580 pages. 2004.

Vol. 3194: R. Camacho, R. King, A. Srinivasan (Eds.), Inductive Logic Programming. XI, 361 pages. 2004.

Vol. 3192: C. Bussler, D. Fensel (Eds.), Artificial Intelligence: Methodology, Systems, and Applications. XIII, 522 pages. 2004.

Vol. 3191: M. Klusch, S. Ossowski, V. Kashyap, R. Unland (Eds.), Cooperative Information Agents VIII. XI, 303 pages. 2004.

Vol. 3187: G. Lindemann, J. Denzinger, I.J. Timm, R. Unland (Eds.), Multiagent System Technologies. XIII, 341 pages. 2004.

Vol. 3176: O. Bousquet, U. von Luxburg, G. Rätsch (Eds.), Advanced Lectures on Machine Learning. IX, 241 pages. 2004.

Vol. 3171: A.L.C. Bazzan, S. Labidi (Eds.), Advances in Artificial Intelligence – SBIA 2004. XVII, 548 pages. 2004.

Vol. 3159: U. Visser, Intelligent Information Integration for the Semantic Web. XIV, 150 pages. 2004.

Vol. 3157: C. Zhang, H. W. Guesgen, W.K. Yeap (Eds.), PRICAI 2004: Trends in Artificial Intelligence. XX, 1023 pages. 2004.

Vol. 3155: P. Funk, P.A. González Calero (Eds.), Advances in Case-Based Reasoning. XIII, 822 pages. 2004.

Vol. 3139: F. Iida, R. Pfeifer, L. Steels, Y. Kuniyoshi (Eds.), Embodied Artificial Intelligence. IX, 331 pages. 2004.

Vol. 3131: V. Torra, Y. Narukawa (Eds.), Modeling Decisions for Artificial Intelligence. XI, 327 pages. 2004.

Vol. 3127: K.E. Wolff, H.D. Pfeiffer, H.S. Delugach (Eds.), Conceptual Structures at Work. XI, 403 pages. 2004.

Vol. 3123: A. Belz, R. Evans, P. Piwek (Eds.), Natural Language Generation. X, 219 pages. 2004.

Vol. 3120: J. Shawe-Taylor, Y. Singer (Eds.), Learning Theory. X, 648 pages. 2004.

Vol. 3097: D. Basin, M. Rusinowitch (Eds.), Automated Reasoning. XII, 493 pages. 2004.

Vol. 3071: A. Omicini, P. Petta, J. Pitt (Eds.), Engineering Societies in the Agents World. XIII, 409 pages. 2004.

Vol. 3070: L. Rutkowski, J. Siekmann, R. Tadeusiewicz, L.A. Zadeh (Eds.), Artificial Intelligence and Soft Computing - ICAISC 2004. XXV, 1208 pages. 2004.

Vol. 3068: E. André, L. Dybkjær, W. Minker, P. Heisterkamp (Eds.), Affective Dialogue Systems. XII, 324 pages. 2004.

Vol. 3067: M. Dastani, J. Dix, A. El Fallah-Seghrouchni (Eds.), Programming Multi-Agent Systems. X, 221 pages. 2004.

Vol. 3066: S. Tsumoto, R. Słowiński, J. Komorowski, J.W. Grzymała-Busse (Eds.), Rough Sets and Current Trends in Computing. XX, 853 pages. 2004.

Vol. 3065: A. Lomuscio, D. Nute (Eds.), Deontic Logic in Computer Science. X, 275 pages. 2004.